Regulation of Crop Quality and Stress Responses

Regulation of Crop Quality and Stress Responses

Editors

Mingxun Chen
Lixi Jiang
Yuan Guo

MDPI

Basel • Beijing • Wuhan • Barcelona • Belgrade • Novi Sad • Cluj • Manchester

Editors

Mingxun Chen
Northwest A&F University
Yangling
China

Lixi Jiang
Zhejiang University
Hangzhou
China

Yuan Guo
Northwest A&F University
Yangling
China

Editorial Office
MDPI
St. Alban-Anlage 66
4052 Basel, Switzerland

This is a reprint of articles from the Special Issue published online in the open access journal *Plants* (ISSN 2223-7747) (available at: https://www.mdpi.com/journal/plants/special_issues/regulation_crop_quality_stress_responses).

For citation purposes, cite each article independently as indicated on the article page online and as indicated below:

Lastname, A.A.; Lastname, B.B. Article Title. *Journal Name* **Year**, *Volume Number*, Page Range.

ISBN 978-3-0365-9084-4 (Hbk)
ISBN 978-3-0365-9085-1 (PDF)
doi.org/10.3390/books978-3-0365-9085-1

Cover image courtesy of Yuan Guo

Contents

About the Editors

Mingxun Chen

Mingxun Chen is a Professor of Crop Science at the College of Agronomy, Northwest A&F University, the People's Republic of China. He completed his Ph.D. in Crop Science at Zhejiang University in China. Dr. Chen has long been engaged in research on the regulation mechanism of high-yield and abiotic stress resistance and the innovation and utilization of germplasm resources in rapeseed. He has published more than 40 articles in international journals, and he is an editor, editorial board member, and reviewer for several scientific journals.

Lixi Jiang

Lixi Jiang is a Professor of Crop Science at the College of Agriculture and Biotechnology, Zhejiang University, the People's Republic of China. He completed his Ph.D. in Plant Breeding at the University of Göttingen in Germany. Dr. Jiang has mainly engaged in basic research on rapeseed and Arabidopsis thaliana. His research topics include: (1) the molecular biology mechanism of seed oil formation in cruciferous plants, and (2) the developmental mechanism of cruciferous plant flowers, and the molecular mechanisms of male sterility for the utilization of crop heterosis. He has published more than 60 articles in international journals, and he is an editor, editorial board member, and reviewer for several scientific journals.

Yuan Guo

Yuan Guo is an Assistant Professor of Crop Science at the College of Agronomy, Northwest A&F University, the People's Republic of China. She completed her Ph.D. in Plant Breeding at Kiel University in Germany. Dr. Guo has mainly engaged in research on the synergistic improvement of rapeseed quality and abiotic stress resistance, and rapeseed genetic breeding. She has published more than 10 articles in international journals.

Preface

The quality formation and stress resistance of crops are critical to ensure food security. Due to its fixation, from seed germination, seedling establishment, and vegetative and reproductive growth to finally obtaining mature seeds to complete the life cycle, crops are vulnerable to adverse temperature, light, moisture and other stressful conditions throughout their life, resulting in yield reduction. Changes in physical, chemical, and biological environmental factors in nature all have impacts on the processes of crop physiology, biochemistry, and growth and development. Moreover, global climate change is leading to the frequent occurrence of extreme climate conditions, which puts forward higher requirements for the adaptability of crops to biotic and abiotic stresses. In recent years, with the rapid development of our understanding of stress signaling pathways, it has been proposed that stress can significantly affect crop quality. The nutrients in crops, such as unsaturated fatty acids, anthocyanins, and melatonin, can not only meet the nutritional needs of humans but also affect the stress resistance of crops. This Special Issue captures some of the latest research on the mechanisms of crop response to light, temperature, moisture, and other related abiotic stresses, as well as the metabolic regulation of crop quality. We are grateful to all authors for their contributions as well as the reviewers for their professional suggestions and decisions. We are also thankful to the editorial team of Plants for this opportunity to collaborate.

Mingxun Chen, Lixi Jiang, and Yuan Guo
Editors

plants

MDPI

Review

Pea Breeding for Resistance to Rhizospheric Pathogens

Osman Z. Wohor [1,2,*], Nicolas Rispail [1], Chris O. Ojiewo [3] and Diego Rubiales [1]

1 Instituto de Agricultura Sostenible, CSIC, Avenida Menéndez Pidal s/n, 14004 Córdoba, Spain
2 Savanna Agriculture Research Institute, CSIR, Nyankpala, Tamale Post TL52, Ghana
3 International Maize and Wheat Improvement Center (CIMMYT), ICRAF House, United Nations Avenue—Gigiri, Nairobi P.O. Box 1041-00621, Kenya
* Correspondence: owohor@ias.csic.es

Abstract: Pea (*Pisum sativum* L.) is a grain legume widely cultivated in temperate climates. It is important in the race for food security owing to its multipurpose low-input requirement and environmental promoting traits. Pea is key in nitrogen fixation, biodiversity preservation, and nutritional functions as food and feed. Unfortunately, like most crops, pea production is constrained by several pests and diseases, of which rhizosphere disease dwellers are the most critical due to their long-term persistence in the soil and difficulty to manage. Understanding the rhizosphere environment can improve host plant root microbial association to increase yield stability and facilitate improved crop performance through breeding. Thus, the use of various germplasm and genomic resources combined with scientific collaborative efforts has contributed to improving pea resistance/cultivation against rhizospheric diseases. This improvement has been achieved through robust phenotyping, genotyping, agronomic practices, and resistance breeding. Nonetheless, resistance to rhizospheric diseases is still limited, while biological and chemical-based control strategies are unrealistic and unfavourable to the environment, respectively. Hence, there is a need to consistently scout for host plant resistance to resolve these bottlenecks. Herein, in view of these challenges, we reflect on pea breeding for resistance to diseases caused by rhizospheric pathogens, including fusarium wilt, root rots, nematode complex, and parasitic broomrape. Here, we will attempt to appraise and harmonise historical and contemporary knowledge that contributes to pea resistance breeding for soilborne disease management and discuss the way forward.

Keywords: rhizosphere; soilborne disease; pea; breeding; fusarium; broomrape; rhizotrons; pathogens; resistance

Citation: Wohor, O.Z.; Rispail, N.; Ojiewo, C.O.; Rubiales, D. Pea Breeding for Resistance to Rhizospheric Pathogens. *Plants* **2022**, *11*, 2664. https://doi.org/10.3390/plants11192664

Academic Editor: Mingxun Chen

Received: 19 September 2022
Accepted: 6 October 2022
Published: 10 October 2022

1. Introduction

Pea (*Pisum sativum* L.) is one of the oldest domesticated crops in the world. Pea is a self-pollinating diploid (2n = 14) with a haploid genome size of 4.5 Gbp. Pea served as a model crop in the hybridization work of Mendel, leading to the postulate on heritability [1]. Its centre of origin is primarily in the Near East with secondary diversification in the Mediterranean, Middle East and East Africa [2]. The taxonomy of the genus *Pisum* has been widely debated, but is generally accepted to contain three main species, *P. sativum*, *P. fulvum* Sibth and Sm. and *P. abyssinicum* A. Br. [1]. However, molecular evidence suggests that *P. sativum* subsp. *elatius* and *P. fulvum* are the two wild species from which domesticated forms are derived. The domesticated *P. sativum* and *P. abyssinicum* are considered derivatives of *P. sativum* subsp. *elatius* in two independent domestication events [3]. Therefore, pea has a very rich genetic diversity due to its typical broad wild progenitors.

This genetic diversity and germplasm reservoir are vital for pea breeding, thus a large collection is maintained in gene banks and is well-studied and preserved. These genetic materials include approximately 98,000 pea accessions distributed in about 25 gene banks, of which some 58,000 are unique accessions (https://www.genesys-pgr.org (accessed on 18 September 2022)). Despite the availability of these large collections, less than one per cent is made up of wild relatives [4]. Yet, wild species are a valuable reservoir of resistance traits, particularly useful for pre-breeding and disease resistance breeding. Hence, it is critical to properly characterize and preserve them to maximise their utilization [5,6]. Pea wild relatives have already been successfully explored and used in pea breeding [7–9]. So, resistance breeding can be accelerated by the introgression of desirable wild alleles, complemented with the adoption of novel techniques and tools for pea precision breeding [10,11].

The cultivated forms of pea are grouped into green pea for human consumption, dry pea and fodder pea for animal feed [12]. Pea is the fourth most important grain legume worldwide, following soybean, peanut and dry bean. The current annual production estimates for the year 2020 are around 14.6 million metric tons (MT) of dry pea and 19.8 million MT of green pea with yield averages of 2.0 t/ha and 7.9 t/ha, respectively, projected to increase in the coming years. The leading producers are Canada, the Russian Federation, China, India, Ukraine, the United States of America, Australia, Ethiopia and Tanzania [13]. The current crop productivity must be increased to feed the continual population growth envisaged to reach 10 billion by 2050, despite the expected reduction in arable land. It is consequently crucial to explore smart agriculture and suitable land use to ensure climate change mitigation and food security [14,15]. Accordingly, pea is a legume crop candidate in this race for food security owing to its multipurpose low input requirement, nitrogen-use efficiency, soil economy amendment and biodiversity attributes. Moreover, it has a nutritional purpose with valuable sources of dietary fibre, high proteins (25%), mineral nutrients and many health benefits [16]. Likewise, the mutualistic association of pea with the N_2-fixing soilborne rhizobium reduces chemical fertilizer inputs, attaining yields with minimal impact on the ecosystem [17,18]. Pea and related legumes as pre-crops in a rotation programme can also provide important benefits for the environment by liberating nitrogen for the succeeding crops, and can serve as a buffer to cereal crop farmers in terms of price instability and crop failure [19–21].

Despite these benefits, the pea crop can be constrained by various diseases that severely affect yield and seed quality [4,12]. The most challenging disease limitations are the rhizospheric diseases found in the vicinity of host plant roots within the bulk soil, where they incubate and infect their hosts [22]. Yet, the available management techniques are limited—those available are either not economical or unhealthy to the environment, and most research efforts against pea rhizospheric diseases only present an incomplete resistance. Therefore, this review aims to consolidate the progress from various research findings on rhizospheric pathogens and serve as a resource for sustainable pea soilborne disease management and future advancements in pea resistance breeding. We delve into the most important pea rhizospheric diseases, including fusarium wilt (*Fusarium oxysporum* f. sp. *pisi*), root rot complex (fusarium, aphanomyces, thielaviopsis root rot, seed-based rhizoctonia and pythium rots), parasitic broomrapes (*Orobanche crenata*), and nematodes complex (cyst, root-knot and root lesion nematodes). Here, soilborne diseases and rhizospheric diseases are synonymously used to refer to diseases found incubating and surviving in the soil niche—with emphasis on those diseases that are closely associated with or infecting the host pea root zone (rhizosphere).

2. Host Pea-Rhizosphere-Microbial Interactions and Stress Amelioration

The rhizosphere harbours a great diversity of microorganisms involved in plant–microbe and plant–rhizosphere–microbe interactions. These microorganisms include plant pathogenic, beneficial, antagonistic, and synergistic associations [23]. The selective release of exudates from host plants activates and sustains specific rhizobacterial communities at the locality of the host's rhizosphere [24]. For instance, the successive rotational cultivation of pea and other pulses with cereals modified soil structure and increased the diversity of the rhizosphere microbial community [25], suggesting that legumes pose a much stronger influence on the selection of their rhizosphere than cereals. Accordingly, pea plants influence the configuration of microbial populations in the rhizosphere systems (Figure 1).

Figure 1. Rhizosphere–microbiota configuration in pea roots showcasing a multitude of interactions. Represented here are mutual associations such as rhizobia bacteria and host pea nodules—fixing nitrogen for pea growth while pea maintains bacteria nourishment, and beneficial associations such as earthworms soil burrow activities—improving soil aeration and fertility for pea, and antagonistic associations such as parasitic broomrapes and parasitic nematodes affect pea production. (Illustration made in ©BioRender—biorender.com).

Advanced high-throughput sequencing tools can identify host plant–rhizosphere–microbial associations to provide clarity regarding the resistance mechanisms in legume root diseases [26]. This high-throughput information has been used to map rhizosphere related traits in pea. For instance, 16S rRNA gene amplicon sequencing and quantitative polymerase chain reaction (qPCR) techniques were used to analyse a pea microbial community—the abundance of proteobacteria, rhizobacteria spp., and discovered plant-growth-promoting genes indicated that pea plants shape their rhizosphere for nutrient uptake and stress amelioration [27]. Pea nodules were also profiled using both responsive genetic constructs and hormones, suggesting that enhanced cytokinin during rhizobium symbiosis is associated with bacterial penetration and the subsequent differentiation of bacteroid within plant cells [28]. Hence, plant growth regulators could enhance microbial activity. Moreover, a study established that a supernumerary chromosome of *Nectria haematococca* carries pea-pathogenicity-related genes and a trait for pea rhizosphere competitiveness. These genes can enable nutrient absorption and stress tolerance in pea [29]. The role of rhizosphere traits in delineating toxic compounds has also been elucidated, postulating that the soil–microbial–plant continuum can mitigate aluminium toxicity through the solubilization of soil phosphates, thereby increasing nutrient availability in the pea rhizosphere [30]. A study of pre-penetration mechanisms of *Fop* revealed that pea root's secretion of the toxin phytoalexin-Pisatin into the rhizosphere could reduce pathogen pressure [31].

In tomatoes, rhizosphere traits were used to identify beneficial bacterial genes involved in the metabolism of plant polysaccharides, iron, sulphur, trehalose, and vitamins, whose genetic variation was linked to specific quantitative trait loci (QTLs) [32]. The information on these beneficial genes can be exploited to facilitate the mapping of pea rhizosphere traits. Furthermore, plant-growth-enabling microbes that form an active part of organic matter or biofertilizers can enhance crop yield and complement plant resilience against pathogens [33,34]. The notable mutual association between pea root nodules and the bacteria *Rhizobium leguminosarium* for N_2 fixation and the release of root exudates, which promote the infection of parasitic broomrapes, is well-established [35]. A recent study suggested that prolonged organic soil amendment can harness phosphate-solubilizing microbes in the rhizosphere. These beneficial bacterial endophytes within the pea rhizosphere can improve insoluble nutrient uptake [36]. The rhizosphere deposition of host-root exudates disseminates complex extra-cellular DNA molecules, and antimicrobial proteins to neutralise pathogenic threats and avoid host tissue intrusion [37]. However, some dispositions of secreted effectors by rhizosphere pathogens may inhibit plant defences by affecting RNA helicase involved in root defence and development [38]. An understanding of this linkage and the increasing evidence of genetic variation controlling pea–microbial interactions can be exploited for pea breeding. Current insights into the genetic basis of host pea–rhizosphere interplay are key for defence against disease complexes, providing opportunities for pea resistance breeding [39]. Therefore, the resistance against rhizospheric diseases can be achieved by the indirect selection of rhizosphere associated traits. These can be used as pointers and their correlated attributes incorporated into breeding programmes [40,41]. Hence, a holistic understanding of the entire soil ecosystem is needed for resistance breeding strategies to unravel the obscure and complex defence mechanisms against rhizospheric diseases [26,42].

3. Pea Rhizospheric Diseases

Novel plant diseases will continue to emerge due to changing climate conditions, farming systems and feeding modes of pathogens and parasites. As a result, previously non-economic pathogens may become important in range and severity [43,44]. The outbreak of any disease in the rhizosphere depends not only on the pathogen, but also on its interaction with the host and the environment, in the so-called pathogen–host–environment disease triangle [45]. Control strategies must be formulated to disrupt this triangle disease balance, either by altering the environment, reducing pathogen pressure or increasing host resistance [46]. Unfortunately, rhizospheric pathogens are more difficult to manage than their aboveground counterparts since they have the ability to survive on plant debris and soil in modified forms [26]. These pathogens can be harnessed and characterised with modern tools to estimate their historical emergence, spread and evolution in the rhizosphere in order to improve diagnostics and the design of efficient control measures [47,48].

The production of grain legumes such as pea is severely affected by soil-infecting pathogens. Pea rhizospheric or soilborne diseases can cause yield losses of between 80 and 100% when not controlled [49,50]. The resemblance of soilborne disease symptoms to other biotic factors further complicates the diagnosis strategy. Thus, the use of chemical, biocontrol and cultural means of control is crucial, although insufficient in most instances. Hence, it is necessary to find the most efficient, economic and environmentally friendly approaches using resistant cultivars [51]. However, breeding for resistance to rhizosphere diseases is complex, as the quantitative nature of resistance is often partial in nature and difficult to select. On the other hand, monogenic resistance is easy to select and has been achieved in some soilborne diseases of legumes, but the rapid evolution of pathogens can easily break down the existing levels of resistance. So, the understanding and evaluation of soilborne diseases by efficient identification and screening techniques is a requirement for the implementation of effective control strategies and resistance breeding [49,52]. Therefore, in this review, some of the important pea rhizospheric diseases are deliberated.

3.1. *Fusarium* Wilt

The genus *Fusarium* encompasses many, mostly soilborne, species reported to affect animals, humans and plants alike, including endophytes, saprobes and pathogens [53]. The genus was initially classified into 70 species on the basis of morphology, biology and phylogenetic criteria [54]. A more recent evaluation reduced their number to nine species, of which *F. oxysporum* is the most important, broadly affecting many crops [39]. Within *F. oxysporum*, the forma specialis *pisi* (*Fop*) primarily infects pea [55,56] and grass pea [57]. Similar to the other f. sp. of *F. oxysporum*, *Fop* has the ability to colonise and infect the soil, organic material residues, host roots and shoots leading to vascular wilt and yield penalties [57,58]. Four races have been described for *Fop* so far, of which races 5 and 6 are mostly found within the Americas, while races 1 and 2 are found worldwide [59,60]. Races 3 and 4 were initially defined as distinctive variants, prior to their reclassification as aggressive variants of near wilt *Fop* race 2 [61]. This explains why there are four races. Generally, symptoms of the disease are typically progressive from older to younger leaves/stems resulting in stunting, yellowing, necrosis to wilting, and finally, plant death [55,62,63]. *Fop* can be disseminated over long distances through the transport of contaminated soil samples by animals, farm technicians, machinery or infected crop seed lots [64,65]. Fungi show survival plasticity in most soil conditions, through the formation of thick-walled chlamydospores that can hibernate in the soil for years. This persistence is further enhanced through aggressive saprophytic mode in host plant debris and via released microconidia and macroconidia, which aid new infections and dispersal [60,66]. Therefore, a clear understanding of the disease cycle and mode of spread is required for the efficient implementation of diagnostic and control strategies.

The disease cycle (Figure 2) initiates in the soil, where the spores, in the form of chlamydospores, lie in wait for unsuspecting hosts. Pea roots emit unknown signals that trigger spore germination in the presence of conducive microclimatic factors. After germination, germlings grow towards pea roots by chemotropism. Once in contact with the host roots, the elongating spore attaches to the root and then enters through root tips, root hairs and wounds. The invading hypha then directly penetrates the root epidermis without the formation of a differentiated penetrative structure. The penetrating hypha then advances through the root cortex until it reaches the vascular tissues and enters the xylem vessels through the pith. Upon reaching the vascular stele, the fungus modifies into an endo-phytic mode of colonization using the xylem vessels to colonize the upper part of the host plant. At this stage, the fungus remains exclusively within the xylem vessels, taking nourishment from the host. The extensive growth of the fungus in the xylem leads to the interruption of the water flux, causing the distinctive wilt symptom manifestation on leaves and ultimate death of the host. Upon plant death, it begins to grow out of the xylem vessel to reach the surface of the dead tissues, where it produces chlamydospores, which disperses into the soil, causing looming infestation in subsequent seasons [55,57,58].

Once the disease cycle and dynamics are well-understood and proper screening methods are identified, managing the disease becomes simple. Fortunately, the fungi can easily be recovered from infected soil and plant tissues and maintained on an axenic medium, such as potato dextrose broth/agar, for its morphological and molecular characterisation. While this has aided the identification and diagnosis of the pathogen, its control strategies continue to be a huge challenge [67,68]. The disease can be managed by enforcing policy compliance via host plant material quarantine at border points of entry from hotspot zones to contain the pathogen [69]. A number of agronomic and cultural practices, such as crop rotation, fallow cropping and fertilization, can improve soil integrity to suppress *Fop* by reducing the pathogen inoculum [57]. For instance, micro-fertilizers inhibit the production of the mycotoxin, fusaric acid, which is a key pathogenic element in *Fop* infection [70,71].

Biological control through the use of natural enemies and suppressive organisms can be used to control *Fop*. The use of bioagents such as *Trichoderma* spp. and *Bacillus*-based antimicrobials have been reported as potent mediators for managing *Fop*, and also stimulates systemic host plant resistance and development [72,73]. Further, chemical control such as soil fumigation with ammonium bicarbonate or plant treatment chemicals can be effective for *Fop* control [74]. In the absence of resistant cultivars, the integrated application of these control methods would be more effective on *Fop*.

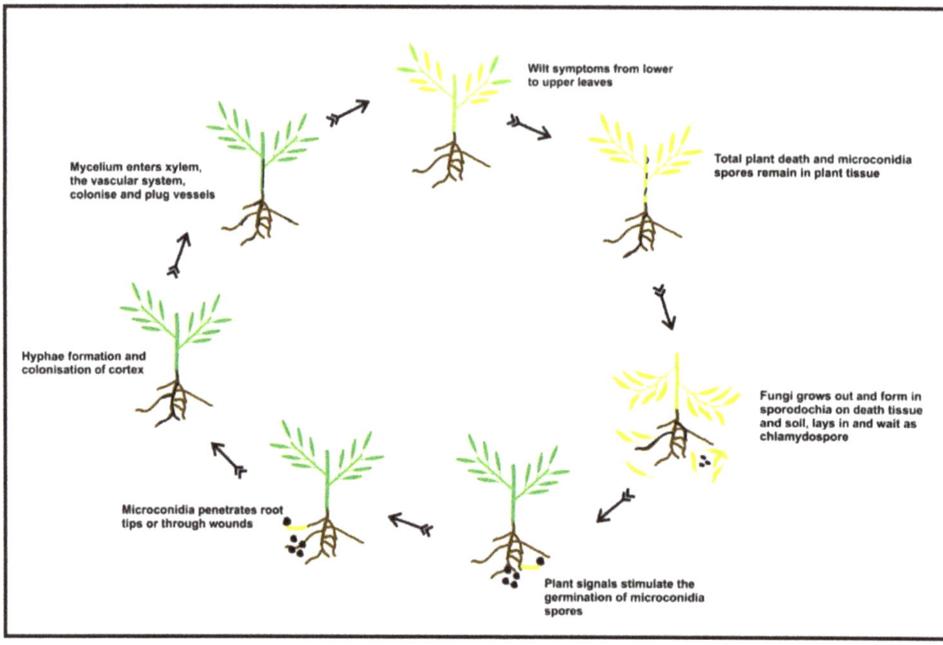

Figure 2. A pictorial presentation of fusarium wilt life cycle.

The identification of resistance sources within germplasms and a combination of practical phenotypic evaluation standards and host resistance genotyping can improve sustainable *Fop* control strategies in the long term [57,75,76]. So, the knowledge of the fungal patho-systems, genetics and physiological variations can inform sustainable resistance breeding strategies [52,64]. Qualitative resistance for *Fop* race 1, 5 and 6 have been identified and deployed into pea varieties using classical breeding [62,77]. Resistance to race 1 is located in the pea linkage group (LG) III in the vicinity of a SCAR marker Y15_999_*Fw*, which could be useful for improving selection for race 1-resistant cultivars through marker-assisted selection (MAS) [78,79]. Resistance to *Fop* race 2, which causes late symptoms, regarded as "near wilt" in the field, appears to be quantitative in nature in both pea and grass pea [55,68,80]. Accordingly, the resistance identified so far against race 2 is more complex, with two minor loci identified in LG III (*Fnw3.1* and *Fnw3.2*) and a major allele at LG IV (*Fnw4.1*) providing a basis for MAS in pea breeding against this race of *Fop* [81]. Pea association mapping for agronomic and quality traits revealed 71 significant marker associations linked to 25 valuable traits, including disease resistance [82]. In addition, a recent genome-wide association study (GWAS) in grass pea identified 17 single-nucleotide polymorphisms (SNP) associated with *Fop* resistance, seven of which were assigned to pea chromosomes 1, 6, and 7 by sequence homology hits [57,83]. This knowledge can improve the use of molecular strategies for *Fop* resistance breeding. Further studies showed that, host resistance is a result of the development of physical and chemical barriers within pea root tissues, leading to cell wall and xylem reinforcement to block pathogen growth [60]. Furthermore, these studies found that some pea accessions harboured a constitutive pre-penetration resistance inhibiting *Fop* germination, which might contribute to reduce pathogenic pressure [31]. These mechanisms can be exploited in pea breeding against *Fop*. Although monogenic resistance is easier to handle by breeding, there is the risk of breakdown by the constant mutation of the pathogen that already led to the emergence of the distinct races of *Fop* [49,79]. Thus, it is essential to continuously search for novel sources of resistance to complement and reinforce the resistance of elite cultivars, hence the need to pay attention to durable resistance [55,68].

3.2. The Rots Complex

Root, seed and seedling rots are a group of soilborne pathogens that affect pea production in many areas. Pea root rots are mainly caused by *Aphanomyces euteiches*, *Fusarium solani*, and *Thielaviopsis* spp., whereas seed and seedling rots are caused by *Rhizoctonia* spp. and *Pythium* spp. [63]. Some viruses and bacteria are also cited to provoke root rots [34]. In this paper, they are together referred to as the rots complex. These rot complexes can be instigated by a single pathogen or a cocktail of pathogens causing seed/seedling decay, root and foot decay, and necrotic wilts [84]. Consequently, the rots complex leads to severely stunted host growth, loss of vigour and yield injury to crops [85]. They also stimulate pre-emergence and post-emergence damping-off and other root rot pestilences [50]. Commonly used management practices, such as soil treatments, adjustment of planting time, seed vigour and quality, seed treatment with fungicides and crop rotation have not been broadly successful [86]. Therefore, the deployment of integrated techniques in the rhizosphere with host-plant-associated microbes could promote the control of the rots complex [87].

3.2.1. Aphanomyces Root Rot

The soilborne oomycete *Aphanomyces euteiches* is a highly specialised legume pathogen causing root rots, which result in economic yield losses. This pathogen is considered an important disease in pea-growing economies and affects the plant at all developmental stages [88]. Infection is initiated by zoospores and oospores stimulated to germinate by pea root exudates; they penetrate pea roots and colonise host tissues, forming a network of mycelia [89]. The mycelia take assimilates from the host plant, causing root damage and a yield penalty of up to 86% in heavily infested pea fields [90]. These effects are further aggravated through the action of other pathogenic enzymes when found in associations with root rots [49]. The root-rot-afflicted plants become dwarfed and water-stressed due to the progressive development of watery lesions and dark brown roots with cortex decay, resulting in subsequent yellowing and wilting of the upper part of the host plant [91]. After the host roots and tissues decay, spores are re-injected into the soil and some remain on the plant material debris, all serving as an inoculum awaiting another infection cycle. These spores are so hardy that they can remain in the soil for years in the absence of the host. The survival of the pathogen is further prolonged by secondary and volunteer weeds serving as alternative hosts that also increase the inoculum bank in the soil [63,90,92]. In addition to their longevity, the pathogenic spores can be disseminated over long distances by running water and other farm operations [87].

The resistance to *A. euteiches* in pea is of quantitative nature with few cultivars containing good levels of partial resistance, and there are no efficient control methods [93,94]. However, several applications of biocontrol products with antagonistic soil bacterial strains such as *Bacillus* spp., *Pseudomonas fluorescens*, *Pantoea agglomerans*, and *Lysobacter capsici* were found to suppress aphanomyces root rot in pea [95]. Although biological control agents have better efficacy under control conditions than under field trials, the integration of a mixed inoculant of different strains could be more effective than a sole application, improving their large-scale efficacy [96]. For instance, the combined application of *Lumbricus terrestris* and *Bacillus velezensis* was found to reduce *A. euteiches* infection in pea, with the response being attributed to soil disturbance and direct antagonistic feeding [97]. The efficacy of some biological agents is comparable to chemical applications. In turn, the application of the endophyte fungus *Clonostachys rosea* was found to reduce pathogenic aphanomyces intensity by 76%, similar to fungicide treatments [84]. Although certain chemicals can be effective, their cost and environmental concerns make them less preferred. The fungicides metalaxyl-M and fosetyl-AL were reported to increase seedling emergence and delay the infection of *A. euteiches* in field pea [84,90], while ethaboxam fungicides were found to suppress *A. euteiches* of pea in North America [98]. On the other hand, cultural methods of soil testing can help to avoid infested areas, and the use of *Brassicaceae* and *Poaceae* families in rotation management practices can minimize the spread of *A. euteiches* in pea fields [89,90].

Advanced breeding techniques for resistance against *A. euteiches* are promising. For example, molecular marker technology was used to identify and release eight F8-derived recombinant inbred lines (RIL) of pea with improved partial resistance to *A. euteiches* and acceptable agronomic attributes [99]. QTL association studies revealed partial resistance genes against *A. euteiches* in pea, reporting a reliable QTL (*Aph1*) in LG IVb, which explains 47% of the variation as a potential option for pea improvement [100]. In a meta-analysis of partial resistance genes to *A. euteiches* in four main sources of resistance in pea, seven highly consistent genomic regions with potential for MAS were identified, and candidate genes underlying six meta-QTL regions were found in collinearity between pea and *Medicago truncatula* genomes [101]. QTL validation was also performed to confirm *A. euteiches* resistance in different pea backgrounds in a backcross to generate near-isogenic line (NIL) populations. These allowed for the development of several breeding lines carrying distinct levels of resistance by marker-assisted backcrossing [102]. Similar association studies also identified 11 markers significantly associated with *A. euteiches* resistance, confirming and refining the location of previously identified QTLs, and uncovering four novel resistance QTLs [103–105]. In addition, one significant SNP was mapped to the major QTL *Ae-Ps7.6*, associated with both *A. euteiches* resistance and pea root system architecture traits [104]. Recently, another stable and major QTL was mapped to an approximately 20.0 cM region on pea chromosome 4, which was identified as the most consistent region conferring partial resistance to *A. euteiches* [106]. These efforts together with some transcriptomic pathway studies that identified expressed candidate genes at the cellular level [92], guided the improvements of *A. euteiches* resistance towards precision and marker-assisted breeding.

3.2.2. Fusarium Root Rot

Fusarium root rot has a wide host range, attacking the host at the cotyledon and tap root zones within ground clearance down to the rhizosphere. In pea, it can be caused by *Fusarium solani* f. sp. *pisi* (*Fsp*) and *Fusarium avenaceum* complex [61]. These fungi are distinct from *Fop*, though they sometimes combine to form a complex and associate with other diseases to infect pea [63,107,108]. Although *Fsp* was initially described as the main causal agent of fusarium root rot in pea, *F. avenaceum* seems to be gradually gaining more prominence over *Fsp*, as *F. avenaceum* was found to constitute 45- 48% of recovered isolates from analysed infected samples of pulse crop residues [109]. The infecting fungal chlamydospores can lay dormant in the soil for many years until their germination is activated by exudates from imbibed host seeds and nourishment obtained from the germinating seedling. Fusarium rot fungi can also freely form cocoon-like colonies in the soil rhizosphere, which increase the pathogen fitness and longevity in the soil [85]. The infection of the pathogen is through the stomata of epi-hypocotyl zones and downstream into the root system. Pathogen severity and spread are further exacerbated by conducive soil moisture, low soil fertility gradient and other pathogenic stresses [65,85,110]. Fusarium root-rot-infected plants display root lesions, vascular tissue decolourisation and root dysfunctions similar to other root rot complexes [39,111].

Control management involves the cultural use of extensive rotation regimes to minimise the inoculum bank, use of good agronomic practices to improve soil fertility and root growth, the avoidance of soil compaction, and the use of good quality seeds [56,110,112]. Some biocontrol agents, such as rhizosphere mycoflora and *Bacillus* spp., have been used as potential solutions to suppress *Fsp* [73,113]. Interestingly, *Pseudomonas* spp. and *Bacillus* spp. can produce volatiles with an antibiotic effect on *Fsp*, whilst *Pythium oligandrum*, *Trichoderma* spp. and *Streptomyces* spp. exhibit hyper parasitism and mycoparasitism against the fusarium root rot [114]. Currently, there are no effective chemical fungicides for the control of *Fsp* [110].

Resistance to *Fsp* has been reported as a quantitative genetic trait with some partial resistance obtained in pea germplasms. The accessions with pigmented flowers were observed to have a tendency of greater partial resistance to *Fsp* than white-flowered cultivars [115]. A pea RIL population was used to further identify one major QTL and five minor QTLs for *Fsp* resistance and one QTL against *F. avenaceum*. The major QTL, *Fsp-Ps*$_{2.1}$, was located within a 1.2 cm interval on chromosome 6 and explained 44–53% of the total variance, while the five minor QTLs were more loosely located on chromosome 1, 4, 5 and 7 [65]. Similarly, four QTLs associated with resistance to *F. avenaceum* in a pea RL population were identified by QTL analysis. These markers identified a key QTL on chromosome 7 that explained 21.7% of the variance in resistance [116]. Another study identified a QTL associated with resistance to *Fsp*, where the QTL was found flanking markers *AA416* and *AB60* on LG VII with 39% explained variance [117]. A more recent study using SNP-derived markers from differentially expressed genes and two RIL populations detected additional QTLs on chromosomes 2 and 3 and confirmed two minor QTLs on chromosome 5 [34]. The application of these identified loci is expected to improve pea resistance breeding. In addition, an extensive RNASeq approach comparing eight tolerant and susceptible pea accessions to *Fsp* identified more than 42,000 differentially expressed genes (DEG) in response to *Fsp* [118]. These DEGs could complement pea breeding efforts for resistance to fusarium root rot.

3.2.3. Black Root Rot

The causal pathogen of black root rot, *Thielaviopsis basicola*, has a wide host range affecting various crops. On a global scale, black root rot is of minor importance [119] but can gain prominence over time with increased inoculum when not controlled. *T. basicola* forms a complex with other soil microbes to exacerbate injury to host roots. Host infection is initiated by pathogenic spores in soil and plant debris, which attack host seeds and root surfaces to spread and cause harm to plants. The spores infect root hairs, and the germ tube penetrates the root cells. The hyphae differentiate into feeding structures (haustoria-like) to absorb nutrients from host cells and disrupt water flux, causing cortical cell death [63]. Field infection is aided and disseminated by farm tools, water flow downstream and aggravated by high soil moisture with high temperatures. *T. basicola* causes pea roots to turn dark-brown or to develop dark lesions and necrosis at the root base, leading to water deficit, stunting growth and plant dysfunction [63,108].

A recent phylogenetic analysis showed that *T. basicola* includes cryptic sister species, and it was proposed to rename them as *Berkeleyomyces basicola* and *B. rouxiae* [120]. This suggests that the pathogen could be evolving, and highlights that a more detailed understanding is needed to establish control strategies. The main management strategy is to avoid the spread of the pathogen into non-infected areas. Furthermore, crop rotations and biocontrol measures can limit the pathogen build up [63]. Although little is understood about the management of the pathogen in the pea crop, lessons can be learned from other crops. In bean fields, cover cropping and green manure are reported to reduce root rot intensity [121]. Compost and other organic amendments are also useful in controlling root rots [108]. Chemicals have not been feasible against this disease, hence breeding for resistance is deemed economically appropriate but it is not considered a major concern in pea [63]. However, there are concerns about the emergence of new variant of *T. basicola* that may become a major threat to pea in the future due to changing climate conditions. *T. basicola* resistance is controlled by a single dominant gene in tobacco [122], and similar resistance genes (R-genes) were found to reduce root lesions in tree cotton (*Gossypium arboreum*). Three resistance QTLs, which together explained a total phenotypic variation of 32.7%, have been identified in *G. hirsutum*. A subsequent synteny analysis of these significant QTL regions with Arabidopsis revealed a total of 624 genes, including 22 pathogen defence genes and 36 stress-related genes that could correspond with *T. basicola* resistant QTLs [123]. These associated genes and novel disease discovery mechanisms are valuable for understanding the resistance of *T. basicola* in crops including legumes. Although the mechanism of resistance for *T. basicola* in pea is not sufficiently studied, it is anticipated that a similar mode of quantitative resistance may be present in pea as observed in other crop species.

3.2.4. Rhizoctonia Root Rot

Rhizoctonia root rot is caused by *Rhizoctonia* spp. which are sporeless and largely consist of hyphae, hyphal propagules and sclerotia. They are classified under filamentous hymenomycetes with an asexual mode of reproduction comprising a host of unrelated species. To distinguish them, *Rhizoctonia* spp. isolates are ascribed to distinct anastomosis groups (AG) via a somatic incompatibility test. The most important group is composed of isolates of *Rhizoctonia solani* [124]. Most AG groups of *R. solani* can cause root rots in pea, but the AG4 is the most frequent and most virulent [63]. The economic merit of this pathogen spans most pea-growing areas, and it is harmful to a broad host range of plants. In severe cases, the effects of the pathogen induce yield losses from about 75% to total crop failure [125]. The germination of *R. solani* sclerotia is triggered by host roots or imbibed seed exudates. The elongating hyphae then enter the seedling by means of soft spots or wounds. The appressorium attaches to host cells and supports pathogen growth, where the pathogen secretion further weakens the host cells to enable colony growth and host root tissue invasion. This eventually causes cell death and the production of sclerotia mass as inoculum for the next pathogenic cycle [61,63,125]. *R. solani* causes soggy lesions on juvenile shoots and roots, adventitious shoots formation, as well as the damping-off and dieback of roots and seedlings. This is exacerbated by high-moisture regimes, poor soil and drainage conditions [63]. *R. solani* is mostly responsible for poor germination and growth due to seed/seedling infections in the rhizosphere. The effects of the disease are visible in the whole plant, and pathogen coexistence with other complexes further complicates diagnosis [126,127]. *R. solani* is reported to be the prevalent isolate causing severe root rots in pea in the Colombia and Washington basins of the USA. The recovery of the pathogen from plant samples and molecular identification by laboratory techniques are key to the pathogen diagnosis [125,128]. *R. solani* can live for lengthy periods in the soil and plant materials as infectious hyphae or hibernate in saprophytic mode [124]. The pathogen exhibits opportunism to non-host species and weeds, further expanding its host scope in a diverse manner through secondary hosts [63].

Complete control strategies of *R. solani* are lacking, hence it is more practical to foster preventive and resistance strategies [125]. Consequently, it is paramount to use eco-friendly techniques to manage the pathogen. Thus, several measures for the sustainable improvement of pea against this rot complex are well-established [52]. Resistance to seedling rot caused by *R. solani* is linked with seedling epicotyl thickness and age, with younger seedlings being more susceptible to infection [63,125]. Thus, good husbandry practices with the use of certified disease-free seeds, seed treatment with fungicides or starter macro-nutrients are useful control strategies. This would improve seedling vigour and limit the inoculum of *R. solani* [129].

In a long rotation and no-till programme, a low population of *R. solani* was observed in legumes compared to cereal cultures, suggesting that long rotations prior to planting can reduce the pathogen activity in pea [130]. The most significant biocontrol study undertaken on pea is the combination of *Rhizobium leguminosarum* and *Trichoderma* spp. isolates to manage *R. solani* [127]. These beneficial mycoparasitism processes mostly involve the synthesis of cell wall lytic enzymes. Promising studies on other crops include the use of bio-formulation of rhizobacteria mixtures as seed treatment significantly suppressed *R. solani* in sunflower [131], *Trichoderma* species were effectively used for *R. solani* management in chickpea [132], and the *Bacillus* spp.-mediated synthesis of selenium nano-particles was also found to be efficient to attenuate *R. solani* in faba bean [133]. Furthermore, *R. solani* fungal genome mycovirus sequence breakthroughs have advanced control methods [134,135]. These fungal genome sequences have already steered the identification of a new effective candidate genus (Betapartitivirus) to help understand the patho-dynamics and enhance control strategies [136]. These mycoviruses have the ability to reduce the mycelial growth of many fungi [135]. Some microbial markers associated with pea roots have also been elucidated and show promise as a *R. solani* control [129]. Thus, they could be useful as biocontrol agents to control *R. solani* in pea. Chemical controls are mostly limited but the use of bavistin 50 WP, and provax-200 fungicides show a complete inhibition of *R. solani* colony growth in pea [137]. Fumigants with the active ingredients of thiophthalimide (captan), quinone outside inhibitors (azoxystrobin), succinate dehydrogenase inhibitors (fluxapyroxad), demethylation inhibitors (ipconazole), aromatic hydrocarbons (tolclofos-methyl) and phenylpyrroles (fludioxonil) have been used as seed treatments to manage *Rhizoctonia* spp. These compounds are mostly single-site inhibitors, so it is recommended to combine them with other treatments to reduce the risk of pathogen resistance [63]. The integration of management techniques could be more efficient against *R. solani* since host plant resistance is limited [26]. It has been demonstrated that canola lines expressing pea defence gene *DRR206* confer resistance to *R. solani*. These constitutively mediated defence genes are often effective against different pathogen species and can be exploited for *R. solani* resistance in pea [138]. In a study of *R. solani* AG2-2IIIB strain, it was shown that lysin motif (LysM) effector protein contributes to virulence through the evasion of chitin-triggered immunity, and the obstruction of this protein has the potential to protect pea against *R. solani* [139]. In the model legume, *M. truncatula*, ethylene-mediated signalling and the overexpression of isoflavonoid proteins have been identified to improve resistance against *R. solani*. These expressed defence genes can protect legumes against root pathogens [140]. The understanding of the role of these disease response proteins can be used to complement pea resistance breeding against *R. solani*.

3.2.5. Pythium Seed/Seedling Rot

This root rot is caused by the oomycetes *Pythium* spp., including *P. ultimum* and *P. aphanidermatun*, which broadly affect pea. The disease is severe under conducive moist and poor soil conditions, causing poor germination, bare patches, and a watery-brown discoloured root system [84,141]. The pathogen produces oospores that reside in the soil and on plant tissues, infecting pea at the pre-emergence and post-emergence stages to cause seed/seedling rot and damping-off [63]. The host plant attracts the zoospores, which encyst on the root surface. The cyst–spore germinates and the mycelia invade and propagate within the root tissue, taking up nutrients to produce sporangia and more oospores to aid their survival [142]. The produced propagules can be spread by animals and farm operation tools and can remain in the soil for several infection cycles [50,130]. Their survival structures exist as saprophytes and persevere in the soil as colonizers of organic matter. Therefore, a thorough understanding of the disease pathway will be essential to establish successful control strategies. This can be elucidated using analyses of the genetic relatedness among isolates from diseased plants, water and soil samples [142].

Control of this pythium complex is difficult due to the interwoven nature of many species. Damping-off can be controlled using integrated disease management strategies involving cultural, biocontrol and host-resistance breeding [39,143,144]. For cultural management practices, it is important to avoid poorly drained soils, maintain good soil health, good tillage, and use disease-free seeds [133,144]. Biological control using rhizobium strains has been shown to offer protection against pythium damping-off in pea [84]. Other biological agents of *Bacillus* spp., *Trichoderma* spp., *Pseudomonas* spp., and *Streptomyces* spp., are appropriate beneficial species that are useful against pythium rots at small scale, but they may not be viable for larger areas [63]. Apart from biocontrol agents, compost has been reported to be effective against pythium rots in vegetables and legumes. Soils with enforced organic matter content also suppress the disease manifestation [112,144]. Chemical treatment with appropriate fumigants, including mefenoxam and ethaboxam, can be effective against pea [63] if long-term effects are discounted.

In pea, resistance to *Pythium* spp. is not readily available. Pea cultivars with large seeds have been reported to give vigorous seedling stands, exhibiting better resistance than those with small seeds [145]. Consequently, using quality seeds with maximum vigour is crucial for healthy seedling emergence to reduce the risk of infection [63,146]. In soybean, partial resistance has been achieved through the identification of two QTLs against *P. ultimum*, explaining 13% and 16% of the phenotypic variance, respectively [147], and two other QTLs (0.8534 and 0.6955 heritability) against *P. aphanidermatun* [148]. An additional GWAS on soybean lines was used to identify 7 SNP markers and 5 QTLs (9.7–16% of variance) associated with partial resistance to *Pythium* spp [149]. Similarly, in snap bean, polygenic resistance was identified in a major QTL associated with *P. ultimum* resistance, which explained 25-49% of variability [150]. Although it is difficult to identify race-specific resistance genes against *Pythium* spp., quantitative gene expression approaches have proven valuable for developing pythium complex resistant varieties [151]. The motif Arg-Xaa-Leu-Arg (RxLR), encoding protein effectors in *Pythium* spp. that exhibited a suppressive activity, was identified and shown to protect host plants by promoting cell death [152]. This knowledge is useful for developing potential candidates for pre-breeding efforts in pea genetic resistance.

3.3. The Nematode Complex

Soil nematodes are micro-worms with unsegmented bodies, less than 2 mm in diameter. The genus comprises up to 20 genera, some of which are associated with the rhizosphere of field pea roots and related legumes [61,63]. They have a wide host range, infesting and causing injury to many plant species. The most economically damaging nematodes on pea are the *Heterodera goettingiana* (cyst), *Meloidogyne* spp. (root knot) and *Pratylenchus* spp. (root lesion) [153,154], making up the pea nematode complex under consideration.

The infection cycle of nematodes is initiated by free-living eggs and juveniles in the rhizosphere of an ideal host or volunteer host plants. Eggs hatch and juveniles detect and penetrate host roots, guided by host plant signals. They then feed on the root tissues and grow to form a colony, which causes the most damage. The infection cycle is similar for most nematodes, except the pea cyst, where the mode of feeding and colony structure of females undergo shedding. Pea cyst nematode eggs also requires a host stimulus to hatch, and they are very mobile. Contrary to cysts, root knot nematodes do not depend on host stimuli to hatch and are largely spot feeders, whilst females of root lesion nematodes do not shed at root penetration and are highly migratory. In general, female nematodes are the most destructive and the producers of eggs for their sustained survival advantage. Thus resistance techniques should be targeted at limiting female populations since males are absent in most species [153], except in lesion nematodes, which assume a sexual mode of reproduction. The nematode complex can continue their unabated infection spree on available hosts and thrive for many years [61,63,155–157].

Nematode-affected plants show signs of dwarfism, patches of necrosis and yellowing, loss of vigour approaching those of soil health deficiencies and a reduction in host root system function [158]. The affected areas also differ in intensity by soil fertility gradients, soil type, soil gravimetric contents, host plant resilience and the parasitic-ecosystem dynamics [155]. In pea, parasitic nematodes generally interfere with rhizobium activity, which reduces nitrogen fixation and leads to yield penalty. In addition, they can serve as vectors of other diseases, exposing plants to further stresses [63]. Nematodes are detected through keen field observations, soil sampling and analysis to estimate the economic threshold per gram of soil. The minimum threshold of 3 to 15 eggs per gram of soil is enough to engulf an entire pea field, resulting in serious yield penalties of up to 50% if not controlled [63].

Control strategies for these soilborne parasitic nematode complexes are limited, thus there is a need to prevent the movement of the pathogen into non-infected areas by adhering to field phytosanitary and cultural protocols. Given the difficulty to completely control nematode complexes, the overall goal of the applied control strategies is usually to prevent parasitic nematode populations from reaching economic injury levels [159]. Cultural methods of tillage and crop management have been recommended to reduce nematode incidence [130,159,160]. Some fungi are nematophagous and they prey on nematode eggs as food, which can be exploited for biocontrol management [161]. Mycorrhizal fungi have also been reported to inhibit nematode entry by aiding hormonal balance in the host to improve nutrient uptake, and different strategies of biocontrol have been reported [161,162]. For chemical control, fumigation with dichloropropene-based fungicides and non-volatile nematicides can provide a good control of these nematodes to improve pea yields [163]. Host resistance efforts, using infested fields and controlled pot experiments show promise [157,164–166]. However, for decades, the interest in plant parasitic nematodes has mainly focused on biological control and host–parasite interactions [154].

3.3.1. Pea Cyst Nematodes

Cyst nematodes represented by isolates of *Heterodera goettingiana* is an important economic parasite of pea, which can survive for over 10 years without a viable host. Cyst nematode populations are widespread in Europe, Russia and the Mediterranean belt [63]. Besides peas, many legume species are suitable hosts of this nematode, including faba bean, vetch and black medick [167]. Infection is promoted by non-aerated soil, a suitable moist climate and host plant vulnerability. In addition to depriving host plants of nutrients and water, they also reduce the plant's natural defences and synergise with other harmful microbes to injure plants [161]. Host exudates trigger pathogen activity and root penetration, where juveniles feed on root cells and moult. Upon its entry into the plant roots, the nematode intracellularly migrates to the vascular bundle, where it selects a cell to become its initial feeding site, and subsequently develops into a syncytium [168]. Cysts are often found within feeding sites, but they can be transported by running water, farm equipment and other soil fauna, which increase the infection and spread of the pathogen [63].

The best management strategies are to use fallow systems, host resistance and trap crops to activate nematode activity leading to suicidal natural death and predation [45]. This makes it important to study nematode morphological diversity, population dynamics and ecology to promote sustainable management strategies [169]. Late planting may reduce the infection rate of the pathogen due to reduced moisture conditions towards the end of the season [166]. It was reported [63] that the utilization of longer rotation systems in the absence of host plants for up to 6 years can drastically reduce cyst nematode populations. However, for the success of rotations, a better understanding of the host range of nematodes is required to prevent incubation in the absence of the main host plant [167]. The incorporation of 10% aldicarb nematicides into infected soil before planting has been reported to control *H. goettingiana* in pea [170]. The nematicide oxamyl also improved pea growth against *H. goettingiana* [171]. Pea-recommended herbicides such as prometryne are reported to hinder egg-hatching ability and affect female *H. goettingiana* development [172]. This could protect plants against nematodes in the absence of resistant cultivars.

The screening of a pea collection under controlled conditions led to the identification of potential sources of resistance. The results show that five *Pisum abyssinicum* accessions, a *P. arvense* accession, and a *P. elatius* accession demonstrated a moderate resistance to *H. goettingiana* [173]. Subsequently, a histological observation of some pea accessions showed that resistance to *H. goettingiana* was caused by the rapid degradation of cysts through a hypersensitive response and associated cell death. The resistance reaction was also connected with lignification and suberisation processes surrounding the necrosed area of infection [174]. Reverse transcriptomic expression analysis revealed that the polygalacturonase-inhibiting protein improves pea defence against *H. goettingiana*. The gene *Pspgip1* was expressed in infected cortical cells and localised in cells bordering cyst-induced syncytia in resistant pea roots, confirming that this gene was key in preventing *H. goettingiana* establishment [175]. Further molecular studies suggested that the involvement of lipoxygenase and the polygalacturonase-inhibiting protein in cyst nematode induced resistance in pea [174,176]. Although these findings improved the understanding of the cyst nematode resistance reaction, they have not been effectively exploited to create resistant elite cultivars through breeding [63,177].

3.3.2. Root Knot Nematodes

Root knot disease is caused by *Meloidogyne* spp. These nematodes have a wide host range, being capable of infesting most crop species, including legumes. Amongst the

Meloidogyne spp. complex, the *M. incognita* (Figure 3) is nature's most flourishing plant pathogen. Infected plants become malnourished, stressed with a loss of vigour, and could prematurely shed inflorescence [153,155]. *Meloidogyne* spp. are obligate endoparasites and stationary feeders but can be spread by farm activities and running water [45]. They possess a feeding conduit, which operates as a molecular sieve for ingesting assimilates. The accumulation of these assimilated compounds forms the characteristic giant cell chambers (nodule-like) on host roots. Their effect reduces host crop physiological functions through root system deprivation of essential nutrients.

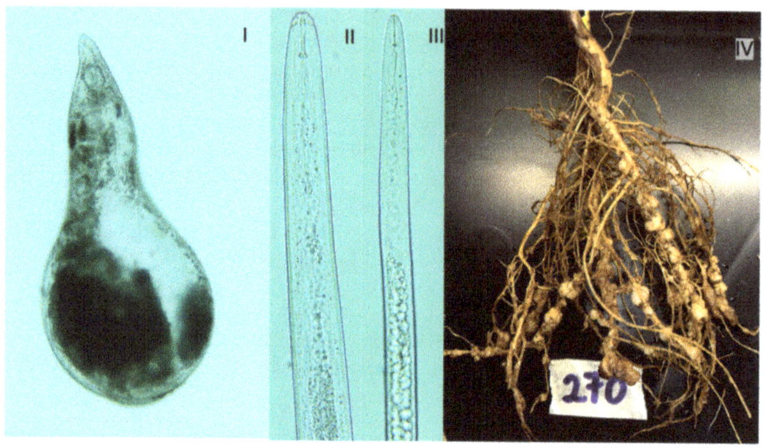

Figure 3. Microscopic view of (**I**) female *M. incognita*, (**II**) male *M. incognita*, (**III**) juvenile *M. incognita* (source P. Castillo, IAS-CSIC) collected from (**IV**) galled root symptoms on soybean plant (O.Z. Wohor IAS-CSIC).

These giant cells manifesting in all *Meloidogyne* spp. indicate that nematode–host-plant-associated genetic factors are involved in the formation process [153], thus enabling the exchange of biochemical compounds between the pathogen and host plant. The *Meloidogyne* spp. infection complex in association with *F. oxysporum* increases fusarium wilt severity and other biotic stresses [178]. Therefore, it is essential to understand the genetics and dynamics of the host–parasite associations to identify management approaches.

The management of root knot nematodes requires integrated strategies, including cultural methods of prolonged fallowing [157], avoiding host plant monocultures and abstaining from areas with secondary host weeds such as alfalfa [63,167]. The beneficial associations of some microbes can be exploited to enhance the availability of nutrients and useful minerals to improve plants' health and immunity to induce resistance to root knot pathogens [114,179]. Some of these microbial isolates from rice rhizosphere are reported as biocontrol agents against root knot nematodes [180]. Panth et al. [114] reported findings of important biocontrol agents such as *Gliocladium catenulatum*, which produce toxins effective against nematodes, *Purpureocillium lilacirium* as parasites on nematodes, and some beneficial *Mycorrhizae* spp. that compete with nematodes for space and resources. In chemical control, low-risk organic chemicals from essential oils of *Ocimum basilicum* at 125 mg/L were effective against *M. incognita*, resulting in a 70% mortality rate in a bioassay test conducted under controlled conditions [181]. The use of chemical fumigants such as 1,3-dichloropropene alone or in combination with methyl isothiocyanate has great potential for the management of root knot nematodes in pea [163].

Host resistance to root knot nematodes is considered a viable long-term goal, although little resistance has been found and released for cultivation so far [63]. A number of criteria have been used to screen for resistance, such as the rating of infected roots on a 1–5 scale, gall severity index [166], or a 0–4 index scale and the reproduction index, calculated as the ratio of total nematode population count per host on the initial inoculated population count [182]. The use of reproductive index led to the identification of pea accessions with a moderate-to-high level of partial resistance to *M. incognita* and *M. javanica* [63,182–184]. The advanced gene manipulation techniques of using nematode protein effectors contributed to elucidate resistance signalling pathways against *M. incognita*. These pathways can modify the host cell wall and regulate stress signalling and hypersensitive response [185]. In *Arabidopsis thaliana*, it was revealed that the actin-depolymerizing factor (ADF) is upregulated in the giant feeding cells during host infection and the knockdown of a specific ADF isotype inhibits *Meloidogyne* spp. proliferation. Hence, limiting the expression of this gene in host plants can prevent the development of *Meloidogyne* spp. [186]. The draft genome of *M. incognita* provided insights into some parasitic adaptive genes and their effects on immune-competent hosts [187]. These may be applicable through gene expression studies to identify *Meloidogyne* spp. resistance sources to promote pea breeding.

3.3.3. Root Lesion Nematodes

Root lesion is elicited by *Pratylenchus* spp. The most prolific and common root lesion nematodes affecting pea are *P. neglectus* and *P. thornei*. Root lesion nematodes are difficult to manage due to the high invasion of both juveniles and adults at point of entry and inside root tissue [63]. The juvenile stage 4 (J4) can assume a dormant non-feeding form in a self-regulatory mode under unsuitable conditions, while waiting for the appropriate conditions to rejuvenate. All the larval stages are active feeders on root cells, and they reproduce in a sexual mode. Quantitatively, one pathogen per gram of soil is capable of rapidly ravaging pea fields [188], validating their aggressiveness and spread in the rhizosphere.

Their persistence is further strengthened by their parthenogenic nature since females can produce eggs in the absence of males [189]. They are migratory and can navigate between feeding sites for nourishment and spread exudates and pathogens between host plants. These exudates and movements to and from feeding sites cause characteristic lesion symptoms in roots. Unlike root knot nematodes, they do not exhibit the usual root bulging signs. Instead, they degenerate the root epidermal cells causing serious root damage and yield losses [190,191].

Early detection is key for the prevention and management of this nematode. So, there is the need to frequently examine pea fields to avoid new infections and establish standard disease management strategies to prevent the spread of lesion pathogens [45,166]. Although crop rotation can be efficient for controlling lesion nematodes, it is less effective against root lesions due to their persistence and ability to infest both cereals and legumes [63,130,192]. Sterilization methods using gamma irradiation (7.5 k Gy), aerated steam (80 °C) and methyl bromide fumigation successfully eliminated *P. thornei* in vertisols in Australia [193]. The non-volatile nematicides (aldicarb, oxamyl, carbofuran, thionazin and fenamiphos) provide good control against these nematodes [163,194]; however, these chemical sterilizations may not be economically feasible in large pea fields [63]. Therefore, the combination of different disease management techniques should be applied to efficiently reduce root lesions severity.

Disrupting the ideal environments of the pathogen and searching for host resistance is needed to control *Pratylenchus* spp. [63,165]. Despite limited available information on resistance against *Pratylenchus* spp., some progress has been made. For instance, a screening for low pathogenic growth and reproduction index reported moderate levels of resistance in pea accessions against *P. nanus* [188]. Resistance can also be determined from soil samples by comparing the initial pathogen population density over the end of season population density. This is more representative than using lesion symptoms [166]. Furthermore, moderate levels of resistance against *P. thornei* and *P. neglectus* were found in chickpea cultivars and wild relatives [189,195]. The broader characterisation of the resistance mechanisms acting in these legumes could have a positive impact on pea breeding against root-lesion nematodes.

3.4. Root Parasitic Weeds: Broomrapes

Broomrapes are soilborne root parasitic plants that constrain crop production. They are widely distributed in the temperate regions of the Mediterranean and Middle East [196]. The family *Orobancheae* consists of about 150 species, some of which can broadly affect important crops, with the species *Orobanche cumana* and *O. crenata* being specific to a few plant genera, whilst *Phelipanche aegyptiaca* and *P. ramosa* are broadly found in nature. The legume-damaging species are *O. crenata*, which is distributed widely; *P. aegyptiaca*, which is restricted to the Eastern Mediterranean and the Middle East; and *O. foetida*, so far limited to Tunisia and Morocco [196,197]. Between them, *O. crenata* is the most important root parasitic weed constraining pea production since it is capable of causing up to 100% yield losses if not controlled [198]. *O. crenata* is a holoparasite without chlorophyll and therefore feeds solely on host plants for all of its nourishment. Germination is usually initiated by seed preconditioning under conducive environments triggered by strigolactones and other chemical signals typically produced by host roots [199]. Upon germination, the emerging radicle grows towards the plant roots by chemotropism. This is guided by the concentration gradients of the host stimuli. The germination radicle then attaches to the root and penetrates the host root vascular system for nourishment and development. This leads to the formation of specific nodules and tubercles on the root surface from which the broomrape shoots are differentiated and emerge from the soil to flower, allowing seed formation and dispersal to continue its endless lifecycle (Figure 4) [200–203]. Broomrape seeds can remain dormant in the soil rhizosphere for many years in the absence of a host.

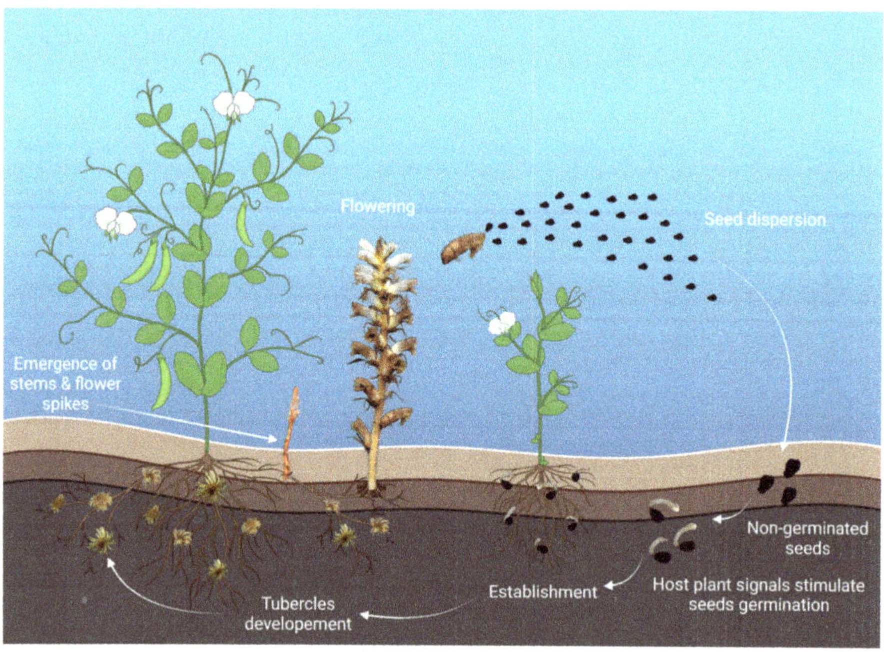

Figure 4. Broomrape infection life cycle, exhibiting below- and above-ground scenarios. (Illustration made in ©BioRender—biorender.com).

A series of broomrape management strategies has been proposed, including field sanitation and containment, prevention/avoidance of spread, and agronomic and chemical treatments to reduce seed banks in the soil [197,203,204]. Cultural control measures have shown promise against *O. crenata* through crop rotations/trap cropping or intercropping with allelopathic crops and field sanitation. Intercropping pea with oat, fenugreek, or berseem clover has been shown to reduce *O. crenata* infection [205,206]. Control techniques that trigger suicidal seed germination have long been suggested but have not reached a commercial stage due to difficulties in the formulation and delivery of germination stimulants or trap crop use [202].

Chemical control of broomrape by foliar applications of glyphosate at low rates is recommended for faba bean and vetches, but pea is highly sensitive to glyphosate-based herbicides [200]. However, pea has better tolerance to pre-emergence and post-emergence imidazolinone herbicide treatments, although no complete control is provided, and treatments are mostly less effective on earlier sowing dates [207]. The interspecific association between the host pea and parasite makes the selective use of herbicides practically impossible. Moreover, the efficacy of herbicides requires repeated applications that are often not cost-efficient for farmers, given the low-input cultivation system of pea [196].

Biological control has also been proposed to protect pea against broomrapes. Some rhizospheric-associated beneficial microorganisms such as arbuscular mycorrhizal fungi, *Azospirillum* spp., *Azotobacter* spp., *Bacillus* spp., *Pseudomonas* spp. and other rhizobacteria can provide protection to pea by suppressing *O. crenata* germlings [208]. These rhizobacteria can reduce the parasitism of *O. crenata* on pea roots via induced chemical and mechanical blockages in the host xylem [209]. Likewise, some arbuscular mycorrhizal fungi (AMF) can reduce the seed germination rate of *O. crenata* [35], and *O. cumana* seed establishment [210]. The use of non-pathogenic *F. oxysporum* strains has also been reported to control *O. cumana* and *P. aegyptiaca* [85]. Despite these positive results at the laboratory scale, biological controls are yet to be used in large-scale commercial applications. Since there is no single efficient control mechanism for managing *O. crenata*, the most promising mechanism is integrated pathogen management with different control strategies and breeding for resistant cultivars [204].

Breeding approaches for resistance to *O. crenata* mostly target host resistance and herbicide resistance [211]. Broomrape resistance breeding is challenging due to the low levels of resistance detected in existing pea germplasms and the low heritability of the resistance [200,212]. Therefore, highly variable germplasms and an understanding of the parasitic host specificity and population dynamics are needed for implementing efficient breeding strategies. Fortunately, unlike reported biotypes in *O cumana* [210], there is no clear proof of the existence of biotypes for *O. crenata* species thus far [213]. Therefore, the effectiveness of *O. crenata* resistance found in a host genotype may not diverge from the environment in which it is grown.

The genetic base for resistance to broomrapes is suggested to involve a complex inheritance governed by minor genes with small effects [214,215]. Resistance was confirmed to be largely quantitatively inherited and steered the identification of partial resistance in wild pea and landraces [200]. The achieved resistance has since been successfully bred into pea cultivars [9,216]. This highlights the importance of wild relatives as sources of useful alleles in pea breeding. As an alternative to a host resistance control strategy, *O. crenata* resistance may be achieved by breeding for early maturity lines, which have the advantage of escape to outcompete the parasite [200,217].

The current use of applied breeding is leveraging biotechnological tools to develop efficient markers to help breeders speed up cultivar release to farmers [4,211]. Molecular markers were utilised in an $F_{2\text{-}3}$ population to detect two QTLs (*Ocp1* and *Ocp2*) for broomrape resistance, which together explained 20% of phenotypic variance [214]. A subsequent study on a RIL of the same biparental population elucidated four useful QTLs associated with field traits and broomrape resistance mechanisms assessed in vitro. The QTLs individually explained about 10 to 17% and 8 to 37% of the phenotypic variance for field and in vitro attributes, respectively [215]. These molecular markers provide the basis for linked trait association for marker-assisted selection (MAS). However, further saturation of the detected loci is often required to bridge the gap between alleles and the genomic regions flanking markers to enable marker-assisted breeding (MAB) [197,218]. The use of microarrays, transcriptomics and proteomics provided quality gene expression indicators to elucidate *O. crenata* resistance. For example, a gene expression profiling of *M. truncatula* against *O. crenata* using a transcriptome pathway found many functional genes and validated 35 associated defence genes acting at the early penetration and late tubercle necrosis stages [218]. A proteomics analysis using gel electrophoresis and mass spectrometry untangles 49 differential gene spots of defence- and stress-related proteins against broomrape infection in model *M. truncatula* [219]. This confirmed previous proteomic studies that reported 22 expressed gene spots related to defence response against *O. crenata* in pea [220]. Therefore, the recent advances in omics and molecular breeding have the potential to change the paradigm of pea breeding [4] and elucidate the molecular basis of *O. crenata* resistance.

4. Breeding Enabling Approaches for Disease Resistance

Progress has been made in identifying sources of resistance to pea rhizosphere root diseases. The effects of these diseases are determined by multiple soil microbial factors and host plant attributes [129]. Conventional breeding approaches are expected to remain viable in disease screening for many decades and beyond. However, these traditional methods need to be complemented with novel genomic and biotechnological approaches for better results. The breeding strategy for pea is similar to other self-pollinated crops. These can be implemented by germplasm assembly and the selection of ideal parents for breeding new cultivars. The development of improved cultivars for resistance against a single pathogen is often a simple process. However, this requires a good source of resistance with efficient and extensive screening approaches to provide sufficient selection pressure [201,221].

4.1. Phenotyping

Phenotyping entails the morphological description of visually developing plant parts, physiology, stress resistance and agronomic parameters linked to yield. Phenotyping is crucial to plant breeding since it is the main basis for selecting lines for developing new cultivars [14]. Novel genomic sequencing technologies have become more affordable, resulting in quality reference genomes and huge genomic datasets, yet phenotyping remains a limiting factor in accessing these gains [222,223]. Disease resistance screening is performed by subjecting breeding lines to uniform disease pressure. This enables the discrimination of contrasting lines into resistant, susceptible and intermediate classes. The selection of desired traits requires an ideal environment to permit full phenotypic expression of resistant alleles [224]. Traditional plant phenotyping based on visual observations and manual data capture is predisposed to evaluation errors [225]. So, breeders are required to utilize the understanding of genotype and environment relations to improve phenotypic accuracy and reduce these errors [226]. Currently, efficient and simple specialized systems are beginning to accelerate pea resistance breeding under optimal growth conditions [227].

4.1.1. Field and Controlled Condition Phenotyping

Under field and controlled conditions, detailed descriptions of screening methods for legume root diseases [39,49,145], parasitic weeds [228], nematodes [177], and their application in pea breeding have been reported. Field screening enables the simultaneous evaluation of genetic materials at a large scale under natural environmental conditions, although it allows less control of the pathogen and the environmental factors. Under controlled conditions, disease resistance assessments are more accurate, and environmental factors, such as light, water, nutrients and temperature, are better controlled than under field conditions [229]. Here, seedlings, in vitro cultures, detached plant parts and young/short cycle plants are mostly preferred. However, there is often a poor correlation between the results obtained in the greenhouse or laboratory and those collected under field conditions due to field plasticity, and the high genotype by environment interaction under field conditions [44,49]. Thus, refinements in screening methods are continuously being implemented to improve accuracy. For instance, Bani et al. [55] modified the root-dip method of inoculating soilborne fungal pathogens on pea roots and incorporated a two-way rating scale of disease severity at the whole-plant and leaf scales. This method improved fusarium wilt disease examination and provided a comprehensive description of pea resistance to *Fop*. Similarly, refinement of the pot and in vitro screenings revealed many potential pea lines with some levels of resistance against broomrapes based on avoidance, low host induction, suicide/necrosis of germlings, and exudate/germination inhibition mechanisms [8,230,231]. Recently, a non-inversive greenhouse system was shown to be useful for screening field pea against aphanomyces root rots to elucidate biogenic markers for the pathogen control [232], thus providing an advanced and efficient phenotyping regime for pea disease assessment.

4.1.2. High-Throughput Phenotyping

Nowadays, low-throughput screening is considered a bottleneck to phenotyping and often requires specialised breeders' expertise, thus high-throughput techniques are beginning to replace visual screening [224,233]. High-throughput phenotyping gives the opportunity to overcome visual assessment bias and difficulty to access plant traits. This would improve selection intensity and allow cost-efficient precision screening of large numbers of samples [234]. Currently, intelligent above- and below-ground vehicles are equipped with remote sensors to efficiently capture quantitative and geographic data across broad areas to improve breeding programmes [235]. The modern use of smart cameras, unmanned aerial vehicles (UAV), near-infrared reflectance spectroscopy (NIRS)/infrared systems, X-ray tomography, artificial intelligence (AI), machine learning (ML) data [236,237], and other specialised precision intelligent robotic systems are starting to deliver reliable phenotyping [238]. The advent of 5G technology could further accelerate rapid data capture, data transfer and interoperability between intelligent platforms, which would aid real-time cloud data storage. This is expected to eliminate unreliable data storage and multiple data transfer misrepresentations—transposition/inversion and substitution of figures manually inputted between systems [239]. The application of high-throughput imagery phenotyping has elucidated early pea vigour traits and improved pea breeding [240]. Likewise, novel infrared imaging technique has been exploited to evaluate differential pea and *M. truncatula* lines against their respective *F. oxysporum* pathogens to discriminate between resistant and susceptible plants. This could be useful for obtaining first-hand disease information before the development of disease symptoms in host plants [67,241]. Consequently, high-throughput phenotyping tools have overcome previous time-consuming constraints. These innovative tools coupled with industrial-scale genotyping could be used to mine germplasms for important traits. This can improve scientists' use of genomics, bioinformatics and biotechnological methods [223,242,243].

4.1.3. Innovative Rhizotrons for Rhizosphere Phenotyping

In the past, conventional phenotypic techniques to assess root disease, including soil excavation and profiling soil sections served their purpose, albeit being disruptive to plants [244] and presenting a high heterogeneity and interference ambiguity [245]. To circumvent these difficulties, researchers have, over the years, developed several innovative systems to achieve the in vitro characterisation of root development and root responses to stresses [246]. These innovative tools specifically designed for root examinations are broadly termed as rhizotrons or mini-rhizotrons (Figure 5). They are basically a plant roots observatory system for continuous monitoring and non-destructive sampling of the rhizosphere functions at different developmental stages of crops [247]. Rhizotron observatories can be useful for improving pea-breeding programmes. However, light capture bias in most enclosed systems may hinder comparative analysis with field data [248]. So, rhizotrons have to be accurate to mimic soil conditions to enable results to be representative of field scenarios [246]. Rhizotrons come in various configurations with assorted materials, forms, shapes and sizes, comprising basic wooden boxes, soil trenches with plastic insertions, glass walls to recent opaque imaging tools, and other underground observatory facilities [249,250].

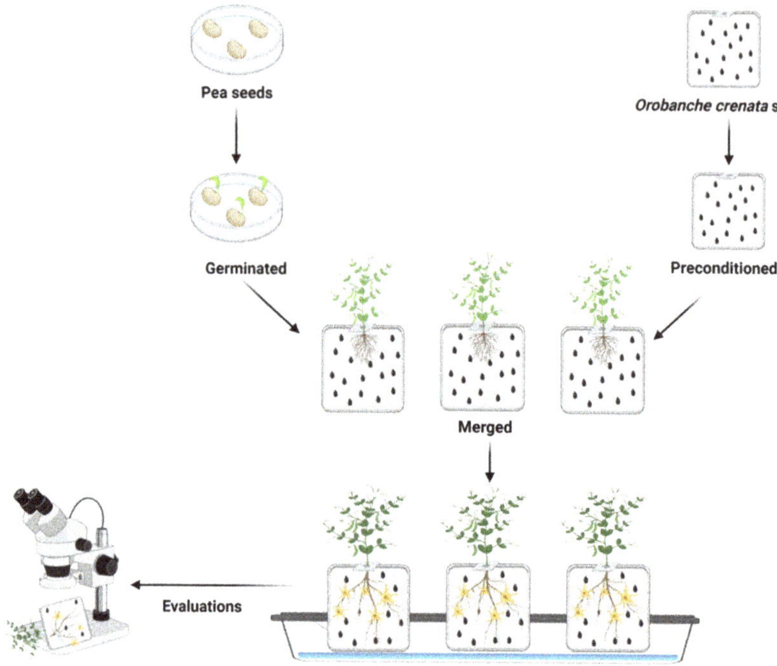

Figure 5. A mini-rhizotron set up for *O. crenata* screening, indicating pre-germination of host plant seeds and preconditioning of parasitic seeds to merger for association and evaluation. (Illustration made in ©BioRender—biorender.com).

Recently, breakthroughs in the non-disruptive examination of plant roots have been made by innovative high-throughput phenotyping methods and simple root analytic tools [247]. Topical innovative soil-filled rhizoboxes have been established for rhizospheric exploration. These have been used to determine root physicochemical properties based on hyperspectral imaging [251]. Likewise, an innovative minirhizotron was presented as 'SoilCam' for automated root system monitoring and root imagery analysis to promote crop performance [252]. Rhizotrons can be integrated into speed-breeding protocols [14,253,254] to improve pea resistance breeding against rhizosphere pathogens. An example of cheap rhizotrons for practical pea root examinations are simple Petri dishes filled with soil media sandwiched between fiberglass paper and the host roots, which have been adopted for determining the resistance response against broomrapes in pea [8,212,217] and other legumes [255,256]. Similar minirhizotrons are effective for evaluating resistance against parasitic *Striga hermonthica* in cereals [257–259], and assessing the biocontrol of broomrapes using rhizobium strains [208–210]. Therefore, rhizotrons have great implications for pea improvements towards the control of *O. crenata* and may be useful for the study of rhizobacterium dynamics, parasitic nematodes, and other microfauna interactions with pea roots.

4.2. Trait Discovery and Pre-Breeding for Resistance Breeding

Gene discovery and pre-breeding are crucial for the continuity of pea breeding programmes. The success of breeding programmes is dependent on broader genetic pools with a high genetic diversity and environmental adaptability. Germplasms from gene banks are a great source of genetic diversity, which is key for crop improvement [11,196,223]. This variability can be sourced from wild relatives, landraces, breeding lines, and mutants, which are very useful for pre-breeding, and can stimulate breeding. Understanding the genetics of cross-compatibility among germplasm can reduce breeding barriers to enable the full utilisation of pea genetic diversity in breeding. All *Pisum* sp. are readily crossable to the cultigen, *P. sativum* ssp. *sativum*. Therefore, the use of pea wild relatives with appropriate breeding regimes can generate multitudes of pre-bred genotypes, which can further be utilised in mainstream breeding for continuous accelerated gains [6]. Although selection in breeding is labour-intensive, breeders need to evaluate their decisions with scientific accuracy and economic merit to improve genetic gain [260,261].

4.3. Genotyping: Genomic Tools and Genetic Breeding Approaches

Genomic approaches using high-throughput genomic information in the areas of genome sequencing, data resequencing, genome-wide markers, genetic maps, QTLs, diagnostic markers, and omics strategies (transcriptomics, proteomics, metabolomics biomarkers), assist with and direct multiple breeding strategies [4,233,262]. The genetic revolution provided by the next-generation sequencing (NGS) platforms ensures the development of approaches, such as genotyping by sequence (GBS), diversity array technology sequencing (DArTseq), ribonucleic acid sequencing (RNA-Seq), whole-genome sequencing (WGS), among others, which have improved the quality of marker technologies [263,264]. This led to the discovery of extensive single-nucleotide polymorphic (SNPs) markers [265] with a huge potential for pea improvements [233]. These novel platforms have already guided the identification of heritable QTLs contributing to phenotypic variance in pea resistance breeding [102,266–268]. These high-throughput techniques also enabled the quantitative elucidation of nematode population composition [269], and pea genetic diversity studies [270,271].

The upscaling of QTL mapping techniques is useful but mostly confined to detecting gene variants in bi-parental segregants. Therefore, GWAS based on the non-random linkage of loci in haplotypes enhances high-resolution mapping of quantitative traits. Thus, improving and complementing traditional bi-parental mapping and validation of QTL alleles to facilitate MAS [223,272]. The efficiency of GWAS is determined by LD, population structure and genetic diversity, and has the power to uncover causative genes missed in QTL mapping populations [104,273]. Thus, GWAS has successfully aided the identification of novel variant–trait associations for breeding valuable disease resistance traits in pea and other legumes [82,83,104,274,275]. Marker–trait association analysis has been used to uncover significant linkage among pea inbred lines for resistance to *Fop*-R1. Five candidate genes were identified, and three of those markers (*Fw_Trap_480, Fw_Trap_340, Fw_Trap_220*) were tightly linked to the *Fw* locus at 1.2 cM, thus offering the potential for MAS in pea [276]. The associated genetics with nanopore sequencing (SQK-RAD004) have been used to manipulate genomic regions controlling pea pod colour and other traits, which are expected to accelerate the pan-genome variation in pea Mendelian traits [277]. An association study was also used to elucidate the linkage between aphanomyce-resistant genes and late flowering, and flower colour variants in pea [103]. Novel QTL approaches can be applied to reduce the limitations of GWAS since a marker may not be in LD with the causal genes in succeeding generations [278]. This can be achieved by targeting haplotypes to accelerate trait introgression and QTL stacking via nested association mapping (NAM) with multiple association studies (QTL+GWAS). The use of advanced–backcross loci (AB-QTL) for allele delivery and expressed loci (eQTL) analysis can help explain the variation of SNPs in gene expression and pan-genomics for dissecting entire sets of gene families [263]. These advanced QTL techniques can enhance trait associations to narrow the genetic distance between markers and alleles for MAB.

The advances in proteomic pathways revealed gene expressions associated with disease resistance in pea. This has improved the understanding of the genetic basis for broomrapes and *Fop* infection studies [220,279,280]. Recently, genomic selection (GS) has become valuable in predicting the breeding values of cultivars without phenotypic information, using genomic data obtained from the prediction models of a training population. This has been successfully explored in pea predictive breeding [281] and selection for pea yield and abiotic stress tolerance [264,282]. Genomic selection can be used to identify multiple traits to improve pea rhizosphere disease-resistance mechanisms through genomic-assisted breeding [283]. Hence, these genetic tools and procedures improve the capacity of breeders for the uptake of biotechnology and reduce the gap between genomics, molecular and conventional breeding strategies [245,284]. For instance, the post-genome reverse genetics technique of gene silencing (RNAi) and 'targeted induced local lesions IN genomes' (TILLING) for gene deletion and point mutation can confirm gene function to accelerate a selection of desirable traits [4,285]. This has facilitated the characterisation of mutant nodulation traits in pea [286]. The construction of ethylmethan sulfonate reference mutant populations and data bases using phenotypic and sequence data have improved the availability of pea TILLING genes. These mutants can be scanned with the BLAST tool to locate similar gene families for different useful traits [287]. These mutation libraries are reported to improve virus-induced gene silencing (VIGS) implementation in pea [288]. VIGS of a yeast protein *MtSTP13* has been found to repress pathogenesis-related gene expression and enhance powdery mildew susceptibility in *M. truncatula*, while its transient overexpression improves the resistance of pea against powdery mildew [289].

The novel revolutionary gene editing CRISPR/Cas9 and single-guided RNA sequence (sgRNA) are powerful optimised tools for gene functional studies in many crops [14]. For example, the legume sugar transport protein gene *STP13* and its derivative protein products present in many plants, including legumes, have been found to contribute to basal resistance against both biotrophic and necrotrophic pathogens, which is exacerbated by the specific mutation of one of its amino acids. Hence, the knockout of this gene to modify the amino acid through CRISPR/Cas9 would eliminate its function and prevent or slow pathogen infection in pea [289]. Similarly, the use of transgenic canola lines expressing the pea defence gene (*DRR*206) was able to control *R. solani* infection [138]. This further reiterates the importance of transgenes in disease control. *Agrobacterium tumefaciens*-mediated gene transformation (T-DNA) techniques have led to the introduction of useful genes from wild relatives and unrelated sources into cultivated lines. These have been used to transform pea roots to induce defence-related mechanisms in response to colonization by a model non-pathogenic *Fusarium oxysporum* (*Fo*47) [290]. This T-DNA mediated technique could be applied to transform pea donor parents with disease-resistance traits, and the transgene can be used to create elite inbred lines to broaden genetic pools. Although some of these novel approaches are still in the discovery phase, they are expected to facilitate pea breeding in the near future and improve the accuracy of disease-resistance evaluations. Decisively, the availability of the genome sequence of pea is improving the application of high-throughput genotyping and genetic tools, thus enhancing association mapping studies to allow breeders to gain better insight into pea genetics to select important traits and their use in MAS [18,291]. This high-resolution pea genome could also serve as a model to strengthen the study of pea-phylogeny-related species.

5. Concluding Remarks

Soilborne pathogens are difficult to manage, and available control methods are not cost-effective nor environmentally friendly. Considerable efforts have been made to improve pea resistance against soilborne pathogens with some levels of incomplete resistance accumulated in cultivars. However, available sources of resistance are limited, and screening methods are time-consuming, reinforcing the need to improve phenotyping accuracy at an affordable cost by adopting novel technologies. This should be complemented with molecular techniques, although these techniques still require accurate phenotypic data to enhance their results. It is crucial to improve the current phenotyping bottlenecks to complement the available sophisticated genomic technologies for pea rhizospheric disease management. The current advances in powerful high-throughput phenotyping platforms would be useful for verifying genetic data. Correspondingly, the recent accompanying cost reduction in precision molecular genetic tools for genome sequencing and marker-trait discoveries are strengths for future pea resistance breeding. There is a great deal of available information and genetic resources as assets for enhancing pea breeding efforts. Remarkably, the optimised pea reference genome released in 2019 is positively impacting pea breeding strategies at the molecular level. This is expected to foster more precise breeding and plant performance indices to heighten complex trait discovery for genetic gain. Furthermore, different breeding programmes can collaborate to share variable breeding materials and enable the gene pyramiding of useful traits and implement different hybridization regimes with recurrent selections to stabilize acquired resistance genes.

At the same time, there is the need to complement these genetic gains with integrated disease management techniques to sustain and prevent such resistance from deterioration. It is paramount to continue to scout for soil amendment disease suppressers to improve soilborne disease breeding efforts. Management efforts that seek to prevent the establishment of soilborne diseases are the ideal control strategies. Thus, disease surveillance through quarantine, prohibition of planting materials from infested areas, seed treatment and approved seed sources, farmer and technician training and education, laboratory diagnostics, and collaboration among institutions and scientists for quick response to disease outbreaks is necessary. The way forward for the complex nature of rhizospheric diseases calls for more robust multidisciplinary approaches that employ all aspects of biological and socio-scientific understanding and policy directives to enhance environmental sustainability and food security.

Author Contributions: Writing—original draft, O.Z.W.; writing—review and editing, O.Z.W.; N.R.; C.O.O. and D.R.; funding acquisition, D.R. All authors have read and agreed to the published version of the manuscript.

Funding: This research was funded by the projects PID2020-114668RB-100 from the Spanish Research Agency (AEI) MCIN/AEI/ 10.13039/501100011033 and Tropical Legumes III project by International Crops Research Institute for the Semi-Arid Tropics (ICRISAT).

Conflicts of Interest: The authors declare no conflict of interest.

References

1. Smýkal, P.; Aubert, G.; Burstin, J.; Coyne, C.J.; Ellis, N.T.; Flavell, A.J.; Ford, R.; Hýbl, M.; Macas, J.; Neumann, P.; et al. Pea (*Pisum sativum* L.) in the genomic era. *Agronomy* **2012**, *2*, 74–115. [CrossRef]
2. Rubiales, D.; Ambrose, M.J.; Domoney, C.; Burstin, J. Pea. In *Genetics, Genomics and Breeding of Cool Season Grain Legumes*; Pérez-de-la-Vega, M., Torres, A.M., Cubero, J.I., Kole, C., Eds.; Science Publishers: Enfield, NH, USA, 2011; pp. 1–49, ISBN 978-1-57808-765-5.
3. Trněný, O.; Brus, J.; Hradilová, I.; Rathore, A.; Das, R.R.; Kopecký, P.; Coyne, C.J.; Reeves, P.; Richards, C.; Smýkal, P. Molecular evidence for two domestication events in the pea crop. *Genes* **2018**, *9*, 535. [CrossRef] [PubMed]
4. Pandey, A.K.; Rubiales, D.; Wang, Y.; Fang, P.; Sun, T.; Liu, N.; Xu, P. Omics resources and omics-enabled approaches for achieving high productivity and improved quality in pea (*Pisum sativum* L.). *Theor. Appl. Genet.* **2021**, *134*, 755–776. [CrossRef] [PubMed]
5. Coyne, C.J.; Kumar, S.; von Wettberg, E.J.B.; Marques, E.; Berger, J.D.; Redden, R.J.; Ellis, T.N.; Brus, J.; Zablatzká, L.; Smýkal, P. Potential and limits of exploitation of crop wild relatives for pea, lentil, and chickpea improvement. *Legum. Sci.* **2020**, *2*, e36. [CrossRef]
6. Pratap, A.; Das, A.; Kumar, S.; Gupta, S. Current Perspectives on Introgression Breeding in Food Legumes. *Front. Plant Sci.* **2021**, *11*, 589189. [CrossRef] [PubMed]
7. Cobos, M.J.; Satovic, Z.; Rubiales, D.; Fondevilla, S. *Er3* gene, conferring resistance to powdery mildew in pea, is located in pea LGIV. *Euphytica* **2018**, *214*, 203. [CrossRef]
8. Rubiales, D.; Fondevilla, S.; Fernández-Aparicio, M. Development of pea breeding lines with resistance to *Orobanche crenata* derived from pea landraces and wild *Pisum* spp. *Agronomy* **2021**, *11*, 36. [CrossRef]
9. Rubiales, D.; Osuna-Caballero, S.; González-Bernal, M.J.; Cobos, M.J.; Flores, F. Pea Breeding Lines Adapted to Autumn Sowings in Broomrape Prone Mediterranean Environments. *Agronomy* **2021**, *11*, 769. [CrossRef]
10. Das, A.; Parihar, A.K.; Saxena, D.; Singh, D.; Singha, K.D.; Kushwaha, K.P.S.; Chand, R.; Bal, R.S.; Chandra, S.; Gupta, S. Deciphering genotype-by- Environment interaction for targeting test environments and rust resistant genotypes in field pea (*Pisum sativum* L.). *Front. Plant Sci.* **2019**, *10*, 825. [CrossRef]
11. Varshney, R.K.; Barmukh, R.; Roorkiwal, M.; Qi, Y.; Kholova, J.; Tuberosa, R.; Reynolds, M.P.; Tardieu, F.; Siddique, K.H. Breeding custom-designed crops for improved drought adaptation. *Adv. Genet.* **2021**, *2*, e202100017. [CrossRef]

12. Rubiales, D.; González-Bernal, M.J.; Warkentin, T.; Bueckert, R.; Vaz Patto, M.C.; McPhee, K.; McGee, R.; Smýkal, P. Advances in pea breeding. In *Achieving Sustainable Cultivation of Vegetables*; Hochmuth, G., Ed.; Burleig Dodds Science Publishing Limited: Cambridge, UK, 2019; pp. 575–606. [CrossRef]
13. FAOSTAT, Statistical Database. Food and Agriculture Organization of the United Nations. Available online: http://www.fao.org/faostat (accessed on 29 June 2022).
14. Hickey, L.T.; Hafeez, A.N.; Robinson, H.; Jackson, S.A.; Leal-Bertioli, S.C.M.; Tester, M.; Gao, C.; Godwin, I.D.; Hayes, B.J.; Wulff, B.B. Breeding crops to feed 10 billion. *Nat. Biotechnol.* **2019**, *37*, 744–754. [CrossRef]
15. Varshney, R.K.; Bohra, A.; Roorkiwal, M.; Barmukh, R.; Cowling, W.A.; Chitikineni, A.; Lam, H.M.; Hickey, L.T.; Croser, J.S.; Bayer, P.E.; et al. Fast-forward breeding for a food-secure world. *Trends Genet.* **2021**, *37*, 1124–1136. [CrossRef]
16. Smýkal, P.; Varshney, R.K.; Singh, V.K.; Coyne, C.J.; Domoney, C.; Kejnovský, E.; Warkentin, T. From Mendel's discovery on pea to today's plant genetics and breeding: Commemorating the 150th anniversary of the reading of Mendel's discovery. *Theor. Appl. Genet.* **2016**, *129*, 2267–2280. [CrossRef]
17. Graham, P.H.; Vance, C.P. Update on Legume Utilization Legumes: Importance and Constraints to Greater Use. *Plant Physiol.* **2003**, *131*, 872–877. [CrossRef]
18. Kreplak, J.; Madoui, M.A.; Cápal, P.; Novák, P.; Labadie, K.; Aubert, G.; Gali, K.K. A reference genome for pea provides insight into legume genome evolution. *Nat. Genet.* **2019**, *51*, 1411–1422. [CrossRef]
19. Zander, P.; Amjath-Babu, T.S.; Preissel, S.; Reckling, M.; Bues, A.; Schläfke, N.; Kuhlman, T.; Bachinger, J.; Uthes, S.; Stoddard, F.; et al. Grain legume decline and potential recovery in European agriculture: A review. *Agron. Sustain. Dev.* **2016**, *36*, 26. [CrossRef]
20. Ojiewo, C.; Monyo, E.; Desmae, H.; Boukar, O.; Mukankusi-Mugisha, C.; Thudi, M.; Pandey, M.K.; Saxena, R.K.; Gaur, P.M.; Chaturvedi, S.K.; et al. Genomics, genetics and breeding of tropical legumes for better livelihoods of smallholder farmers. *Plant Breed.* **2019**, *138*, 487–499. [CrossRef]
21. Döring, T.F.; Rosslenbroich, D.; Giese, C.; Athmann, M.; Watson, C.; Vágó, I.; Kátai, J.; Tállai, M.; Bruns, C. Disease suppressive soils vary in resilience to stress. *Appl. Soil Ecol.* **2020**, *149*, 103482. [CrossRef]
22. Vishwakarma, K.; Kumar, N.; Shandilya, C.; Mohapatra, S.; Bhayana, S.; Varma, A. Revisiting Plant–Microbe Interactions and Microbial Consortia Application for Enhancing Sustainable Agriculture: A Review. *Front. Microbiol.* **2020**, *11*, 560406. [CrossRef]
23. Müller, D.B.; Vogel, C.; Bai, Y.; Vorholt, J.A. The plant microbiota: Systems-level insights and perspectives. *Annu. Rev. Genet.* **2016**, *50*, 211–234. [CrossRef]
24. Latati, M.; Aouiche, A.; Tellah, S.; Laribi, A.; Benlahrech, S.; Kaci, G.; Ouarem, F.; Ounane, S.M. Intercropping maize and common bean enhances microbial carbon and nitrogen availability in low phosphorus soil under Mediterranean conditions. *Eur. J. Soil Biol.* **2017**, *80*, 9–18. [CrossRef]
25. Hamel, C.; Gan, Y.T.; Sokolski, S.; Bainard, L.D. High frequency cropping of pulses modifies soil nitrogen level and the rhizosphere bacterial microbiome in 4-year rotation systems of the semiarid prairie. *Appl. Soil Ecol.* **2018**, *126*, 47–56. [CrossRef]
26. Wille, L.; Messmer, M.M.; Studer, B.; Hohmann, P. Insights to plant–microbe interactions provide opportunities to improve resistance breeding against root diseases in grain legumes. *Plant Cell Environ.* **2019**, *42*, 20–40. [CrossRef]
27. Chaudhari, D.; Rangappa, K.; Das, A.; Layek, J.; Basavaraj, S.; Kandpal, B.K.; Shouche, Y.; Rahi, P. Pea (*Pisum sativum* L.) plant shapes its rhizosphere microbiome for nutrient uptake and stress amelioration in acidic soils of the North-East region of India. *Front. Microbiol.* **2020**, *11*, 968. [CrossRef]
28. Dolgikh, E.A.; Kusakin, P.G.; Kitaeva, A.B.; Tsyganova, A.V.; Kirienko, A.N.; Leppyanen, I.V.; Dolgikh, A.V.; Ilina, E.L.; Demchenko, K.N.; Tikhonovich, I.A.; et al. Mutational analysis indicates that abnormalities in rhizobial infection and subsequent plant cell and bacteroid differentiation in pea (*Pisum sativum*) nodules coincide with abnormal cytokinin responses and localization. *Ann. Bot.* **2020**, *125*, 905–923. [CrossRef]
29. Rodriguez-Carres, M.; White, G.; Tsuchiya, D.; Taga, M.; VanEtten, H.D. The supernumerary chromosome of *Nectria haematococca* that carries pea-pathogenicity-related genes also carries a trait for pea rhizosphere competitiveness. *Appl. Environ. Microbiol.* **2008**, *74*, 3849–3856. [CrossRef]
30. Belimov, A.A.; Shaposhnikov, A.I.; Syrova, D.S.; Kichko, A.A.; Guro, P.V.; Yuzikhin, O.S.; Azarova, T.S.; Sazanova, A.L.; Sekste, E.A.; Litvinskiy, V.A.; et al. The role of symbiotic microorganisms, nutrient uptake and rhizosphere bacterial community in response of pea (*Pisum sativum* L.) genotypes to elevated Al concentrations in soil. *Plants* **2020**, *9*, 1801. [CrossRef]
31. Bani, M.; Cimmino, A.; Evidente, A.; Rubiales, D.; Rispail, N. Pisatin involvement in the variation of inhibition of *Fusarium oxysporum* f. sp. *pisi* spore germination by root exudates of *Pisum* spp. germplasm. *Plant Pathol.* **2018**, *67*, 1046–1054. [CrossRef]
32. Oyserman, B.O.; Flores, S.S.; Griffioen, T.; Pan, X.; van der Wijk, E.; Pronk, L.; Lokhorst, W.; Nurfikari, A.; Paulson, J.N.; Movassagh, M.; et al. Disentangling the genetic basis of rhizosphere microbiome assembly in tomato. *Nat. Commun.* **2022**, *13*, 3228. [CrossRef]
33. Whipps, J.M. Microbial interactions and biocontrol in the rhizosphere. *J. Exp. Bot.* **2001**, *52*, 487–511. [CrossRef]
34. Williamson-Benavides, B.A.; Sharpe, R.M.; Nelson, G.; Bodah, E.T.; Porter, L.D.; Dhingra, A. Identification of Root Rot Resistance QTLs in Pea Using *Fusarium solani* f. sp. *pisi*-Responsive Differentially Expressed Genes. *Front. Genet.* **2021**, *12*, 1379. [CrossRef] [PubMed]
35. Fernández-Aparicio, M.; García-Garrido, J.M.; Ocampo, J.A.; Rubiales, D. Colonisation of field pea roots by arbuscular mycorrhizal fungi reduces *Orobanche* and *Phelipanche* species seed germination. *Weed Res.* **2010**, *50*, 262–268. [CrossRef]

36. Harkes, P.; Van Steenbrugge, J.J.M.; Van Den Elsen, S.J.J.; Suleiman, A.K.A.; De Haan, J.J.; Holterman, M.H.M.; Helder, J. Shifts in the active rhizobiome paralleling low *Meloidogyne chitwoodi* densities in fields under prolonged organic soil management. *Front. Plant Sci.* **2020**, *10*, 1697. [CrossRef] [PubMed]
37. Hawes, M.; Allen, C.; Turgeon, B.G.; Curlango-Rivera, G.; Minh Tran, T.; Huskey, D.A.; Xiong, Z. Root border cells and their role in plant defense. *Annu. Rev. Phytopathol.* **2016**, *54*, 143–161. [CrossRef] [PubMed]
38. Camborde, L.; Kiselev, A.; Pel, M.J.C.; Le Ru, A.; Jauneau, A.; Pouzet, C.; Dumas, B.; Gaulin, E. An oomycete effector targets a plant RNA helicase involved in root development and defense. *New Phytol.* **2021**, *233*, 2232–2248. [CrossRef] [PubMed]
39. Wille, L.; Messmer, M.M.; Bodenhausen, N.; Studer, B.; Hohmann, P. Heritable Variation in Pea for Resistance Against a Root Rot Complex and Its Characterization by Amplicon Sequencing. *Front. Plant Sci.* **2020**, *11*, 542153. [CrossRef]
40. Tosi, M.; Mitter, E.K.; Gaiero, J.; Dunfield, K. It takes three to tango: The importance of microbes, host plant, and soil management to elucidate manipulation strategies for the plant microbiome. *Can. J. Microbiol.* **2020**, *66*, 413–433. [CrossRef]
41. Dubey, S.; Sharma, S. Rhizospheric Engineering by Plant-Mediated Indirect Selection of Microbiome for Agricultural Sustainability. *Crit. Rev. Plant Sci.* **2021**, *40*, 379–397. [CrossRef]
42. Hartmann, A.; Rothballer, M.; Schmid, M. Lorenz Hiltner, a pioneer in rhizosphere microbial ecology and soil bacteriology research. *Plant Soil* **2008**, *312*, 7–14. [CrossRef]
43. Lucas, M.R.; Huynh, B.L.; da Silva Vinholes, P.; Cisse, N.; Drabo, I.; Ehlers, J.D.; Roberts, P.A.; Close, T.J. Association studies and legume synteny reveal haplotypes determining seed size in *Vigna unguiculata*. *Front. Plant Sci.* **2013**, *4*, 95. [CrossRef]
44. Niks, R.E.; Parlevliet, J.E.; Lindhout, P.; Bai, Y. *Breeding Crops with Resistance to Diseases and Pests*; Wageningen Academic Publishers: Wageningen, The Netherlands, 2019.
45. Rubiales, D. Developing pest- and disease-resistant cultivars of grain legumes. In *Achieving Sustainable Cultivation of Grain Legumes. Volume 1: Advances in Breeding and Cultivation Techniques*; Sivasankar, S., Bergvinson, D., Gaur, P., Kumar, S., Beebe, S., Tamò, M., Eds.; Burleig Dodds Science Publishing: Cambridge, UK, 2018; ISBN 978-1-78676-136-1.
46. Raaijmakers, J.M.; Paulitz, T.C.; Steinberg, C.; Alabouvette, C.; Moënne-Loccoz, Y. The rhizosphere: A playground and battlefield for soilborne pathogens and beneficial microorganisms. *Plant Soil* **2009**, *321*, 341–361. [CrossRef]
47. O'Keeffe, K.R.; Carbone, I.; Jones, C.D.; Mitchell, C.E. Plastic potential: How the phenotypes and adaptations of pathogens are influenced by microbial interactions within plants. *Curr. Opin. Plant Biol.* **2017**, *38*, 78–83. [CrossRef]
48. Lyu, D.; Msimbira, L.A.; Nazari, M.; Antar, M.; Pagé, A.; Shah, A.; Monjezi, N.; Zajonc, J.; Tanney, C.A.; Backer, R.; et al. The coevolution of plants and microbes underpins sustainable agriculture. *Microorganisms* **2021**, *9*, 1036. [CrossRef]
49. Infantino, A.; Kharrat, M.; Riccioni, L.; Coyne, C.J.; McPhee, K.E.; Grünwald, N.J. Screening techniques and sources of resistance to root diseases in cool season food legumes. *Euphytica* **2006**, *147*, 201–221. [CrossRef]
50. Foster, K.; You, M.P.; Nietschke, B.; Edwards, N.; Barbetti, M.J. Soilborne root disease pathogen complexes drive widespread decline of subterranean clover pastures across diverse climatic zones. *Crop Pasture Sci.* **2017**, *68*, 33–44. [CrossRef]
51. Rubiales, D.; Fondevilla, S.; Chen, W.; Gentzbittel, L.; Higgins, T.J.V.; Castillejo, M.A.; Singh, K.B.; Rispail, N. Achievements and Challenges in Legume Breeding for Pest and Disease Resistance. *CRC. Crit. Rev. Plant Sci.* **2015**, *34*, 195–236. [CrossRef]
52. Zitnick-Anderson, K.; del Río Mendoza, L.E.; Forster, S.; Pasche, J.S. Associations among the communities of soil-borne pathogens, soil edaphic properties and disease incidence in the field pea root rot complex. *Plant Soil* **2020**, *457*, 339–354. [CrossRef]
53. Di Pietro, A.; Madrid, M.P.; Caracuel, Z.; Delgado-Jarana, J.; Roncero, M.I.G. *Fusarium oxysporum*: Exploring the molecular arsenal of a vascular wilt fungus. *Mol. Plant Pathol.* **2003**, *4*, 315–325. [CrossRef]
54. Leslie, J.F.; Summerell, B.A. *The Fusarium Laboratory Manual*, 1st ed.; Blackwell Publishing: Ames, IA, USA, 2006; Volume 22, pp. 3–274.
55. Bani, M.; Rubiales, D.; Rispail, N. A detailed evaluation method to identify sources of quantitative resistance to *Fusarium oxysporum* f. sp. *pisi* race 2 within a *Pisum* spp. germplasm collection.. *Plant Pathol.* **2012**, *61*, 532–542. [CrossRef]
56. Willsey, T.; Patey, J.; Vucurevich, C.; Chatterton, S.; Carcamo, H. Evaluation of foliar and seed treatments for integrated management of root rot and pea leaf weevil in field pea and faba bean. *Crop Prot.* **2021**, *143*, 105538. [CrossRef]
57. Sampaio, A.M.; De Sousa Araújo, S.; Rubiales, D.; Patto, M.C.V. Fusarium wilt management in legume crops. *Agronomy* **2020**, *10*, 1073. [CrossRef]
58. Kraft, J.M. Fusarium wilt of peas (a review). *Agronomie* **1994**, *14*, 561–567. [CrossRef]
59. Mcphee, K.E.; Tullu, A.; Kraft, J.M.; Muehlbauer, F.J. Resistance to Fusarium Wilt Race 2 in the *Pisum* Core Collection. *J. Am. Soc. Hortic. Sci.* **1999**, *124*, 28–31. [CrossRef]
60. Bani, M.; Pérez-de-Luque, A.; Rubiales, D.; Rispail, N. Physical and chemical barriers in root tissues contribute to quantitative resistance to *Fusarium oxysporum* f. sp. *pisi* in Pea. . *Front. Plant Sci.* **2018**, *9*, 199. [CrossRef]
61. Kraft, J.M.; Pfleger, F.L. *Compendium of Pea Diseases and Pests*, 2nd ed.; American Phytopathological Society (APS Press): St. Paul, MN, USA, 2001.
62. Shubha, K.; Dhar, S.; Choudhary, H.; Dubey, S.C.; Sharma, R.K. Identification of resistant sources and inheritance of Fusarium wilt resistance in garden pea (*Pisum sativum* ssp. *hortense*). *Indian J. Hortic.* **2016**, *73*, 356–361. [CrossRef]
63. Harveson, R.M.; Pasche, J.S.; Porter, L.; Chen, W.; Burrows, M. *Compendium of Pea Diseases and Pests*, 3rd ed.; American Phytopathological Society: St. Paul, MN, USA, 2021; ISBN 978-0-89054-655-0.
64. Leslie, J.F.; Anderson, L.L.; Bowden, R.L.; Lee, Y.W. Inter- and intra-specific genetic variation in *Fusarium*. *Int. J. Food Microbiol.* **2007**, *119*, 25–32. [CrossRef]

65. Coyne, C.J.; Porter, L.D.; Boutet, G.; Ma, Y.; McGee, R.J.; Lesné, A.; Baranger, A.; Pilet-Nayel, M.L. Confirmation of Fusarium root rot resistance QTL Fsp-Ps 2.1 of pea under controlled conditions. *BMC Plant Biol.* **2019**, *19*, 98. [CrossRef]
66. Jha, U.C.; Bohra, A.; Pandey, S.; Parida, S.K. Breeding, genetics, and genomics approaches for improving Fusarium wilt resistance in major grain legumes. *Front. Genet.* **2020**, *11*, 1001. [CrossRef] [PubMed]
67. Rispail, N.; Bani, M.; Rubiales, D. Resistance reaction of *Medicago truncatula* genotypes to *Fusarium oxysporum*: Effect of plant age, substrate and inoculation method. *Crop Pasture Sci.* **2015**, *66*, 506–515. [CrossRef]
68. Sampaio, A.M.; Vitale, S.; Turrà, D.; Di Pietro, A.; Rubiales, D.; van Eeuwijk, F.; Vaz Patto, M.C. A diversity of resistance sources to *Fusarium oxysporum* f. sp. *pisi* found within grass pea germplasm. *Plant Soil.* **2021**, *463*, 19–38. [CrossRef]
69. Leslie, J.F.; Summerell, B.A. Fusarium laboratory workshops–A recent history. *Mycotoxin Res.* **2006**, *22*, 73. [CrossRef]
70. Wang, R.; Huang, J.; Liang, A.; Wang, Y.; Mur, L.A.J.; Wang, M.; Guo, S. Zinc and copper enhance cucumber tolerance to fusaric acid by mediating its distribution and toxicity and modifying the antioxidant system. *Int. J. Mol. Sci.* **2020**, *21*, 3370. [CrossRef]
71. Wang, M.; Liu, W.; Yan, J.; Sun, P.; Chen, F.; Jiang, B.; Xie, D.; Lin, Y.; Peng, Q.; He, X. A Transcriptomic Analysis of Gene Expression in Chieh-Qua in Response to Fusaric Acid Stress. *Horticulturae* **2021**, *7*, 88. [CrossRef]
72. Kraft, J.M.; Papavizas, G.C. Use of host resistance, Trichoderma, and fungicides to control soilborne diseases and increase seed yields of peas. *Plant Dis.* **1983**, *11*, 1234–1237. [CrossRef]
73. Riaz, R.; Khan, A.; Khan, W.J.; Jabeen, Z.; Yasmin, H.; Naz, R.; Nosheen, A.; Hassan, M.N. Vegetable associated *Bacillus* spp. suppress the pea (*Pisum sativum* L.) root rot caused by *Fusarium solani*. *Biol. Control* **2021**, *158*, 104610. [CrossRef]
74. Summerell, B.A.; Leslie, J.F. Fifty years of Fusarium: How could nine species have ever been enough? *Fungal Divers.* **2011**, *50*, 135–144. [CrossRef]
75. Saremi, H.; Okhovvat, S.M.; Ashrafi, S.J. Fusarium diseases as the main soil borne fungal pathogen on plants and their control management with soil solarization in Iran. *Afr. J. Biotechnol.* **2011**, *10*, 18391–18398. [CrossRef]
76. Dita, M.; Barquero, M.; Heck, D.; Mizubuti, E.S.G.; Staver, C.P. Fusarium wilt of banana: Current knowledge on epidemiology and research needs toward sustainable disease management. *Front. Plant Sci.* **2018**, *871*, 1468. [CrossRef]
77. Grajal-Martin, M.J.; Muehlbauer, F.J. Genomic location of the Fw gene for resistance to Fusarium wilt race 1 in peas. *J. Hered.* **2002**, *93*, 291–293. [CrossRef]
78. McClendon, M.T.; Inglis, D.A.; McPhee, K.E.; Coyne, C.J. DNA markers linked to fusarium wilt race 1 resistance in pea. *J. Am. Soc. Hortic. Sci.* **2002**, *127*, 602–607. [CrossRef]
79. Okubara, P.A.; Keller, K.E.; McClendon, M.T.; Inglis, D.A.; McPhee, K.E.; Coyne, C.J. Y15_999 Fw, a dominant SCAR marker linked to the Fusarium wilt race 1 (Fw) resistance gene in pea. *Pisum Genet.* **2015**, *37*, 30–33.
80. Wohor, Z.O.; Rispail, N.; Rubiales, D. Evaluation of a *Pisum* spp. Germplasm Collection for the Resistance to *Fusarium oxysporum* f. sp. *pisi* Schlecht and *Orobanche crenata* Forsk. In Proceedings of the 9th International Conference on Legume Genetics and Genomics ICLGG, Dijon, France, 13 May 2019; Volume 71.
81. McPhee, K.E.; Inglis, D.A.; Gundersen, B.; Coyne, C.J. Mapping QTL for Fusarium wilt Race 2 partial resistance in pea (*Pisum sativum*). *Plant Breed.* **2012**, *131*, 300–306. [CrossRef]
82. Cheng, P.; Holdsworth, W.; Ma, Y.; Coyne, C.J.; Mazourek, M.; Grusak, M.A.; Fuchs, S.; McGee, R.J. Association mapping of agronomic and quality traits in USDA pea single-plant collection. *Mol. Breed.* **2015**, *35*, 1–13. [CrossRef]
83. Sampaio, A.M.; Alves, M.L.; Pereira, P.; Valiollahi, E.; Santos, C.; Šatović, Z.; Rubiales, D.; Araújo, S.D.S.; van Eeuwijk, F.; Vaz Patto, M.C. Grass pea natural variation reveals oligogenic resistance to *Fusarium oxysporum* f. sp. *pisi*. *Plant Genome* **2021**, *14*, e20154. [CrossRef]
84. Xue, A.G. Biological control of pathogens causing root rot complex in field pea using *Clonostachys rosea* strain ACM941. *Phytopathology* **2003**, *93*, 329–335. [CrossRef]
85. Smith, S.N. An Overview of Ecological and Habitat Aspects in the Genus *Fusarium* with Special Emphasis on the Soil- Borne Pathogenic Forms. *Plant Pathol.* **2007**, *16*, 97–120.
86. Zitnick-Anderson, K.; Simons, K.; Pasche, J.S. Detection and qPCR quantification of seven *Fusarium* species associated with the root rot complex in field pea. *Can. J. Plant Pathol.* **2018**, *40*, 261–271. [CrossRef]
87. Tu, J.C. Management of root rot diseases of peas, beans, and tomatoes. *Can. J. Plant Pathol.* **1992**, *14*, 92–99. [CrossRef]
88. Sivachandra Kumar, N.T.; Cox, L.; Armstrong-Cho, C.; Banniza, S. Optimization of zoospore production and inoculum concentration of *Aphanomyces euteiches* for resistance screening of pea and lentil. *Can. J. Plant Pathol.* **2020**, *42*, 419–428. [CrossRef]
89. Hossain, S.; Bergkvist, G.; Berglund, K.; Mårtensson, A.; Persson, P. Aphanomyces pea root rot disease and control with special reference to impact of Brassicaceae cover crops. *Acta Agric. Scand. Sect. B Soil Plant Sci.* **2012**, *62*, 477–487. [CrossRef]
90. Wu, L.; Chang, K.F.; Conner, R.L.; Strelkov, S.; Fredua-Agyeman, R.; Hwang, S.F.; Feindel, D. *Aphanomyces euteiches*: A Threat to Canadian Field Pea Production. *Engineering* **2018**, *4*, 542–551. [CrossRef]
91. Billard, E.; Quillévéré-Hamard, A.; Lavaud, C.; Pilet-Nayel, M.L.; Le May, C. Testing of life history traits of a soilborne pathogen in vitro: Do characteristics of oospores change according the strains of *Aphanomyces euteiches* and the host plant infected by the pathogen? *J. Phytopathol.* **2019**, *167*, 313–320. [CrossRef]
92. Gaulin, E.; Madoui, M.A.; Bottin, A.; Jacquet, C.; Mathé, C.; Couloux, A.; Wincker, P.; Dumas, B. Transcriptome of *Aphanomyces euteiches*: New Oomycete putative pathogenicity factors and metabolic pathways. *PLoS ONE* **2008**, *3*, e1723. [CrossRef] [PubMed]

93. Pilet-Nayel, M.L.; Muehlbauer, F.J.; McGee, R.J.; Kraft, J.M.; Baranger, A.; Coyne, C.J. Consistent quantitative trait loci in pea for partial resistance to *Aphanomyces euteiches* isolates from the United States and France. *Phytopathology* **2005**, *95*, 1287–1293. [CrossRef] [PubMed]
94. Sharma, A.; Rani, M.; Lata, H.; Thakur, A.; Sharma, P.; Kumar, P.; Jayswal, D.K.; Rana, R.S. Global dimension of root rot complex in garden pea: Current status and breeding prospective. *Crop Prot.* **2022**, *158*, 106004. [CrossRef]
95. Godebo, A.T.; Germida, J.J.; Walley, F.L. Isolation, identification, and assessment of soil bacteria as biocontrol agents of pea root rot caused by *Aphanomyces euteiches*. *Can. J. Soil Sci.* **2020**, *100*, 206–216. [CrossRef]
96. Godebo, A.T.; Wee, N.M.J.; Yost, C.K.; Walley, F.L.; Germida, J.J. A Meta-Analysis to Determine the State of Biological Control of Aphanomyces Root Rot. *Front. Mol. Biosci.* **2021**, *8*, 777042. [CrossRef]
97. Lagerlöf, J.; Ayuke, F.; Heyman, F.; Meijer, J. Effects of biocontrol bacteria and earthworms on *Aphanomyces euteiches* root-rot and growth of peas (*Pisum sativum*) studied in a pot experiment. *Acta Agric. Scand. Sect. B–Soil Plant Sci.* **2020**, *70*, 427–436. [CrossRef]
98. Hossain, S.; Bergkvist, G.; Glinwood, R.; Berglund, K.; Mårtensson, A.; Hallin, S.; Persson, P. Brassicaceae cover crops reduce Aphanomyces pea root rot without suppressing genetic potential of microbial nitrogen cycling. *Plant Soil* **2015**, *392*, 227–238. [CrossRef]
99. McGee, R.J.; Coyne, C.J.; Pilet-Nayel, M.L.; Moussart, A.; Tivoli, B.; Baranger, A.; Hamon, C.; Vandemark, G.; McPhee, K. Registration of pea germplasm lines partially resistant to *Aphanomyces* root rot for breeding fresh or freezer pea and dry pea types. *J. Plant Regist.* **2012**, *6*, 203–207. [CrossRef]
100. Pilet-Nayel, M.L.; Muehlbauer, F.J.; McGee, R.J.; Kraft, J.M.; Baranger, A.; Coyne, C.J. Quantitative trait loci for partial resistance to Aphanomyces root rot in pea. *Theor. Appl. Genet.* **2002**, *106*, 28–39. [CrossRef]
101. Hamon, C.; Coyne, C.J.; McGee, R.J.; Lesné, A.; Esnault, R.; Mangin, P.; Hervé, M.; Le Goff, I.; Deniot, G.; Roux-Duparque, M.; et al. QTL meta-analysis provides a comprehensive view of loci controlling partial resistance to *Aphanomyces euteiches* in four sources of resistance in pea. *BMC Plant Biol.* **2013**, *13*, 45. [CrossRef]
102. Lavaud, C.; Lesné, A.; Piriou, C.; Le Roy, G.; Boutet, G.; Moussart, A.; Poncet, C.; Delourme, R.; Baranger, A.; Pilet-Nayel, M.L. Validation of QTL for resistance to *Aphanomyces euteiches* in different pea genetic backgrounds using near-isogenic lines. *Theor. Appl. Genet.* **2015**, *128*, 2273–2288. [CrossRef]
103. Desgroux, A.; L'Anthoëne, V.; Roux-Duparque, M.; Rivière, J.P.; Aubert, G.; Tayeh, N.; Moussart, A.; Mangin, P.; Vetel, P.; Piriou, C.; et al. Genome-wide association mapping of partial resistance to *Aphanomyces euteiches* in pea. *BMC Genom.* **2016**, *17*, 124. [CrossRef]
104. Desgroux, A.; Baudais, V.N.; Aubert, V.; Le Roy, G.; de Larambergue, H.; Miteul, H.; Aubert, G.; Boutet, G.; Duc, G.; Baranger, A.; et al. Comparative genome-wide-association mapping identifies common loci controlling root system architecture and resistance to *Aphanomyces euteiches* in pea. *Front. Plant Sci.* **2018**, *8*, 2195. [CrossRef]
105. Bonhomme, M.; Fariello, M.I.; Navier, H.; Hajri, A.; Badis, Y.; Miteul, H.; Samac, D.A.; Dumas, B.; Baranger, A.; Jacquet, C.; et al. A local score approach improves GWAS resolution and detects minor QTL: Application to *Medicago truncatula* quantitative disease resistance to multiple *Aphanomyces euteiches* isolates. *Heredity* **2019**, *123*, 517–531. [CrossRef]
106. Wu, L.; Fredua-Agyeman, R.; Hwang, S.F.; Chang, K.F.; Conner, R.L.; McLaren, D.L.; Strelkov, S.E. Mapping QTL associated with partial resistance to Aphanomyces root rot in pea (*Pisum sativum* L.) using a 13.2 K SNP array and SSR markers. *Theor. Appl. Genet.* **2021**, *134*, 2965–2990. [CrossRef]
107. Hadwiger, L.A. Pea-Fusarium solani interactions contributions of a system toward understanding disease resistance. *Phytopathology* **2008**, *98*, 372–379. [CrossRef]
108. Baćanović-Šišić, J.; Šišić, A.; Schmidt, J.H.; Finckh, M.R. Identification and characterization of pathogens associated with root rot of winter peas grown under organic management in Germany. *Eur. J. Plant Pathol.* **2018**, *151*, 745–755. [CrossRef]
109. Fernandez, M.R.; Huber, D.; Basnyat, P.; Zentner, R.P. Impact of agronomic practices on populations of *Fusarium* and other fungi in cereal and noncereal crop residues on the Canadian Prairies. *Soil Tillage Res.* **2008**, *100*, 60–71. [CrossRef]
110. Bodah, E.T.; Porter, L.D.; Chaves, B.; Dhingra, A. Evaluation of pea accessions and commercial cultivars for fusarium root rot resistance. *Euphytica* **2016**, *208*, 63–72. [CrossRef]
111. Singh, B.P.; Singh, G.; Krishna, K.; Nayak, S.C.; Srinivasa, N. *Management of Fungal Pathogens in Pulses: Current Status and Future Challenges*; Springer Nature: Cham, Switzerland, 2020; Volume 11224, pp. 35947–35948. [CrossRef]
112. Mazzola, M. Mechanisms of natural soil suppressiveness to soilborne diseases. *Antonie Van Leeuwenhoek Int. J. Gen. Mol. Microbiol.* **2002**, *81*, 557–564. [CrossRef]
113. Jha, P.K.; Jalali, B.L. Biocontrol of pea root rot incited by *Fusarium solani* f. sp. *pisi* with rhizosphere mycoflora. *Indian Phytopathol.* **2006**, *59*, 41–43.
114. Panth, M.; Hassler, S.C.; Baysal-Gurel, F. Methods for Management of Soilborne Diseases in Crop Production. *Agriculture* **2020**, *10*, 16. [CrossRef]
115. Grünwald, N.J.; Coffman, V.A.; Kraft, J.M. Sources of Partial Resistance to Fusarium Root Rot in the *Pisum* Core Collection. *Plant Dis.* **2003**, *87*, 1197–1200. [CrossRef]
116. Li, W.J.; Feng, J.; Chang, K.F.; Conner, R.L.; Hwang, S.F.; Strelkov, S.E.; Gossen, B.D.; McLaren, D.L. Microsatellite DNA markers indicate quantitative trait loci controlling resistance to pea root rot caused by *Fusarium avenaceum* (Corda ex Fries) Sacc. *Plant Pathol. J.* **2012**, *11*, 114–119. [CrossRef]

117. Feng, J.; Hwang, R.; Chang, K.F.; Conner, R.L.; Hwang, S.F.; Strelkov, S.E.; Gossen, B.D.; McLaren, D.L.; Xue, A.G. Identification of microsatellite markers linked to quantitative trait loci controlling resistance to Fusarium root rot in field pea. *Can. J. Plant Sci.* **2011**, *91*, 199–204. [CrossRef]
118. Williamson-Benavides, B.A.; Sharpe, R.M.; Nelson, G.; Bodah, E.T.; Porter, L.D.; Dhingra, A. Identification of *Fusarium solani* f. sp. *pisi* (*Fsp*) Responsive Genes in *Pisum sativum*. *Front. Genet.* **2020**, *11*, 950. [CrossRef]
119. Grünwald, N.J.; Chen, W.; Larsen, R.C. Pea Diseases and their Management. In *Diseases of Fruits and Vegetables*; Springer: Dordrech, The Netherlands, 2004; Volume 2, pp. 301–331. [CrossRef]
120. Nel, W.J.; Duong, T.A.; Wingfield, B.D.; Wingfield, M.J.; de Beer, Z.W. A new genus and species for the globally important, multihost root pathogen *Thielaviopsis basicola*. *Plant Pathol.* **2018**, *67*, 871–882. [CrossRef]
121. Abawi, G.S.; Widmer, T.L. Impact of soil health management practices on soilborne pathogens, nematodes and root diseases of vegetable crops. *Appl. Soil Ecol.* **2000**, *15*, 37–47. [CrossRef]
122. Hood, M.E.; Shew, H.D. Pathogenesis of Thielaviopsis basicola on a susceptible and a resistant cultivar of burley tobacco. *Phytopathology* **1996**, *86*, 38–44. [CrossRef]
123. Niu, C.; Lister, H.E.; Nguyen, B.; Wheeler, T.A.; Wright, R.J. Resistance to *Thielaviopsis basicola* in the cultivated a genome cotton. *Theor. Appl. Genet.* **2008**, *117*, 1313–1323. [CrossRef] [PubMed]
124. Keijer, J. The initial steps of the infection process in *Rhizoctonia solani*. In *Rhizoctonia Species: Taxonomy, Molecular Biology, Ecology, Pathology and Disease Control*; Springer: Dordrecht, The Netherlands, 1996; pp. 149–162.
125. Sharma-Poudyal, D.; Paulitz, T.C.; Porter, L.D.; Sharma-Poudyal, D. Characterization and pathogenicity of Rhizoctonia and Rhizoctonia-like spp. From pea crops in the Columbia Basin of Oregon and Washington. *Plant Dis.* **2015**, *99*, 604–613. [CrossRef] [PubMed]
126. Beniwal, S.P.S.; Ahmed, S.; Gorfu, D. Wilt/root rot diseases of chickpea in Ethiopia. *Trop. Pest Manag.* **1992**, *38*, 48–51. [CrossRef]
127. Ketta, H.; Elkhateeb, N.; Saleh, M.; Kamel, S. Efficiency Assessment of Combinations Between *Rhizobium leguminosarum* and *Trichoderma* spp. for Controlling of Pea (*Pisum sativum* L.) Damping-off Disease. *Egypt. J. Phytopathol.* **2021**, *49*, 1–14. [CrossRef]
128. Uwaremwe, C.; Yue, L.; Liu, Y.; Tian, Y.; Zhao, X.; Wang, Y.; Xie, Z.; Zhang, Y.; Cui, Z.; Wang, R. Molecular identification and pathogenicity of *Fusarium* and *Alternaria* species associated with root rot disease of wolfberry in Gansu and Ningxia provinces, China. *Plant Pathol.* **2021**, *70*, 397–406. [CrossRef]
129. Wille, L.; Kurmann, M.; Messmer, M.M.; Studer, B.; Hohmann, P. Untangling the Pea Root Rot Complex Reveals Microbial Markers for Plant Health. *Front. Plant Sci.* **2021**, *12*, 737820. [CrossRef]
130. Flower, K.C.; Hüberli, D.; Collins, S.J.; Thomas, G.; Ward, P.R.; Cordingley, N. Progression of plant-parasitic nematodes and foliar and root diseases under no-tillage with different crop rotations. *Soil Tillage Res.* **2019**, *191*, 18–28. [CrossRef]
131. El_Komy, M.H.; Hassouna, M.G.; Abou-Taleb, E.M.; Al-Sarar, A.S.; Abobakr, Y. A mixture of *Azotobacter*, *Azospirillum*, and *Klebsiella* strains improves root-rot disease complex management and promotes growth in sunflowers in calcareous soil. *Eur. J. Plant Pathol.* **2020**, *156*, 713–726. [CrossRef]
132. Khan, M.R.; Ashraf, S.; Rasool, F.; Salati, K.M.; Mohiddin, F.A.; Haque, Z. Field performance of Trichoderma species against wilt disease complex of chickpea caused by *Fusarium oxysporum* f. sp. *ciceri* and *Rhizoctonia solani*. *Turk. J. Agric. For.* **2014**, *38*, 447–454. [CrossRef]
133. Hashem, A.H.; Abdelaziz, A.M.; Askar, A.A.; Fouda, H.M.; Khalil, A.M.A.; Abd-Elsalam, K.A.; Khaleil, M.M. Bacillus megaterium-mediated synthesis of selenium nanoparticles and their antifungal activity against *Rhizoctonia solani* in faba bean plants. *J. Fungi* **2021**, *7*, 195. [CrossRef]
134. Cubeta, M.A.; Thomas, E.; Dean, R.A.; Jabaji, S.; Neate, S.M.; Tavantzis, S.; Toda, T.; Vilgalys, R.; Bharathan, N.; Fedorova-Abrams, N.; et al. Draft genome sequence of the plant-pathogenic soil fungus *Rhizoctonia solani* anastomosis group 3 strain Rhs1AP. *Genome Announc.* **2014**, *2*, e01072-14. [CrossRef]
135. Zhong, J.; Chen, C.Y.; Gao, B.D. Genome sequence of a novel mycovirus of Rhizoctonia solani, a plant pathogenic fungus. *Virus Genes* **2015**, *51*, 167–170. [CrossRef]
136. Chen, Y.; Su, J.E.; Qin, X.Y.; Fan, Z.Y.; Zhang, X.H.; Yu, Q.; Xia, Z.Y.; Zou, C.M.; Zhao, G.K.; Lin, Z.L. A novel putative betapartitivirus isolated from the plant-pathogenic fungus *Rhizoctonia solani*. *Arch. Virol.* **2020**, *165*, 1697–1701. [CrossRef]
137. Akhter, W.; Bhuiyan, M.K.A.; Sultana, F.; Hossain, M.M. Integrated effect of microbial antagonist, organic amendment and fungicide in controlling seedling mortality (*Rhizoctonia solani*) and improving yield in pea (*Pisum sativum* L.). *Comptes Rendus Biol.* **2015**, *338*, 21–28. [CrossRef]
138. Wang, Y.; Fristensky, B. Transgenic canola lines expressing pea defense gene DRR206 have resistance to aggressive blackleg isolates and to *Rhizoctonia solani*. *Mol. Breed.* **2001**, *8*, 263–271. [CrossRef]
139. Dölfors, F.; Holmquist, L.; Dixelius, C.; Tzelepis, G. A LysM effector protein from the basidiomycete *Rhizoctonia solani* contributes to virulence through suppression of chitin-triggered immunity. *Mol. Genet. Genom.* **2019**, *294*, 1211–1218. [CrossRef]
140. Liu, Y.; Hassan, S.; Kidd, B.N.; Garg, G.; Mathesius, U.; Singh, K.B.; Anderson, J.P. Ethylene signaling is important for isoflavonoid-mediated resistance to *Rhizoctonia solani* in roots of *Medicago truncatula*. *Mol. Plant-Microbe Interact.* **2017**, *30*, 691–700. [CrossRef]
141. Schroeder, K.L.; Martin, F.N.; de Cock, A.W.A.M.; Lévesque, C.A.; Spies, C.F.J.; Okubara, P.A.; Paulitz, T.C. Molecular detection and quantification of pythium species: Evolving taxonomy, new tools, and challenges. *Plant Dis.* **2013**, *97*, 4–20. [CrossRef]
142. Kageyama, K. Molecular taxonomy and its application to ecological studies of *Pythium* species. *J. Gen. Plant Pathol.* **2014**, *80*, 314–326. [CrossRef]

143. Khalil, S.A.M.; Nehal, S.-M.; Nadia, G.-G.; Mokhtar, M.-K. Field approaches of chemical inducers and bioagents for controlling root diseases incidence of pea (*Pisum sativum* L.) under field conditions. *Plant Pathol. J.* **2020**, *19*, 166–175. [CrossRef]
144. Wu, W.; Ogawa, F.; Ochiai, M.; Yamada, K.; Fukui, H. Common strategies to control pythium disease. *Rev. Agric. Sci.* **2020**, *8*, 58–69. [CrossRef]
145. Kraft, J.M.; Haware, M.P.; Jiménez-Díaz, R.M.; Bayaa, B.; Harrabi, M. Screening techniques and sources of resistance to root rots and wilts in cool season food legumes. *Euphytica* **1993**, *73*, 27–39. [CrossRef]
146. Alcala, A.V.C.; Paulitz, T.C.; Schroeder, K.L.; Porter, L.D.; Derie, M.L.; du Toit, L.J. Pythium species associated with damping-off of pea in certified organic fields in the Columbia Basin of central Washington. *Plant Dis.* **2016**, *100*, 916–925. [CrossRef]
147. Klepadlo, M.; Balk, C.S.; Vuong, T.D.; Dorrance, A.E.; Nguyen, H.T. Molecular characterization of genomic regions for resistance to *Pythium ultimum* var. *ultimum* in the soybean cultivar Magellan. *Theor. Appl. Genet.* **2019**, *132*, 405–417. [CrossRef]
148. Urrea, K.; Rupe, J.; Chen, P.; Rothrock, C.S. Characterization of seed rot resistance to *Pythium aphanidermatum* in soybean. *Crop Sci.* **2017**, *57*, 1394–1403. [CrossRef]
149. Lin, F.; Wani, S.H.; Collins, P.J.; Wen, Z.; Li, W.; Zhang, N.; McCoy, A.G.; Bi, Y.; Tan, R.; Zhang, S.; et al. QTL mapping and GWAS for identification of loci conferring partial resistance to *Pythium sylvaticum* in soybean (*Glycine max* (L.) Merr). *Mol. Breed.* **2020**, *40*, 1–11. [CrossRef]
150. Navarro, F.; Sass, M.E.; Nienhuis, J. Identification and confirmation of quantitative trait loci for root rot resistance in snap bean. *Crop Sci.* **2008**, *48*, 962–972. [CrossRef]
151. Arora, H.; Sharma, A.; Sharma, S.; Haron, F.F.; Gafur, A.; Sayyed, R.Z.; Datta, R. Pythium damping-off and root rot of *Capsicum annuum* L.: Impacts, diagnosis, and management. *Microorganisms* **2021**, *9*, 823. [CrossRef]
152. Ai, G.; Yang, K.; Ye, W.; Tian, Y.; Du, Y.; Zhu, H.; Li, T.; Xia, Q.; Shen, D.; Peng, H.; et al. Prediction and characterization of RXLR effectors in Pythium species. *Mol. Plant-Microbe Interact.* **2020**, *33*, 1046–1058. [CrossRef]
153. Trudgill, D.L.; Blok, V.C. Apomictic, Polyphagous Root-Knot Nematodes: Exceptionally Successful and Damaging Biotrophic Root Pathogens. *Annu. Rev. Phytopathol.* **2001**, *39*, 53–77. [CrossRef] [PubMed]
154. Mesa-Valle, C.M.; Garrido-Cardenas, J.A.; Cebrian-Carmona, J.; Talavera, M.; Manzano-Agugliaro, F. Global research on plant nematodes. *Agronomy* **2020**, *10*, 1148. [CrossRef]
155. Castillo, P.; Navas-Cortés, J.A.; Landa, B.B.; Jiménez-Díaz, R.M.; Vovlas, N. Plant-parasitic nematodes attacking chickpea and their in planta interactions with rhizobia and phytopathogenic fungi. *Plant Dis.* **2008**, *92*, 840–853. [CrossRef] [PubMed]
156. Jones, J.T.; Haegeman, A.; Danchin, E.G.J.; Gaur, H.S.; Helder, J.; Jones, M.G.K.; Kikuchi, T.; Manzanilla-López, R.; Palomares-Rius, J.E.; Wesemael, W.M.L.; et al. Top 10 plant-parasitic nematodes in molecular plant pathology. *Mol. Plant Pathol.* **2013**, *14*, 946–961. [CrossRef] [PubMed]
157. Dobosz, R.; Krawczyk, R. Meloidogyne hapla development on growing legume plants–Short Communication. *Plant Prot. Sci.* **2019**, *55*, 274–277. [CrossRef]
158. Singh, S.K.; Hodda, M.; Ash, G.J. Plant-parasitic nematodes of potential phytosanitary importance, their main hosts and reported yield losses. *EPPO Bull.* **2013**, *43*, 334–374. [CrossRef]
159. Kimpinski, J.; Sturz, A.V. Managing crop root zone ecosystems for prevention of harmful and encouragement of beneficial nematodes. *Soil Tillage Res.* **2003**, *72*, 213–221. [CrossRef]
160. Zhang, S.; Cui, S.; McLaughlin, N.B.; Liu, P.; Hu, N.; Liang, W.; Wu, D.; Liang, A. Tillage effects outweigh seasonal effects on soil nematode community structure. *Soil Tillage Res.* **2019**, *192*, 233–239. [CrossRef]
161. Zhang, Y.; Li, S.; Li, H.; Wang, R.; Zhang, K.Q.; Xu, J. Fungi–nematode interactions: Diversity, ecology, and biocontrol prospects in agriculture. *J. Fungi* **2020**, *6*, 206. [CrossRef]
162. Li, X.; Liu, C.; Zhao, H.; Gao, F.; Ji, G.; Hu, F.; Li, H. Similar positive effects of beneficial bacteria, nematodes and earthworms on soil quality and productivity. *Appl. Soil Ecol.* **2018**, *130*, 202–208. [CrossRef]
163. Di Vito, M.; Greco, N. Control of food legume nematodes in the Mediterranean Basin 1. *EPPO Bull.* **1994**, *24*, 489–494. [CrossRef]
164. Sillero, J.C.; Villegas-Fernández, A.M.; Thomas, J.; Rojas-Molina, M.M.; Emeran, A.A.; Fernández-Aparicio, M.; Rubiales, D. Faba bean breeding for disease resistance. *Field Crops Res.* **2010**, *115*, 297–307. [CrossRef]
165. Kosterin, O.E. Prospects of the use of wild relatives for pea breeding. *Russ. J. Genet. Appl. Res.* **2016**, *6*, 233–243. [CrossRef]
166. Zwart, R.S.; Thudi, M.; Channale, S.; Manchikatla, P.K.; Varshney, R.K.; Thompson, J.P. Resistance to Plant-Parasitic Nematodes in Chickpea: Current Status and Future Perspectives. *Front. Plant Sci.* **2019**, *10*, 966. [CrossRef]
167. Vovlas, A.; Santoro, S.; Radicci, V.; Leonetti, P.; Castillo, P.; Palomares-Rius, J.E. Host-suitability of black medick (*Medicago lupulina* L.) and additional molecular markers for identification of the pea cyst nematode Heterodera goettingiana. *Eur. J. Plant Pathol.* **2017**, *149*, 193–199. [CrossRef]
168. Jones, M.G.K. Host cell responses to endoparasitic nematode attack: Structure and function of giant cells and syncytia. *Ann. Appl. Biol.* **1981**, *97*, 353–372. [CrossRef]
169. Munawar, M.; Yevtushenko, D.P.; Castillo, P. Integrative taxonomy, distribution, and host associations of *Geocenamus brevidens* and *Quinisulcius capitatus* from southern Alberta, Canada. *J. Nematol.* **2021**, *53*, 1–17. [CrossRef]
170. Whitehead, A.G.; Bromilow, R.H.; Tite, D.J.; Finch, P.H.; Fraser, J.E.; French, E.M. Incorporation of granular nematicides in soil to control pea cyst-nematode, *Heterodera goettingiana*. *Ann. Appl. Biol.* **1979**, *92*, 81–91. [CrossRef]
171. Green, C.D.; Williamson, K.; Dennis, E.B.; McBurney, T. The effect of oxamyl on the growth of peas attacked by pea cyst nematode. *Ann. Appl. Biol.* **1981**, *97*, 303–309. [CrossRef]

172. Dopierata, U.; Giebel, J. Herbicides can influence the level of pea infestation by *Heterodera goettingiana*. *J. Plant Prot.* **2002**, *42*, 337–341.
173. Di Vito, M.; Perrino, P. Reaction of Pisum spp. to the attacks of *Heterodera goettingiana*. *Nematol. Mediterr.* **1978**, *6*, 113–118.
174. Bleve-Zacheo, T.; Melillo, M.T.; Zacheo, G. Syncytia development in germplasm pea accessions infected with *Heterodera goettingiana*. *Nematol Mediterr.* **1990**, *18*, 93–102.
175. Veronico, P.; Melillo, M.T.; Saponaro, C.; Leonetti, P.; Picardi, E.; Jones, J.T. A polygalacturonase-inhibiting protein with a role in pea defence against the cyst nematode *Heterodera goettingiana*. *Mol. Plant Pathol.* **2011**, *12*, 275–287. [CrossRef]
176. Veronico, P.; Giannino, D.; Melillo, M.T.; Leone, A.; Reyes, A.; Kennedy, M.W.; Bleve-Zacheo, T. A novel lipoxygenase in pea roots. Its function in wounding and biotic stress. *Plant Physiol.* **2006**, *141*, 1045–1055. [CrossRef]
177. Sharma, S.B.; Sikora, R.A.; Greco, N.; Di Vito, M.; Caubel, G. Screening techniques and sources of resistance to nematodes in cool season food legumes. *Euphytica* **1993**, *73*, 59–66. [CrossRef]
178. Castillo, P.; Navas-Cortés, J.A.; Gomar-Tinoco, D.; Di Vito, M.; Jiménez-Díaz, R.M. Interactions between *Meloidogyne artiellia*, the Cereal and Legume Root-Knot Nematode, and *Fusarium oxysporum* f. sp. *ciceris* Race 5 in Chickpea. *Phytopathology* **2003**, *93*, 1513–1523. [CrossRef]
179. Sturz, A.V.; Christie, B.R. Beneficial microbial allelopathies in the root zone: The management of soil quality and plant disease with rhizobacteria. *Soil Tillage Res.* **2003**, *72*, 107–123. [CrossRef]
180. Haque, Z.; Khan, M.R. Identification of multi-facial microbial isolates from the rice rhizosphere and their biocontrol activity against *Rhizoctonia solani* AG1-IA. *Biol. Control* **2021**, *161*, 104640. [CrossRef]
181. Pandey, R.; Kalra, A.; Tandon, S.; Mehrotra, N.; Singh, H.N.; Kumar, S. Essential oils as potent source of nematicidal compounds. *J. Phytopathol.* **2000**, *148*, 501–502. [CrossRef]
182. Sharma, A.; Haseeb, A.; Abuzar, S. Screening of field pea (*Pisum sativum*) selections for their reactions to root-knot nematode (*Meloidogyne incognita*). *J. Zhejiang Univ. Sci. B.* **2006**, *7*, 209–214. [CrossRef] [PubMed]
183. Gautam, N.K.; Marla, S.S.; Mirza, N.; Khan, Z.; Singh, B.; Wankhede, D.P.; Gawade, B.H. Evaluation of field pea accessions for root-knot nematode resistance and possible role of NADP dependent malic enzyme gene in host resistance. *Indian J. Genet. Plant Breed.* **2017**, *77*, 4. [CrossRef]
184. Youssef, M.; El-Nagdi, W. Differential responses of certain field pea and cowpea cultivars to root-knot nematode, *Meloidogyne incognita* for commercial release. *Bull. Natl. Res. Cent.* **2019**, *43*, 178. [CrossRef]
185. Hewezi, T.; Baum, T.J. Manipulation of plant cells by cyst and root-knot nematode effectors. *Mol. Plant-Microbe Interact.* **2013**, *26*, 9–16. [CrossRef]
186. Clément, M.; Ketelaar, T.; Rodiuc, N.; Banora, M.Y.; Smertenko, A.; Engler, G.; Abad, P.; Hussey, P.J.; de Almeida Engler, J. Actin-depolymerizing factor2-mediated actin dynamics are essential for root-knot nematode infection of Arabidopsis. *Plant Cell* **2009**, *21*, 2963–2979. [CrossRef]
187. Abad, P.; Gouzy, J.; Aury, J.M.; Castagnone-Sereno, P.; Danchin, E.G.; Deleury, E.; Perfus-Barbeoch, L.; Anthouard, V.; Artiguenave, F.; Blok, V.C.; et al. Genome sequence of the metazoan plant-parasitic nematode Meloidogyne incognita. *Nat. Biotechnol.* **2008**, *26*, 909–915. [CrossRef]
188. Upadhaya, A.; Yan, G.; Pasche, J. Reproduction Ability and Growth Effect of Pin Nematode, *Paratylenchus nanus*, With Selected Field Pea Cultivars. *Plant Dis.* **2019**, *103*, 2520–2526. [CrossRef]
189. Reen, R.A.; Mumford, M.H.; Thompson, J.P. Novel Sources of Resistance to Root-Lesion Nematode (*Pratylenchus thornei*) in a New Collection of Wild *Cicer* Species (*C. reticulatum* and *C. echinospermum*) to Improve Resistance in Cultivated Chickpea (*C. arietinum*). *Phytopathology* **2019**, *109*, 1270–1279. [CrossRef]
190. Kandel, S.L.; Smiley, R.W.; Garland-Campbell, K.; Elling, A.A.; Huggins, D.; Paulitz, T.C. Spatial distribution of root lesion nematodes (*Pratylenchus* spp.) in a long-term no-till cropping system and their relationship with soil and landscape properties. *Eur. J. Plant Pathol.* **2018**, *150*, 1011–1021. [CrossRef]
191. Smiley, R. Root-lesion Nematodes Affecting Dryland Cereals in the Semiarid Pacific Northwest USA. *Plant Dis.* **2021**, *105*, 3324–3343. [CrossRef]
192. Taylor, S.P.; Hollaway, G.J.; Hunt, C.H. Effect of field crops on population densities of *Pratylenchus neglectus* and *P. thornei* in Southeastern Australia; Part 1: P. neglectus. *J. Nematol.* **2000**, *32*, 591–599.
193. Thompson, J.P. Treatments to eliminate root-lesion nematode (*Pratylenchus thornei* Sher & Allen) from a vertisol. *Nematologica* **1990**, *36*, 123–127.
194. Taylor, S.P.; Vanstone, V.A.; Ware, A.H.; McKay, A.C.; Szot, D.; Russ, M.H. Measuring yield loss in cereals caused by root lesion nematodes (*Pratylenchus neglectus* and *P. thornei*) with and without nematicide. *Aust. J. Agric. Res.* **1999**, *50*, 617–627. [CrossRef]
195. Thompson, J.P.; Reen, R.A.; Clewett, T.G.; Sheedy, J.G.; Kelly, A.M.; Gogel, B.J.; Knights, E.J. Hybridisation of Australian chickpea cultivars with wild *Cicer* spp. increases resistance to root-lesion nematodes (*Pratylenchus thornei* and *P. neglectus*). *Australas. Plant Pathol.* **2011**, *40*, 601–611. [CrossRef]
196. Rubiales, D.; Fernández-Aparicio, M. Parasitic weed: Broomrape. In *Compendium of Pea Diseases and Pests*, 3rd ed.; Harveson, R.M., Ed.; American Phytopathological Society: St. Paul, MN, USA, 2021; pp. 67–69. ISBN 978-0-89054-655-0.
197. Rubiales, D.; Fernández-Aparicio, M. Innovations in parasitic weeds management in legume crops. A review. *Agron. Sustain. Dev.* **2012**, *32*, 433–449. [CrossRef]

198. Fernández-Aparicio, M.; Flores, F.; Rubiales, D. The effect of *Orobanche crenata* infection severity in faba bean, field pea, and grass pea productivity. *Front. Plant Sci.* **2016**, *7*, 1409. [CrossRef]
199. Fernández-Aparicio, M.; Yoneyama, K.; Rubiales, D. The role of strigolactones in host specificity of *Orobanche* and *Phelipanche* seed germination. *Seed Sci. Res.* **2011**, *21*, 55–61. [CrossRef]
200. Rubiales, D.; Moreno, M.T.; Sillero, J.C. Search for resistance to crenate broomrape (*Orobanche crenata* Forsk.) in pea germplasm. *Genet. Resour. Crop Evol.* **2005**, *52*, 853–861. [CrossRef]
201. Rispail, N.; Dita, M.A.; González-Verdejo, C.; Pérez-de-Luque, A.; Castillejo, M.A.; Prats, E.; Román, B.; Jorrín, J.; Rubiales, D. Plant resistance to parasitic plants: Molecular approaches to an old foe. *New Phytol.* **2007**, *173*, 703–712. [CrossRef]
202. Pérez-de-Luque, A.; Moreno, M.T.; Rubiales, D. Host plant resistance against broomrapes (*Orobanche* spp.): Defence reactions and mechanisms of resistance. *Ann. Appl. Biol.* **2008**, *152*, 131–141. [CrossRef]
203. Fernández-Aparicio, M.; Rubiales, D. Advances in understanding plant root response to weedy root parasites. In *Improving Crop Root Function*, 1st ed.; Gregory, P., Ed.; Burleig Dodds Science Publishing Limited: Cambridge, UK, 2021; pp. 215–230. [CrossRef]
204. Rubiales, D.; Fernández-Aparicio, M.; Wegmann, K.; Joel, D.M. Revisiting strategies for reducing the seedbank of *Orobanche* and *Phelipanche* spp. *Weed Res.* **2009**, *49*, 23–33. [CrossRef]
205. Fernández-Aparicio, M.; Sillero, J.C.; Rubiales, D. Intercropping with cereals reduces infection by *Orobanche crenata* in legumes. *Crop Prot.* **2007**, *26*, 1166–1172. [CrossRef]
206. Fernández-Aparicio, M.; Emeran, A.A.; Rubiales, D. Control of Orobanche crenata in legumes intercropped with fenugreek (*Trigonella foenum-graecum*). *Crop Prot.* **2008**, *27*, 653–659. [CrossRef]
207. Rubiales, D. Parasitic plants, wild relatives and the nature of resistance. *New Phytol.* **2003**, *160*, 459–461. [CrossRef] [PubMed]
208. Mabrouk, Y.; Zourgui, L.; Sifi, B.; Delavault, P.; Simier, P.; Belhadj, O. Some compatible *Rhizobium leguminosarum* strains in peas decrease infections when parasitised by *Orobanche crenata*. *Weed Res.* **2007**, *47*, 44–53. [CrossRef]
209. Mabrouk, Y.; Mejri, S.; Belhadj, O. Biochemical mechanisms of induced resistance by rhizobial lipopolysaccharide in pea against crenate broomrape. *Rev. Bras. Bot.* **2016**, *39*, 107–114. [CrossRef]
210. Louarn, J.; Carbonne, F.; Delavault, P.; Bécard, G.; Rochange, S. Reduced Germination of *Orobanche cumana* Seeds in the Presence of Arbuscular Mycorrhizal Fungi or Their Exudates. *PLoS ONE* **2012**, *7*, e49273. [CrossRef]
211. Rubiales, D.; Fernández-Aparicio, M.; Pérez-de-Luque, A.; Castillejo, M.A.; Prats, E.; Sillero, J.C.; Rispail, N.; Fondevilla, S. Breeding approaches for crenate broomrape (*Orobanche crenata* Forsk.) management in pea (*Pisum sativum* L.). *Pest Manag. Sci.* **2009**, *65*, 553–559. [CrossRef]
212. Pérez-de-Luque, A.; Jorrín, J.; Cubero, J.I.; Rubiales, D. *Orobanche crenata* resistance and avoidance in pea (*Pisum* spp.) operate at different developmental stages of the parasite. *Weed Res.* **2005**, *45*, 379–387. [CrossRef]
213. Rubiales, D. Broomrape threat to agriculture. *Outlooks Pest Manag.* **2020**, *31*, 141–144. [CrossRef]
214. Valderrama, M.R.; Román, B.; Satovic, Z.; Rubiales, D.; Cubero, J.I.; Torres, A.M. Locating quantitative trait loci associated with *Orobanche crenata* resistance in pea. *Weed Res.* **2004**, *44*, 323–328. [CrossRef]
215. Fondevilla, S.; Fernández-Aparicio, M.; Satovic, Z.; Emeran, A.A.; Torres, A.M.; Moreno, M.T.; Rubiales, D. Identification of quantitative trait loci for specific mechanisms of resistance to *Orobanche crenata* Forsk. in pea (*Pisum sativum* L.). *Mol. Breed.* **2010**, *25*, 259–272. [CrossRef]
216. Fondevilla, S.; Flores, F.; Emeran, A.A.; Kharrat, M.; Rubiales, D. High productivity of dry pea genotypes resistant to crenate broomrape in Mediterranean environments. *Agron. Sustain. Dev.* **2017**, *37*, 61. [CrossRef]
217. Fernández-Aparicio, M.; Flores, F.; Rubiales, D. Recognition of root exudates by seeds of broomrape (*Orobanche* and *Phelipanche*) species. *Ann. Bot.* **2009**, *103*, 423–431. [CrossRef]
218. Dita, M.A.; Die, J.V.; Román, B.; Krajinski, F.; Küster, H.; Moreno, M.T.; Cubero, J.I.; Rubiales, D. Gene expression profiling of *Medicago truncatula* roots in response to the parasitic plant *Orobanche crenata*. *Weed Res.* **2009**, *49*, 66–80. [CrossRef]
219. Castillejo, M.Á.; Maldonado, A.M.; Dumas-Gaudot, E.; Fernández-Aparicio, M.; Susín, R.; Rubiales, D.; Jorrín, J.V. Differential expression proteomics to investigate responses and resistance to *Orobanche crenata* in *Medicago truncatula*. *BMC Genom.* **2009**, *10*, 294. [CrossRef]
220. Castillejo, M.Á.; Amiour, N.; Dumas-Gaudot, E.; Rubiales, D.; Jorrín, J.V. A proteomic approach to studying plant response to crenate broomrape (*Orobanche crenata*) in pea (*Pisum sativum*). *Phytochemistry* **2004**, *65*, 1817–1828. [CrossRef]
221. Rubiales, D. Can we breed for durable resistance to broomrapes? *Phytopathol. Mediterr.* **2018**, *57*, 170–185.
222. Varshney, R.K.; Kudapa, H.; Pazhamala, L.; Chitikineni, A.; Thudi, M.; Bohra, A.; Gaur, P.M.; Janila, P.; Fikre, A.; Kimurto, P.; et al. Translational Genomics in Agriculture: Some Examples in Grain Legumes. *CRC. Crit. Rev. Plant Sci.* **2015**, *34*, 169–194. [CrossRef]
223. Thudi, M.; Palakurthi, R.; Schnable, J.C.; Chitikineni, A.; Dreisigacker, S.; Mace, E.; Srivastava, R.K.; Satyavathi, C.T.; Odeny, D.; Tiwari, V.K.; et al. Genomic resources in plant breeding for sustainable agriculture. *J. Plant Physiol.* **2021**, *257*, 153351. [CrossRef]
224. Tivoli, B.; Baranger, A.; Avila, C.M.; Banniza, S.; Barbetti, M.; Chen, W.; Davidson, J.; Lindeck, K.; Kharrat, M.; Rubiales, D.; et al. Screening techniques and sources of resistance to foliar diseases caused by major necrotrophic fungi in grain legumes. *Euphytica* **2006**, *147*, 223–253. [CrossRef]
225. Burud, I.; Lange, G.; Lillemo, M.; Bleken, E.; Grimstad, L.; From, P.J. Exploring robots and UAVs as phenotyping tools in plant breeding. *IFAC-PapersOnLine* **2017**, *50*, 11479–11484. [CrossRef]
226. Chen, C.Y.; Butts, C.L.; Dang, P.M.; Wang, M.L. Advances in phenotyping of functional traits. In *Phenomics in Crop Plants: Trends, options and Limitations*; Springer: New Delhi, India, 2015; pp. 163–180.

227. Cazzola, F.; Bermejo, C.J.; Guindon, M.F.; Cointry, E. Speed breeding in pea (*Pisum sativum* L.), an efficient and simple system to accelerate breeding programs. *Euphytica* **2020**, *216*, 178. [CrossRef]
228. Rubiales, D.; Pérez-de-Luque, A.; Fernández-Aparico, M.; Sillero, J.C.; Román, B.; Kharrat, M.; Khalil, S.; Joel, D.M.; Riches, C. Screening techniques and sources of resistance against parasitic weeds in grain legumes. *Euphytica* **2006**, *147*, 187–199. [CrossRef]
229. Furbank, R.T.; Tester, M. Phenomics–technologies to relieve the phenotyping bottleneck. *Trends Plant Sci.* **2011**, *16*, 635–644. [CrossRef]
230. Fernández-Aparicio, M.; Rubiales, D. Characterisation of resistance to crenate broomrape (*Orobanche crenata* Forsk.) in *Lathyrus cicera* L. *Euphytica* **2010**, *173*, 77–84. [CrossRef]
231. Fernández-Aparicio, M.; Moral, A.; Kharrat, M.; Rubiales, D. Resistance against broomrapes (*Orobanche* and *Phelipanche* spp.) in faba bean (*Vicia faba*) based in low induction of broomrape seed germination. *Euphytica* **2012**, *186*, 897–905. [CrossRef]
232. Marzougui, A.; Rajendran, A.; Mattinson, D.S.; Ma, Y.; McGee, R.J.; Garcia-Perez, M.; Ficklin, S.P.; Sankaran, S. Evaluation of biogenic markers-based phenotyping for resistance to Aphanomyces root rot in field pea. *Inf. Process. Agric.* **2022**, *9*, 1–10. [CrossRef]
233. Divyanth, L.G.; Marzougui, A.; Gonzalez-Bernal, M.J.; McGee, R.J.; Rubiales, D.; Sankaran, S. Evaluation of effective class-balancing techniques for CNN-based assessment of Aphanomyces root rot resistance in pea. *Sensors* **2022**, *22*, 7237. [CrossRef]
234. Araus, J.L.; Kefauver, S.C.; Zaman-Allah, M.; Olsen, M.S.; Cairns, J.E. Translating high-throughput phenotyping into genetic gain. *Trends Plant Sci.* **2018**, *23*, 451–466. [CrossRef]
235. Araus, J.L.; Cairns, J.E. Field high-throughput phenotyping: The new crop breeding frontier. *Trends Plant Sci.* **2014**, *19*, 52–61. [CrossRef]
236. Araus, J.L.; Kefauver, S.C. Breeding to adapt agriculture to climate change: Affordable phenotyping solutions. *Curr. Opin. Plant Biol.* **2018**, *45*, 237–247. [CrossRef] [PubMed]
237. Jung, J.; Maeda, M.; Chang, A.; Bhandari, M.; Ashapure, A.; Landivar-Bowles, J. The potential of remote sensing and artificial intelligence as tools to improve the resilience of agriculture production systems. *Curr. Opin. Biotechnol.* **2021**, *70*, 15–22. [CrossRef] [PubMed]
238. Quirós Vargas, J.J.; Zhang, C.; Smitchger, J.A.; McGee, R.J.; Sankaran, S. Phenotyping of plant biomass and performance traits using remote sensing techniques in pea (*Pisum sativum* L.). *Sensors* **2019**, *19*, 2031. [CrossRef] [PubMed]
239. Yao, L.; Van De Zedde, R.; Kowalchuk, G. Recent developments and potential of robotics in plant eco-phenotyping. *Emerg. Top. Life Sci.* **2021**, *5*, 289–300. [CrossRef]
240. Nguyen, G.N.; Norton, S.L.; Rosewarne, G.M.; James, L.E.; Slater, A.T. Automated phenotyping for early vigour of field pea seedlings in controlled environment by colour imaging technology. *PLoS ONE* **2018**, *13*, e0207788. [CrossRef]
241. Rispail, N.; Rubiales, D. Rapid and efficient estimation of pea resistance to the soil-borne pathogen *Fusarium oxysporum* by infrared imaging. *Sensors* **2015**, *15*, 3988–4000. [CrossRef]
242. Bohar, R.; Chitkineni, A.; Varshney, R.K. Genetic molecular markers to accelerate genetic gains in crops. *Biotechniques* **2020**, *69*, 158–160. [CrossRef]
243. Moreira, F.F.; Oliveira, H.R.; Volenec, J.J.; Rainey, K.M.; Brito, L.F. Integrating High-Throughput Phenotyping and Statistical Genomic Methods to Genetically Improve Longitudinal Traits in Crops. *Front. Plant Sci.* **2020**, *11*, 681. [CrossRef]
244. Bates, G.H. A Device for the Observation of Root Growth in the Soil. *Nature* **1937**, *139*, 966–967. [CrossRef]
245. Kuijken, R.C.P.; van Eeuwijk, F.A.; Marcelis, L.F.M.; Bouwmeester, H.J. Root phenotyping: From component trait in the lab to breeding. *J. Exp. Bot.* **2015**, *66*, 5389–5401. [CrossRef]
246. Cabrera, J.; Conesa, C.M.; del Pozo, J.C. May the dark be with roots: A perspective on how root illumination may bias in vitro research on plant–environment interactions. *New Phytol.* **2022**, *233*, 1988–1997. [CrossRef]
247. Yee, M.O.; Kim, P.; Li, Y.; Singh, A.K.; Northen, T.R.; Chakraborty, R. Specialized Plant Growth Chamber Designs to Study Complex Rhizosphere Interactions. *Front. Microbiol.* **2021**, *12*, 507. [CrossRef]
248. Jeudy, C.; Adrian, M.; Baussard, C.; Bernard, C.; Bernaud, E.; Bourion, V.; Busset, H.; Cabrera-Bosquet, L.; Cointault, F.; Han, S.; et al. RhizoTubes as a new tool for high throughput imaging of plant root development and architecture: Test, comparison with pot grown plants and validation. *Plant Methods* **2016**, *12*, 31. [CrossRef]
249. Taylor, H.M.; Upchurch, D.R.; McMichael, B.L. Applications and limitations of rhizotrons and minirhizotrons for root studies. *Plant Soil* **1990**, *129*, 29–35. [CrossRef]
250. Klepper, B.; Kaspar, T.C. Rhizotrons: Their Development and Use in Agricultural Research. *Agron. J.* **1994**, *86*, 745–753. [CrossRef]
251. Bodner, G.; Nakhforoosh, A.; Arnold, T.; Leitner, D. Hyperspectral imaging: A novel approach for plant root phenotyping. *Plant Methods* **2018**, *14*, 84. [CrossRef]
252. Rahman, G.; Sohag, H.; Chowdhury, R.; Wahid, K.A.; Dinh, A.; Arcand, M.; Vail, S. SoilCam: A fully automated minirhizotron using multispectral imaging for root activity monitoring. *Sensors* **2020**, *20*, 787. [CrossRef]
253. Ghosh, S.; Watson, A.; Gonzalez-Navarro, O.E.; Ramirez-Gonzalez, R.H.; Yanes, L.; Mendoza-Suárez, M.; Simmonds, J.; Wells, R.; Rayner, T.; Green, P.; et al. Speed breeding in growth chambers and glasshouses for crop breeding and model plant research. *Nat. Protoc.* **2018**, *13*, 2944–2963. [CrossRef]
254. Bhatta, M.; Sandro, P.; Smith, M.R.; Delaney, O.; Voss-Fels, K.P.; Gutierrez, L.; Hickey, L.T. Need for speed: Manipulating plant growth to accelerate breeding cycles. *Curr. Opin. Plant Biol.* **2021**, *60*, 101986. [CrossRef]

255. Rubiales, D.; Pérez-de-Luque, A.; Joel, D.M.; Alcántara, C.; Sillero, J.C. Characterization of resistance in chickpea to crenate broomrape (*Orobanche crenata*). *Weed Sci.* **2003**, *51*, 702–707. [CrossRef]
256. Fernández-Aparicio, M.; Kisugi, T.; Xie, X.; Rubiales, D.; Yoneyama, K. Low strigolactone root exudation: A novel mechanism of broomrape (*Orobanche* and *Phelipanche* spp.) resistance available for faba bean breeding. *J. Agric. Food Chem.* **2014**, *62*, 7063–7071. [CrossRef]
257. Gurney, A.L.; Grimanelli, D.; Kanampiu, F.; Hoisington, D.; Scholes, J.D.; Press, M.C. Novel sources of resistance to *Striga hermonthica* in *Tripsacum dactyloides*, a wild relative of maize. *New Phytol.* **2003**, *160*, 557–568. [CrossRef]
258. Gurney, A.L.; Slate, J.; Press, M.C.; Scholes, J.D. A novel form of resistance in rice to the angiosperm parasite *Striga hermonthica*. *New Phytol.* **2006**, *169*, 199–208. [CrossRef]
259. Kavuluko, J.; Kibe, M.; Sugut, I.; Kibet, W.; Masanga, J.; Mutinda, S.; Wamalwa, M.; Magomere, T.; Odeny, D.; Runo, S. GWAS provides biological insights into mechanisms of the parasitic plant (Striga) resistance in sorghum. *BMC Plant Biol.* **2021**, *21*, 392. [CrossRef]
260. Esquinas-Alcázar, J. Protecting crop genetic diversity for food security: Political, ethical and technical challenges. *Nat. Rev. Genet.* **2005**, *6*, 946–953. [CrossRef]
261. Bariana, H.S.; Bansal, U.K. Breeding for Disease Resistance. *Encycl. Appl. Plant Sci.* **2016**, *3*, 69–76. [CrossRef]
262. Varshney, R.K.; Glaszmann, J.C.; Leung, H.; Ribaut, J.M. More genomic resources for less-studied crops. *Trends Biotechnol.* **2010**, *28*, 452–460. [CrossRef]
263. Varshney, R.K.; Dubey, A. Novel genomic tools and modern genetic and breeding approaches for crop improvement. *J. Plant Biochem. Biotechnol.* **2009**, *18*, 127–138. [CrossRef]
264. Annicchiarico, P.; Nazzicari, N.; Pecetti, L.; Romani, M.; Ferrari, B.; Wei, Y.; Brummer, E.C. GBS-Based Genomic Selection for Pea Grain Yield under Severe Terminal Drought. *Plant Genome* **2017**, *10*, plantgenome2016.07.0072. [CrossRef]
265. Leonforte, A.; Sudheesh, S.; Cogan, N.O.I.; Salisbury, P.A.; Nicolas, M.E.; Materne, M.; Forster, J.W.; Kaur, S. SNP marker discovery, linkage map construction and identification of QTLs for enhanced salinity tolerance in field pea (*Pisum sativum* L.). *BMC Plant Biol.* **2013**, *13*, 161. [CrossRef] [PubMed]
266. Aryamanesh, N.; Zeng, Y.; Byrne, O.; Hardie, D.C.; Al-Subhi, A.M.; Khan, T.; Siddique, K.H.M.; Yan, G. Identification of genome regions controlling cotyledon, pod wall/seed coat and pod wall resistance to pea weevil through QTL mapping. *Theor. Appl. Genet.* **2014**, *127*, 489–497. [CrossRef] [PubMed]
267. Dachapak, S.; Somta, P.; Naito, K.; Tomooka, N.; Kaga, A.; Srinives, P. Detection of quantitative trait loci for salt tolerance in zombi pea [*Vigna vexillata* (L.) A. Rich]. *Euphytica* **2019**, *215*, 208. [CrossRef]
268. Barilli, E.; Carrillo-Perdomo, E.; Cobos, M.J.; Kilian, A.; Carling, J.; Rubiales, D. Identification of potential candidate genes controlling pea aphid tolerance in a *Pisum fulvum* high-density integrated DArTseq SNP-based genetic map. *Pest Manag. Sci.* **2020**, *76*, 1731–1742. [CrossRef] [PubMed]
269. Du, X.F.; Li, Y.B.; Han, X.; Ahmad, W.; Li, Q. Using high-throughput sequencing quantitatively to investigate soil nematode community composition in a steppe-forest ecotone. *Appl. Soil Ecol.* **2020**, *152*, 103562. [CrossRef]
270. Nasiri, J.; Haghnazari, A.; Saba, J. Genetic diversity among varieties and wild species accessions of pea (*Pisum sativum* L.) based on SSR markers. *Afr. J. Biotechnol.* **2009**, *8*, 3405–3417. [CrossRef]
271. Siol, M.; Jacquin, F.; Chabert-Martinello, M.; Smýkal, P.; Le Paslier, M.C.; Aubert, G.; Burstin, J. Patterns of genetic structure and linkage disequilibrium in a large collection of pea germplasm. *G3* **2017**, *7*, 2461–2471. [CrossRef]
272. Varshney, R.K. Exciting journey of 10 years from genomes to fields and markets: Some success stories of genomics-assisted breeding in chickpea, pigeonpea and groundnut. *Plant Sci.* **2016**, *242*, 98–107. [CrossRef]
273. Kankanala, P.; Nandety, R.S.; Mysore, K.S. Genomics of Plant Disease Resistance in Legumes. *Front. Plant Sci.* **2019**, *10*, 1345. [CrossRef]
274. Le Signor, C.; Aimé, D.; Bordat, A.; Belghazi, M.; Labas, V.; Gouzy, J.; Young, N.D.; Prosperi, J.M.; Leprince, O.; Thompson, R.D.; et al. Genome-wide association studies with proteomics data reveal genes important for synthesis, transport and packaging of globulins in legume seeds. *New Phytol.* **2017**, *214*, 1597–1613. [CrossRef]
275. Zitnick-Anderson, K.; Oladzadabbasabadi, A.; Jain, S.; Modderman, C.; Osorno, J.M.; McClean, P.E.; Pasche, J.S. Sources of Resistance to Fusarium solani and Associated Genomic Regions in Common Bean Diversity Panels. *Front. Genet.* **2020**, *11*, 475. [CrossRef]
276. Kwon, S.J.; Smýkal, P.; Hu, J.; Wang, M.; Kim, S.J.; McGee, R.J.; McPhee, K.; Coyne, C.J. User-friendly markers linked to Fusarium wilt race 1 resistance Fw gene for marker-assisted selection in pea. *Plant Breed.* **2013**, *132*, 642–648. [CrossRef]
277. Shirasawa, K.; Sasaki, K.; Hirakawa, H.; Isobe, S. Genomic region associated with pod color variation in pea (*Pisum sativum*). *G3 Genes Genomes Genet.* **2021**, *11*, jkab081. [CrossRef]
278. Dinglasan, E.; Periyannan, S.; Hickey, L.T. Harnessing adult-plant resistance genes to deploy durable disease resistance in crops. *Essays Biochem.* **2022**, *66*, 571–580.
279. Castillejo, M.Á.; Fernández-Aparicio, M.; Rubiales, D. Proteomic analysis by two-dimensional differential in gel electrophoresis (2D DIGE) of the early response of *Pisum sativum* to *Orobanche crenata*. *J. Exp. Bot.* **2012**, *63*, 107–119. [CrossRef]
280. Castillejo, M.Á.; Bani, M.; Rubiales, D. Understanding pea resistance mechanisms in response to *Fusarium oxysporum* through proteomic analysis. *Phytochemistry* **2015**, *115*, 44–58. [CrossRef]

281. Burstin, J.; Salloignon, P.; Chabert-Martinello, M.; Magnin-Robert, J.B.; Siol, M.; Jacquin, F.; Chauveau, A.; Pont, C.; Aubert, G.; Delaitre, C.; et al. Genetic diversity and trait genomic prediction in a pea diversity panel. *BMC Genom.* **2015**, *16*, 105. [CrossRef]
282. Annicchiarico, P.; Nazzicari, N.; Laouar, M.; Thami-Alami, I.; Romani, M.; Pecetti, L. Development and proof-of-concept application of genome-enabled selection for pea grain yield under severe terminal drought. *Int. J. Mol. Sci.* **2020**, *21*, 2414. [CrossRef]
283. Bohra, A.; Pandey, M.K.; Jha, U.C.; Singh, B.; Singh, I.P.; Datta, D.; Chaturvedi, S.K.; Nadarajan, N.; Varshney, R.K. Genomics-assisted breeding in four major pulse crops of developing countries: Present status and prospects. *Theor. Appl. Genet.* **2014**, *127*, 1263–1291. [CrossRef]
284. Varshney, R.K.; Tuberosa, R. Translational genomics in crop breeding for biotic stress resistance: An introduction. *Transl. Genom. Crop Breed. Vol. I Biot. Stress* **2013**, *1*, 1–9. [CrossRef]
285. Zargar, S.M.; Raatz, B.; Sonah, H.; Muslima, N.; Bhat, J.A.; Dar, Z.A.; Agrawal, G.K.; Rakwal, R. Recent advances in molecular marker techniques: Insight into QTL mapping, GWAS and genomic selection in plants. *J. Crop Sci. Biotechnol.* **2015**, *18*, 293–308. [CrossRef]
286. Sagan, M.; Huguet, T.; Duc, G. Phenotypic characterization and classification of nodulation mutants of pea (*Pisum sativum* L.). *Plant Sci.* **1994**, *100*, 59–70. [CrossRef]
287. Dalmais, M.; Schmidt, J.; Le Signor, C.; Moussy, F.; Burstin, J.; Savois, V.; Aubert, G.; Brunaud, V.; De Oliveira, Y.; Guichard, C.; et al. UTILLdb, a *Pisum sativum* in silicoforward and reverse genetics tool. *Genome Biol.* **2008**, *9*, R43. [CrossRef]
288. Tayeh, N.; Aubert, G.; Pilet-Nayel, M.L.; Lejeune-Hénaut, I.; Warkentin, T.D.; Burstin, J. Genomic tools in pea breeding programs: Status and perspectives. *Front. Plant Sci.* **2015**, *6*, 1037. [CrossRef]
289. Gupta, M.; Dubey, S.; Jain, D.; Chandran, D. The *Medicago truncatula* sugar transport protein 13 and its Lr67res-like variant confer powdery mildew resistance in legumes via defense modulation. *Plant Cell Physiol.* **2021**, *62*, 650–667. [CrossRef]
290. Benhamou, N.; Garand, C. Cytological analysis of defense-related mechanisms induced in pea root tissues in response to colonization by nonpathogenic *Fusarium oxysporum Fo47*. *Phytopathology* **2001**, *91*, 730–740. [CrossRef]
291. Gali, K.K.; Tar'an, B.; Madoui, M.A.; van der Vossen, E.; van Oeveren, J.; Labadie, K.; Berges, H.; Bendahmane, A.; Lachagari, R.V.; Burstin, J.; et al. Development of a sequence-based reference physical map of pea (*Pisum sativum* L.). *Front. Plant Sci.* **2019**, *10*, 323. [CrossRef]

plants MDPI

Article

Identification and Spatial Distribution of Bioactive Compounds in Seeds *Vigna unguiculata* (L.) Walp. by Laser Microscopy and Tandem Mass Spectrometry

Mayya P. Razgonova [1,2,*], Marina O. Burlyaeva [1], Yulia N. Zinchenko [1], Ekaterina A. Krylova [1], Olga A. Chunikhina [1], Natalia M. Ivanova [3], Alexander M. Zakharenko [4,5] and Kirill S. Golokhvast [1,2,4,5,*]

1 N.I. Vavilov All-Russian Institute of Plant Genetic Resources, 190000 Saint-Petersburg, Russia
2 Institute of Life Science and Biomedicine, Far Eastern Federal University, 690922 Vladivostok, Russia
3 Department of Botany, Saint-Petersburg State University, 199034 Saint-Petersburg, Russia
4 Siberian Federal Scientific Centre of Agrobiotechnology RAS, 633501 Krasnoobsk, Russia
5 Laboratory of Supercritical Fluid Research and Application in Agrobiotechnology, Tomsk State University, 634050 Tomsk, Russia
* Correspondence: m.razgonova@vir.nw.ru (M.P.R.); golokhvast@sfsca.ru (K.S.G.)

Abstract: The research presents a comparative metabolomic study of extracts of *Vigna unguiculata* seed samples from the collection of the N.I. Vavilov All-Russian Institute of Plant Genetic Resources. Analyzed samples related to different areas of use in agricultural production, belonging to different cultivar groups *sesquipedalis* (vegetable accessions) and *unguiculata* (grain accessions). Metabolome analysis was performed by liquid chromatography combined with ion trap mass spectrometry. Substances were localized in seeds using confocal and laser microscopy. As a result, 49 bioactive compounds were identified: flavonols, flavones, flavan-3-ols, anthocyanidin, phenolic acids, amino acids, monocarboxylic acids, aminobenzoic acids, fatty acids, lignans, carotenoid, sapogenins, steroids, etc. Steroidal alkaloids were identified in *V. unguiculata* seeds for the first time. The seed coat (palisade epidermis and parenchyma) is the richest in phenolic compounds. Comparison of seeds of varieties of different directions of use in terms of the number of bioactive substances identified revealed a significant superiority of vegetable accessions over grain ones in this indicator, 36 compounds were found in samples from cultivar group *sesquipedalis*, and 24 in *unguiculata*. The greatest variety of bioactive compounds was found in the vegetable accession k-640 from China.

Keywords: *Vigna unguiculata*; tandem mass spectrometry; metabolites; local landrace; vegetable cultivar; grain cultivar; bioactive substances; seed; laser microscopy

Citation: Razgonova, M.P.; Burlyaeva, M.O.; Zinchenko, Y.N.; Krylova, E.A.; Chunikhina, O.A.; Ivanova, N.M.; Zakharenko, A.M.; Golokhvast, K.S. Identification and Spatial Distribution of Bioactive Compounds in Seeds *Vigna unguiculata* (L.) Walp. by Laser Microscopy and Tandem Mass Spectrometry. *Plants* **2022**, *11*, 2147. https://doi.org/10.3390/plants11162147

Academic Editors: Mingxun Chen, Lixi Jiang and Yuan Guo

Received: 22 July 2022
Accepted: 9 August 2022
Published: 18 August 2022

1. Introduction

Vigna unguiculata (L.) Walp. is an important component of farming systems in many parts of the world. It is mainly grown on the continents of Africa, Asia, and South America. In recent years, information has appeared about the successful experience of its cultivation in the southern regions of Russia and Russian Far East [1,2]. *V. unguiculata* is a multipurpose vegetable crop and it is valued for its drought and heat tolerance. It is grown mainly for its seeds (cvg. *unguiculata* = ssp. *unguiculata*) or green vegetable pods (cvg. *sesquipedalis* = ssp. *sesquipedalis*). Food products prepared from *Vigna* are a source of many nutrients: proteins, amino acids, carbohydrates, minerals, fiber, vitamins, and other bioactive compounds [3–7].

This crop has a high level of polyphenols, some profiles of which are not commonly found in other legumes. The main polyphenols are phenolic acid derivatives (148–1176 μg/g) and flavonol glycosides (27–1060 μg/g). A number of varieties contain anthocyanins (875–3860 μg/g) and/or flavan-3-ols (2155–6297 μg/g). Monomers, mainly catechin-7-O-glucoside, predominate among the flavan-3-ols.

There are data on the content of bioactive peptides in *V. unguiculata*; although, their content varies depending on the variety. In addition, there is medical evidence showing significant anti-inflammatory effects and benefits of *V. unguiculata* polyphenols and peptides against cancer, diabetes, and cardiovascular diseases [8]. It holds great promise for wider use in modern food products due to nutritional properties that have a positive impact on health and a range of agronomic advantages over other legumes. The high content of polyphenols in the seeds, which are mainly concentrated in the seed coat, provides additional benefits for the use of phenolic extracts as nutraceutical and functional ingredients in food formulations [9].

The seeds of *V. unguiculata* are rich in bioactive compounds, and they can be used in the development of functional foods necessary for a healthy lifestyle. Such nutrition will serve as a medicine at the same time, because seeds of *V. unguiculata* have anti-inflammatory, immune-boosting, neuroprotective, anti-apoptotic, anti-cancer, antioxidant, anti-mutagenic, and cardioprotective properties [10].

At the present, analysis of the metabolomic composition of plants is applied for various purposes [11–18]. This approach is used to identify relationships between biochemical parameters and genetic characteristics of various crops, to solve various breeding problems, to characterize different groups of crops (variety types, subspecies, and species), to identify genotypes that do not differ morphologically and physiologically, etc. In recent years, interest has arisen in the application of metabolomic data in applied research aimed at solving problems in the food and pharmaceutical industries. Nutritional quality is becoming increasingly important for consumers and food manufacturers.

Comprehensive and diversified approaches are necessary to improve the quality of pods and seeds in different groups of *V. unguiculata* varieties, both vegetable and grain use. The data obtained from the analysis of the metabolome of seeds in the future can complement traditional and molecular genetic breeding methods that are aimed at creating new hybrids, donors of valuable traits, inbred lines, and varieties with high levels of bioactive compounds.

When searching for accessions with the highest nutritional value, and in order to create varieties with improved seed quality, it is important to study the content of bioactive substances, taking into account the specifics of their content in varieties with different directions of use. Microscopic methods, including confocal laser scanning microscopy, provide a unique opportunity to study the tissue localization of phytochemical compounds in plants. Comparison of HPLC data with fluorescent microscopy allows not only visualization of these compounds, but also the understanding of their function in plant organs. Currently, microscopic images are successfully used to clarify the morphological structure of cells and the structure of plant tissue [19], and to visualize the location of different groups of chemicals in plant organs [20–22]. However, there is limited information available on the presence and localization of bioactive compounds in *V. unguiculata* seeds.

The purpose of this research was a comparative metabolomic analysis of bioactive substances in *V. unguiculata* seeds derived from the collection of N.I. Vavilov All-Russian Institute of Plant Genetic Resources. Samples were grown in the field in Primorsky Krai (Russia) at the northern border of the crop area.

The objectives of the study were:

- Analysis of bioactive compounds in seeds using high-performance liquid chromatography (HPLC) and tandem mass spectrometry (MS/MS) methods;
- Visualization of the localization of phytochemical compounds in seed tissues using confocal laser microscopy;
- Identification of differences in the content of bioactive compounds in seeds of vegetable and grain accessions (cultivar groups *sesquipedalis* and *unguiculata*).

2. Materials and Methods

2.1. Materials

The object of the research was the seeds of *V. unguiculata* from the group of varieties (cultivar groups) *sesquipedalis* and *unguiculata* harvested in 2020, grown at the Far East Experiment Station Branch of the Federal Research Center the N.I. Vavilov All-Russian Institute of Plant Genetic Resources (Table 1; Figure 1). Landraces were collected by N.I. Vavilov during the 1929 expedition to China (k-640, k-642) and obtained from the extract in 1921 from the USA (k-6) and in 1985 from Germany (k-1783); modern cultivar "Lyanchihe" (k-632341) was developed in Russia in the Primorsky Territory as a result of selection from samples of Chinese origin. Seeds (k-6) had cherry seed color; k-1783—beige; k-640, k-642, and k-632341—reddish-brown with dark strokes. Seeds for analysis were collected at the stage of industrial ripeness at the same time in 2020.

Table 1. *V. unguiculata* seed material samples.

No	VIR Catalogue Number	Name of Accessions	Country of Origin	Acqdate	Cultivar Groups
1	k-6	Cultivar "Clay"	USA	1921	*unguiculata*
2	k-640	Landrace	China	1929	*sesquipedalis*
3	k-642	Landrace	China	1929	*sesquipedalis*
4	k-1783	Landrace	Germany	1985	*unguiculata*
5	k-632341	Cultivar "Lyanchihe"	Far East, Russia	2018	*sesquipedalis*

k-6 k-1783

Figure 1. *Cont.*

k-640 k-642 k-632341

Figure 1. Samples of *V. unguiculata* grown at the Far East Experiment Station Branch of the Federal Research Center the N.I. Vavilov All-Russian Institute of Plant Genetic Resources. Appearance of the plants and seeds.

2.2. Chemicals and Reagents

HPLC-grade acetonitrile was purchased from Fisher Scientific (Southborough, UK), and MS-grade formic acid was from Sigma-Aldrich (Steinheim, Germany). Ultra-pure water was prepared from a SIEMENS ULTRA clear (SIEMENS water technologies, Munich, Germany), and all other chemicals were analytical grade.

2.3. Maceration

Fractional maceration technique was applied to obtain highly concentrated extracts. From 300 g of the sample, 4 g of *V. unguiculata* was randomly selected for maceration. The total amount of the extractant (ethyl alcohol of reagent grade) was divided into 3 parts, and the grains were consistently infused with the first, second, and third parts. The solid–solvent ratio was 1:20. The infusion of each part of the extractant lasted 7 days at room temperature.

2.4. Liquid Chromatography

HPLC was performed using Shimadzu LC-20 Prominence HPLC (Shimadzu, Kyoto, Japan), equipped with a UV-sensor and a Shodex ODP-40 4E reverse phase column to separate multicomponent mixtures. The gradient elution program was as follows: 0.01–4 min, 100% CH_3CN; 4–60 min, 100–25% CH_3CN; 60–75 min, 25–0% CH_3CN; control washing 75–120 min 0% CH_3CN. The entire HPLC analysis was performed using a UV–VIS detector SPD-20A (Shimadzu, Japan) at wavelengths of 230 and 330 nm, at 30 °C provided with column oven CTO-20A (Shimadzu, Japan) with an injection volume of 20 μL.

2.5. Mass Spectrometry

MS analysis was performed on an ion trap amaZon SL (BRUKER DALTONIKS, Bremen, Germany) equipped with an ESI source in negative ion mode. The optimized parameters were obtained as follows: ionization source temperature: 70 °C, gas flow: 4 L/min, nebulizer gas (atomizer): 7.3 psi, capillary voltage: 4500 V, end plate bend voltage: 1500 V, fragmentary: 280 V, collision energy: 60 eV. An ion trap was used in the scan range m/z 100–1.700 for MS and MS/MS. The mass spectra were recorded in negative and positive ion modes. The capture rate was one spectrum/s for MS and two spectrum/s for MS/MS. Data collection was controlled by Hystar Data Analysis 4.1 software (BRUKER DALTONIKS, Bremen, Germany). All experiments were repeated three times. A four-stage ion separation mode (MS/MS mode) was implemented. After a comparison of the m/z values, retention times, and fragmentation patterns with the MS/MS spectral data retrieved from the cited articles and after a database search (MS2T, MassBank, HMDB). The AmaZon SL ion trap is equipped with dedicated software to manage and interface it with 8 major HPLC system manufacturers. The Compass HyStar software (Version Bruker Compass HyStar 4.1 SR1 (4.1.28.0)) was used for synchronization with the Shimadzu chromatograph.

2.6. Optical Microscopy

The study of the structure of the *V. unguiculata* seed coat by light microscopy was carried out by performing sections of dry seeds. Sections were prepared by hand with a safety razor from the middle part of half the seed in a direction perpendicular to the hilum. Photo fixation of sections and the study of the color of the seed coat were carried out in water, immediately after preparation of the sections.

For the confocal laser scanning microscopy, dry untreated *V. unguiculata* seeds were used. The transverse dissection was performed with an MS-2 sled microtome (Tochmedpribor, Kharkiv, Ukraine). The obtained sliced seeds were placed on microscopic cover glass through immersion oil to reduce light refraction by air gaps. The autofluorescence parameters were determined using confocal microscope (LSM 800, Carl Zeiss Microscopy GmbH, Berlin, Germany). The autofluorescence spectrum was chosen using lambda scan mode of the microscope, which allows to determine the emission maximum in a specific sample and obtain spectral acquisition. The specimen was excited by each laser separately and three main peaks of autofluorescence were revealed: excitation by a violet laser, 405 nm (solid state, diode, 5 mW) with the emission maximum of 400–475 nm (blue); excitation by a blue laser, 488 nm (solid state, diode, 10 mW) with the emission maxima in 500–545 nm (green) and 620–700 nm (red). The used power and detector gain for blue, green, and red channels were 5% and 750 V, 4.5% and 800 V, and 7% and 850 V, respectively. The objective Plan-Apochromat 63×/1.40 Oil DIC M27 with 63× magnification and the software ZEN 2.1 (Carl Zeiss Microscopy GmbH, Germany) were used for image acquisition and processing.

2.7. Statistical Data Processing

Statistical analysis included the compilation of binary matrices for each of the compounds identified in seeds of *V. unguiculata*, in which the "presence" (1) or "absence" (0) of the compound was noted in each of the studied samples. Based on the total matrix, a dendrogram was built, demonstrating the relationship between the studied samples. The method of unweighted pair-group cluster analysis with arithmetic averaging (UPGMA) using the TREECON program was used to construct the dendrogram. The cluster analysis was also carried out and a WPGMC (Median Clustering or Weighted Pair Group Method with Centroid Averaging) dendrogram was plotted in the Statistica 7 program, based on the data of the summary matrix.

In addition, based on the results of a comparative analysis of substances identified in *V. unguiculata* seeds, a Consensus tree was constructed using the Winclada-Nona program using the maximum parsimony criterion.

3. Results and Discussion

3.1. Tandem Mass Spectrometry

V. unguiculata extracts were analyzed using an ion trap coupled to high-performance liquid chromatography to better interpret the diversity of phytochemicals available. Primary analysis of the extracts showed a composition rich in bioactive substances. All experiments were repeated three times. A four-stage ion separation mode (MS/MS mode) was implemented. After a comparison of the m/z values, retention times, and fragmentation patterns with the MS/MS spectral data retrieved from the cited articles and after a database search (MS2T, MassBank, HMDB). Tentative identification showed the presence of 49 bioactive compounds detected by mass spectrometric analysis in *V. unguiculata* extracts. Forty-nine target analytes were successfully identified by comparing fragmentation patterns and retention times, most of which were polyphenols. Other compounds were identified by comparing their MS/MS data with the available literature. All identified compounds, along with molecular formulas, calculated and observed m/z, MS/MS data, and their comparative profile for *V. unguiculata*, are shown in Table 2.

Table 2. Compounds identified from extracts of *V. unguiculata* under positive and negative ionization modes by tandem mass spectrometry.

No	VIR Catalogue Number	Class of Compounds	Identified Compounds	Formula	Mass	Molecular Ion [M-H]-	Molecular Ion [M+H]+	2 Fragmentation MS/MS	3 Fragmentation MS/MS	4 Fragmentation MS/MS	References
	Polyphenols										
1	k-6(583); k-642 (582)	Flavonol	Dihydrokaempferol (Aromadendrin; Katuranin)	$C_{15}H_{12}O_6$	288.25	287		151; 269			*Solanium tuberosum* [23]; *F. glaucescens* [24]; *Camellia kucha* [25]; *Echinops* [26]
2	k-6(583); k-632341 (579)	Flavonol	Quercetin	$C_{15}H_{10}O_7$	302.23	301		179; 273	151;		Potato leaves [27]; *Vigna sinensis* [28]; *Vaccinium macrocarpon* [29]; Propolis [30]
3	k-6(583); k-1783 (585); k-640 (589); k-640 (590)	Flavonol	Dihydroquercetin (Taxifolin; Taxifoliol)	$C_{15}H_{12}O_7$	304.25	303		285; 177	241		*Dracocephalum palmatum* [31]; *Vitis amurensis* [32]; *Rhodiola rosea* [33]
4	k-640 (590)	Flavonol	Myricetin	$C_{15}H_{10}O_8$	318.23	317		273	260; 251		*Vaccinium macrocarpon* [29]; *F. glaucescens* [24]; millet grains [34]; *Sanguisorba officinalis* [35]
5	k-6(584)	Flavonol	Quercetin 3-*O*-glucoside (Isoquercitrin; Hirsutrin)	$C_{21}H_{20}O_{12}$	464.37		303	256; 165	229		Potato [23]; *Vigna sinensis* [28]; Andean blueberry [36]; *Lonicera Henryl* [37]
6	k-640 (590)	Flavone	Acacetin (Linarigenin; Buddle-oflavonol)	$C_{16}H_{12}O_5$	284.26		285	257; 239; 177	248; 237; 216; 173		*Mentha* [38]; *Dracocephalum palmatum* [39]; *Wissadula periplocifolia* [40]
7	k-6(583)	Tetrahydro-xyflavane	Luteoliflavan-eriodictyol-O-hexoside	$C_{36}H_{34}O_{16}$	722.64		723	587; 555; 499	543; 516; 499	499	*C. edulis* [24]
8	k-632341 (579)	Flavan-3-ol	Epiafzelechin ((epi)Afzelechin)	$C_{15}H_{14}O_5$	274.26		275	195; 149	167	150	*Cassia granidis* [41]; *Cassia abbreviata* [42,43]; *A. cordifolia; F. glaucescens; F. herrerae* [24]
9	k-632341 (579); k-632341 (580); k-640 (590); k-642 (582)	Flavan-3-ol	Catechin (D-Catechol)	$C_{15}H_{14}O_6$	290.26	289		245; 205	201	175	*Eucalyptus* [44]; *Vaccinium macrocarpon* [45]; *C. edulis* [24]; *Vigna inguiculata* [46]; *Triticum* [47]
10	k-632341 (579); k-640 (590); k-642 (582)	Flavan-3-ol	(Epi)afzelechin-4′-*O*-glucoside	$C_{21}H_{24}O_{10}$	436.41	435		299; 191; 161	151; 117		*Vigna unguiculata* [46]
11	k-632341 (580)	Flavan-3-ol	(Epi)afzelechin-3-*O*-glucoside	$C_{21}H_{24}O_{10}$	436.41	435		313; 299; 273			*Vigna unguiculata* [46]; *Cassia abbreviata* [42]

Table 2. *Cont.*

No	VIR Catalogue Number	Class of Compounds	Identified Compounds	Formula	Mass	Molecular Ion [M-H]-	Molecular Ion [M+H]+	2 Fragmentation MS/MS	3 Fragmentation MS/MS	4 Fragmentation MS/MS	References
12	k-6(583); k-6(584); k-632341 (579); k-642 (582)	Flavan-3-ol	Chinchonain Ia	$C_{24}H_{20}O_9$	452.41	451		289	245	203	Andean blueberry [36]
13	k-632341 (579); k-632341 (580); k-640 (590)	Flavan-3-ol	(epi)Catechin *O*-hexoside	$C_{21}H_{24}O_{11}$	452.41	451		289; 269; 245	245; 231	227	Andean blueberry [36]
14	k-6(583); k-6(584); k-632341 (579); k-632341 (580); k-640 (589); k-640 (590); k-642 (582)	Anthocyanidin	Delphinidin 3-*O*-glucoside	$C_{21}H_{21}O_{12+}$	465.39	463		300	151; 271	169	Rapeseed petals [48]; *Vigna sinensis* [28]; *Berberis ilicifolia; Berberis empetrifolia; Ribes maellanicum; Ribes cucullatum; Myrteola nummalaria* [49]; *Vigna unguiculata* [50]
15	k-632341 (579); k-632341 (580); k-642 (582)	Anthocyanidin	Delphinidin-3,5-*O*-diglucoside	$C_{27}H_{30}O_{17}$	626.52		626	303; 465	257; 165	229; 157	*Vitis labrusca* [51]; *Solanium nigrum* [52]; Muscadine pomace [53]
16	k-1783 (585)	Lignan	Dimethylmatairesinol (Arctigenin Methyl Ether)	$C_{22}H_{26}O_6$	386.44		387	205			Lignans [54]
17	k-640 (590)	Lignan	Medioresinol	$C_{21}H_{24}O_7$	388.41	387		207; 225; 179			Lignans [54]; *Punica granatum* [55]; *Bituminaria* [56]
18	k-632341 (579)	Lignan	Syringaresinol	$C_{22}H_{26}O_8$	418.44		419	326; 248; 151	298; 254; 218; 174	251; 182; 145	*Triticum aestivum* L. [47]; *Lignans* [54]; *Punica granatum* [55]; *Magnolia thailandica* [57]
19	k-632341 (579)	Hydroxybenzoic acid (Phenolic acid)	Protocatechuic acid	$C_7H_6O_4$	154.12		155	126			*Vigna unguiculata* [6]; *Eucalyptus* [44]; *Eucalyptus Globulus* [58]; *Vaccinium macrocarpon* [45]; *Lonicera japonica* [59]
20	k-640 (590)	Polyphenolic acid	Coumaroyl quinic acid methyl ester	$C_{17}H_{20}O_8$	352.34	351		285; 267; 243	242; 200		*F. glaucescens* [24]
21	k-640 (590)	Derivative of hydroxycinnamic acid	Ferulic acid-*O*-hexoside	$C_{16}H_{20}O_9$	356.32	355		191;209; 174	173		*A. cordifolia* [24]; millet grains [34]; Rapeseed petals [48]; beer [60]; strawberry [61]
22	k-632341 (579); k-640 (589)	Hydroxybenzoic acid	Salvianolic acid D	$C_{20}H_{18}O_{10}$	418.35	417		373	347	303	*Salvia miltiorrhiza* [62]; *Lonicera caerulea* [63]
23	k-640 (590)	Phenolic acid	*Trans*-salvianolic acid J	$C_{27}H_{22}O_{12}$	538.46		539	493; 479; 357	420		*Mentha* [38]

Table 2. *Cont.*

No	VIR Catalogue Number	Class of Compounds	Identified Compounds	Formula	Mass	Molecular Ion [M-H]-	Molecular Ion [M+H]+	2 Fragmentation MS/MS	3 Fragmentation MS/MS	4 Fragmentation MS/MS	References
	Others										
24	k-642 (582)	Non-proteinogenic L-α-amino acid	L-Pyroglutamic acid (Pidolic acid; 5-Oxo-L-Proline)	$C_5H_7NO_3$	129.11		130	112			Potato leaves [27]
25	k-632341 (580)	Aminobenzoic acid	4-Aminobenzoic acid (*p*-aminobenzoic acid)	$C_7H_7NO_2$	137.14		138	119			*Solanum tuberosum* [23]
26	k-632341 (579) k-632341 (580) k-640 (589); k-640 (590) k-642 (582)	Carboxylic acid	Indole-3-carboxylic acid	$C_{10}H_9NO_2$	175.18		176	159; 130			Beer [60]
27	k-632341 (579); k-632341 (580); k-640 (590)	Monocarboxylic acid	Dihydroferulic acid	$C_{10}H_{12}O_4$	196.2	195		159; 129			*A. cordifolia* [24]; Coffee [64]
28	k-632341 (579); k-632341 (580)	Amino acid	L-Tryptophan (Tryptophan; (S)-Tryptophan)	$C_{11}H_{12}N_2O_2$	204.23		205	188	146; 144	118	*Camellia kucha* [25]; *Vigna inguiculata* [6,46]; Rapeseed petals [48]; *Perilla frutescens* [65]
29	k-1783 (585)	Omega-5 fatty acid	Myristoleic acid (Cis-9-Tetradecanoic acid)	$C_{14}H_{26}O_2$	226.36		227	209	139	122	*F. glaucescens* [24]
30	k-642 (582)	Purine	Adenosine	$C_{10}H_{13}N_5O_4$	267.24		268	136			*Lonicera japonica* [59]
31	k-632341 (579)	Omega-3 fatty acid	Linoleic acid (Linolic acid; Telfairic acid)	$C_{18}H_{32}O_2$	280.45	279		261; 205	205		Salviae [66]; *Angelicae sinensis Radix* [67]; *Pinus sylvestris* [68]
32	k-640 (590)	Hydroperoxy fatty acid	Hydroperoxy-octadecadienoic acid	$C_{18}H_{32}O_4$	312.44	311		183; 309			Potato [23]
33	k-6(583); k-640 (589)	Unsaturated monocarboxylic acid	9,10-Dihydroxy-8-oxooctadec-12-enoic acid (oxo-DHODE; oxo-Dihydroxy-octadecenoic acid)	$C_{18}H_{32}O_5$	328.44	327		291; 269; 251; 233; 211; 195; 183	279; 258; 247; 236; 217; 195	177; 161	*Bituminaria* [56]; Broccoli [69]; *Phyllostachys nigra* [70]

Table 2. *Cont.*

No	VIR Catalogue Number	Class of Compounds	Identified Compounds	Formula	Mass	Molecular Ion [M-H]-	Molecular Ion [M+H]+	2 Fragmentation MS/MS	3 Fragmentation MS/MS	4 Fragmentation MS/MS	References
34	k-6(583); k-632341 (580); k-640 (590)	Unsaturated monocarboxylic acid	Trihydroxyoctadecadienoic acid	$C_{18}H_{32}O_5$	328.44	327		211; 183; 127	183; 167; 149		Potato leaves [27]
35	k-640 (589)	Omega-hydroxy-long-chain fatty acid	Hydroxy docosanoic acid	$C_{22}H_{44}O_3$	356.58	355		309	305; 132		*A. cordifolia* [24]
36	k-1783 (585); k-640 (590)	Steroidal alkaloid	Solanidine	$C_{27}H_{43}NO$	397.64		398	185; 272	167		Potato [71,72]
37	k-6(583); k-640 (590)	Long-chain fatty acid	Nonacosanoic acid	$C_{29}H_{58}O_2$	438.77	437		393			*C. edulis* [24]
38	k-1783 (585); k-640 (590)	Steroid	Vebonol	$C_{30}H_{44}O_3$	452.67		453	435; 336; 209	336; 226		*Hylocereus polyrhizus* [73]; *Zostera marina* [74]
39	k-6(583)	Carotenoid	*all-trans*-β-cryptoxanthin caprate		706.2		707	625; 587; 571	527		Sarsaparilla [75]
40	k-640 (590)	Steroidal alkaloid	β-chaconine	$C_{39}H_{63}NO_{10}$	705.92		706	690			
41	k-6(583)	Carotenoid	(all-E)-violaxanthin myristate		810.1		811	794; 748; 723; 675; 622; 602			Carotenoids [76]
42	k-6(583); k-1783 (585); k-640 (589); k-640 (590); k-642 (582)	Steroidal alkaloid	α-chaconine	$C_{45}H_{73}NO_{14}$	852.06		852	706	704; 690		*Solanum tuberosum* [72,77,78]
43	k-640 (589); k-642 (582)	Steroidal alkaloid	α-solanine	$C_{45}H_{73}NO_{15}$	868.96		868	722	560; 398	398; 185	*Solanum tuberosum* [72,77,78]
44	k-6(583); k-640 (589); k-640 (590)	Steroidal alkaloid	Solanidenol chacotriose	$C_{45}H_{73}NO_{15}$	868.96		868	850; 823; 765; 747; 722; 706	704	677	Potato [77]
45	k-1783 (585)	Steroidal alkaloid	Solanidadiene solatriose	$C_{45}H_{73}NO_{15}$	868.96		868	706	722; 398; 560		Potato [77]
46	k-6(583); k-640 (590)	Steroidal alkaloid	Solanidenediol chacotriose	$C_{45}H_{73}NO_{16}$	884.06		884	866; 822; 800; 78; 720; 704	849; 822; 720; 704; 691		Potato [77]

Table 2. *Cont.*

No	VIR Catalogue Number	Class of Compounds	Identified Compounds	Formula	Mass	Molecular Ion [M-H]-	Molecular Ion [M+H]+	2 Fragmentation MS/MS	3 Fragmentation MS/MS	4 Fragmentation MS/MS	References
47	k-6(583); k-640 (589); k-640 (590)	Steroidal alkaloid	Leptinine II	$C_{45}H_{73}NO_{16}$	884.06		884	866; 738; 722	720; 704; 677; 654		*Solanum tuberosum* [77]
48	k-6(583); k-632341 (579); k-632341 (580); k-640 (589); k-640 (590); k-642 (582)	Sapogenin	3-Rhamnose-galactose-glucuronic acid-soyasapogenol B	$C_{48}H_{78}O_{18}$	943.12	941		615; 733; 795; 923	571		*Bituminaria bituminosa* [56]; *Medicago truncatula* [79]
49	k-6(583); k-640 (589) k-640 (590); k-642 (582)	Sapogenin	6-deoxyhexose-hexoside-uronic acid–soyasapogenol A	$C_{48}H_{78}O_{19}$	959.12	957		525; 733; 939	457		*Bituminaria* [56]; *Medicago truncatula* [79]

Separately, it is worth noting the presence of steroidal alkaloids in all presented samples of *V. unguiculata*, which were not previously noted by other authors. In addition, the presence of sapogenins A and B is also interesting, as they previously were identified in soybean (*Glycine* Willd.).

Figure 2 shows an example of decoding the spectrum of the steroidal alkaloid α-chaconine from an ion chromatogram obtained by tandem mass spectrometry. The [M + H]+ ion produces five product ions at *m/z* 706, *m/z* 673, *m/z* 560, *m/z* 437, and *m/z* 398 (Figure 2). A fragment ion at *m/z* 706 gives rise to two daughter ions at *m/z* 560 and *m/z* 398. A fragment ion at *m/z* 560 gives rise to five daughter ions at *m/z* 545, *m/z* 454, *m/z* 398, *m/z* 380, and *m/z* 213. This compound is identified in scientific articles as α-chaconine, for example, in *Solanum tuberosum* [72,77,78,80,81].

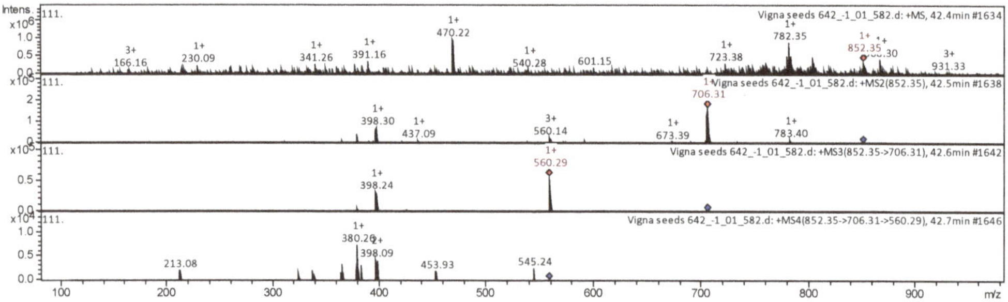

Figure 2. Mass spectrum of the steroidal alkaloid α-chaconine from *V. unguiculata* extract, at *m/z* 852.35.

Table 3 shows the distribution of bioactive substances in accessions of *V. unguiculata*. Based on the tabular data, the sample of vegetable accession k-640 showed the greatest variety of bioactive compounds (29 compounds). In two other vegetable accessions (k-632341 and k-642), 19 and 13 compounds were found, respectively. In grain accessions (k-6 and k-1783), 18 and 7 compounds were identified, respectively.

Table 3. Distribution of bioactive substances in accessions of *V. unguiculata*.

No	Class of Compounds	Identified Compounds	Formula	VIR Catalogue Number				
				k-642	k-632341	k-640	k-1783	k-6
	Polyphenols							
1	Flavonol	Dihydrokaempferol (Aromadendrin; Katuranin)	$C_{15}H_{12}O_6$	■				■
2	Flavonol	Quercetin	$C_{15}H_{10}O_7$		■			■
3	Flavonol	Dihydroquercetin (Taxifolin; Taxifoliol)	$C_{15}H_{12}O_7$			■	■	■
4	Flavonol	Myricetin	$C_{15}H_{10}O_8$			■		
5	Flavonol	Quercetin 3-*O*-glucoside (Isoquercitrin; Hirsutrin)	$C_{21}H_{20}O_{12}$					■
6	Flavan-3-ol	Epiafzelechin ((epi)Afzelechin)	$C_{15}H_{14}O_5$		■			
7	Flavan-3-ol	Catechin (D-Catechol)	$C_{15}H_{14}O_6$	■	■	■		
8	Flavan-3-ol	(Epi)afzelechin-4′-*O*-glucoside	$C_{21}H_{24}O_{10}$	■	■	■		
9	Flavan-3-ol	(Epi)afzelechin-3-*O*-glucoside	$C_{21}H_{24}O_{10}$		■			

Table 3. *Cont.*

No	Class of Compounds	Identified Compounds	Formula	VIR Catalogue Number				
				k-642	k-632341	k-640	k-1783	k-6
10	Flavan-3-ol	Chinchonain Ia	$C_{24}H_{20}O_9$	+	+			+
11	Flavan-3-ol	(epi)Catechin *O*-hexoside	$C_{21}H_{24}O_{11}$		+	+		
12	Flavone	Acacetin (Linarigenin; Buddleoflavonol)	$C_{16}H_{12}O_5$			+		
13	Tetrahydroxyflavan	Luteoliflavan-eriodictyol-O-hexoside	$C_{36}H_{34}O_{16}$					+
14	Anthocyanidin	Delphinidin 3-*O*-glucoside	$C_{21}H_{21}O_{12+}$	+	+	+		+
15	Anthocyanidin	Delphinidin-3,5-*O*-diglucoside	$C_{27}H_{30}O_{17}$	+	+			
16	Lignan	Dimethylmatairesinol (Arctigenin Methyl Ether)	$C_{22}H_{26}O_6$				+	
17	Lignan	Medioresinol	$C_{21}H_{24}O_7$			+		
18	Lignan	Syringaresinol	$C_{22}H_{26}O_8$		+			
19	Hydroxybenzoic acid (Phenolic acid)	Protocatechuic acid	$C_7H_6O_4$		+			
20	Hydroxybenzoic acid (Phenolic acid)	Salvianolic acid D	$C_{20}H_{18}O_{10}$		+	+		
21	Polyphenolic acid	Coumaroyl quinic acid methyl ester	$C_{17}H_{20}O_8$			+		
22	Derivative of hydroxycinnamic acid	Ferulic acid-*O*-hexoside	$C_{16}H_{20}O_9$			+		
23	Phenolic acid	*Trans*-salvianolic acid J	$C_{27}H_{22}O_{12}$			+		
	Others							
24	Non-proteinogenic L-α-amino acid	L-Pyroglutamic acid (Pidolic acid; 5-Oxo-L-Proline)	$C_5H_7NO_3$	+				
25	Aminobenzoic acid	4-Aminobenzoic acid (*p*-aminobenzoic acid)	$C_7H_7NO_2$		+			
26	Monocarboxylic acid	Dihydroferulic acid	$C_{10}H_{12}O_4$		+	+		
27	Carboxylic acid	Indole-3-carboxylic acid	$C_{10}H_9NO_2$	+	+	+		
28	Amino acid	L-Tryptophan (Tryptophan; (S)-Tryptophan)	$C_{11}H_{12}N_2O_2$		+			
29	Omega-5 fatty acid	Myristoleic acid (Cis-9-Tetradecanoic acid)	$C_{14}H_{26}O_2$				+	
30	Purine	Adenosine	$C_{10}H_{13}N_5O_4$	+				
31	Omega-3 fatty acid	Linoleic acid (Linolic acid; Telfairic acid)	$C_{18}H_{32}O_2$		+			
32	Hydroperoxy fatty acid	Hydroperoxy-octadecadienoic acid	$C_{18}H_{32}O_4$			+		
33	Unsaturated monocarboxylic acid	9,10-Dihydroxy-8-oxooctadec-12-enoic acid (oxo-DHODE; oxo-Dihydroxy-octadecenoic acid)	$C_{18}H_{32}O_5$			+		+

Table 3. *Cont.*

No	Class of Compounds	Identified Compounds	Formula	VIR Catalogue Number				
				k-642	k-632341	k-640	k-1783	k-6
34	Unsaturated monocarboxylic acid	Trihydroxyoctadecadienoic acid	$C_{18}H_{32}O_5$		■	■		■
35	Omega-hydroxy-long-chain fatty acid	Hydroxy docosanoic acid	$C_{22}H_{44}O_3$			■		
36	Long-chain fatty acid	Nonacosanoic acid	$C_{29}H_{58}O_2$			■		■
37	Steroid	Vebonol	$C_{30}H_{44}O_3$			■	■	
38	Carotenoid	*all-trans*-β-cryptoxanthin caprate						■
39	Carotenoid	(all-E)-violaxanthin myristate						■
40	Steroidal alkaloid	Solanidine	$C_{27}H_{43}NO$			■	■	
41	Steroidal alkaloid	β-chaconine	$C_{39}H_{63}NO_{10}$			■		
42	Steroidal alkaloid	α-chaconine	$C_{45}H_{73}NO_{14}$	■		■	■	■
43	Steroidal alkaloid	α-solanine	$C_{45}H_{73}NO_{15}$	■		■		
44	Steroidal alkaloid	Solanidenol chacotriose	$C_{45}H_{73}NO_{15}$			■		■
45	Steroidal alkaloid	Solanidadiene solatriose	$C_{45}H_{73}NO_{15}$				■	
46	Steroidal alkaloid	Solanidenediol chacotriose	$C_{45}H_{73}NO_{16}$			■		■
47	Steroidal alkaloid	Leptinine II	$C_{45}H_{73}NO_{16}$			■		■
48	Sapogenin	3-Rhamnose-galactose-glucuronic acid-soyasapogenol B	$C_{48}H_{78}O_{18}$	■	■	■		■
49	Sapogenin	6-deoxyhexose-hexoside-uronic acid–soyasapogenol A	$C_{48}H_{78}O_{19}$	■		■		■

Comparison of samples by the presence or absence of identified substances by different statistical methods did not reveal clear relationships with their origin and belonging to a certain group of varieties (cultivar groups) (Figures 3–5). However, some samples differed from others in the presence of specific substances that were found only in them (Figure 6). Thus, only the grain accession "Clay" (k-6, USA) bred at the beginning of the last century contained the flavonol quercetin 3-*O*-glucoside, carotenoids *all-trans*-β-cryptoxanthin caprate, and *(all-E)*-violaxanthin myristate, tetrahydroxyflavan luteoliflavan-eriodictyol-*O*-hexoside. The steroidal alkaloid solanidadiene solatriose, omega-5 fatty acid myristoleic acid, and lignan dimethylmatairesinol were found in a local grain accession from Germany (k-1783). In addition, unlike other studied samples, only k-1783 lacked the anthocyanidin delphinidin 3-*O*-glucoside and sapogenin 3-rhamnose-galactose-glucuronic acid-soyasapogenol B. Vegetable accessions from China (k-640; k-642), and Primorsky Krai Russia (k-632341) also differed from each other in the content of some bioactive substances. Only k-642 had L-pyroglutamic acid and adenosine. Only k-632341 had aminobenzoic acid, protocatechuic acid, L-tryptophan, *(epi)*afzelechin and *(epi)*afzelechin-3-*O*-glucoside, omega-3 fatty acid linoleic acid. The lignan syringaresinol were not identified only in k-632341. The flavone acacetin, the steroidal alkaloid β-chaconine, and the phenolic acid *Trans*-salvianolic acid J were identified in k-640.

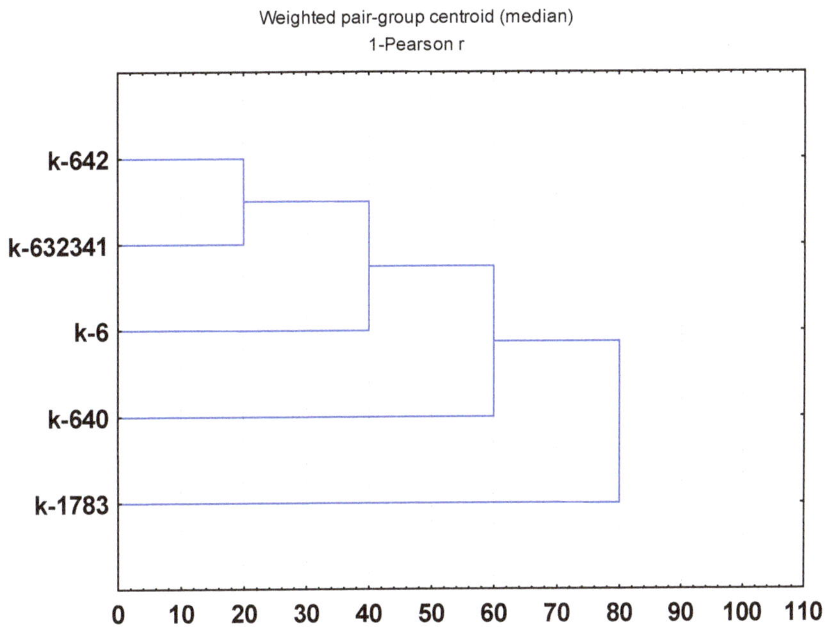

Figure 3. Dendrogram WPGMC (Median Clustering or Weighted Pair Group Method with Centroid Averaging), plotting on the basis of a comparative analysis of substances identified in *V. unguiculata* seeds.

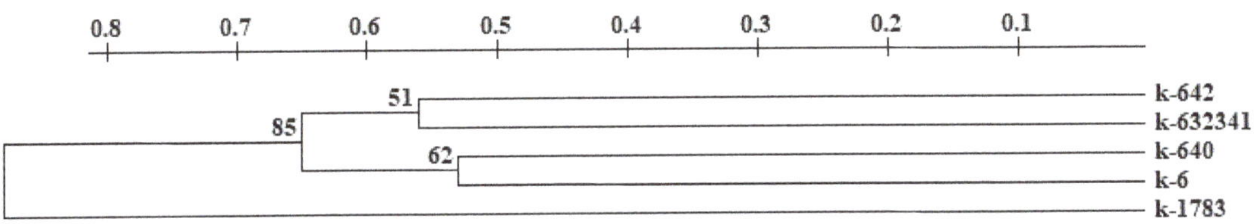

Figure 4. Dendrogram UPGMA (Unweighted Group Average or Unweighted Pair Group Method with Arithmetic Averaging), constructed on the basis of a comparative analysis of substances identified in *V. unguiculata* seeds.

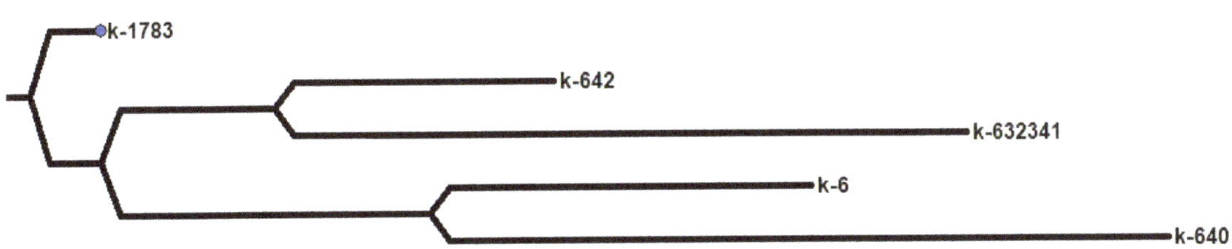

Figure 5. Consensus tree built using the criterion of maximum parsimony based on the results of a comparative analysis of substances identified in *V. unguiculata* seeds (Ci = 77, consistency index—the proportion of homoplasia in the total number of changes in traits, L = 63, Ri = 33 retention index—the number of synapomorphies determined by the data).

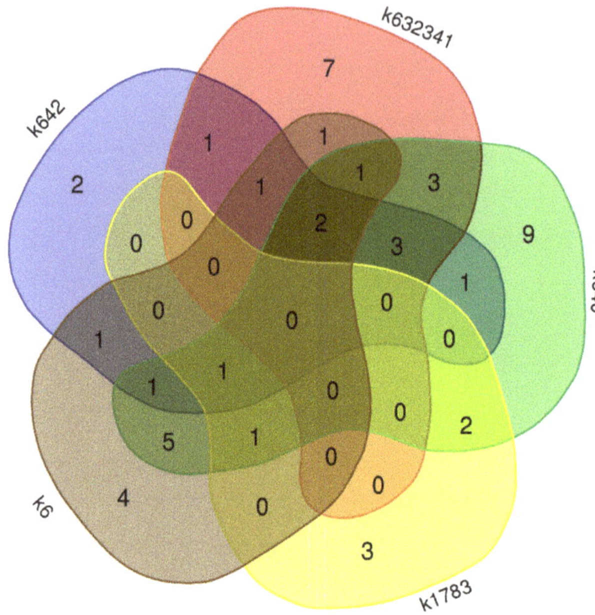

Figure 6. Venn diagram of bioactive compounds between *V. unguiculata* accessions.

Despite the absence of a significant difference in the presence or absence of various bioactive compounds in seeds between grain and vegetable accessions, the samples from these groups differed in the number of identified compounds. Grain accessions had much fewer (24) bioactive substances than in vegetable samples (36) (Tables 2 and 3). A comparison of accessions of different directions of use in terms of the number of identified classes of compounds also showed their smaller number in grain accessions (12) than in vegetable samples (23) (Figure 7).

It should be noted that indole-3-carboxylic acid, catechin, *(Epi)*afzelechin-4′-*O*-glucoside were found in all vegetable accessions and were not identified in grain accessions (Table 3).

Two compounds belonging to the group of anthocyanins were identified in vegetable accessions and one in grain accessions, and delphinidin 3-*O*-glucoside was not found only in k-1783 and k-640.

Five flavonoid compounds were identified: dihydrokaempferol in vegetable accession k-642 and grain accession k-6, quercetin 3-*O*-glucoside in k-6; myricetin in k-640; quercetin in k-632341; and dihydroquercetin in k-640 and in all grain accessions. Flavone acacetin has only been identified in vegetable cultivar k-640. A total of six flavan-3-ols were identified: catechin and *(epi)*afzelechin-4′-*O*-glucoside were found in all vegetable cultivars, *(epi)*afzelechin and *(epi)*afzelechin-3-*O*-glucoside were found only in vegetable cultivar k-632341, chinchonain Ia—in two vegetable accessions (k-642; k-632341) and one grain accession (k-6), *(epi)*catechin *O*-hexoside—in two vegetable accessions (k-632341; k-640). Tetrahydroxyflavan had only been identified in grain accession k-6 (luteoliflavan-eriodictyol-*O*-hexoside).

A class of phenolic acids has also been identified: trans-salvianolic acid J and coumaroyl quinic acid methyl ester were in k-640, salvianolic acid D was identified in k-632341 and k-640; protocatechuic acid was in k-632341, and a derivative of hydroxycinnamic acid (ferulic acid-*O*-hexoside) was found in k-640.

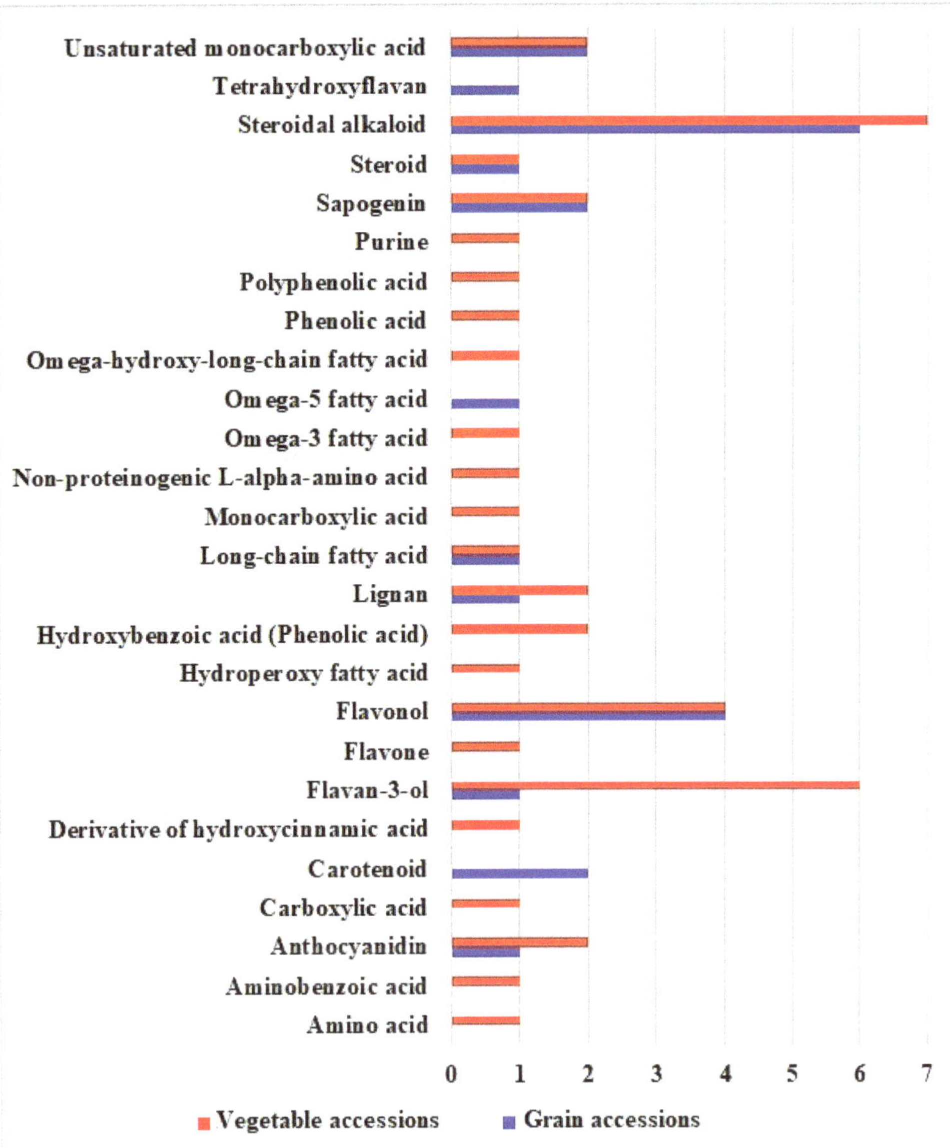

Figure 7. The content of bioactive substances in the seeds of grain and vegetable accessions. 1–7—number of identified substances.

In addition, amino acids (L-tryptophan in k-632341), monocarboxylic acid (dihydroferulic acid in k-632341 and k-640), aminobenzoic acid (4-aminobenzoic acid in k-632341), carboxylic acid (indole-3-carboxylic acid in all vegetable accessions), non-proteinogenic L-α-amino acid (L-pyroglutamic acid in k-642), unsaturated monocarboxylic acid (9,10-dihydroxy-8-oxooctadec-12-enoic acid in k-6, and k-640), and trihydroxyoctadecadienoic acid (in k-640, k-632341, and k-6) were identified in *V. unguiculata* seeds.

From hydroperoxy fatty acids, hydroperoxy-octadecadienoic acids (k-640) were found; from long-chain fatty acids—nonacosanoic acid (k-6, k-640); from omega-3 fatty acids—linoleic acid (k-632341); from omega-5 fatty acid—myristoleic acid (k-1783); and from omega-hydroxy-long-chain fatty acid—hydroxy docosanoic acid (k-640).

In addition, three lignans (dimethylmatairesinol in k-1783, medioresinol in k-640, and syringaresinol in k-640), two carotenoids (*all-trans*-β-cryptoxanthin caprate, *(all-E)*-violaxanthin myristate, only in k-6), purine (Adenosine in k-642), and two sapogenins (3-Rhamnose-galactose-glucuronic acid-soyasapogenol B were identified in all accessions, except for k-1783; 6-deoxyhexose-hexoside-uronic acid-soyasapogenol A was found in k-642, k-640, and k-6.

Eight compounds from the group of steroidal alkaloids were also identified: solanidine (k-640, k-1783), β-chaconine (k-640), α-chaconine (in all samples except k-632341), α-solanine (k-642, k-640), solanidenol chacotriose (k-640; k-6), solanidadiene solatriose (k-1783), and solanidenediol chacotriose and leptinine II (k-640; k-6).

3.2. Confocal Laser Scanning Microscopy

Laser microscopy exploits the ability of the chemicals to fluoresce when excited by a laser and allows certain groups of chemical compounds in the plant tissues to be located. Our study allowed us to find out the spatial arrangement of phenolic compounds in the seed coat and cotyledons of *V. unguiculata*, based on the autofluorescence.

According to the literature data, the blue fluorescence in plants is mainly due to the presence of phenolic hydroxycinnamic acids [82]. The main fluorescent component is ferulic acid, but other hydroxycinnamic (*p*-coumaric, caffeic) acids can also contribute to it [83]. Moreover, lignin is a well-known source of blue fluorescence in plants. It has a wide emission range due to the presence of multiple fluorophore types within the molecule and can be observed when excited by UV and visible light [84]. Previous studies have shown that the lignin content of legume seed coat is low [85,86], and the cotyledons are poorly lignified [87]. Therefore, we suppose that most of the blue fluorescence in *V. unguiculata* seeds comes from hydroxycinnamic acids.

The blue-light-induced green autofluorescence in the range of 500–545 nm can be explained by the presence of flavins and flavonols (myricetin, quercetin, and kaempferol) and their derivatives [88–90]. The emission in the red spectrum mainly occurs due to the presence of anthocyanins and anthocyanidins [91–93].

The seed coat consists of cells of the palisade layer (palisade epidermis), hypoderma, and parenchyma. Sample k-6, with cherry-colored seeds, had a small number of flavonols and flavins in the cell walls of the palisade epidermis and in the cell cavities of this layer (Figure 8c). Anthocyanins and anthocyanidins accumulated mostly only in the parenchyma of the seed coat (Figure 8d). Phenolic acids were the most abundant in cotyledons, especially in their outer layer (Figure 8b).

According to confocal microscopy, the cavities and cell walls of the palisade epidermis of accession k-1783, which had a beige seed color, contained more flavonols and flavins than accession k-6 (Figure 9c). The same substances were found in the hypoderma and in several rows of parenchyma cells adjacent to it, as well as in the cells of the cotyledons. Anthocyanins and anthocyanidins were located in small inclusions in the cell cavities of the palisade epidermis and hypoderma and in the upper cells of the parenchymal layer (Figure 9d). Quite significant differences between the two grain samples can be noted in the location of phenolic acids. Phenolic acids were almost exclusively in the cotyledons in accession k-6 (Figure 8b), and the same acids were in the cotyledons and in the cells of the parenchyma of the seed coat in accession k-1783 (Figure 9b).

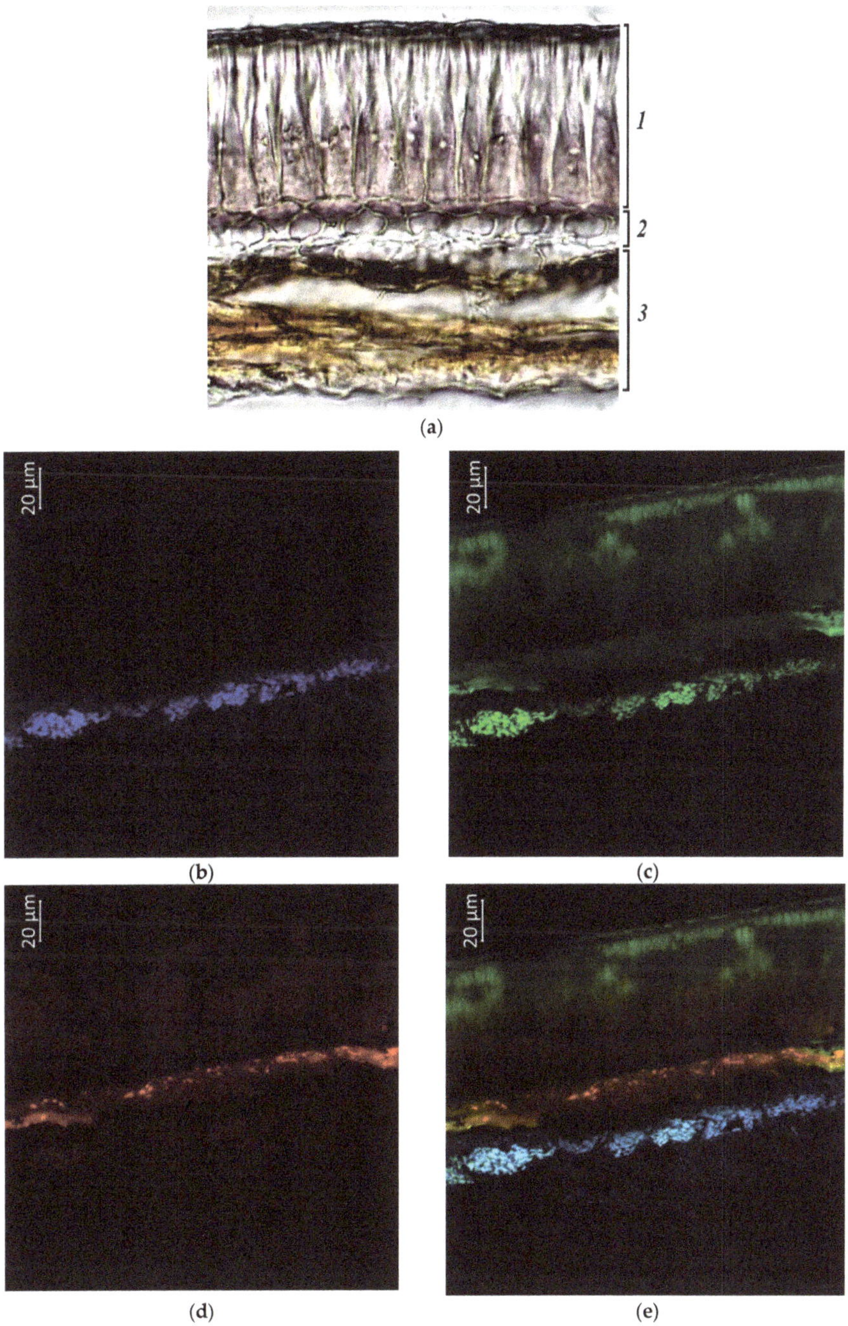

Figure 8. *V. unguiculata* accession k-6: (**a**)—seed coat structure, light microscopy (1—palisade layer, 2—hypoderma, 3—parenchyma); (**b**–**e**) transverse section of the seed, confocal microscopy, (**b**)—excitation 405 nm with the emission in 400–475 nm (blue), (**c**)—excitation 488 nm with the emission in 500–545 nm (green), (**d**)—excitation 488 nm with the emission in 620–700 nm (red), (**e**)—merged.

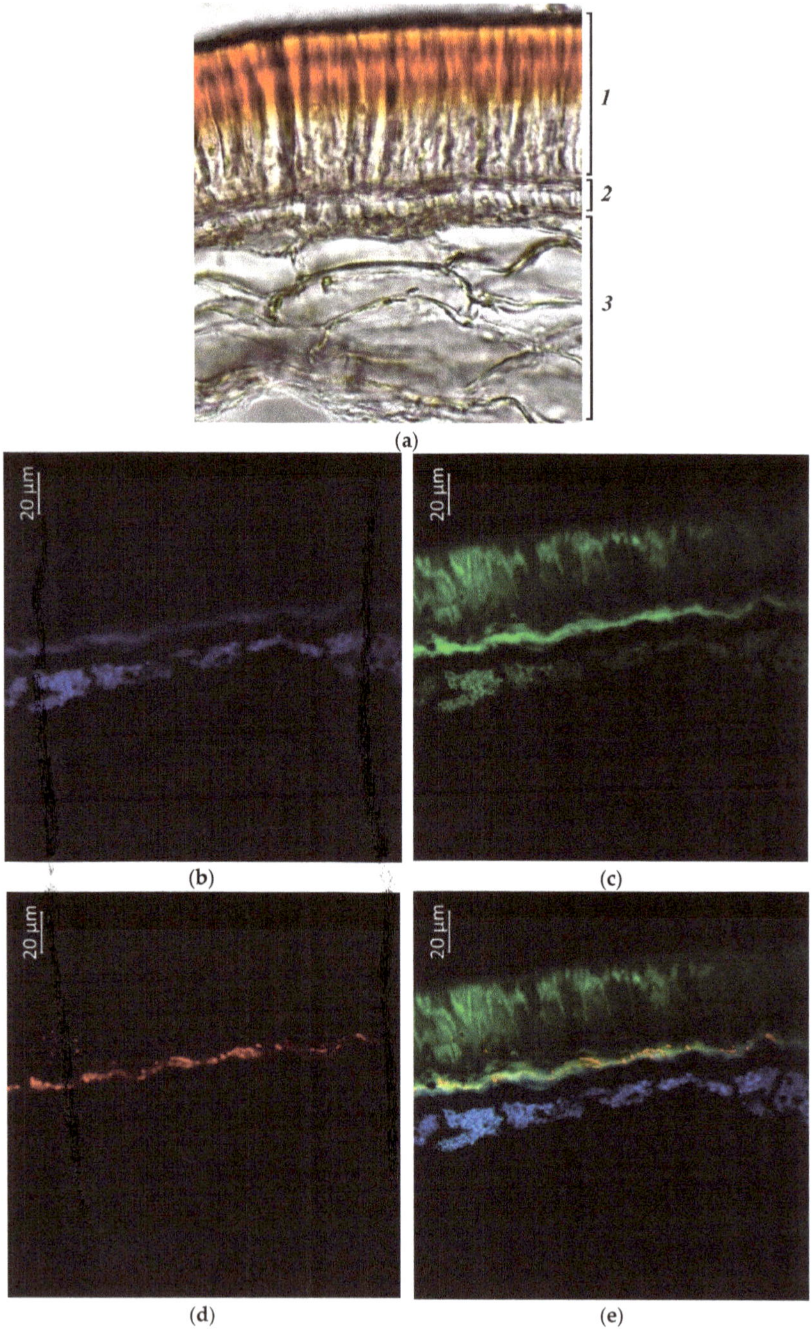

Figure 9. *V. unguiculata* accession k-1783: (**a**)—seed coat structure, light microscopy (1—palisade layer, 2—hypoderma, 3—parenchyma); (**b**–**e**)—transverse section of the seed, confocal microscopy, (**b**)—excitation 405 nm with the emission in 400–475 nm (blue), (**c**)—excitation 488 nm with the emission in 500–545 nm (green), (**d**)—excitation 488 nm with the emission in 620–700 nm (red), (**e**)—merged.

Despite the light beige color of the seeds, the k-1783 accession had much more bioactive substances in the seed coat than the k-6 sample with dark cherry seeds.

Photographs showed a more intense coloration of the palisade epidermis and underlying layers of the seed coat in vegetable accessions (k-640, k-642, and k-632341), in contrast to grain accessions (Figures 10–12). The seeds of these specimens were reddish-brown in color and had longitudinal dark streaks running parallel to the hilum. Anthocyanins and anthocyanidins were present in the palisade epidermis, in the hypoderma, and in the cells of the parenchyma adjacent to the hypodermis. Among vegetable samples, k-640 had a less bright red color, and k-632341 had a stronger red color (characterized by dark cherry pods at the stage of technical ripeness). The green color, indicating the presence of flavonols and flavins, was the most intense for k-642, less for k-640, and k-632341. These compounds were located in all vegetable accessions both in the palisade epidermis, hypoderma, parenchyma, and in the cotyledons (Figures 10c, 11c, and 12c).

Phenolic acids in all vegetable samples were concentrated to a greater extent in the cotyledons, and to a lesser extent in the seed coat (Figures 10b, 11b and 12b). It should be noted that in k-642 and k-632341, the hypoderma and parenchyma of the seed coat were more strongly colored blue than in k-640. Based on the photographs, we can conclude that the highest content of the studied substances was found in accession k-642.

A large number of researchers have shown the important role of the seed coat in supplying the embryo with nutrients during development [94,95]. It plays a significant role in the regulation of seed dormancy and germination, and it is a rich source of many valuable substances. It contains a wide range of compounds: flavonoids, proteins, peptides, amino acids, alkaloids, terpenoids, steroids, etc. [96]. Many components of the seed coat play an important role in seed protection. As a result of our study of seeds using confocal microscopy, it was found that in most samples, the largest number of bioactive substances was in the palisade epidermis of the seed coat, and fewer compounds were found in the hypoderma, parenchyma, and cotyledons. Moreover, phenolic acids were localized mainly in the cotyledons, less in the parenchyma of the seed coat. Flavonols and flavins were located mainly in the palisade epidermis, and less in the cells of the hypoderma, parenchyma of the seed coat, and cotyledons. Anthocyanins and anthocyanidins in vegetable samples (k-640, k-642, k-632341) were present not only in the palisade epidermis, but also in parenchyma cells. Anthocyanins were found mainly in parenchyma cells and hypodermis in grain accessions (k-6, k-1783)

Polyphenolic compounds, including phenolic acids and their derivatives, tannins, and flavonoids, represent the largest group of natural plant nutrients. They determine the color of fruits and seeds and play an important role in disease resistance [97,98]. Most of the phenolic compounds found in legume seeds are also located in the seed coat. In soybean and common bean, the concentration of phenolic compounds such as flavonoids and anthocyanins correlate with seed coat color [99,100]. In our study, according to the results of tandem mass spectrometry, anthocyanidins, which determined the color of cherry and red-brown seeds, were identified in *V. unguiculata* seeds. The darkest seeds (cherry) of accession k-6 had the smallest number of different polyphenols, compared with the seeds of vegetable accessions (k-640, k-642, and k-632341), colored red-brown. It can be assumed that the color of the seed coat of k-6 also depends on the carotenoid (*(all-trans)*-β-cryptoxanthin caprate), which was found only in its seeds.

A comparison of the results obtained using confocal laser microscopy with data on the content of bioactive substances in seeds identified using tandem mass spectrometry provides additional information. Laser microscopy makes it possible to visualize the comparative concentration of substances in plant tissues, which is an additional characteristic in the study of bioactive substances in samples. If, according to the data of tandem mass spectrometry, k-640 had the greatest variety of identified substances, then k-642 was distinguished by the concentration of anthocyanidins and flavonols in the cells of the seed coat.

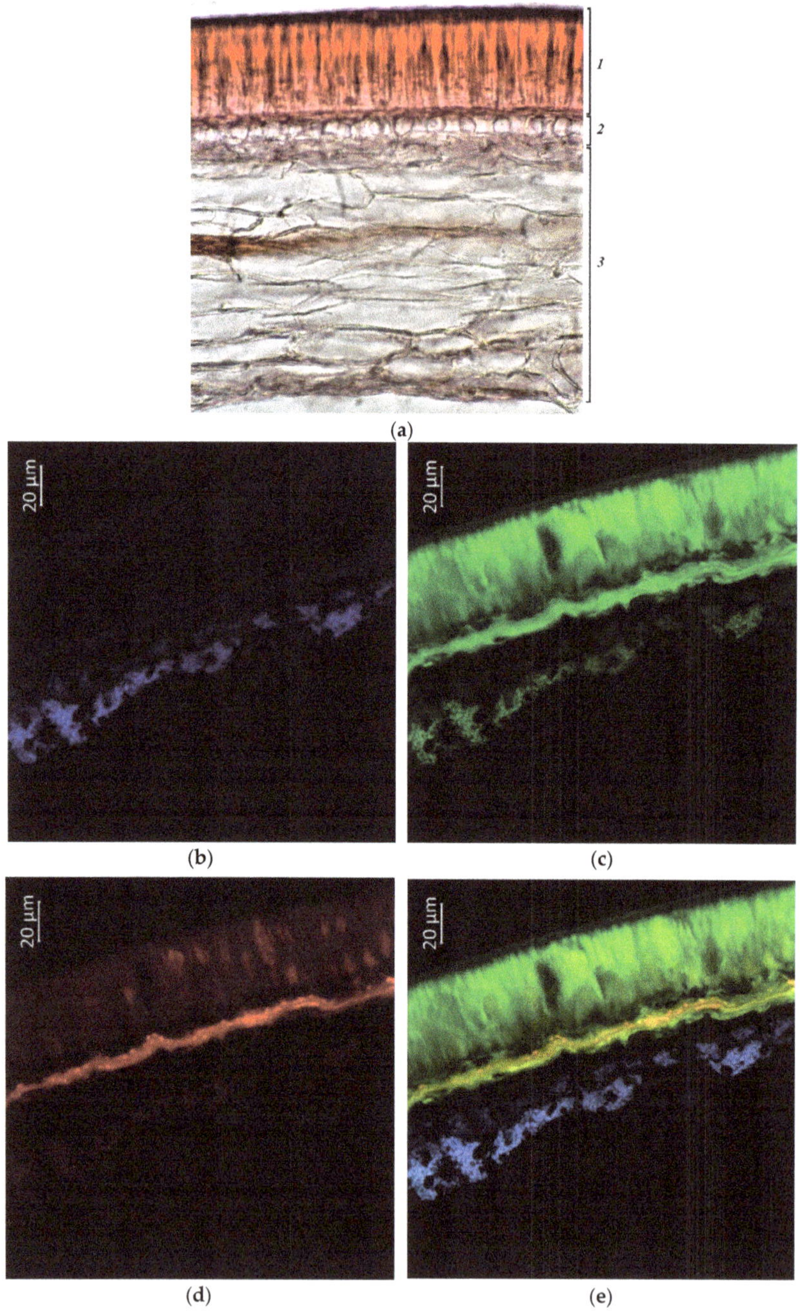

Figure 10. *V. unguiculata* accession k-640: (**a**)—seed coat structure, light microscopy (1—palisade layer, 2—hypoderma, 3—parenchyma); (**b**–**e**)—transverse section of the seed, confocal microscopy, (**b**)—excitation 405 nm with the emission in 400–475 nm (blue), (**c**)—excitation 488 nm with the emission in 500–545 nm (green), (**d**)—excitation 488 nm with the emission in 620–700 nm (red), (**e**)—merged.

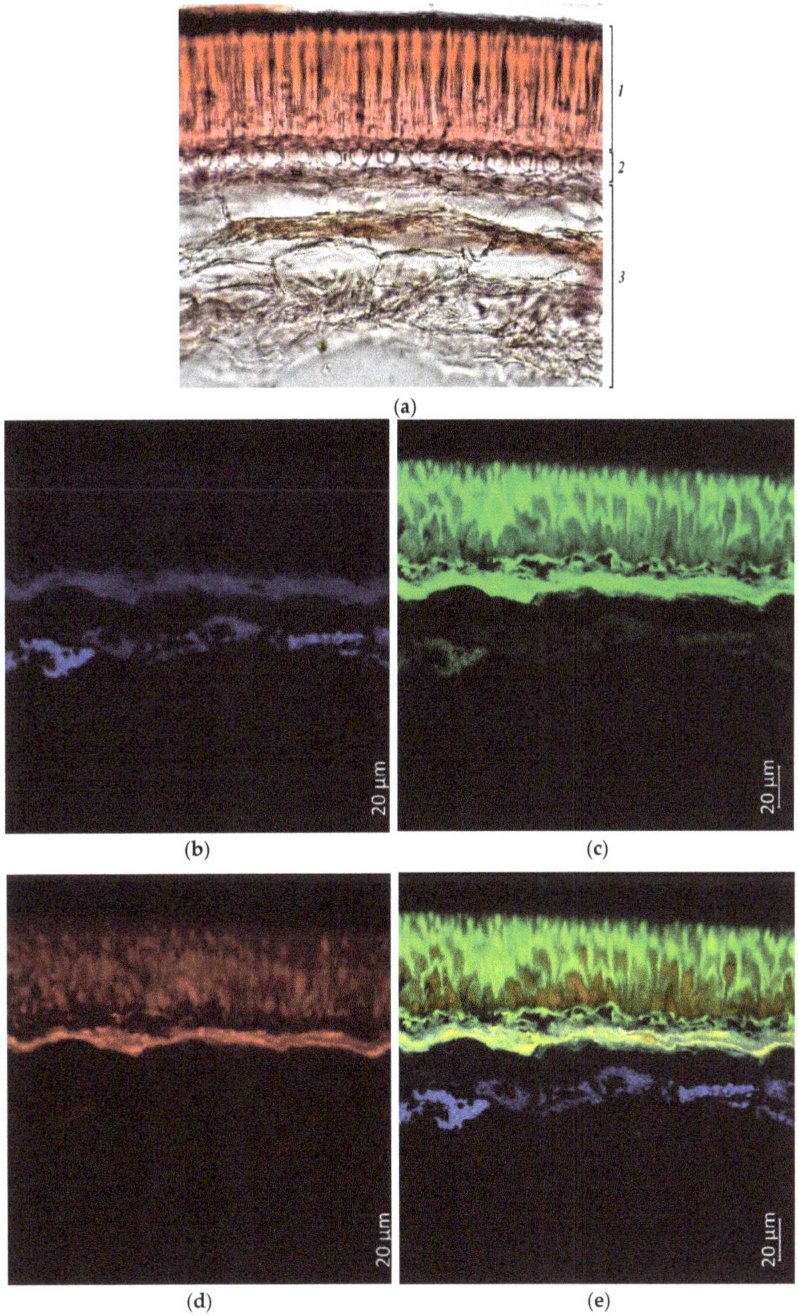

Figure 11. *V. unguiculata* accession k-642: (**a**)—seed coat structure, light microscopy (1—palisade layer, 2—hypoderma, 3—parenchyma); (**b**–**e**) transverse section of the seed, confocal microscopy, (**b**)—excitation 405 nm with the emission in 400–475 nm (blue), (**c**)—excitation 488 nm with the emission in 500–545 nm (green), (**d**)—excitation 488 nm with the emission in 620–700 nm (red), (**e**)—merged.

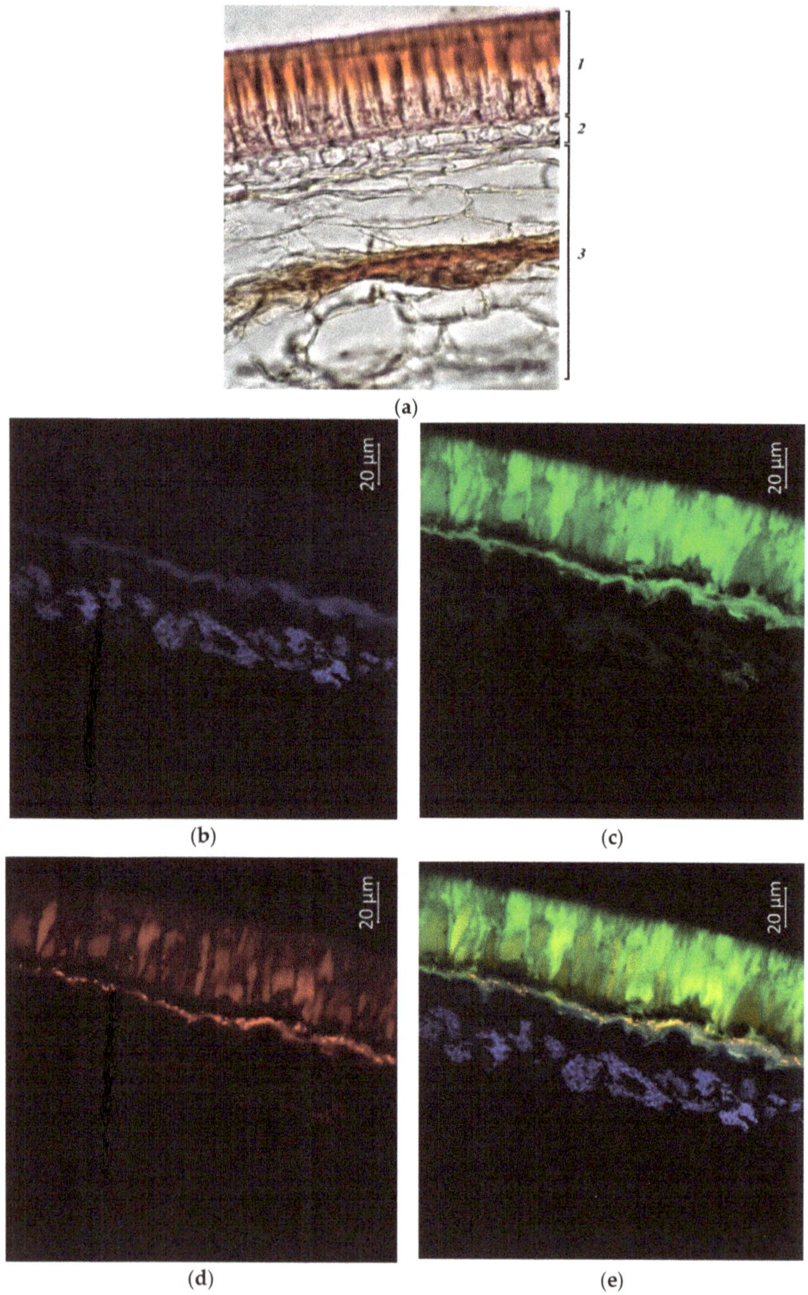

Figure 12. *V. unguiculata* accession k-632341: (**a**)—seed coat structure, light microscopy (1—palisade layer, 2—hypoderma, 3—parenchyma); (**b**–**e**)—transverse section of the seed, confocal microscopy, (**b**)—excitation 405 nm with the emission in 400–475 nm (blue), (**c**)—excitation 488 nm with the emission in 500–545 nm (green), (**d**)—excitation 488 nm with the emission in 620–700 nm (red), (**e**)—merged.

4. Conclusions

The seeds of both vegetable and grain accessions of *V. unguiculata* are rich in bioactive compounds: phenols, polyphenols, flavonols, flavones, anthocyanins, amino acids, carotenoids, omega-3 and 5 fatty acids, sapogenins, steroids, etc. We identified 49 bioactive substances, most of which belonged to the class of polyphenols. For the first time, steroidal alkaloids were found in *V. unguiculata* seeds, and they were present in all studied samples. Most of the bioactive substances were localized in the palisade epidermis; the smaller part was in the hypoderma and parenchyma of the seed coat and cotyledons. Seeds of vegetable accessions differed from seeds of grain accessions in a large number of bioactive substances, 36 and 24, respectively. Comparison of accessions of different directions of use in terms of the number of classes of compounds also showed their smaller number in grain accessions (13) than in vegetable accessions (22).

Given these differences, it is the most effective to include seeds from accessions of different uses in different diets. Vegetable accessions differed in the content of a larger number of compounds related to flavan-3-ol, anthocyanidin, lignan, phenolic acid, etc., only carotenoid was encountered in grain varieties. Further research involving a larger number of samples will provide new data on the regularities of the content of substances in accessions from different use groups and apply them in the development of dietary recommendations, as well as in the selection of varieties with improved seed quality.

Author Contributions: Conceptualization, M.O.B. and M.P.R.; methodology, M.O.B., Y.N.Z. and M.P.R.; software, Y.N.Z. and A.M.Z.; validation, M.O.B., E.A.K. and K.S.G.; formal analysis, N.M.I. and O.A.C.; investigation, K.S.G. and M.P.R.; resources, K.S.G. and M.O.B.; data curation, O.A.C.; writing—original draft preparation—M.P.R. and M.O.B.; writing—review and editing M.O.B., Y.N.Z. and E.A.K.; visualization, M.P.R., M.O.B. and E.A.K.; supervision, K.S.G.; project administration, M.O.B. and K.S.G. All authors have read and agreed to the published version of the manuscript.

Funding: The work was supported by the budget project of VIR No. 0481-2022-0002 "Revealing the possibilities of legumes gene pool to optimize their breeding and diversify their use in various sectors of the national economy".

Institutional Review Board Statement: Not applicable.

Informed Consent Statement: Not applicable.

Data Availability Statement: The data presented in the current study are available in the article.

Conflicts of Interest: The authors declare no conflict of interest.

References

1. Burlyaeva, M.O.; Gurkina, M.V.; Chebukin, P.A.; Perchuk, I.N.; Miroshnichenko, E.V. New varieties of vegetable cowpea (*Vigna unguiculata* subsp. *sesquipedalis* (L.) Verdc.) and prospects of their cultivation in southern Russia. *Veg. Crops Russ.* **2019**, *5*, 33–37. (In Russian) [CrossRef]
2. Burlyaeva, M.O.; Gurkina, M.V.; Miroshnichenko, E.V. Application of multivariate analysis to identify relationships among useful agronomic characters of cowpea and differentiation of cultivars for vegetable and grain uses. *J. Proc. Appl. Bot.* **2021**, *182*, 36–47. [CrossRef]
3. Bouchenak, M.; Lamri-Senhadji, M.Y. Nutritional quality of legumes, and their role in cardiometabolic risk prevention: A review. *J. Med. Food* **2013**, *16*, 185–198. [CrossRef] [PubMed]
4. Gonçalves, A.; Goufo, P.; Barros, A.; Domínguez-Perles, R.; Trindade, H.; Rosa, E.A.S.; Ferreira, L.; Rodrigues, M. Cowpea (*Vigna unguiculata* L. Walp), a renewed multipurpose crop for a more sustainable agri-food system: Nutritional advantages and constraints. *J. Sci. Food Agric.* **2016**, *96*, 2941–2951. [CrossRef] [PubMed]
5. Singh, B.; Singh, J.P.; Shevkani, K.; Singh, N.; Kaur, A. Bioactive constituents in pulses and their health benefits. *J. Food Sci. Technol.* **2017**, *54*, 858–870. [CrossRef]
6. Lazaridi, E.; Ntatsi, G.; Fernandez, J.A.; Karapanos, I.; Carnide, V.; Savvas, D.; Bebeli, P.J. Phenotypic diversity and evaluation of fresh pods of cowpea landraces from Southern Europe. *J. Sci. Food Agric.* **2017**, *97*, 4326–4333. [CrossRef]
7. Collado, E.; Klug, T.V.; Artés-Hernández, F.; Aguayo, E.; Artés, F.; Fernández, J.A.; Gómez, P.A. Quality changes in nutritional traits of fresh-cut and then microwaved cowpea seeds and pods. *Food Bioprocess Technol.* **2019**, *12*, 338–346. [CrossRef]
8. Awika, J.M.; Duodu, K.G. Bioactive polyphenols and peptides in cowpea (*Vigna unguiculata*) and their health promoting properties: A review. *J. Func. Foods* **2017**, *38*, 686–697. [CrossRef]

9. Avanza, M.V.; Álvarez-Rivera, G.; Cifuentes, A.; Mendiola, J.A.; Ibáñez, E. Phytochemical and Functional Characterization of Phenolic Compounds from Cowpea (*Vigna unguiculata* (L.) Walp.) Obtained by Green Extraction Technologies. *Agronomy* **2021**, *11*, 162. [CrossRef]
10. Fasuan, T.O.; Chukwu, C.T.; Uchegbu, N.N.; Olagunju, T.M.; Asadu, K.S.; Nwachukwu, M.C. Effects of pre-harvest synthetic chemicals on post-harvest bioactive profile and phytoconstituents of white cultivar of *Vigna unguiculata* grains. *J. Food Process. Preserv.* **2022**, *46*, e16187. [CrossRef]
11. Grierson, C.S.; Barnes, S.R.; Chase, M.W.; Clarke, M.; Grierson, D.; Edwards, K.J.; Jellis, G.J.; Jones, J.D.; Knapp, S.; Oldroyd, G.; et al. One hundred important questions facing plant science research. *New Phytol.* **2011**, *192*, 6–12. [CrossRef] [PubMed]
12. Frank, T.; Engel, K. Metabolomic Analysis of Plants and Crops. In *Metabolomics in Food and Nutrition*; Elsevier: Amsterdam, The Netherlands, 2013; pp. 148–191.
13. Puzanskiy, R.K.; Yemelyanov, V.V.; Gavrilenko, T.A.; Shishova, M.F. The perspectives of metabolomics studies of potato plants. *Vavilov J. Genet. Breed.* **2017**, *21*, 112–123. [CrossRef]
14. Solovyeva, A.E.; Shelenga, T.V.; Shavarda, A.L.; Burlyaeva, M.O. Comparative analysis of wild and cultivated *Lathyrus*, L. spp. according to their primary and secondary metabolite contents. *Vavilov J. Genet. Breed.* **2019**, *23*, 667–674. [CrossRef]
15. Solovyeva, A.E.; Shelenga, T.V.; Shavarda, A.L.; Burlyaeva, M.O. Comparative analysis of wild and cultivated Lathyrus, L. species to assess their content of sugars, polyols, free fatty acids, and phytosterols. *Vavilov J. Genet. Breed.* **2020**, *24*, 730–737. [CrossRef]
16. Perchuk, I.; Shelenga, T.; Gurkina, M.; Miroshnichenko, E.; Burlyaeva, M. Composition of Primary and Secondary Metabolite Compounds in Seeds and Pods of Asparagus Bean (*Vigna unguiculata* (L.) Walp.) from China. *Molecules* **2020**, *25*, 3778. [CrossRef]
17. Loskutov, I.G.; Shelenga, T.V.; Konarev, A.V.; Khoreva, V.I.; Kerv, Y.A.; Blinova, E.V.; Gnutikov, A.A.; Rodionov, A.V.; Malyshev, L.L. Assessment of oat varieties with different levels of breeding refinement from the Vavilov Institute's collection applying the method of metabolomic profiling. *Proc. Appl. Bot. Genet. Breed.* **2022**, *183*, 104–117. [CrossRef]
18. Porokhovinova, E.A.; Shelenga, T.V.; Kerv, Y.A.; Khoreva, V.I.; Konarev, A.V.; Yakusheva, T.V.; Pavlov, A.V.; Slobodkina, A.A.; Brutch, N.B. Features of Profiles of Biologically Active Compounds of Primary and Secondary Metabolism of Lines from VIR Flax Genetic Collection, Contrasting in Size and Color of Seeds. *Plants* **2022**, *11*, 750. [CrossRef]
19. Meijering, E. Cell segmentation: 50 years down the road [life sciences]. *IEEE Signal Process. Mag.* **2012**, *29*, 140–145. [CrossRef]
20. Hutzler, P.; Fischbach, R.; Heller, W.; Jungblut, T.P.; Reuber, S.; Schmitz, R.; Veit, M.; Weissenböck, G.; Schnitzler, J.-P. Tissue localization of phenolic compounds in plants by confocal laser scanning microscopy. *J. Exp. Bot.* **1998**, *49*, 953–965. [CrossRef]
21. Saboori-Robat, E.; Joshi, J.; Pajak, A.; Solouki, M.; Mohsenpour, M.; Renaud, J.; Marsolais, F. Common Bean (*Phaseolus vulgaris* L.) Accumulates Most S-Methylcysteine as Its γ-Glutamyl Dipeptide. *Plants* **2019**, *8*, 126. [CrossRef]
22. Razgonova, M.; Zinchenko, Y.; Pikula, K.; Tekutyeva, L.; Son, O.; Zakharenko, A.; Kalenik, T.; Golokhvast, K. Spatial Distribution of Polyphenolic Compounds in Corn Grains (*Zea mays* L. var. Pioneer) Studied by Laser Confocal Microscopy and High-Resolution Mass Spectrometry. *Plants* **2022**, *11*, 630. [CrossRef] [PubMed]
23. Oertel, A.; Matros, A.; Hartmann, A.; Arapitsas, P.; Dehmer, K.J.; Martens, S.; Mock, H.P. Metabolite profiling of red and blue potatoes revealed cultivar and tissue specific patterns for anthocyanins and other polyphenols. *Planta* **2017**, *246*, 281–297. [CrossRef]
24. Hamed, A.R.; El-Hawary, S.S.; Ibrahim, R.M.; Abdelmohsen, U.R.; El-Halawany, A.M. Identification of Chemopreventive Components from *Halophytes* Belonging to Aizoaceae and Cactaceae Through LC/MS–Bioassay Guided Approach. *J. Chromatogr. Sci.* **2021**, *59*, 618–626. [CrossRef] [PubMed]
25. Qin, D.; Wang, Q.; Li, H.; Jiang, X.; Fang, K.; Wang, Q.; Li, B.; Pan, C.; Wu, H. Identification of key metabolites based on non-targeted metabolomics and chemometrics analyses provides insights into bitterness in Kucha [*Camellia kucha* (Chang et Wang) Chang]. *Food Res. Int.* **2020**, *138*, 109789. [CrossRef] [PubMed]
26. Seukep, A.J.; Zhang, Y.-L.; Xu, Y.-B.; Guo, M.-Q. In Vitro Antibacterial and Antiproliferative Potential of *Echinops lanceolatus* Mattf. (*Asteraceae*) and Identification of Potential Bioactive Compounds. *Pharmaceuticals* **2020**, *13*, 59.
27. Rodriguez-Perez, C.; Gomez-Caravaca, A.M.; Guerra-Hernandez, E.; Cerretani, L.; Garcia-Villanova, B.; Verardo, V. Comprehensive metabolite profiling of *Solanum tuberosum* L. (potato) leaves T by HPLC-ESI-QTOF-MS. *Molecules* **2018**, *112*, 390–399. [CrossRef]
28. Chang, Q.; Wong, Y.-S. Identification of Flavonoids in Hakmeitau Beans (*Vigna sinensis*) by High-Performance Liquid Chromatography−Electrospray Mass Spectrometry (LC-ESI/MS). *J. Agric. Food Chem.* **2004**, *52*, 6694–6699. [CrossRef]
29. Rafsanjany, N.; Senker, J.; Brandt, S.; Dobrindt, U.; Hensel, A. In Vivo Consumption of Cranberry Exerts ex Vivo Antiadhesive Activity against FimH-Dominated Uropathogenic Escherichia coli: A Combined in Vivo, ex Vivo, and in Vitro Study of an Extract from *Vaccinium macrocarpon*. *J. Agric. Food Chem.* **2015**, *63*, 8804–8818. [CrossRef]
30. Belmehdi, O.; Bouyahya, A.; József, J.E.K.Ő.; Cziáky, Z.; Zengin, G.; Sotkó, G.; Elbaaboua, A.; Senhaji, N.S.; Abrini, J. Synergistic interaction between propolis extract, essential oils, and antibiotics against Staphylococcus epidermidis and methicillin resistant Staphylococcus aureus. *Int. J. Second. Metab.* **2021**, *8*, 195–213. [CrossRef]
31. Okhlopkova, Z.M.; Razgonova, M.P.; Pikula, K.S.; Zakharenko, A.M.; Piekoszewski, W.; Manakov, Y.A.; Ercisli, S.; Golokhvast, K.S. *Dracocephalum palmatum* S. and *Dracocephalum ruyschiana* L. originating from Yakutia: A High-Resolution Mass Spectrometric Approach for the Comprehensive Characterization of Phenolic Compounds. *Appl. Sci.* **2022**, *12*, 1766. [CrossRef]

32. Razgonova, M.; Zakharenko, A.; Pikula, K.; Manakov, Y.; Ercisli, S.; Derbush, I.; Kislin, E.; Seryodkin, I.; Sabitov, A.; Kalenik, T.; et al. LC-MS/MS Screening of Phenolic Compounds in Wild and Cultivated Grapes Vitis amurensis Rupr. *Molecules* **2021**, *26*, 3650. [CrossRef] [PubMed]
33. Zakharenko, A.M.; Razgonova, M.P.; Pikula, K.S.; Golokhvast, K.S. Simultaneous determination of 78 compounds of *Rhodiola rosea* extract using supercritical CO_2-extraction and HPLC-ESI-MS/MS spectrometry. HINDAWY. *Biochem. Res. Int.* **2021**, *2021*, 9957490. [CrossRef] [PubMed]
34. Chandrasekara, A.; Shahidi, F. Determination of antioxidant activity in free and hydrolyzed fractions of millet grains and characterization of their phenolic profiles by HPLC-DAD-ESI-MSn. *J. Funct. Foods* **2011**, *3*, 144–158. [CrossRef]
35. Kim, S.; Oh, S.; Noh, H.B.; Ji, S.; Lee, S.H.; Koo, J.M.; Choi, C.W.; Jhun, H.P. In Vitro Antioxidant and Anti-Propionibacterium acnes Activities of Cold Water, Hot Water, and Methanol Extracts, and Their Respective Ethyl Acetate Fractions, from *Sanguisorba officinalis* L. Roots. *Molecules* **2018**, *23*, 3001. [CrossRef] [PubMed]
36. Aita, S.E.; Capriotti, A.L.; Cavaliere, C.; Cerrato, A.; Giannelli Moneta, B.; Montone, C.M.; Piovesana, S.; Lagana, A. Andean Blueberry of the Genus Disterigma: A High-Resolution Mass Spectrometric Approach for the Comprehensive Characterization of Phenolic Compounds. *Separations* **2021**, *8*, 58. [CrossRef]
37. Jaiswal, R.; Muller, H.; Muller, A.; Karar, M.G.E.; Kuhnert, N. Identification and characterization of chlorogenic acids, chlorogenic acid glycosides and flavonoids from *Lonicera henryi* L. (Caprifoliaceae) leaves by LC–MSn. *Phytochemistry* **2014**, *108*, 252–263. [CrossRef]
38. Xu, L.L.; Xu, J.J.; Zhong, K.R.; Shang, Z.P.; Wang, F.; Wang, R.F.; Liu, B. Analysis of non-volatile chemical constituents of *Menthae Haplocalycis* herba by ultra-high performance liquid chromatography—High resolution mass spectrometry. *Molecules* **2017**, *22*, 1756. [CrossRef]
39. Olennikov, D.N.; Chirikova, N.K.; Okhlopkova, Z.M.; Zulfugarov, I.S. Chemical Composition and Antioxidant Activity of Tánara Ótó (*Dracocephalum palmatum* Stephan), a Medicinal Plant Used by the North-Yakutian Nomads. *Molecules* **2013**, *18*, 14105–14121. [CrossRef]
40. Teles, Y.C.F.; Souza, M.S.R.; de Fatima Vanderlei de Souza, M. New Sulphated Flavonoids: Biosynthesis, Structures, and Biological Activities. *Molecules* **2018**, *22*, 480. [CrossRef]
41. Marcia Fuentes, J.A.; Lopez-Salas, L.; Borras-Linares, I.; Navarro-Alarcon, M.; Segura-Carretero, A.; Lozano-Sanchez, J. Development of an Innovative Pressurized Liquid Extraction Procedure by Response Surface Methodology to Recover Bioactive Compounds from Carao Tree Seeds. *Foods* **2021**, *10*, 398. [CrossRef]
42. Thomford, N.E.; Dzobo, K.; Chopera, D.; Wonkam, A.; Maroyi, A.; Blackhurst, D.; Dandara, C. In vitro reversible and time-dependent CYP450 inhibition profiles of medicinal herbal plant extracts *Newbouldia laevis* and *Cassia abbreviata*: Implications for herb-drug interactions. *Molecules* **2016**, *21*, 891. [CrossRef] [PubMed]
43. Sobeh, M.; Mahmoud, M.F.; Abdelfattah, M.A.O.; Cheng, H.; El-Shazly, A.M.; Wink, M. A proanthocyanidin-rich extract from *Cassia abbreviata* exhibits antioxidant and hepatoprotective activities in vivo. *J. Ethnopharmacol.* **2018**, *213*, 38–47. [CrossRef] [PubMed]
44. Santos, S.A.O.; Vilela, C.; Freire, C.S.R.; Neto, C.P.; Silvestre, A.J.D. Ultra-high performance liquid chromatography coupled to mass spectrometry applied to the identification of valuable phenolic compounds from Eucalyptus wood. *J. Chromatogr. B* **2013**, *938*, 65–74. [CrossRef] [PubMed]
45. Abeywickrama, G.; Debnath, S.C.; Ambigaipalan, P.; Shahidi, F. Phenolics of Selected Cranberry Genotypes (*Vaccinium macrocarpon* Ait.) and Their Antioxidant Efficacy. *J. Agric. Food Chem.* **2016**, *64*, 9342–9351. [CrossRef]
46. Ojwang, L.O.; Yang, L.; Dykes, L.; Awika, J. Proanthocyanidin profile of cowpea (*Vigna unguiculata*) reveals catechin-O-glucoside as the dominant compound. *Food Chem.* **2013**, *130*, 35–43. [CrossRef]
47. Sharma, M.; Sandhir, R.; Singh, A.; Kumar, P.; Mishra, A.; Jachak, S.; Singh, S.P.; Singh, J.; Roy, J. Comparison analysis of phenolic compound characterization and their biosynthesis genes between two diverse bread wheat (*Triticum aestivum*) varieties differing for chapatti (unleavened flat bread) quality. *Front. Plant Sci.* **2016**, *7*, 1870. [CrossRef]
48. Yin, N.-W.; Wang, S.-X.; Jia, L.-D.; Zhu, M.-C.; Yang, J.; Zhou, B.-J.; Yin, J.-M.; Lu, K.; Wang, R.; Li, J.-N.; et al. Identification and Characterization of Major Constituents in Different-Colored Rapeseed Petals by UPLC−HESI-MS/MS. *J. Agric. Food Chem.* **2019**, *67*, 11053–11065. [CrossRef]
49. Ruiz, A.; Hermosín-Gutiérrez, I.; Vergara, C.; von Baer, D.; Zapata, M.; Hitschfeld, A.; Obando, L.; Mardones, C. Anthocyanin profiles in south Patagonian wild berries by HPLC-DAD-ESI-MS/MS. *Food Res. Int.* **2013**, *51*, 706–713. [CrossRef]
50. Ha, T.J.; Lee, M.H.; Park, C.H.; Pae, S.B.; Shim, K.B.; Ko, J.M.; Shin, S.O.; Baek, I.Y.; Park, K.Y. Identification and Characterization of Anthocyanins in Yard-Long Beans (*Vigna unguiculata* ssp. *sesquipedalis* L.) by High-Performance Liquid Chromatography with Diode Array Detection and Electrospray Ionization/Mass Spectrometry (HPLC-DAD-ESI/MS) Analysis. *J. Agric. Food Chem.* **2010**, *58*, 2571–2576.
51. Lago-Vanzela, E.S.; Da-Silva, R.; Gomes, E.; Garcia-Romero, E.; Hermosin-Gutierres, E. Phenolic Composition of the Edible Parts (Flesh and Skin) of Bordô Grape (*Vitis labrusca*) Using HPLC−DAD−ESI-MS/MS. *J. Agric. Food Chem.* **2011**, *59*, 13136–13146. [CrossRef]
52. Chhon, S.; Jeon, J.; Kim, J.; Park, S.U. Phenolic Accumulation of Anthocyanins through Overexpression of AtPAP1 in *Solanum nigrum* Lin. (Black Nightshade). *Biomolecules* **2020**, *10*, 277. [CrossRef] [PubMed]

53. Anari, Z.; Mai, C.; Sengupta, A.; Howard, L.; Brownmiller, C.; Wickramasinghe, S.R. Combined Osmotic and Membrane Distillation for Concentration of Anthocyanin from Muscadine Pomace Biomolecules. *J. Food Sci.* **2019**, *84*, 2199–2208. [CrossRef] [PubMed]
54. Eklund, P.C.; Backman, M.J.; Kronberg, L.A.; Smeds, A.I.; Sjoholm, R.E. Identification of lignans by liquid chromatography-electrospray ionization ion-trap mass spectrometry. *J. Mass Spectrom.* **2008**, *43*, 97–107. [CrossRef] [PubMed]
55. Bonzanini, F.; Bruni, R.; Palla, G.; Serlataite, N.; Caligiani, A. Identification and distribution of lignans in *Punica granatum* L. fruit endocarp, pulp, seeds, wood knots and commercial juices by GC–MS. *Food Chem.* **2009**, *117*, 745–749. [CrossRef]
56. Llorent-Martinez, E.J.; Spinola, V.; Gouveia, S.; Castilho, P.C. HPLC-ESI-MSn characterization of phenolic compounds, terpenoid saponins, and other minor compounds in *Bituminaria bituminosa*. *Ind. Crops Prod.* **2015**, *69*, 80–90. [CrossRef]
57. Monthong, W.; Pitchuanchom, S.; Nuntasaen, N.; Pompimon, W. (+)-Syringaresinol Lignan from New Species *Magnolia Thailandica*. *Am. J. Appl. Sci.* **2011**, *8*, 1268–1271. [CrossRef]
58. Pan, M.; Lei, Q.; Zang, N.; Zhang, H. A Strategy Based on GC-MS/MS, UPLC-MS/MS and Virtual Molecular Docking for Analysis and Prediction of Bioactive Compounds in *Eucalyptus Globulus* Leaves. *Int. J. Mol. Sci.* **2019**, *20*, 3875. [CrossRef] [PubMed]
59. Cai, Z.; Wang, C.; Zou, L.; Liu, X.; Chen, J.; Tan, M.; Mei, Y.; Wei, L. Comparison of Multiple Bioactive Constituents in the Flower and the Caulis of *Lonicera japonica* Based on UFLC-QTRAP-MS/MS Combined with Multivariate Statistical Analysis. *Molecules* **2019**, *24*, 1936. [CrossRef]
60. Quifer-Rada, P.; Vallverdu-Queralt, A.; Martinez-Huelamo, M.; Chiva-Blanch, G.; Jauregui, O.; Estruch, R.; Lamuela-Raventos, R. A comprehensive characterization of beer polyphenols by high resolution mass spectrometry (LC-ESI-LTQ-Orbitrap-MS). *Food Chem.* **2015**, *169*, 336–343. [CrossRef]
61. Alvarez-Fernandez, M.; Cerezo, A.B.; Canete-Rodriguez, A.M.; Troncoso, A.M.; Garcia-Parrilla, M.C. Composition of nonanthocyanin polyphenols in alcoholic-fermented strawberry products using LC-MS (QTRAP), high-resolution MS (UHPLC-Orbitrap-MS), LC-DAD, and antioxidant activity. *J. Agric. Food Chem.* **2015**, *63*, 2041–2051. [CrossRef]
62. Jiang, R.-W.; Lau, K.-M.; Hon, P.-M.; Mak, T.C.W.; Woo, K.-S.; Fung, K.-P. Chemistry and Biological Activities of Caffeic Acid Derivatives from *Salvia miltiorrhiza*. *Curr. Med. Chem.* **2005**, *12*, 237–246. [CrossRef]
63. Razgonova, M.P.; Tikhonova, N.G.; Sabitov, A.S.; Mikhailova, N.M.; Luchko, S.R.; Zakharenko, A.M.; Pikula, K.S.; Golokhvast, K.S. Identification of phenolic constituents in *Lonicera caerulea* L. by HPLC with diode array detection electrospray ionisation tandem mass spectrometry. In *BIO Web of Conferences*; EDP Sciences: Les Ulis, France, 2021; Volume 32, p. 02010.
64. Lang, R.; Dieminger, N.; Beusch, A.; Lee, Y.M.; Dunkel, A.; Suess, B.; Skurk, T.; Wahl, A.; Hauner, H.; Hofmann, T. Bioappearance and pharmacokinetics of bioactives upon coffee consumption). *Anal. Bioanal. Chem.* **2013**, *405*, 8487–8503. [CrossRef] [PubMed]
65. Zhou, X.-J.; Yan, L.-L.; Yin, P.-P.; Shi, L.-L.; Zhang, J.-H.; Liu, J.-H.; Ma, C. Structural characterisation and antioxidant activity evaluation of phenolic compounds from cold-pressed *Perilla frutescens* var. arguta seed flour. *Food Chem.* **2014**, *164*, 150–157. [CrossRef] [PubMed]
66. Yang, S.T.; Wu, X.; Rui, W.; Guo, J.; Feng, Y.F. UPLC/Q-TOF-MS analysis for identification of hydrophilic phenolics and lipophilic diterpenoids from *Radix Salviae Miltiorrhizae*. *Acta Chromatogr.* **2015**, *27*, 711–728. [CrossRef]
67. Huang, Y.; Yao, P.; Leung, K.W.; Wang, H.; Kong, X.P.; Wang, L.; Dong, T.T.X.; Chen, Y.; Tsim, K.W.K. The Yin-Yang Property of Chinese Medicinal Herbs Relates to Chemical Composition but Not Anti-Oxidative Activity: An Illustration Using Spleen-Meridian Herbs. *Front. Pharmacol.* **2018**, *9*, 1304. [CrossRef]
68. Ekeberg, D.; Flate, P.-O.; Eikenes, M.; Fongen, M.; Naess-Andresen, C.F. Qualitative and quantitative determination of extractives in heartwood of Scots pine (*Pinus sylvestris* L.) by gas chromatography. *J. Chromatogr. A* **2006**, *1109*, 267–272. [CrossRef] [PubMed]
69. Park, S.K.; Ha, J.S.; Kim, J.M.; Kang, J.Y.; Lee, D.S.; Guo, T.J.; Lee, U.; Kim, D.-O.; Heo, H.J. Antiamnesic Effect of Broccoli (*Brassica oleracea* var. italica) Leaves on Amyloid Beta (Aβ)1-42-Induced Learning and Memory Impairment. *J. Agric. Food Chem.* **2016**, *64*, 3353–3361. [CrossRef]
70. Van Hoyweghen, L.; De Bosscher, K.; Haegeman, G.; Deforce, D.; Heyerick, A. *In vitro* inhibition of the transcription factor NF-κB and cyclooxygenase by Bamboo extracts. *Phytother. Res.* **2014**, *28*, 224–230. [CrossRef]
71. Bianco, G.; Schmitt-Kopplin, P.; De Benedetto, G.; Kettrup, A.; Cataldi, T.R.I. Determination of glycoalkaloids and relative aglycones by nonaqueous capillary electrophoresis coupled with electrospray ionization-ion trap mass spectrometry. *Electrophoresis* **2002**, *23*, 2904–2912. [CrossRef]
72. Hossain, M.B.; Brunton, N.P.; Rai, D.K. Effect of Drying Methods on the Steroidal Alkaloid Content of Potato Peels, Shoots and Berries. *Molecules* **2016**, *21*, 403. [CrossRef]
73. Wu, Y.; Xu, J.; He, Y.; Shi, M.; Han, X.; Li, W.; Zhang, X.; Wen, X. Metabolic Profiling of Pitaya (*Hylocereus polyrhizus*) during Fruit Development and Maturation. *Molecules* **2019**, *24*, 1114. [CrossRef]
74. Razgonova, M.P.; Tekutyeva, L.A.; Podvolotskaya, A.B.; Stepochkina, V.D.; Zakharenko, A.M.; Golokhvast, K. *Zostera marina* L.: Supercritical CO2-extraction and Mass Spectrometric Characterization of Chemical Constituents Recovered from Eelgrass. *Separations* **2022**, *24*, 1114. [CrossRef]
75. Delgado-Pelayo, R.; Homero-Mendez, D. Identification and Quantitative Analysis of Carotenoids and Their Esters from Sarsaparilla (*Smilax aspera* L.) Berries. *J. Agric. Food Chem.* **2012**, *60*, 8225–8232. [CrossRef]
76. Mercadante, A.Z.; Rodrigues, D.B.; Petry, F.C.; Barros Mariutti, L.R. Carotenoid esters in foods—A review and practical directions on analysis and occurrence. *Food Res. Int.* **2017**, *99*, 830–850. [CrossRef] [PubMed]

77. Shakya, R.; Navarre, D.A. LC-MS Analysis of Solanidane Glycoalkaloid Diversity among Tubers of Four Wild Potato Species and Three Cultivars (*Solanum tuberosum*). *J. Agric. Food Chem.* **2008**, *56*, 6949–6958. [CrossRef]
78. Steinert, K.; Hovelmann, Y.; Huber, F.; Humpf, H.-U. LC-MS Identification of Novel Iso-Esculeoside B from Tomato Fruits and LC−MS/MS-Based Food Screening for Major Dietary Steroidal Alkaloids Focused on Esculeosides. *J. Agric. Food Chem.* **2020**, *68*, 14492–14501. [CrossRef] [PubMed]
79. Pollier, J.; Morreel, K.; Geelen, D.; Goossens, A. Metabolite Profiling of Triterpene Saponins in *Medicago truncatula* Hairy Roots by Liquid Chromatography Fourier Transform Ion Cyclotron Resonance Mass Spectrometry. *J. Nat. Prod.* **2011**, *74*, 1462–1476. [CrossRef]
80. Deuber, H.; Guignard, C.; Hoffmann, L.; Evers, D. Polyphenol and glycoalkaloid contents in potato cultivars grown in Luxembourg. *Food Chem.* **2012**, *135*, 2814–2824.
81. Huang, W.; Serra, O.; Dastmalchi, K.; Jin, L.; Yang, L.; Stark, R.E. Comprehensive MS and Solid-state NMR Metabolomic Profiling Reveals Molecular Variations in Native Periderms from Four *Solanum tuberosum* Potato Cultivars. *J. Agric. Food Chem.* **2017**, *65*, 2258–2274. [CrossRef] [PubMed]
82. Corcel, M.; Devaux, M.-F.; Guillon, F.; Barron, C. Identification of tissular origin of particles based on autofluorescence multispectral image analysis at the macroscopic scale. In Proceedings of the EPJ Web of Conferences, Crete, Greece, 17–29 August 2017; p. 05012.
83. Lichtenthaler, H.K.; Schweiger, J. Cell wall bound ferulic acid, the major substance of the blue-green fluorescence emission of plants. *J. Plant Physiol.* **1998**, *152*, 272–282. [CrossRef]
84. Donaldson, L. Softwood and hardwood lignin fluorescence spectra of wood cell walls in different mounting media. *IAWA J.* **2013**, *34*, 3–19. [CrossRef]
85. Brillouet, J.M.; Riochet, D. Cell wall polysaccharides and lignin in cotyledons and hulls of seeds from various lupin (*Lupinus,* L.) species. *J. Sci. Food Agric.* **1983**, *34*, 861–868. [CrossRef]
86. Krzyzanowski, F.C.; Franca Neto, J.D.B.; Mandarino, J.M.G.; Kaster, M. Evaluation of lignin content of soybean seed coat stored in a controlled environment. *Rev. Bras. De Sementes* **2008**, *30*, 220–223. [CrossRef]
87. Brillouet, J.M.; Carré, B. Composition of cell walls from cotyledons of *Pisum sativum*, *Vicia faba* and *Glycine max*. *Phytochemistry* **1983**, *22*, 841–847. [CrossRef]
88. Sudo, E.; Teranishi, M.; Hidema, J.; Taniuchi, T. Visualization of flavonol distribution in the abaxial epidermis of onion scales via detection of its autofluorescence in the absence of chemical processes. *Biosci. Biotechnol. Biochem.* **2009**, *73*, 2107–2109. [CrossRef]
89. Monago-Maraña, O.; Durán-Merás, I.; Galeano-Díaz, T.; de la Peña, A.M. Fluorescence properties of flavonoid compounds. Quantification in paprika samples using spectrofluorimetry coupled to second order chemometric tools. *Food Chem.* **2016**, *196*, 1058–1065. [CrossRef]
90. Roshchina, V.V.; Kuchin, A.V.; Yashin, V.A. Application of Autofluorescence for Analysis of Medicinal Plants. *Spectrosc. Int. J.* **2017**, *2017*, 7159609. [CrossRef]
91. Talamond, P.; Verdeil, J.-L.; Conéjéro, G. Secondary metabolite localization by autofluorescence in living plant cells. *Molecules* **2015**, *20*, 5024–5037. [CrossRef]
92. Collings, D.A. Anthocyanin in the vacuole of red onion epidermal cells quenches other fluorescent molecules. *Plants* **2019**, *8*, 596. [CrossRef]
93. Mackon, E.; Ma, Y.; Jeazet Dongho Epse Mackon, G.C.; Li, Q.; Zhou, Q.; Liu, P. Subcellular Localization and Vesicular Structures of Anthocyanin Pigmentation by Fluorescence Imaging of Black Rice (*Oryza sativa* L.) Stigma Protoplast. *Plants* **2021**, *10*, 685. [CrossRef]
94. Weber, H.; Borisjuk, L.; Wobus, U. Molecular physiology of legume seed development. *Annu. Rev. Plant Biol.* **2005**, *56*, 253–279. [CrossRef] [PubMed]
95. Moïse, J.A.; Han, S.; Gudynaitę-Savitch, L.; Johnson, D.A.; Miki, B.L. Seed coats: Structure, development, composition, and biotechnology. *In Vitro Cell. Dev. Biol.-Plant* **2005**, *41*, 620–644. [CrossRef]
96. Ndakidemi, P.A.; Dakora, F.D. Legume seed flavonoids and nitrogenous metabolites as signals and protectants in early seedling development. *Funct. Plant Biol.* **2003**, *30*, 729–745. [CrossRef]
97. Salunkhe, D.K.; Jadhav, S.J.; Kadam, S.S.; Chavan, J.K. Chemical, biochemical, and biological significance of polyphenols in cereals and legumes. *Crit. Rev. Food Sci. Nutr.* **1982**, *17*, 277–305. [CrossRef] [PubMed]
98. Dixon, R.A.; Paiva, N.L. Stress-induced phenylpropanoid metabolism. *Plant Cell* **1995**, *7*, 1085–1097. [CrossRef] [PubMed]
99. Benitez, E.R.; Funatsuki, H.; Kaneko, Y.; Matsuzawa, Y.; Bang, S.W.; Takahashi, R. Soybean maturity gene effects on seed coat pigmentation and cracking in response to low temperatures. *Crop Sci.* **2004**, *44*, 2038–2042. [CrossRef]
100. Nakamura, T.; Yang, D.; Kalaiselvi, S.; Uematsu, Y.; Takahashi, R. Genetic analysis of net-like cracking in soybean coats. *Euphytica.* **2003**, *133*, 179–184. [CrossRef]

plants

Article

Combination of Genomics, Transcriptomics Identifies Candidate Loci Related to Cold Tolerance in Dongxiang Wild Rice

Dianwen Wang †, Yulong Xiao †, Hongping Chen, Cheng Huang, Ping Chen, Dazhou Chen, Wei Deng and Jilin Wang *

Rice National Engineering Research Center (Nanchang), Rice Research Institute, Jiangxi Academy of Agricultural Sciences, Nanchang 330200, China

* Correspondence: wangjilin1982@163.com; Tel.: +86-133-8753-2293

† These authors contributed equally to this study.

Abstract: Rice, a cold-sensitive crop, is a staple food for more than 50% of the world's population. Low temperature severely compromises the growth of rice and challenges China's food safety. Dongxiang wild rice (DXWR) is the most northerly common wild rice in China and has strong cold tolerance, but the genetic basis of its cold tolerance is still unclear. Here, we report quantitative trait loci (QTLs) analysis for seedling cold tolerance (SCT) using a high-density single nucleotide polymorphism linkage map in the backcross recombinant inbred lines that were derived from a cross of DXWR, and an indica cultivar, GZX49. A total of 10 putative QTLs were identified for SCT under 4 °C cold treatment, each explaining 2.0–6.8% of the phenotypic variation in this population. Furthermore, transcriptome sequencing of DXWR seedlings before and after cold treatment was performed, and 898 and 3413 differentially expressed genes (DEGs) relative to 0 h in cold-tolerant for 4 h and 12 h were identified, respectively. Gene ontology and Kyoto encyclopedia of genes and genomes (KEGG) analysis were performed on these DEGs. Using transcriptome data and genetic linkage analysis, combined with qRT-PCR, sequence comparison, and bioinformatics, *LOC_Os08g04840* was putatively identified as a candidate gene for the major effect locus *qSCT8*. These findings provided insights into the genetic basis of SCT for the improvement of cold stress potential in rice breeding programs.

Keywords: Dongxiang wild rice; seedling cold tolerance; quantitative trait locus; transcriptomics; differentially expressed genes

Citation: Wang, D.; Xiao, Y.; Chen, H.; Huang, C.; Chen, P.; Chen, D.; Deng, W.; Wang, J. Combination of Genomics, Transcriptomics Identifies Candidate Loci Related to Cold Tolerance in Dongxiang Wild Rice. *Plants* **2022**, *11*, 2329. https://doi.org/10.3390/plants11182329

Academic Editors: Lixi Jiang, Mingxun Chen and Yuan Guo

Received: 27 July 2022
Accepted: 2 September 2022
Published: 6 September 2022

1. Introduction

Rice (*Oryza sativa* L.) is the staple food for half of the world's population. Increased rice yield has become an important issue for the global economy and food security due to rapid population growth and a reduction in arable land. The growth and development of rice are sensitive to temperature fluctuations, and the optimum temperature for rice cultivation is 25–30 °C [1]. Double-cropping early rice often encounters cold weather in late spring, and suffers from chilling stress during its seedling stage, causing chlorosis, reduction of growth rate and tillering, and low seedling vigor [1–4]. Thus, improving cold tolerance (CT) in cultivars to promote high and stable rice yield has been a major goal of rice breeding [5].

CT in rice is a quantitative trait that is controlled by quantitative trait loci (QTL) and is largely influenced by the environment. Numerous loci were studied for cold tolerance by QTL mapping and association analysis in rice, including more than 80 QTLs related to seedling cold tolerance (SCT) [4,6–16]. Among these loci, only a few QTLs have been thoroughly researched and cloned using map-based cloning strategies, and most of the functional mechanisms remain largely unknown [16]. *COLD1* encodes a Ca^{2+} signaling regulator that interacts with *qCTS9*, encoding a novel Brassinosteroid Insensitive-1 protein, and co-regulates cold tolerance in rice [17,18]. Functional interaction between *bZIP73* and *bZIP71* can significantly enhance cold tolerance in rice seedlings [19]. The substitution of a single nucleotide at position 343 from A to G in *qPSR10* can effectively enhance cold tolerance in the rice seedling stage [20]. *HAN1* encodes a biologically active jasmonyl-L-isoleucine (JA-Ile) to the inactive form 12-hydroxy-JA-Ile (12OH-JA-Ile) oxidase, which is mediated by fine-tuning JA cold tolerance of rice [21]. *OsWRKY115* (*qCT7*) transcription factor positively regulates the cold tolerance of rice seedlings [22]. In this context, a large effort is still required to explore multiple beneficial alleles for the improvement of cold tolerance in rice.

The development of DNA sequencing technology has supported the possibility of identifying functional genes at the genome-wide level [23–25]. RNA-sequencing (RNA-seq), characterizing the genome-wide gene expression profile, is an efficient means to detect stimuli-responsive genes at the genome-wide level [26]. Transcriptome analysis of seedlings before and after cold treatment showed that there are many genes involved in response to cold stress at the seedling stage, such as transport, metabolism, signal transduction, and transcriptional regulation [27–29]. In particular, comparative transcriptome analysis of germplasms with distinct cold tolerance phenotypes provides important clues for revealing key genes and mechanisms controlling cold tolerance in rice [30–32]. Moreover, the strategy combining transcriptome analysis with QTL mapping was explored to accelerate the identification of candidate genes, including *LOC_Os07g22494*, for a CT-related QTL [33–35]. Jiangxi Dongxiang wild rice (DXWR) is the northernmost wild rice in China (28°14′ N), and its seedlings are extremely cold-tolerant [36]. Several QTLs and candidate genes associated with cold tolerance have been identified in DXWR [14,37,38]. However, the genetic basis for the regulation of cold tolerance in DXWR remains unclear.

In this present study, a high-density single nucleotide polymorphism (SNP) linkage map was constructed in the backcross recombinant inbred line (BRIL) population, which was derived from a backcross of DXWR and GZX49 to detect QTLs associated with SCT. At the same time, we cold-treated DXWR seedlings and performed transcriptome sequencing to analyze differentially expressed genes. For a novel major effect, *qSCT8* candidate gene analysis was performed, and *LOC_Os08g04840* encoding an MYB transcription factor was used as the *qSCT8* candidate gene. These findings provide new insights into the genetic basis of SCT in rice.

2. Results

2.1. Development and Construction of High-Density Bin Map of BRILs

To construct the backcross recombinant inbred lines (BRILs), the F_1 plant derived from a cross between Ganzaoxian49 (GZX49) and Dongxiang wild rice (DXWR) was backcrossed as a receptor with GZX49. The BC_1F_1 seeds were harvested and single-seed descent to obtain a BRIL population comprising 132 lines (Figure 1A). The population was genotyped using the genotyping-by-sequencing (GBS) method, and generated a total of 70.76 Gb of raw data, approximately 536.09 Mb each line, with an overall effective mean depth coverage of 1.97-fold. Based on the Rice Genome Annotation Project (version 7), a total of 55,036 SNPs were detected in the population and were evenly distributed throughout the genome (Figure 1B, Table S1). According to the high-density genotype of the BRIL population, the recombination breakpoint of each line is determined. In addition, a bin map was constructed based on the recombination breakpoints of each line. The bin map contains 1658 recombination bins distributed on 12 chromosomes, and the average size of the bin is 220.4 kb, with a size range of 12.8 kb to 4.97 Mb. The genetic linkage map was further constructed using a bin map, the total length of the linkage map was 2341.9 cM, and the average interval between adjacent bins was 1.44 cM (Figure 1C, Table S1).

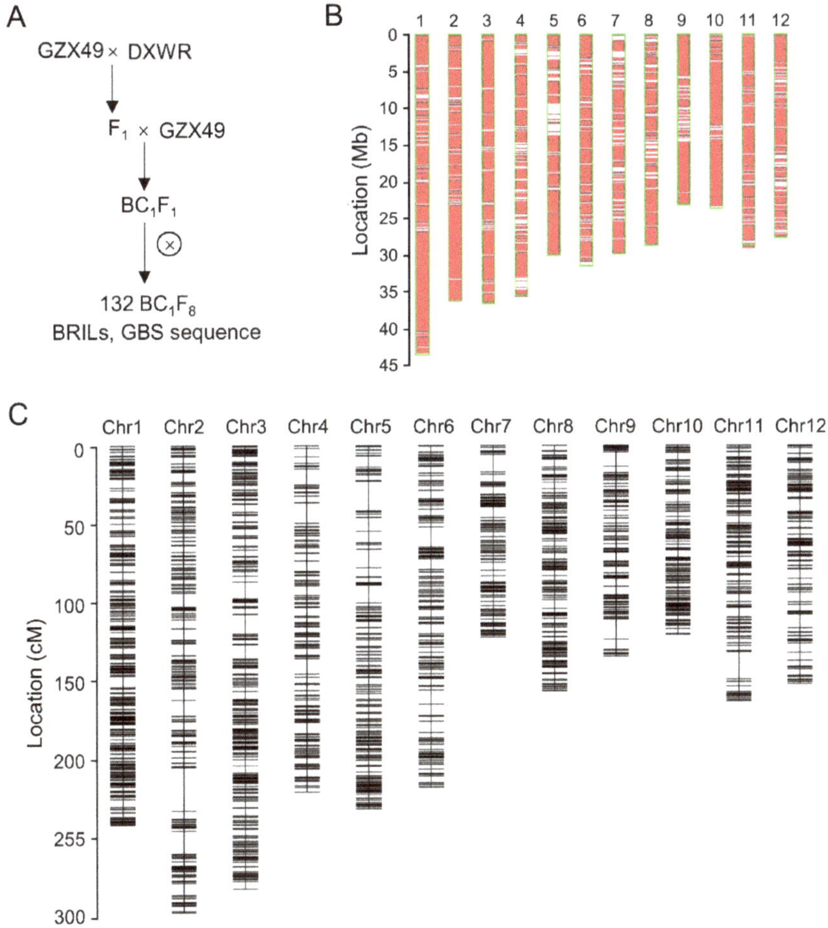

Figure 1. Development and construction of high-density bin map of BRILs. (**A**) Flowchart of developing backcross recombinant inbred line (BRIL) population. ⊗ represents self-cross. (**B**) Polymorphic SNPs between DXWR and GZX49 distributed on chromosomes. (**C**) Genetic linkage map constructed by population BRILs.

2.2. QTL analysis of Seedling Cold Tolerance

Seedling cold tolerance (SCT) was measured for the two parental lines and the BRILs at the seedling stage. The seedlings grew to trifoliate (Figure 2A), were treated at a low temperature of 4 °C for 84 h in the incubator, and then resumed growth at 28 °C for 7 days; the results showed that the survival rate of DXWR was still 100% after cold treatment, while the survival rate of GZX49 was almost zero (Figure 2B). This indicated that DXWR had strong SCT compared with GZX49. The survival rate of seedlings after cold treatment in the BRIL population showed a continuous distribution from 0 to 100% (Figure 2C,D). These results suggested SCT was in quantitative inheritance controlled by polygenes or QTLs, and there was a large genetic variation for SCT among the BRILs.

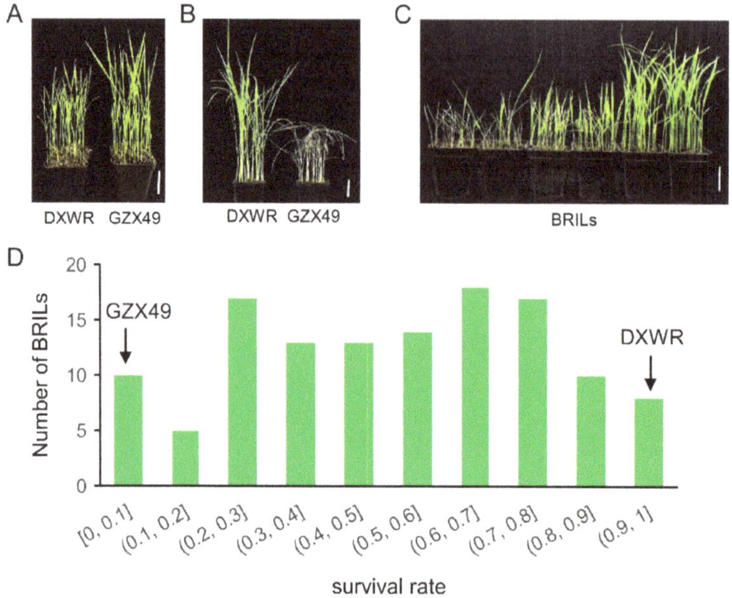

Figure 2. Phenotypic identification of seedling for cold tolerance. (**A**,**B**) represent the phenotypes of GZX49 and DXWR before and after cold treatment, respectively. (**C**) Phenotypes of some lines of the BRIL population after cold treatment. The scale bar of (**A**–**C**) means 5 cm. (**D**) Frequency distribution of seedling survival rate after cold treatment. Arrows indicate the means of parental lines GZX49 and DXWR.

The ridge regression analysis for the QTL detection was performed in the BRIL with the bin genotypes. A total of 10 QTLs for SCT were identified and found to explain 48.9% of the phenotypic variance, distributed on all chromosomes except chromosomes 5, 6, and 10 (Figure 3A, Table S2). The majority (9/10) of the QTLs for SCT had a positive effect, indicating that alleles from DXWR increased SCT (Table S2). The phenotypic variation explained (PVE) by each QTL ranged from 2.0% to 6.8% (Table S2), confirming that the SCT was a complex quantitative trait controlled by multiple genes. Among these QTLs, 8 QTLs correspond to bin intervals less than 500 kb, and 5 QTLs correspond to bin intervals that were smaller (less than 200 kb) (Table S2). The detection of QTLs in relatively small intervals was of great significance for further fine mapping and candidate gene mapping of these QTLs. Based on the mapping interval of two QTLs, *qSCT1* and *qSCT2*, it was found that these two QTLs contained the two reported SCT-related genes *OsMYB3R-2* [39] and *OsTPP1* [40], respectively. Using qRT-PCR technology to detect the expression of *OsMYB3R-2* under 4 °C of treatment, we found that the expression of *OsMYB3R-2* was significantly up-regulated in DXWR but not significantly changed in GZX49 (Figure 3B). Sequence comparison of *OsMYB3R-2* between DXWR and GZX49 revealed in its promoter region (2-kb upstream of the predicted start codon), 17 SNPs and 4 Indels (Figure 3C). Thus, *OsMYB3R-2* was a possible candidate gene for *qSCT1*. Similarly, qRT-PCR technology and sequence comparison were conducted to screen out the degree of response of *OsTPP1* expression to cold stress and the causal polymorphisms of *OsTPP1* (Figure 3D,E). Thus, *OsTPP1* is a possible candidate gene for *qSCT2*. Additionally, these results indicate that the QTLs are entirely effective.

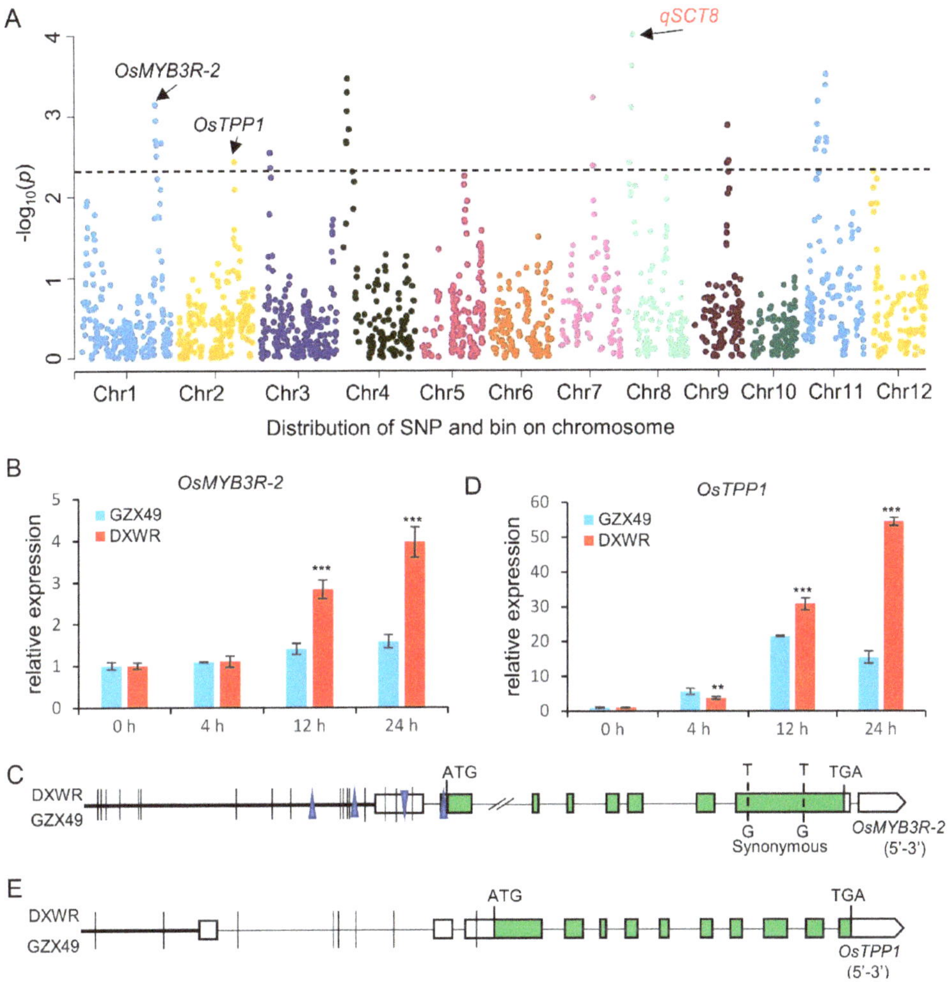

Figure 3. QTL analysis of the BRIL population. (**A**) Manhattan plots of the loci for seedling cold tolerance (SCT). The *x*-axis represents single nucleotide polymorphism (SNP) along each numbered chromosome; the *y*-axis represents the negative logarithm of the *p*-value (-log10 p) for the SNP association. Horizontal dashed lines in the plots indicate the declaration thresholds. (**B**,**D**) Expression levels of the *OsMYB3R-2* and *OsTPP1* in GZX49 and DXWR after cold stress measured by qRT-PCR, respectively. The results were statistically analyzed using Student's *t*-test (** $p < 0.01$, *** $p < 0.005$). Transcription levels relative to 0 h, which was set to 1, are presented as the mean and SE of triplicates. *LOC_Os03g13170* (Ubiquitin) is the control gene. (**C**,**E**) Sequence comparison of *OsMYB3R-2* and *OsTPP1* among GZX49 and DXWR. The vertical bars and triangle represent SNPs and nucleotide deletion, respectively.

2.3. Transcriptome Analysis of Differentially Expressed Genes in DXWR under Cold Stress

To identify the cold stress-responsive genes contained in DXWR, DXWR seedlings were treated at 4 °C, and transcriptome sequencing of aboveground tissues treated for 0 h, 4 h, and 12 h, respectively, was performed. There were 898 and 3413 differentially expressed genes (DEGs) relative to 0 h at 4 °C for 4 h and 12 h, respectively (Figure 4A). Among them, 291 and 607 genes were up- and down-regulated at 4 h, respectively, while

1294 and 2119 genes were up- and down-regulated at 12 h, respectively, by 4 °C treatment (Figure 4A). Additionally, there were 1047 DEGs between 4 h and 12 h of treatment at 4 °C, including 610 up-regulated genes and 437 down-regulated genes (Figure 4A). Among the DEGs at 4 h, 93.4% (273/291) of the up-regulated genes and 86.7% (526/607) of the down-regulated genes were repeatedly detected in the DEGs of 12 h, respectively (Figure 4B,C). In addition, among the differentially expressed genes at 12 h, 79% (1021/1293) of up-regulated genes and 75.2% (1593/2119) of down-regulated genes had no detectable differences at 4 h (Figure 4B,C). This indicated that the response mechanism of DXWR to cold stress was mainly at 12 h of cold treatment.

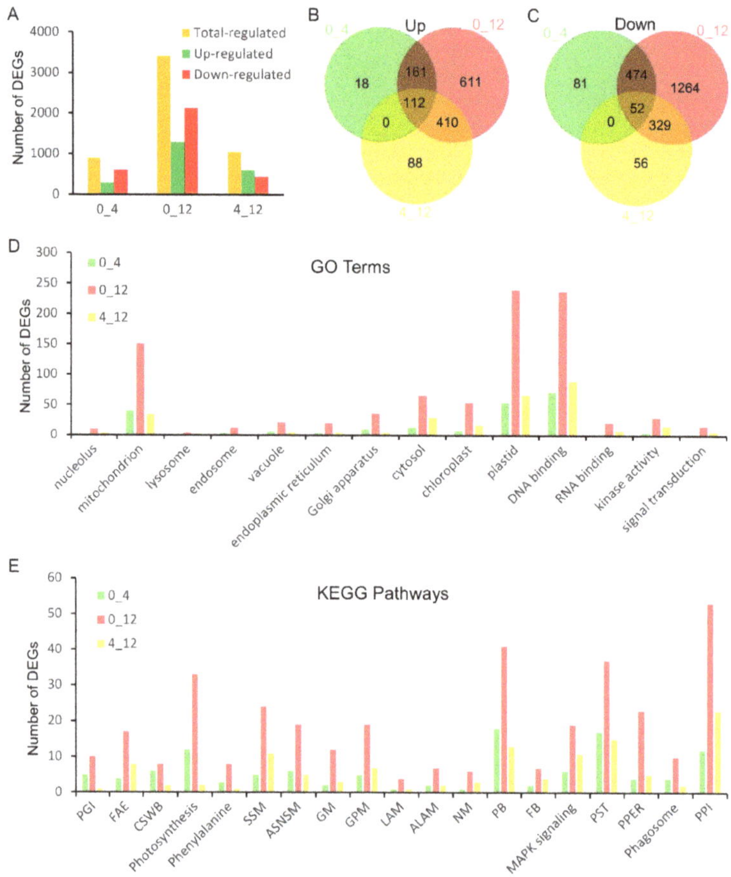

Figure 4. Transcriptome analysis of the genetic mechanism of DXWR in response to cold stress. (**A**) Differentially expressed gene statistics. (**B**,**C**) are Venn diagrams of up-regulated and down-regulated genes, respectively. (**D**,**E**) are gene ontology (GO) functional enrichment histogram and Kyoto encyclopedia of genes and genomes (KEGG) pathway enrichment histogram, respectively. 0_4, 0_12, and 4_12 represent the comparison of DXWR seedling treatment at 4 °C for 0 h and 4 h, 0 h and 12 h, and 4 h and 12 h, respectively. PGI, FAE, CSWB, SSM, ASNSM, GM, GPM, LAM, ALAM, NM, PB, FB, PST, PPER, and PPI in subfigure (**E**) represent pentose and glucuronate interconversions, fatty acid elongation, cutin suberine and wax biosynthesis, starch and sucrose metabolism, amino sugar and nucleotide sugar metabolism, glycerolipid metabolism, gycerophospholipid metabolism, linoleic acid metabolism, alpha-linolenic acid metabolism, nitrogen metabolism, phenylpropanoid biosynthesis, flavonoid biosynthesis, plant hormone signal transduction, protein processing in endoplasmic reticulum, and plant–pathogen interaction, respectively.

Furthermore, the DEGs were classified according to gene ontology (GO) and Kyoto encyclopedia of genes and genomes (KEGG) terms. DEGs at 4 h and 12 h were enriched in 14 GO terms, mainly containing mitochondrion, cytosol, chloroplast, plastid, and DNA binding (Figure 4D). KEGG analysis showed that DEGs at 4 h and 12 h displayed enrichment in diverse metabolic pathways, including photosynthesis metabolism, phenylalanine metabolism, starch and sucrose metabolism, amino sugar and nucleotide sugar metabolism, and glycerolipid metabolism (Figure 4E). In addition, the DEGs were classified as plant–pathogen interaction, plant hormone signal transduction, protein processing in the endoplasmic reticulum, and MAPK signaling pathway (Figure 4E).

2.4. Candidate Gene Analysis for *qSCT8*

To identify the possible novel genes for STC within the QTLs, *qSCT8* was thoroughly analyzed because its *p*-value is the most significant. *qSCT8* was first identified within a 638 kb interval. Using Rice Genome Annotation Project (uga.edu), it was found that the interval of *qSCT8* contains about 90 genes. According to the transcriptome sequencing data of DXWR, it was found that among the 90 genes, 18 genes were detected to be expressed at the seedling stage, of which only *LOC_Os08g04840*, encodes an MYB family transcription factor, was up-regulated after cold treatment (Figure 5A). The expression of *LOC_Os08g04840* during 4 °C cold stress in DXWR and GZX49 was further analyzed by qRT-PCR technology, and it was found that the expression of *LOC_Os08g04840* did not change significantly in DXWR and GZX49 after 4 h of treatment compared with 0 h, but it was significantly up-regulated by 2-fold after 12 h of treatment. It was up-regulated by 2-fold, and finally, DXWR was up-regulated by 3.5-fold, and GZX49 was only 2.3-fold after 24 h of treatment (Figure 5B). This indicated that the response intensity of *LOC_Os08g04840* in DXWR was higher than that of GZX49 in the late 4 °C cold treatment. Furthermore, sequence comparison revealed that *LOC_Os08g04840* had substantial variation among rice varieties. In its promoter region (2-kb upstream of the predicted start codon), 4 SNPs were found between DXWR and GZX49. In the coding region, there were no nucleotide variations. Thus, *LOC_Os08g04840* was putatively a possible candidate gene for *qSCT8*.

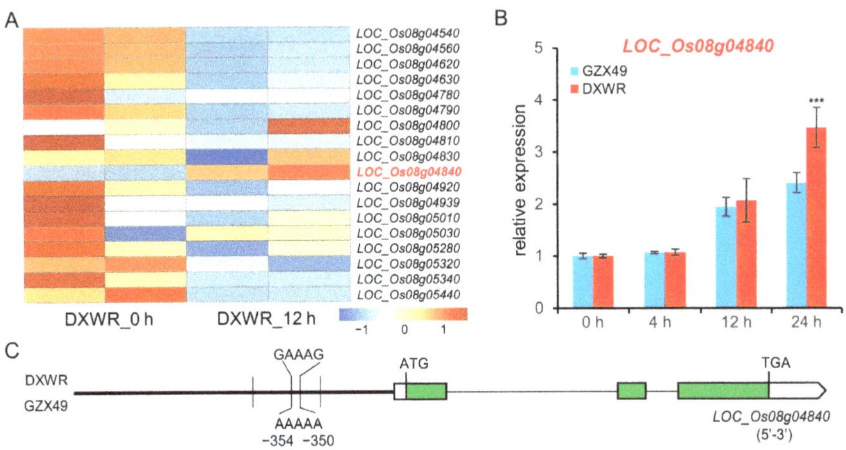

Figure 5. *qSCT8* candidate gene analysis. (**A**) Clustering heat maps of the relative expression levels of genes within the qSCT8 localization interval determined using RNA-seq data. Standard scores (Z-scores) were used as the numerical signs to evaluate the standard deviations from the mean of the corresponding samples. (**B**) Expression levels of the *LOC_Os08g04840* in GZX49 and DXWR after cold stress measured by qRT-PCR. The results were statistically analyzed using Student's *t*-test (*** $p < 0.005$). (**C**) Sequence comparison of *LOC_Os08g04840* among GZX49 and DXWR. Vertical bars represent SNPs.

3. Discussion

In recent years, direct seeding of rice has gained popularity due to its time and labor savings and low input. However, the transplanting of seedlings was carried out in a greenhouse, so the genes related to cold tolerance in the seedling stage of rice were lost due to domestication, which leads to major problems such as poor chlorosis, reduced growth rate and tillering, and low seedling vigor in direct-seeded rice when it encounters a cold spring [1,2,41]. Therefore, cultivating rice varieties with strong cold tolerance at the seedling stage is a necessary condition for the application of direct seeding cultivation of rice. DXWR survives heavy snow and can naturally overwinter in its habitat (Figure 6A). The cold tolerance of DXWR is stronger than the main variety LongJing31 in northeast China (Figure 6B). However, most of the identified cold tolerance genes are from japonica rice, which limits the development of cold tolerance breeding in cultivated rice [4,16,18,42–44]. Therefore, the identification of SCT-related genes and analysis of the genetic basis of cold tolerance in DXWR has important scientific value for improving the low-temperature seedling vigor ability of rice.

Figure 6. Cold tolerance phenotype of DXWR. (**A**) The phenotype of DXWR encountering heavy snow weather in its habitat. (**B**) Phenotypes of DXWR, Longjing 31, and GZX49 after 4 days of treatment at 0 °C at the seedling stage. LJ31 stands for LongJing31. The scale bar of (**B**) means 5 cm.

In this study, a high-generation genetic population was constructed using DXWR/GZX49, and a high-density linkage map was constructed using this population, which contained 1658 bins with an average physical interval of 220.4 kb (Table S1). Using this population, 10 QTLs affecting seedling survival after 78 h of cold treatment at 4 °C were identified, and these QTLs explained 2.06–6.8% of the phenotypic variation range. By comparing the positions of these QTLs, 6 QTLs overlapped with the reported QTLs related to cold tolerance (Table S2). The location of *qSCT1* was very close to that of the *QTL3* [44], and the identified cold tolerance-related gene *OsMYB3R-2* was included in this mapping interval [39]. The *qSCT2* mapping region contained the identified cold tolerance gene *OsTPP1* [40]. *qSCT4* was mapped near *qCTB4-1*, and two cold tolerance-related genes, *CTB4a* and *CTB2*, have been cloned in this interval [45,46]. *qSCT7* and *qSCT9* contain intervals that overlap with the cloned cold tolerance genes *qCT7* [22] and *qCTS-9* [17], respectively. Two QTL *qSCT11.1* and *qSCT11.2* were identified on chromosome 11, which were detected as *qCTS11-2* and *qCTS11-4*, respectively, in previous reports [43]. The results of the above comparison reflected the accuracy of this study and also illustrated the complexity of genetic regulation of cold tolerance in the rice seedling stage. In addition, this study showed that DXWR was extremely cold tolerant due to its inclusion of multiple cold-tolerant genes. These results not only reinforced those of previous studies but also reflected the complexity of developing strong cold-tolerant rice varieties.

Many rice cold tolerance genes have been identified using reverse genetics, which is important for understanding the regulatory mechanisms of rice cold tolerance [16]. However, only cold tolerance genes with natural allelic variation can be directly used for breeding improvement. Cold stress could induce the up-regulated expression of *OsMYB3R-2*; meanwhile, overexpression of *OsMYB3R-2* in *Oryza sativa japonica* 'Zhonghua 10' could significantly enhance the cold tolerance of its seedlings [39]. Low-temperature stress could activate the activity of *OsMAPK3*, and the active *OsMAPK3* could phosphorylate *OsbHLH002* and make it accumulate without ubiquitinase degradation. *OsbHLH002* can promote the expression of *OsTPP1*, thereby increasing trehalose content and resistance to low-temperature damage [40]. Our study found that the $OsMYB3R\text{-}2^{DXWR}$ and $OSTPP1^{DXWR}$ allele was significantly more up-regulated than $OsMYB3R\text{-}2^{GZX49}$ and $OSTPP1^{GZX49}$ under cold stress conditions (Figure 3B,D), indicating that $OsMYB3R\text{-}2^{DXWR}$ and $OSTPP1^{DXWR}$ may be a natural allelic variation with strong cold tolerance, which can be directly used for breeding improvement. This further indicates that DXWR contains many excellent cold tolerance alleles, which can provide important genetic resources for the improvement of rice cold tolerance breeding.

In order to further analyze the cold tolerance characteristics of Dongxiang wild rice at the level of gene expression, transcriptomes were performed on DXWR during cold treatment in this study, and more than 3000 cold stress response genes were identified (Figure 4A–C). GO terms and KEGG pathways analysis were performed on these genes, and it was found that these genes have been involved in multiple biological processes (Figure 4D,E). This further illustrated the complexity of DXWR resistance to cold stress.

Combined QTL-mapping and transcriptomes analysis, compared to the approaches for identifying candidate genes using only traditional QTL-mapping or high-throughput expression profiling, takes less time, reduces labor costs, and increases the selection veracity of the target regions or candidate genes [35,47]. In this study, a novel cold-tolerant QTL, *qSCT8*, was first identified within a 638 kb interval (Table S2 and Figure 3A). The genes *LOC_Os08g04840*, coding an MYB family transcription, that responded to cold stress in the *qSCT8* mapping region were screened out by transcriptome, and qRT-PCR was used to further verify *LOC_Os08g04840* which showed differences in expression before and after cold treatment (Figure 5A,B). In particular, the two G/A variant at the -350 and -354 site upstream of the start codon may cause MADS combined element (AAAAAAAAAGAAAG) defects in $LOC\_Os08g04840^{GZX49}$ (Figure 5C) [48]. Transcription factors of MYB and MADS family have been reported to regulate cold tolerance [39,49–52]. The interaction mode of MADS—MYB might play an important role in the signal transmission of cold tolerance in DXWR. This presumption might provide a new idea for the follow-up study on the genetic mechanism of cold tolerance in the DXWR seedling stage.

4. Materials and Methods

4.1. Plant Materials

A backcross recombinant inbred line (BRIL) population was developed by single-seed descent from a backcross (BC_1F_1) of Dongxiang wild rice (DXWR) as donor and *indica* cultivar GZX49 as the recurrent parent. This population consisted of 132 lines, which were backcrossed for 1 generation with GZX49 as the parent, and then selfed for 8 generations (Figure 1A). The BRIL population was grown at the experimental field of Jiangxi Academy of Agricultural Sciences in 2019 at Nanchang (28.57′ N, 115.9′ E), China.

4.2. Evaluation of Cold Tolerance at the Seedling Stage

The harvested BRIL seeds were incubated at 40 °C for approximately 36 h to break dormancy and soaked in deionized water at 30 °C for approximately 60 h for germination. A total of 30 germinated seeds from each line were selected and sown in soil in pots and cultivated under a 12 h light/12 h dark cycle at 28 °C/26 °C with 80% humidity. When most of the seedlings grew to trifoliate, weak seedlings were removed, and the rest were treated at a low temperature of 4 °C for 84 h in the incubator and then resumed growth at 28 °C for 7 days. Survival rates were determined after 14-day treatments by counting the surviving plants (leaf contains about 40% green part) and the dead plants. All lines were subjected to three independent replicates.

4.3. DNA Extraction and SNP Genotyping

Genomic DNA extraction was carried out using the plant genomic DNA extraction Kit (TIANGEN, Beijing, China), following the manufacturer's instructions. RNase A was then added to digest RNA. The quality and concentrations of DNA were detected using a NanoDrop 2000 (Thermo Fisher Scientifc, Waltham, MA, USA), while DNA integrity was examined by electrophoresis on 1% agarose gels. To prepare the reduced representation libraries for sequencing, the GBS protocol was carried out according to the method reported by Elshire et al. [53]. In brief, the genomic DNA was first digested by restriction enzymes. In this case, *EcoRI* and *MseI* were selected to effectively reduce genome complexity. Barcode adapters were designed and modified according to the standard Illumina adapter design for paired-end read libraries. The ligation reaction was incubated for 1 h at 22 °C with T4 DNA ligase (Thermo Scientific, Madison, WI, USA) and inactivated at 65 °C for 20 min. The ligation products from each sample were pooled in a single tube, and the products were amplified with 10 cycles of PCR. The amplified library was purified using a QIA quick PCR purification kit (Qiagen, Hilden, Germany), quantified on an Agilent 2100 Bioanalyzer (Agilent Technologies, Palo Alto, CA, USA), and finally sequenced on an Illumina HiSeq3000 instrument (Illumina, San Diego, CA, USA), which generated 150 bp paired-end reads.

The sequencing reads were aligned to the Nipponbare genome sequence (IRGSPv7) using SOAP2 software (version 2.20) [54]. SNP calling was performed with realSFS based on the Bayesian estimation of site frequency at every site. All SNPs were filtered using a Practical Extraction and Report Language (PERL) script based on the following criteria: loci with >50% missing data and minor allele frequency less than 5% in the population. In the end, we obtained 55,036 SNPs. Based on SNP genotyping, the bin was defined by a unique overlapping recombination segment across the BRILs according to a previously reported approach [55]. A bin without breakpoints was generated using the R/qtl package function *fill.geno* with the "argmax" method. The high-density bin map was constructed as previously described [56]. Briefly, the sliding window approach was adopted to evaluate a group of consecutive SNPs for genotyping and determination of recombination breakpoints along the chromosomes of each individual. Blocks with a length less than 250 kb in which the number of sequenced SNPs was fewer than 5 were masked as missing data to avoid false double recombination. Genotypes of bins for regions at the transitions between two different genotype blocks were imputed using the R/qtl package. The genetic linkage map was constructed using the R/qtl package function *est.map* with the Haldane map method [57]. Finally, we used 48,339 SNPs to construct bin maps with 1658 bins. The genotype data of the BRIL population are available on request.

4.4. QTL Analysis

The QTL analysis of the phenotypic data with bin maps with 1658 bins in the BILs was performed using the linear ridge regression method to reduce the multicollinearity among markers, as described previously [58]. A significance level of $p < 0.005$ was set as the threshold in the BILs to declare the presence of a putative QTL in a given bin. If several adjacent bins showed p-values lower than the threshold, the QTL was tentatively located in the bin (peak bin) with the lowest p-value [58]. The phenotypic variance explained by each QTL was decomposed using the "relaimpo" package of R ("lmg" function). QTL nomenclature followed the principles suggested by a previous report [59].

Gene annotations for a given peak bin were obtained from the Rice Genome Annotation Project Database (http://rice.plantbiology.msu.edu/, accessed on 8 June 2021).

4.5. RNA-Seq Analysis

Six RNA samples, including cold treatment and controls, were used for RNA sequencing. The aerial parts of 20 seedlings of the corresponding treatment were mixed for each sample in duplicate. In total, 3 μg of RNA per sample was used as the input material for the RNA sample preparations. The sequencing libraries were prepared with the NEBNext Ultra RNA Library Prep Kit for Illumina (NEB, Ipswich, MA, USA) following the manufacturer's recommendations; index codes were added to attribute sequences to each sample. Clustering of the index-coded samples was performed on a cBot Cluster Generation System with the TruSeq PE Cluster Kit v3-cBot-HS (Illumina) according to the manufacturer's instructions. After cluster generation, the libraries were sequenced with an Illumina sequencing platform (Hiseq 2500), and 125-bp or 150-bp paired-end reads were generated. The reads were aligned to the reference transcript sequence (http://rice.plantbiology.msu.edu/, accessed on 8 December 2021), and the number of reads covered from the start to the end of each gene was counted. We used the RSEM v1.2.31 tool to quantify gene expression levels. Due to the influence of sequencing depth and gene length, the gene expression value of RNA-seq is generally not represented by read counts, and TPM (Transcripts Per Million mapped reads) is often used as a standardized value. TPM has successively corrected the gene length and sequencing depth. The sum of the TPM of each sample is the same, and the TPM value can reflect the ratio of reads of a gene in the comparison so that the value can be directly compared between samples. Difference analysis was used to find the reasons for the differences in different samples and the degree of their influence on the differences. Statistical analysis of sample expression data was performed to screen genes with significantly different expression levels in different states. The difference analysis is divided into three steps: first, normalize the original read counts, mainly to correct the sequencing depth; then, the statistical model is used to calculate the hypothesis test probability (*p*-value); finally, the multiple hypothesis test correction is performed. Obtain the false discovery rate (FDR) value. For samples with biological replicates, DESeq2 v1.10.1 was selected for differential gene expression analysis. The above RNA-seq-related experiments were completed by Wuhan Genoseq Bioinformatics Technology Co., Ltd. The original RNA-seq data set has been deposited in NCBI (PRJNA871989).

4.6. *Quantitative Real-Time PCR*

Total RNAs were isolated with the TRIzol kit (Invitrogen, Carlsbad, CA, USA) according to the manufacturer's instructions. The RNA was treated with DNase I (Invitrogen), and approximately 3 µg of total RNA was used to synthesize first-strand cDNA using oligo $(dT)_{18}$ as a primer (Promega, Shanghai, China). The trans-intron ACTIN primer (ACTIN-M) was used to detect whether the reverse transcribed cDNA still has genomic DNA, and the cDNA without genomic DNA was used for subsequent Quantitative real-time PCR. Quantitative real-time (qRT) PCR was performed using gene-specific primers and the FastStart Universal SYBR Green Master (Roche) on a real-time PCR ViiA7 system (Applied Biosystems). Genes *ubiquitin* (*LOC_Os03g13170*) and *actin1* (*LOC_Os03g50885*) not differentially expressed in RNAseq were used as the internal control. The relative quantification method ($2^{-\triangle\triangle CT}$) was used to evaluate gene expression level [60]. As similar expression results were observed regardless of which control genes were used, the *ubiquitin* of expressions was applied for the relative expression analyses for every assayed sample. At least three biological replicates, each containing four technical, were performed for each experiment.

The primers were designed according to the Nipponbare reference genome by Primer3 (http://redb.ncpgr.cn/modules/redbtools/primer3.php, accessed on 8 April 2022). The sequences were analyzed using Sequencer 5.0 (Gene Codes Corporation, Ann Arbor, MI, USA). All primers were synthesized at Sangon Biotech (Shanghai, China) and are listed in Table S3.

5. Conclusions

In conclusion, Dongxiang wild rice (DXWR) has strong cold tolerance. A BRIL population of DXWR/GZX49 with a high-density bin map was used to identify several QTLs and potential candidate genes for seedling cold tolerance in this study. Cold stress-responsive genes in DXWR were detected by transcriptome sequencing. Moreover, using transcriptome data and genetic linkage analysis, combined with qRT-PCR, sequence comparison, and bioinformatics, the candidate gene of the major effect locus *qSCT8* was identified as

LOC_Os08g04840. These findings might provide insights into the genetic basis of SCT for the improvement of cold stress potential in rice breeding programs.

Supplementary Materials: The following supporting information can be downloaded at: https://www.mdpi.com/article/10.3390/plants11182329/s1, Table S1: Distribution of SNP and bin on chromosome; Table S2: QTLs detected for seedling cold tolerance in BRIL population; Table S3: Primers used in the study.

Author Contributions: D.W. and J.W. designed and conceived the research; D.W., Y.X., J.W., C.H., P.C., W.D. and H.C. conducted the experiments; J.W., D.C. and H.C. developed the population; D.W., Y.X. and J.W. analyzed data and wrote the paper. All authors have read and agreed to the published version of the manuscript.

Funding: This research is funded by the National Natural Science Foundation of China (31960076), the Jiangxi Modern Agricultural Scientific Research Collaborative Innovation Special Project (JXXTCXB-SJJ202204), the Jiangxi Technological Innovation Guidance Program (20212BDH81023), the Project of Discovery of Favorable Genes of Wild Rice and Breeding of Green and Efficient Varieties of Jiangxi Province (20213AAF01001) and the Doctoral Startup Fund Project of Jiangxi Academy of Agricultural Sciences (20191CBS002).

Institutional Review Board Statement: Not applicable.

Informed Consent Statement: Not applicable.

Data Availability Statement: Data is contained within the article and Supplementary Material.

Acknowledgments: We thank Xiaohua Tu and Xianmei Wang for assistance with phenotypic evaluation. We are grateful to Chaopu Zhang and Kun Xie for assistance in the data analysis.

Conflicts of Interest: The authors declare no conflict of interest.

References

1. Zhang, Q.; Chen, Q.; Wang, S.; Hong, Y.; Wang, Z. Rice and cold stress: Methods for its evaluation and summary of cold tolerance-related quantitative trait loci. *Rice* **2014**, *7*, 24. [CrossRef] [PubMed]
2. Moraes de Freitas, G.P.; Basu, S.; Ramegowda, V.; Thomas, J.; Benitez, L.C.; Braga, E.B.; Pereira, A. Physiological and transcriptional responses to low-temperature stress in rice genotypes at the reproductive stage. *Plant Signal. Behav.* **2019**, *14*, e1581557. [CrossRef] [PubMed]
3. Jena, K.K.; Kim, S.M.; Suh, J.P.; Yang, C.I.; Kim, Y.G. Identification of cold-tolerant breeding lines by quantitative trait loci associated with cold tolerance in rice. *Crop Sci.* **2012**, *52*, 517–523. [CrossRef]
4. Li, J.; Zhang, Z.; Chong, K.; Xu, Y. Chilling tolerance in rice: Past and present. *J. Plant Physiol.* **2022**, *268*, 153576. [CrossRef]
5. Qian, Q. Smart super rice. *Sci. China Life Sci.* **2017**, *60*, 1460–1462. [CrossRef]
6. Dai, L.; Lin, X.; Ye, C.; Ise, K.; Saito, K.; Kato, A.; Xu, F.; Yu, T.; Zhang, D. Identification of quantitative trait loci controlling cold tolerance at the reproductive stage in yunnan landrace of rice, Kunmingxiaobaigu. *Breed. Sci.* **2004**, *54*, 253–258. [CrossRef]
7. Saito, K.; Hayano-Saito, Y.; Kuroki, M.; Sato, Y. Map-based cloning of the rice cold tolerance gene Ctb1. *Plant Sci.* **2010**, *179*, 97–102. [CrossRef]
8. Andaya, V.C.; Mackill, D.J. Mapping of QTLs associated with cold tolerance during the vegetative stage in rice. *J. Exp. Bot.* **2003**, *54*, 2579–2585. [CrossRef]
9. Andaya, V.C.; Tai, T.H. Fine mapping of the qCTS12 locus, a major QTL for seedling cold tolerance in rice. *Theor. Appl. Genet.* **2006**, *113*, 467–475. [CrossRef]
10. Han, L.; Qiao, Y.; Zhang, S.; Zhang, Y.; Cao, G.; Kim, J.; Lee, K.; Koh, H. Identification of quantitative trait loci for cold response of seedling vigor traits in rice. *J. Genet. Genom.* **2007**, *34*, 239–246. [CrossRef]
11. Lou, Q.; Chen, L.; Sun, Z.; Xing, Y.; Li, J.; Xu, X.; Mei, H.; Luo, L. A major QTL associated with cold tolerance at seedling stage in rice (*Oryza sativa* L.). *Euphytica* **2007**, *158*, 87–94. [CrossRef]
12. Suh, J.P.; Lee, C.K.; Lee, J.H.; Kim, J.J.; Kim, S.M.; Cho, Y.C.; Park, S.H.; Shin, J.C.; Kim, Y.G.; Jena, K.K. Identification of quantitative trait loci for seedling cold tolerance using RILs derived from a cross between japonica and tropical japonica rice cultivars. *Euphytica* **2012**, *184*, 101–108. [CrossRef]
13. Shirasawa, S.; Endo, T.; Nakagomi, K.; Yamaguchi, M.; Nishio, T. Delimitation of a QTL region controlling cold tolerance at booting stage of a cultivar, 'Lijiangxintuanheigu', in rice, *Oryza sativa* L. *Theor. Appl. Genet.* **2012**, *124*, 937–946. [CrossRef]
14. Mao, D.; Yu, L.; Chen, D.; Li, L.; Zhu, Y.; Xiao, Y.; Zhang, D.; Chen, C. Multiple cold resistance loci confer the high cold tolerance adaptation of Dongxiang wild rice (*Oryza rufipogon*) to its high-latitude habitat. *Theor. Appl. Genet.* **2015**, *128*, 1359–1371. [CrossRef]

15. Endo, T.; Chiba, B.; Wagatsuma, K.; Saeki, K.; Ando, T.; Shomura, A.; Mizubayashi, T.; Ueda, T.; Yamamoto, T.; Nishio, T. Detection of QTLs for cold tolerance of rice cultivar 'Kuchum' and effect of QTL pyramiding. *Theor. Appl. Genet.* **2016**, *129*, 631–640. [CrossRef]
16. Lv, Y.; Hussain, M.A.; Luo, D.; Tang, N. Current understanding of genetic and molecular basis of coldtolerance in rice. *Mol. Breed.* **2019**, *39*, 159. [CrossRef]
17. Zhao, J.; Zhang, S.; Dong, J.; Yang, T.; Mao, X.; Liu, Q.; Wang, X.; Liu, B. A novel functional gene associated with cold tolerance at the seedling stage in rice. *Plant Biotechnol. J.* **2017**, *15*, 1141–1148. [CrossRef]
18. Ma, Y.; Dai, X.; Xu, Y.; Luo, W.; Zheng, X.; Zeng, D.; Pan, Y.; Lin, X.; Liu, H.; Zhang, D.; et al. COLD1 confers chilling tolerance in rice. *Cell* **2015**, *160*, 1209–1221. [CrossRef]
19. Liu, C.; Ou, S.; Mao, B.; Tang, J.; Wang, W.; Wang, H.; Cao, S.; Schläppi, M.R.; Zhao, B.; Xiao, G.; et al. Early selection of bZIP73 facilitated adaptation of japonica rice to cold climates. *Nat. Commun.* **2018**, *9*, 3302. [CrossRef]
20. Xiao, N.; Gao, Y.; Qian, H.; Gao, Q.; Wu, Y.; Zhang, D.; Zhang, X.; Yu, L.; Li, Y.; Pan, C.; et al. Identification of genes related to cold tolerance and a functional allele that confers cold tolerance. *Plant Physiol.* **2018**, *177*, 1108–1123. [CrossRef]
21. Mao, D.; Xin, Y.; Tan, Y.; Hu, X.; Bai, J.; Liu, Z.-y.; Yu, Y.; Li, L.; Peng, C.; Fan, T.; et al. Natural variation in the *HAN1* gene confers chilling tolerance in rice and allowed adaptation to a temperate climate. *Proc. Nalt. Acad. Sci. USA* **2019**, *116*, 3494–3501. [CrossRef] [PubMed]
22. Liu, H.; Yang, L.; Xu, S.; Lyu, M.J.; Wang, J.; Wang, H.; Zheng, H.; Xin, W.; Liu, J.; Zou, D. OsWRKY115 on qCT7 links to cold tolerance in rice. *Theor. Appl. Genet.* **2022**, *135*, 2353–2367. [CrossRef] [PubMed]
23. Hufford, M.B.; Xu, X.; van Heerwaarden, J.; Pyhäjärvi, T.; Chia, J.-M.; Cartwright, R.A.; Elshire, R.J.; Glaubitz, J.C.; Guill, K.E.; Kaeppler, S.M.; et al. Comparative population genomics of maize domestication and improvement. *Nat. Genet.* **2012**, *44*, 808–811. [CrossRef] [PubMed]
24. Qi, J.; Liu, X.; Shen, D.; Miao, H.; Xie, B.; Li, X.; Zeng, P.; Wang, S.; Shang, Y.; Gu, X.; et al. A genomic variation map provides insights into the genetic basis of cucumber domestication and diversity. *Nat. Genet.* **2013**, *45*, 1510–1515. [CrossRef] [PubMed]
25. Li, Y.-B.; Zhao, J.-Z.; Pu, W.-F.; Jia, H.; Peng, H.; Zhong, D.; Wang, S.-K. Catalytic effect analysis of metallic additives on light crude oil by TG and DSC tests. *J. Therm. Anal. Calorim.* **2013**, *113*, 579–587. [CrossRef]
26. Wang, L.; Li, P.; Brutnell, T.P. Exploring plant transcriptomes using ultra high-throughput sequencing. *Brief. Funct. Genom.* **2010**, *9*, 118–128. [CrossRef] [PubMed]
27. Bai, B.; Wu, J.; Sheng, W.-T.; Zhou, B.; Zhou, L.-J.; Zhuang, W.; Yao, D.-P.; Deng, Q.-Y. Comparative analysis of anther transcriptome profiles of two different rice male sterile lines genotypes under cold stress. *Int. J. Mol. Sci.* **2015**, *16*, 11398–11416. [CrossRef]
28. Zhao, J.; Zhang, S.; Yang, T.; Zeng, Z.; Huang, Z.; Liu, Q.; Wang, X.; Leach, J.; Leung, H.; Liu, B. Global transcriptional profiling of a cold-tolerant rice variety under moderate cold stress reveals different cold stress response mechanisms. *Physiol Plant.* **2015**, *154*, 381–394. [CrossRef]
29. Chawade, A.; Lindlof, A.; Olsson, B.; Olsson, O. Global expression profiling of low temperature induced genes in the chilling tolerant japonica rice Jumli Marshi. *PLoS ONE* **2013**, *8*, e81729. [CrossRef]
30. Zhang, T.; Zhao, X.; Wang, W.; Pan, Y.; Huang, L.; Liu, X.; Zong, Y.; Zhu, L.; Yang, D.; Fu, B. Comparative transcriptome profiling of chilling stress responsiveness in two contrasting rice genotypes. *PLoS ONE* **2012**, *7*, e43274. [CrossRef]
31. Dametto, A.; Sperotto, R.A.; Adamski, J.M.; Blasi, É.A.R.; Cargnelutti, D.; de Oliveira, L.F.V.; Ricachenevsky, F.K.; Fregonezi, J.N.; Mariath, J.E.A.; da Cruz, R.P.; et al. Cold tolerance in rice germinating seeds revealed by deep RNAseq analysis of contrasting indica genotypes. *Plant Sci.* **2015**, *238*, 1–12. [CrossRef]
32. Guan, S.; Xu, Q.; Ma, D.; Zhang, W.; Xu, Z.; Zhao, M.; Guo, Z. Transcriptomics profiling in response to cold stress in cultivated rice and weedy rice. *Gene* **2019**, *685*, 96–105. [CrossRef]
33. Kadambari, G.; Vemireddy, L.R.; Srividhya, A.; Nagireddy, R.; Jena, S.S.; Gandikota, M.; Patil, S.; Veeraghattapu, R.; Deborah, D.; Reddy, G.E.; et al. QTLSeq-based genetic analysis identifies a major genomic region governing dwarfness in rice (*Oryza sativa* L.). *Plant Cell Rep.* **2018**, *37*, 677–687. [CrossRef]
34. Takagi, H.; Abe, A.; Yoshida, K.; Kosugi, S.; Natsume, S.; Mitsuoka, C.; Uemura, A.; Utsushi, H.; Tamiru, M.; Takuno, S.; et al. QTL-seq: Rapid mapping of quantitative trait loci in rice by whole genome resequencing of DNA from two bulked populations. *Plant J.* **2013**, *74*, 174–183. [CrossRef]
35. Liu, F.; Xu, W.; Song, Q.; Tan, L.; Liu, J.; Zhu, Z.; Fu, Y.; Su, Z.; Sun, C. Microarray-assisted fine-mapping of quantitative trait loci for cold tolerance in rice. *Mol. Plant* **2013**, *6*, 757–767. [CrossRef]
36. Zhang, F.; Cui, F.; Zhang, L.; Wen, X.; Luo, X.; Zhou, Y.; Li, X.; Wan, Y.; Zhang, J.; Xie, J. Development and identification of a introgression line with strong drought resistance at seedling stage derived from *Oryza sativa* L. mating with *Oryzarufipogon Griff*. *Euphytica* **2014**, *200*, 1–7. [CrossRef]
37. Zhao, J.; Qin, J.J.; Song, Q.; Sun, C.Q.; Liu, F.X. Combining QTL mapping and expression profile analysis to identify candidate genes of cold tolerance from Dongxiang common wild rice (*Oryza rufipogon Griff*). *J. Integr. Agric.* **2016**, *15*, 1933–1943. [CrossRef]
38. Liang, Y.S.; Zheng, J.; Yan, C.; Li, X.X.; Liu, S.F.; Zhou, J.J.; Qin, X.J.; Nan, W.B.; Yang, Y.Q.; Zhang, H.M. Locating QTLs controlling overwintering trait in Chinese perennial Dongxiang wild rice. *Mol. Genet. Genom.* **2018**, *293*, 81–93. [CrossRef]
39. Ma, Q.; Dai, X.; Xu, Y.; Guo, J.; Liu, Y.; Chen, N.; Xiao, J.; Zhang, D.; Xu, Z.; Zhang, X.; et al. Enhanced tolerance to chilling stress in OsMYB3R-2 transgenic rice is mediated by alteration in cell cycle and ectopic expression of stress genes. *Plant Physiol.* **2009**, *150*, 244–256. [CrossRef]

40. Zhang, Z.; Li, J.; Li, F.; Liu, H.; Yang, W.; Chong, K.; Xu, Y. OsMAPK3 Phosphorylates OsbHLH002/OsICE1 and inhibits its ubiquitination to activate OsTPP1 and enhances rice chilling tolerance. *Dev. Cell* **2017**, *43*, 731–743.e5. [CrossRef]
41. Liu, C.T.; Wang, W.; Mao, B.G.; Chu, C.C. Cold stress tolerance in rice: Physiological changes, molecular mechanism, and future prospects. *Yi Chuan* **2018**, *40*, 171–185. [CrossRef] [PubMed]
42. Lv, Y.; Guo, Z.; Li, X.; Ye, H.; Li, X.; Xiong, L. New insights into the genetic basis of natural chilling and cold shock tolerance in rice by genome-wide association analysis. *Plant Cell Environ.* **2016**, *39*, 556–570. [CrossRef]
43. Wang, D.; Liu, J.; Li, C.; Kang, H.; Wang, Y.; Tan, X.; Liu, M.; Deng, Y.; Wang, Z.; Liu, Y.; et al. Genome-wide association mapping of cold tolerance genes at the seedling stage in rice. *Rice* **2016**, *9*, 61. [CrossRef]
44. Song, J.; Li, J.; Sun, J.; Hu, T.; Wu, A.; Liu, S.; Wang, W.; Ma, D.; Zhao, M. Genome-wide association mapping for cold tolerance in a core collection of rice (*Oryza sativa* L.) landraces by using high-density single nucleotide polymorphism markers from specific-locus amplified fragment sequencing. *Front. Plant Sci.* **2018**, *9*, 875. [CrossRef]
45. Zhang, Z.; Li, J.; Pan, Y.; Li, J.; Zhou, L.; Shi, H.; Zeng, Y.; Guo, H.; Yang, S.; Zheng, W.; et al. Natural variation in CTB4a enhances rice adaptation to cold habitats. *Nat. Commun.* **2017**, *8*, 14788. [CrossRef]
46. Li, J.; Zeng, Y.; Pan, Y.; Zhou, L.; Zhang, Z.; Guo, H.; Lou, Q.; Shui, G.; Huang, H.; Tian, H.; et al. Stepwise selection of natural variations at CTB2 and CTB4a improves cold adaptation during domestication of japonica rice. *New Phytol.* **2021**, *231*, 1056–1072. [CrossRef]
47. Peng, Z.; Zhao, C.; Li, S.; Guo, Y.; Xu, H.; Hu, G.; Liu, Z.; Chen, X.; Chen, J.; Lin, S.; et al. Integration of genomics, transcriptomics and metabolomics identifies candidate loci underlying fruit weight in loquat. *Hortic. Res.* **2022**, *9*, uhac037. [CrossRef]
48. Jin., J.; Tian., F.; Yang., D.; Meng., Y.; Kong., L.; Luo., J.; Gao., G. PlantTFDB 4.0: Toward a central hub for transcription factors and regulatory interactions in plants. *Nucleic Acids Res.* **2017**, *45*, 1040–1045. [CrossRef]
49. Lombardo, F.; Kuroki, M.; Yao, S.-G.; Shimizu, H.; Ikegaya, T.; Kimizu, M.; Ohmori, S.; Akiyama, T.; Hayashi, T.; Yamaguchi, T.; et al. The superwoman1-cleistogamy2 mutant is a novel resource for gene containment in rice. *Plant Biotechnol. J.* **2017**, *15*, 97–106. [CrossRef]
50. Chen, R.; Ma, J.; Luo, D.; Hou, X.; Ma, F.; Zhang, Y.; Meng, Y.; Zhang, H.; Guo, W. CaMADS, a MADS-box transcription factor from pepper, plays an important role in the response to cold, salt, and osmotic stress. *Plant Sci.* **2019**, *280*, 164–174. [CrossRef]
51. Jia, J.; Zhao, P.; Cheng, L.; Yuan, G.; Yang, W.; Liu, S.; Chen, S.; Qi, D.; Liu, G.; Li, X. MADS-box family genes in sheepgrass and their involvement in abiotic stress responses. *BMC Plant Biol.* **2018**, *18*, 42. [CrossRef]
52. Wang, X.; Niu, Y.; Zheng, Y. Multiple functions of MYB transcription factors in abiotic stress responses. *Int. J. Mol. Sci.* **2021**, *22*, 6125. [CrossRef] [PubMed]
53. Elshire, R.J.; Glaubitz, J.C.; Sun, Q.; Poland, J.A.; Kawamoto, K.; Buckler, E.S.; Mitchell, S.E. A robust, simple genotyping-by-sequencing (GBS) approach for high diversity species. *PLoS ONE* **2011**, *6*, e19379. [CrossRef] [PubMed]
54. Li, R.; Yu, C.; Li, Y.; Lam, T.W.; Yiu, S.M.; Kristiansen, K.; Wang, J. SOAP2: An improved ultrafast tool for short read alignment. *Bioinformatics* **2009**, *25*, 1966–1967. [CrossRef]
55. Paran, I.; Zamir, D. Quantitative traits in plants: Beyond the QTL. *Trends Genet.* **2003**, *19*, 303–306. [CrossRef]
56. Xie, W.; Feng, Q.; Yu, H.; Huang, X.; Zhao, Q.; Xing, Y.; Yu, S.; Han, B.; Zhang, Q. Parent-independent genotyping for constructing an ultrahigh-density linkage map based on population sequencing. *Proc. Nalt. Acad. Sci. USA* **2010**, *107*, 10578–10583. [CrossRef]
57. Broman, K.W.; Wu, H.; Sen, S.; Churchill, G.A. R/qtl: QTL mapping in experimental crosses. *Bioinformatics* **2003**, *19*, 889–890. [CrossRef]
58. Sun, W.; Zhou, Q.; Yao, Y.; Qiu, X.; Xie, K.; Yu, S. Identification of genomic regions and the isoamylase gene for reduced grain chalkiness in rice. *PLoS ONE* **2015**, *10*, e0122013. [CrossRef]
59. McCouch, S.R. Cgsnl, Gene nomenclature system for rice. *Rice* **2008**, *1*, 72–84. [CrossRef]
60. Livak, K.J.; Schmittgen, T.D. Analysis of relative gene expression data using real-time quantitative PCR and the 2(-Delta Delta C(T)) Method. *Methods* **2001**, *25*, 402–408. [CrossRef]

plants

MDPI

Article

Effects of Nitrogen Deficiency on the Metabolism of Organic Acids and Amino Acids in *Oryza sativa*

Ling-Hua Chen [1], Zu-Xin Cheng [2], Ming Xu [2], Zhi-Jian Yang [2,3] and Lin-Tong Yang [4,*]

1 College of Jinshan, Fujian Agriculture and Forestry University, Fuzhou 350002, China
2 Fujian Engineering Technology Research Center of Breeding and Utilization for Special Crops, College of Agriculture, Fujian Agriculture and Forestry University, Fuzhou 350002, China
3 Key Laboratory of Crop Biotechnology, College of Agriculture, Fujian Agriculture and Forestry University, Fuzhou 350002, China
4 College of Resources and Environment, Fujian Agriculture and Forestry University, Fuzhou 350002, China
* Correspondence: talstoy@163.com or talstoy@fafu.edu.cn

Abstract: Organic acids metabolism and nitrogen (N) metabolism in rice seedlings and the relationship between them are not fully understood. In this study, rice (*Oryza sativa* L. ssp. Indica) variety "Huanghuazhan" was used as the experimental material, and three N levels (5 mM, 1 mM, and 0 mM NH_4NO_3) were set by the hydroponic method for different levels of N treatment. Our results showed that the increased content of malate in rice leaves caused by reducing N level was related to the increased synthesis of malate (the activity of leaf PEPC increased)and the decreased degradation of malate (the activity of leaf NADP-ME decreased), while the increased contents of citrate and isocitrate in rice leaves caused by reducing N level might not be caused by the increased biosynthesis, but due to the decrease in degradation of citrate and isocitrate (the activities of leaf CS, ACO, and NADP-IDH decreased). The increased content of malate in rice roots caused by reducing N level might be related to the increased biosynthesis and the decreased degradation of root malate (the activities of root NAD-MDH and PEPC increased, while the activity of NADP-ME decreased). Compared to the control (5 mM NH_4NO_3), the increased content of citrate in rice roots caused by reducing N level might be related to the increased biosynthesis rather than the decreased degradation of citrate, due to the higher activities of CS and ACO in rice roots under 0 mM N and 1mM N treatment when compared to that of the control ones. At the same time, the increased content of isocitrate in roots was related to the increased isomerization of isocitrate (the activity of root ACO increased) and the decreased degradation of isocitrate (the activity of root NADP-IDH decreased). With the reducing N level, the activities of N metabolism-related enzymes, such as nitrate reductase (NR), glutamine synthetase (GS), and glutamate synthase (GOGAT), decreased in rice leaves and roots, resulting in the decreased contents of total free amino acids (TFAAs) and soluble proteins in rice seedlings, and finally led to the growth inhibition. Our results showed that the dynamics of organic acids metabolism caused by reducing N level were different in rice leaves and roots. In conclusion, there was a close correlation between organic acids metabolism and N metabolism in rice leaves and roots under N-limited conditions; furthermore, such a correlation was more obvious in rice leaves than that of roots.

Keywords: rice; nitrogen deficiency; organic acid; nitrogen metabolism; growth

Citation: Chen, L.-H.; Cheng, Z.-X.; Xu, M.; Yang, Z.-J.; Yang, L.-T. Effects of Nitrogen Deficiency on the Metabolism of Organic Acids and Amino Acids in *Oryza sativa*. *Plants* **2022**, *11*, 2576. https://doi.org/10.3390/plants11192576

Academic Editors: Mingxun Chen, Lixi Jiang and Yuan Guo

Received: 3 September 2022
Accepted: 27 September 2022
Published: 29 September 2022

1. Introduction

Nitrogen (N) is a macro-nutrient for plant growth and development. It is an important component of functional molecules, such as nucleic acid, amino acid, protein, chlorophyll, and some plant hormones. It is one of the main factors limiting plant growth and crop yield [1]. N is the most abundant nutrient element in plants, accounting for about 2–4% of the dry weight of plants [2]. Although some organic forms of N, such as urea, amino acids, and amides, can also be absorbed by plants, ammonium and nitrate are the two main forms of N obtained by plant roots from soil solutions. These N forms are scarce in natural soil. When the total N absorbed by plant roots from the rhizosphere is less than the N required for plant growth, N deficiency will occur. In the absence of artificial N fertilizer input, N deficiency is one of the important factors restricting the crop yield and quality of agricultural products. Organic acids are the intermediate products of major carbon metabolism in plant cells and participate in various biochemical pathways, such as glycolysis, the tricarboxylic acid (TCA) cycle, photorespiration, the glyoxylic acid cycle, or the photosynthetic C4 cycle [3]. The intermediate products of organic acid metabolism, such as oxaloacetic acid and 2-ketoglutarate, provide the carbon skeletons for transamination and amino acid synthesis [4–6]. It is important to study the relationship between N level and organic acid metabolism and N metabolism for understanding plant N metabolism and guiding the application of N fertilization.

Application of N fertilizer to the field can regulate the flowering period of rice and increase the number of panicles, the number of grains per panicle, and the rice yield [7], while N deficiency can lead to the degradation of photosynthetic pigments in crop leaves, the yellowing of leaves, the reduction in carbon dioxide assimilation rate, and the inhibitionof the growth and yield of crops. The phenomenon that N deficiency reduces the photosynthetic rate and affects plant growth has been reported in many crops, including tea, citrus, corn, cucumber, sunflower, tobacco, rice, and barley [2,8–11].

The carbon (C) metabolism and N assimilation are interrelated. N assimilation requires energy and a carbon skeleton, which is provided by carbohydrate metabolism. Nitrate is an important source of inorganic N in plants. Changes in exogenous nitrate concentration could affect the rapid change in plant metabolism. For example, it could induce the synthesis of nitrate-assimilating enzymes and accelerate glycolysis and the TCA cycle to convert starch into organic acids to assimilate ammonium [12]. Excess nitrate may be accumulated into vacuoles, but excessive ammonium absorbed by crop roots and ammonium produced by nitrate through the action of nitrate reductase (NR) and nitrite reductase (NiR) have toxic effects on plants. Ammonium assimilation requires an adequate carbon skeleton and promotes carbon flow into the TCA cycle [13]. Inorganic N is assimilated and incorporated into glutamine, glutamate, asparagine, and aspartic acid, which are important N carriers in plants. These primary amino acids are the main compounds in the total free amino acid pool of many plants and can be used to synthesize other amides, such as urea, other amino acids, and amines [12,13]. In this process, the glutamine synthetase/glutamate synthase (GS/GOGAT) cycle is the main pathway of nitrogen assimilation in plants. GS catalyzes the reaction of NH_4^+ with glutamate to produce glutamine. GOGAT subsequently catalyzes the transfer of amide groups from glutamine to 2-ketoglutarate (2-OG), generating two glutamate molecules [14]. Among them, 2-ketoglutarate, the main carbon receptor in the GS/GOGAT pathway, is synthesized through the combined action of phosphoenolpyruvate (PEP) carboxylase (PEPC, EC 4.1.1.31), pyruvate kinase (PK), pyruvate dehydrogenase (PDH), aconitase (ACO, EC 4.2.1.3), and NADP-isocitrate dehydrogenase (NADP-IDH, EC1.1.1.42) [15].

The organic-acid-metabolism-related enzymes PEPC, NAD-malate dehydrogenase (NAD-MDH, EC 1.1.1.37), NADP-malic enzyme (NADP-ME, EC 1.1.1.40), citrate synthase (CS, EC 4.1.3.7), ACO, and NADP-IDH are considered to play an important role in balancing C and N metabolism [16,17]. PEPC catalyzes irreversible β-carboxylation of PEP in the presence of HCO_3^- and Mg_2^+ to yield oxaloacetate, which is used to synthesize malate by NAD-MDH [18,19]. The degradation of malate is mediated by cytosolic NADP-ME in plant cells [20,21]. In the cytoplasm and mitochondrion, PEP is converted to acetyl-CoA by the sequential actions of PK and PDH. Aceytl-CoA and oxaloacetate are then combined by CS to generate citrate in the mitochondrion [22]. The citrate is reversibly isomerized to isocitrate via *cis*-aconitate by ACO [23] Isocitrate is then catalyzed to 2OG by NADP-IDH [24]. It was found on the model plant *Arabidopsis thaliana* that the interaction of C-N metabolism could significantly affect the expression levels of more than 300 genes ($p < 0.05$). Among them, the expression levels of PEPC, NAD-MDH, NADP-ME, and NADP-IDH were found to be altered by N deficiency [5]. Studies have shown that plant physiological metabolism, energy, and protein synthesis are found to be significantly affected by the interaction between C/N signals [25,26]. Specific metabolites of C/N metabolism, such as glucose, sucrose, or nitrate, serve as signals to regulate genes encoding enzymes involved in many important biological processes, including photosynthesis, carbon metabolism, nitrogen metabolism, and metabolite allocation [5]. In the study of N-deficiency transcriptome in rice, Shao et al. reported that N deficiency treatment upregulated the expression level of *PEPC* in rice leaves and increased the content of malic acid in leaves [2]. Similarly, it was also found that N-deficiency treatment increased the contents of glycolic acid and succinic acid in tea leaves and citric acid, malic acid, isocitric acid, succinic acid, and fumaric acid in tea roots [9].

Rice is an important grain crop in the world. Nitrogen fertilizer plays an important role in its growth and yield [27]. Meanwhile, increasing the soil nitrogen supply can significantly improve the soil carbon pool [28]. Previous studies have shown that under N deficiency, the activities of enzymes related to N metabolism (such as GS, GOGAT, and glutamate dehydrogenase (GDH)) were affected to varying degrees in rice [14]. Therefore, in order to further understand the mode of C and N metabolism in rice under N reduction conditions, this study carried out research on organic acid metabolism and N metabolism of rice leaves and roots under low-N and N-deficiency treatments. The results of this research can be used for rice N management, high yield, and high-quality cultivation and sustainable development of farmland.

2. Results

2.1. Effects of Different Nitrogen Treatments on Rice Growth and Nitrogen Content in Leaves and Roots

Different nitrogen levels significantly affected the growth of rice seedlings (Figure 1). Compared to the control (5 mM NH_4NO_3), under the treatment of 0 mM NH_4NO_3, rice leaves turned yellow and plant growth was inhibited. After the treatment of different N levels, the fresh weight of the plants was measured. Compared to the control, both low-N (1 mM NH_4NO_3) and N-deficiency (0 mM NH_4NO_3) treatments significantly reduced the fresh weight of rice seedlings (Figure 2A). The shoot height of rice seedlings was measured. Compared to the control, N-deficiency and low-N treatment significantly reduced the shoot height of rice seedlings (Figure 2B). After oven-drying, N content was measured. The result showed that the leaf N and root N of rice seedlings were significantly reduced by either low N (1 mMNH_4NO_3) or N deficiency (0 mM NH_4NO_3) (Figure 2C,D).

Figure 1. Phenotypeof *O. sativa* seedlings under different N levels. Rice seedlings were transplanted to the Hogland nutrient solution containing 5 mM (control), 1 mM (low N), and 0 mM NH_4NO_3 (N deficiency). Seedlings were kept under a 14 h light/10 h dark regime with white photo-illumination of 150 μmol m^{-2} s^{-1} as well as a relative humidity of 68% in a light incubator (28 °C). Fifteen days after transplanting, leaf and root samples were collected.

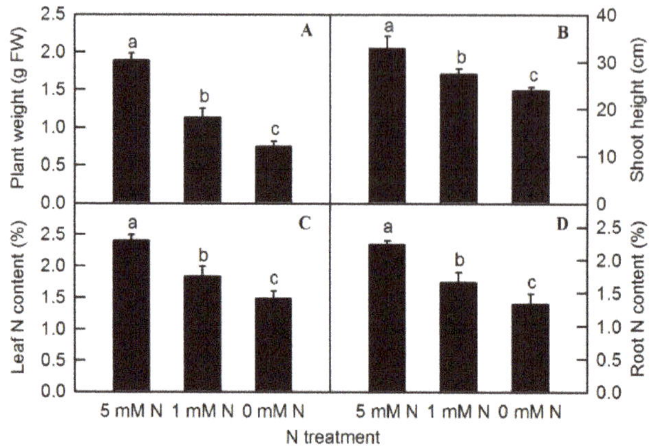

Figure 2. Effects of N deficiency on plant weight (**A**), shoot height (**B**), leaf N content (**C**), and root N content (**D**) in *O. sativa*. Bars represent means ± SD (n = 4). Difference among the treatments was analyzed by Duncan's multiple range test. Different letters above the bar indicate a significant difference at $p < 0.05$.

2.2. Effects of Different N Treatments on the Contents of Organic Acids, TFAAs, and Soluble Protein in Rice Leaves and Roots

Different N levels affected the organic acid content of rice leaves and roots. The contents of malate, citrate in leaves, and malate in roots increased significantly with decreasing N supply (Figure 3A,B,F). Compared to the control, both low N and N deficiency significantly increased the contents of leaf isocitrate (Figure 3C) and citric acid in roots (Figure 3G), while the contents of isocitrate in roots were significantly higher under N deficiency than those under low nitrogen and control treatment (Figure 3H). With the reduction in N supply, the contents of leaf TFAA (Figure 3D), leaf-soluble protein (Figure 3E), and root TFAA (Figure 3I) showed a significant downward trend, while the content of root-soluble protein had no significant difference among different N treatments (Figure 3J). All the metabolites measured above were notably lower in the root than in the shoot (Figure 3). Furthermore, it can be seen from the above results that reducing the N supply increased the contents of malate, citrate, and isocitrate in rice leaves and roots, while it decreased the contents of TFAA and leaf-soluble protein in rice, which indicated that the change trend of organic acids and TFAA content in response to reduced N availability was similar between rice leaves and roots.

2.3. Effects of Different NTreatments on the Activities of Enzymes Related to Organic Acid Metabolism in Rice Leaves and Roots

Compared to the control, the activities of NAD-MDH, NADP-ME, NADP-IDH, ACO, and CS in rice leaves were significantly decreased by both low-N and N-deficiency treatments (Figure 4A–C,E,F), but their activities were not significantly different between low-N and N-deficiency treatments. With the decrease in nutrient liquid nitrogen concentration, the PEPC enzyme activity of rice leaves showed a significant upward trend (Figure 4D). In rice roots, compared to the control, low-N and N-deficiency treatments significantly increased the enzyme activities of NAD-MDH, PEPC, and CS in the root system (Figure 4G,J,L), and significantly decreased the enzyme activities of NADP-ME and NADP-IDH in rice roots (Figure 4H,I). At the same time, there was no significant difference in NAD-MDH, PEPC, CS, NADP-ME, and NADP-IDH activities between low-N and N-deficiency treatments in rice roots (Figure 4G–J,L). With the decreased N supply, root ACO showed a significant upward trend (Figure 4K). Interestingly, all the activities of enzymes related to organic acids metabolism measured above were notably lower in the root than in the shoot (Figure 4).

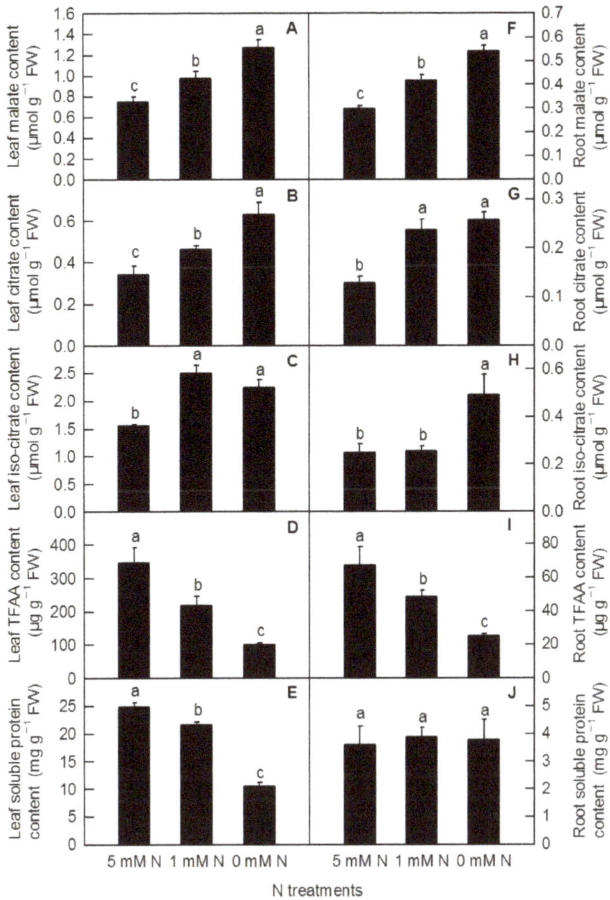

Figure 3. Effects of N deficiency on the contents of organic acids, TFAA, and soluble proteins in *O. sativa* leaves (**A–E**) and roots (**F–J**). Bars represent means ± SD (n = 3). Difference among the treatments was analyzed by Duncan's multiple range test. Different letters above the bar indicate a significant difference at $p < 0.05$.

2.4. Effects of Different N Treatments on the Activities of Enzymes Related to N Metabolism in Rice Leaves and Roots

Compared to the control, N-deficiency and low-N treatment significantly reduced the activities of leaf NR, leaf GS, leaf GOGAT, root NR, root GS, and root GOGAT inrice seedlings (Figure 5). In addition to no significant change in enzyme activity of leaf GS between low-N and N-deficiency treatments, low-N treatment significantly decreased the activities of the above several enzymes related to nitrogen metabolism (Figure 5B). Similar to the parameters measured above, all the activities of enzymes related to N metabolism measured above were notably lower in the root than in the shoot (Figure 5).

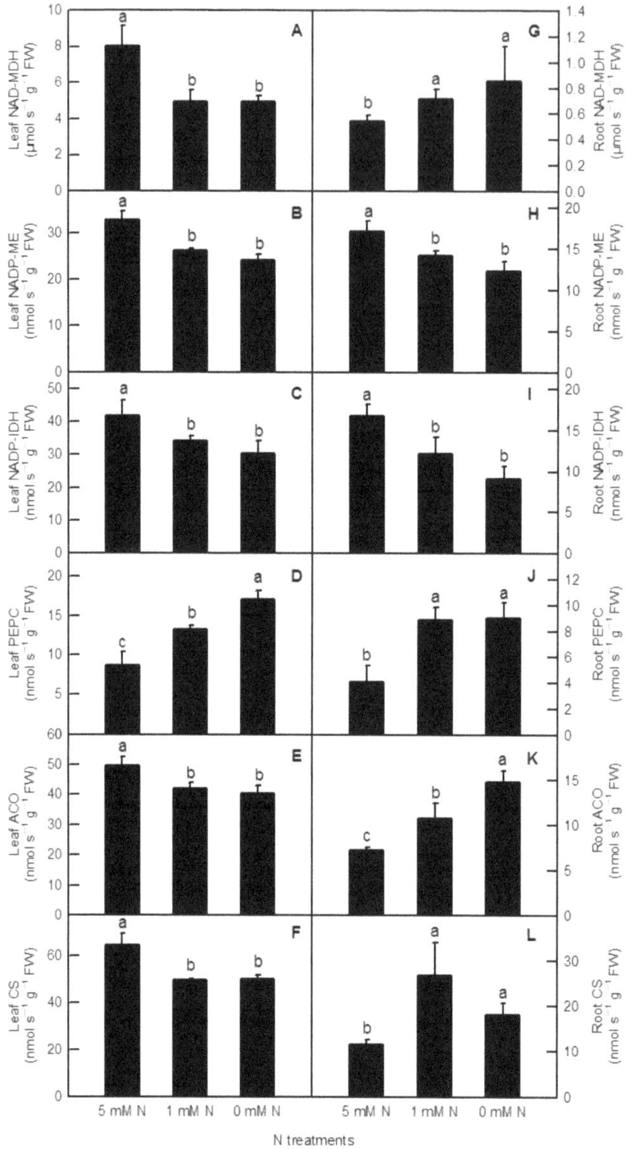

Figure 4. Effects of N deficiency on the activities of enzymes related to organic acid metabolism in *O. sativa* leaves (**A–F**) and roots (**G–L**). Bars represent means ± SD (n = 3). Difference among the treatments was analyzed by Duncan's multiple range test. Different letters above the bar indicate a significant difference at $p < 0.05$.

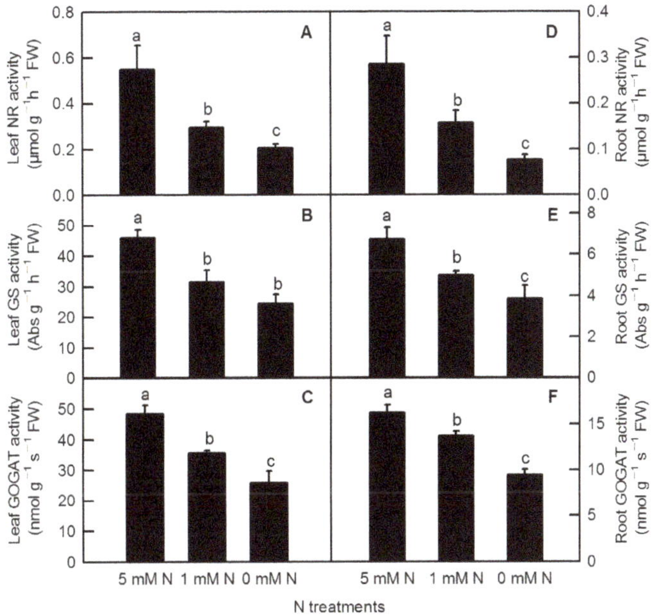

Figure 5. Effects of N deficiency on the activities of enzymes related to N metabolism in *O. sativa* leaves(**A–C**) and roots (**D–F**). Bars represent means ± SD (n = 3). Difference among the treatments was analyzed by Duncan's multiple range test. Different letters above the bar indicate a significant difference at $p < 0.05$.

2.5. CorrelationAnalysis and Principal Component Analysis (PCA) Loading Plots

The contents of leaf organic acids (malate, citrate, and isocitrate) were negatively correlated with all the parameters measured in rice leaf, whereas they were positively correlated with leaf PEPC (Figure 6A). Except for leaf PEPC, the activities of leaf-organic-acids-metabolism-related enzymes were all positively correlated with leaf-N-metabolism-related parameters (leaf N, leaf-soluble protein, leaf TFAA, leaf GOGAT, leaf GS, and leaf NR) (Figure 6A). In contrast, root-N-metabolism-related parameters, such as root GOGAT, root GS, root NR, and root TFAA, were positively correlated with enzymes related to root organic acids catabolism, such as root NADP-ME and NADP-IDH, whereas they were negatively correlated with enzymes related to root organic acids biosynthesis (such as root PEPC and NADP-MDH) (Figure 6B). Interestingly, root-soluble protein had no obvious correlation with other parameters measured in this study in rice root (Figure 6B).

The principal component analysis (PCA) loading plot generated by Sigmaplot 10.0 visualized two loadings against each other to investigate the relationships between the variables. There were three or four replicates for each treatment in PCA. Thirty-two parameters including plant weight, shoot height, tissue N contents, soluble proteins, TFAAs, organic acids, and the activities of their metabolism-related-enzymes, from *O. sativa* leaves and roots, were transformed for PCA analysis (Figure 7). The first two PCs explained 89.28% of the physiological variation in response to different N-supplying levels with PC1 being accounted for by71.56% and PC2 by 17.72%. The factor loadings are listed in Table S1. The PCA result clearly showed that tissue N contents, plant weight, and shoot height were clustered tightly. Parameters related to leaf organic acids metabolism, leaf N metabolism, and root N metabolism were clustered together, whereas parameters related to root organic acids metabolism closely clustered with leaf organic acids metabolism. Furthermore, the interconnection between leaf organic acids metabolism and leaf N metabolism was closer than that between root organic acids metabolism and root N metabolism (Figure 7).

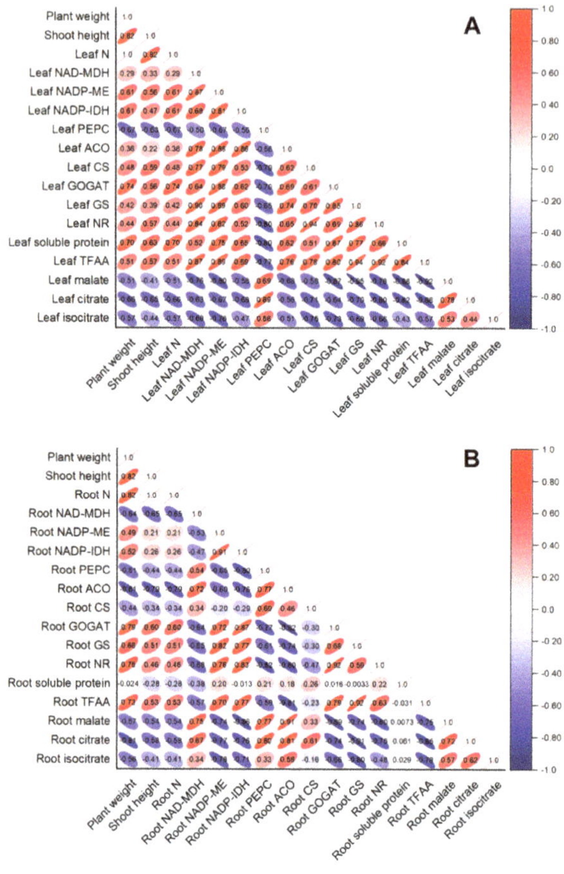

Figure 6. Correlation analysis of variables in rice leaves (**A**) and root (**B**) in response to different N treatments. The direction and color of the ellipse represent the positive or negative relationship between two variables. The number in the ellipse indicates the correlation coefficient between two variables.

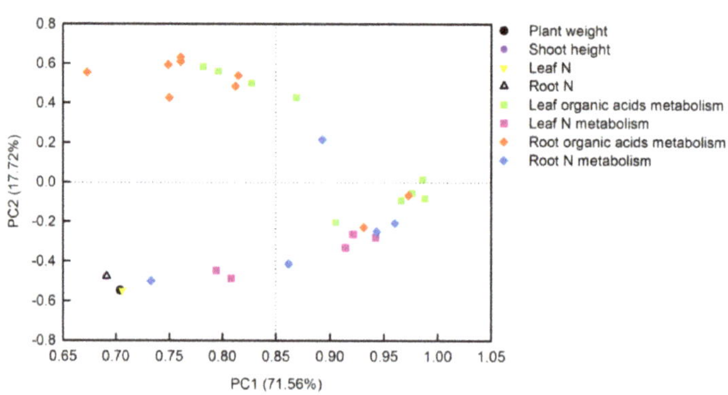

Figure 7. Principal component analysis (PCA) loading plot of the parameters in *O. sativa* leaves sand roots under different N levels. Thirty-two parameters from *O. sativa* leaves and roots were transformed for PCA analysis. The first two PCs explained 89.28% of the physiological variation in response to different N-supplying levels.

3. Discussion

Nfertilizer is an important factor for increasing yield in agricultural production. Both the previous literature and results of the current study showed that N deficiency could lead to the degradation of photosynthetic pigments in rice leaves and the yellowing of leaves, which eventually led to the reduction in photosynthetic rate and affected plant growth (Figures 1 and 2) [2]. The phenomenon that N deficiency induceda reduction in photosynthetic rate and, thus, affected plant growth has been reported in many crops, such as tea, citrus, maize, cucumber, sunflower, tobacco, and barley [2,8–11]. According to the statistics of the Food and Agriculture Organization of the United Nations (FAO), the amount of agricultural nitrogen fertilizer has increased several times in recent decades (FAOSTAT) [29]. With the prominent increase in global population and the decline in arable land, the demand for fertilizer will increase [30]. However, excessive application of fertilizer will lead to eutrophication of the atmosphere and water system, and cause severe environmental problems. Therefore, it is of great significance to evaluate the physiological and biochemical reactions of crops to the nutrient in fertilizers to guide the rational application of fertilizers.

It was found on the model plant *Arabidopsis thaliana* that C/N interaction could significantly affect the differential expression of more than 300 genes ($p < 0.05$) [5]. Among them, the organic acid metabolic enzyme gene *PEPC* was considered to play an important role in balancing C and N metabolism [17]. In the study of transcriptomic analysis inrice under N-deficiency, Shao et al. reported that N deficiency upregulated the expression of the PEPC gene in rice leaves and increased the content of malate in rice leaves [2]. In this study, we found that with the decrease inN level, the contents of malate and citrate increased significantly in rice leaves and roots (Figure 3A,B,F). Compared to the control, both low N and N deficiency significantly increased the content of isocitrate in rice leaves and roots, while the content of root isocitrate was significantly higher under N-deficiency treatment than that under low-N and the control treatment (Figure 3C,G,H). Through the determination of the activities of organic-acid-metabolism-related enzymes, it was found that the increase in malate in rice leaves caused by the reduction inN supply was related to the increased biosynthesis of malate and the decreased degradation of malate, which was consistent with the increased PEPC activity and the decreased NADP-ME activity in rice leaves (Figures 4B,D and 6A). Organic acids are synthesized by photosynthetic products through glycolysis and the TCA cycle, while 2-ketoglutarategenerated by citrate degradation in the cytoplasm is mainly used for NH_4^+assimilation in leaf cells [31,32]. Masumoto et al. reported that *Osppc4* was obtained by the knockdown of a leaf PEPC gene in rice and their comparative analysis of the leaf metabolome showed that *Osppc4* knockdown greatly reduced the content of organic acids, especially malate, thus inhibiting NH_4^+assimilation and subsequent amino acids biosynthesis in the GS/GOGAT cycle, resulting in the slowdown of plant growth [18]. In the current study, the increased contents of citrate and isocitrate in rice leaves caused by N reduction might not be caused by the increased activity of biosynthesis, but due to the reduced degradation of citrate and isocitrate in the TCA cycle, because, compared to the control, the enzyme activities of CS, ACO, and NADP-IDH in rice leaves under nitrogen reduction treatment all showed a downward trend (Figures 4C,E,F and 6A). Similarly, it was also reported that N deficiency led to an increase in the contents of organic acids in potato shoot [33]. As the enzyme activities of NAD-MDH and PEPC increased in rice roots under N deficiency conditions, while the activities of NADP-ME enzymes decreased, the increased content of malate in rice roots caused by low-Nand N-deficiency treatments might be related to the increase in synthesis and reduction in degradation of malate (Figures 3F, 4J,H and 6B). Compared to the control, the increased content of root citrate under low N and N deficiency might be related to the increased biosynthesis of citrate rather than the decreased degradation ofcitrate, because the activities of root CS and ACO were higher in low-N and N-deficiency treatments than that of the control ones (Figures 3G and 4K,L). At the same time, the root isocitrate content and root ACO activity were higher under N-deficiency treatment than those under control and low-N treatment ones, while theactivity of root NADP-IDH under N-deficiency and low-N treatment was higher than that of the control ones, which indicated that the increased isocitrate content in roots was related to the increase in citrate isomerization activity and the decrease in isocitrate degradation (Figures 3H, 4I,K and 6B). It can be seen from the above results that although the change mode of organic acids in rice leaves and roots under reducing N level was basically the same, the change pattern of organic-acids-metabolism-related enzymes caused by reducing N level was different between rice leaves and roots. Our correlation analysis and PCA result showed that parameters related to leaf organic acids metabolism and root organic acids metabolism were tightly clustered together, indicating that organic acids metabolism in the leaf and root were closely interconnected (Figures 6 and 7).

Under N-deficient conditions, the content of ammonium used for amino acids biosynthesis in plants decreased. Therefore, the condensation of the amino and transamination process will decrease, which will inevitably lead to a change in the content of amino acids and proteins in plants [10,25]. In the current study, except for the soluble protein in rice roots, the contents of leaf TFAA, leaf-soluble protein, and root TFAA were decreased by the decreasing N level (Figure 3D,E,I,J). This is consistent with the results obtained in *Citrus* [10,25], tomato [34], tea [9], *Japonica* rice [4], and hybrid rice [35]. NR, GS, and GOGAT are three key enzymes for N metabolism in plants [36]. After nitrate is absorbed by roots, it is first reduced to nitrite by NR in the cytoplasm, and then reduced to NH_4^+ by nitrite reductase (NiR) in the plastid. Through the GS/GOGAT cycle, NH_4^+ produced from nitrate reduction and ammonium absorbed by rice ammonium transporters *Os*AMTs are assimilated into amino acids [27]. With the decreasing N level in nutrient solution, the activities of N-metabolism-related enzymes such as NR, GS, and GOGAT in rice leaves and roots decreased (Figure 5). Xiong et al. showed N deficiency at the tillering stage and N compensation at the spike differentiation stage, and N deficiency treatment could reduce the enzyme activities of NR and GS in the double cropping hybrid rice, while the normal N application and double N compensation groups increased the activities of NR and GS to a certain extent, thereby improving the photosynthetic performance of hybrid rice leaves and rice yield [35]. N deficiency treatment reduced the content of total N, ammonium, glutamine, and TFAAs in potato roots, and reduced the expression of N-metabolism-related genes such as *AMT* (ammonium transporter gene) and *GS* [34]. Similarly, Chen et al. and Huang et al. found that with the reduced N-supplying level in nutrient solution, the activities of NR, GS, GOGAT, and other enzymes related to N metabolism in *Citrus* leaves and roots decreased, resulting in a reduction in total free amino acid and soluble protein content in seedlings, and ultimately resulting in the inhibition of plant growth [10,25]. Interestingly, our PCA result demonstrated that leaf organic acids metabolism had a closer relationship with leaf N metabolism than those between root organic acids metabolism and root N metabolism (Figure 7; Table S1).

4. Materials and Methods

4.1. Plant Material and Treatment

The plant material used in this experiment was rice (*Oryza sativa* L. ssp. Indica) variety "Huanghuazhan" selected by the Rice Research Institute of Guangdong Academy of Agricultural Sciences. Uniform seeds of rice seeds were sowed in the plastic tray containing paddy soil with the content of alkaline hydrolysis N of 143 ± 6.98 mg/kg. When the height of the aboveground part of the rice seedlings was about 11.18 cm, the rice seedlings were transplanted to the Hogland nutrient solution containing 5 mM NH_4NO_3 (control), 1 mM NH_4NO_3 (low N), and 0 mM NH_4NO_3 (N deficiency), and cultured in a light incubator (28 °C). Seedlings were kept under a 14 h light/10 h dark regime with white photo-illumination of 150 μmol m^{-2} s^{-1} as well as a relative humidity of 68%. The nutrient solution was changed every two days. The formula of Hoagland nutrient solution contained the following macroelements (in mM): NH_4NO_3, 5 mM; KH_2PO_4, 1 mM; $MgSO_4$, 2 mM; microelements (in μM):H_3BO_3,5 mM; $MnCl_2$, 2 μM; $ZnSO_4$, 2; $CuSO_4$, 0.5 μM;

$(NH_4)_6Mo_7O_{24}$, 0.065 µM; $FeSO_4$-EDTA, 20 µM. In order to better promote the growth of rice seedlings, a concentration of 0.1 µM Na_2SiO_3 was added to each nutrient solution. After 15 days of culture, leaf and root samples were collected to determine the fresh weight and height of shoots after the phenotype of seedlings were recorded (Figure 1). There were four replicates for fresh weight and shoot height. At the same time, the rice plants were divided into upper and lower parts with scissors and put into aluminum foil bags. After freezing in liquid nitrogen, allthe samples were stored at −80 °C until assayed.

4.2. Determination of N Content in Rice Leaves and Roots

The nitrogen content was determined by the Kjeldahl method [2]. The ground samples of rice leaves and roots were digested with concentrated H_2SO_4-H_2O_2, and then the N contents in rice leaves and roots were analyzed using a Foss Kjeltec 8200 nitrogen analyzer (Hilleroed, Denmark). There were four replicates for the measurement of leaf N and root N.

4.3. Determination of Organic Acids, Total Free Amino Acids (TFAA),and Total Soluble Proteins in Rice Leaves and Roots

The extraction and determination of organic acids in rice leaves and roots were carried out according to the method described by Lu et al. [37]. Organic acids were extracted by 4% perchloric acid and centrifuged to determine the contents of malate, citrate, and isocitrate. The reaction mixture of malate contained 50 mM 3-amino-1-propanol-HCl (pH = 10), 30 mM sodium glutamate-NaOH (pH = 10), 2.7 mM NAD, 1 Uglutamate-oxaloacetate transaminase (GOT, EC 2.6.1.1), 10 U NAD-malate dehydrogenase (NAD-MDH, EC 1.1.37), and 50 µL of extracted supernatant. The reaction system of citrate contained 100 mM Tris-HCl (pH = 7.6), 0.2 mM NADH, 7 U lactate dehydrogenase (LDH, EC 1.1.1.27), 14 U NAD-MDH, 0.5 U citrate lyase (EC 4.1.3.6), and an appropriate amount of extracted supernatant. The reaction system of isocitrate contained 100 mMTris-HCl (pH = 7.6), 3.3 mM $MnSO_4$, 0.2 mM NADP, 0.1 U NADP isocitrate dehydrogenase (NADP-IDH, EC 1.1.1.42), and an appropriate amount of extracted supernatant. There were three biological replicates for each treatment.

TFAAs were determined according to the method of Li et al. [38]. The total free amino acids were determined by the ninhydrin method. About 0.1 g of root and leaf samples were ground and extracted with 1.6 mL of 10% acetic acid. After centrifugation at 12,000× *g* for 10 min at 4 °C, 0.9 mL of ultrapure water, 1.5 mL of hydrated ninhydrin solution, and 0.25 mL of 1% ascorbic acid were added to a 5 mL reaction tube containing 0.1 mL of supernatant, mixed well, and reacted in a boiling water bath for 15 min. After reaction, 1.25 mL of 60% ethanol was added to the mixture, and after mixing, the absorbance value at 570 nm wavelength was measured with an Ultraviolet/Visible spectrophotometer. Leucine was used to make the standard curve and the concentration of leucine for the standard curve ranged from 0 µg/mL to 2.5 µg/mL. Each treatment was performed for three biological replicates. The total soluble protein content was determined by the Coomassie brilliant blue method described by Bradford [39]. Rice samples (about 0.1 g) were extracted by grinding with 1.6 mL of 50 mM Na_2HPO_4-KH_2PO_4 (pH = 7.0) and centrifuged at 3000× *g* for 10 min at 4 °C. Twenty-five microliters of supernatant and 975 µLof Coomassie brilliant blue solution were mixed and allowed to react at 25 °C for 5 minutes, and then the absorbance at 595 nm was determined. At the same time, bovine serum albumin (BSA) was used to prepare standard curves under the same conditions, and each treatment had three biological replicates.

4.4. Determination of Enzymes Related to Organic Acid Metabolism and N Metabolism in Rice Leaves and Roots

The extraction and activity measurement of organic-acid-metabolism-related enzymes in rice leaves and roots were carried out according to the method described by Lu et al. [37]. The extraction solution contained 50 mM HEPES KOH (pH = 7.5), 10 mM $MgCl_2$, 2 mM EDTA-Na_2, 10 mM dithiothreitol (DTT), 1% (*v/v*) Triton X-100, 5% water-insoluble PVPP, 1% (*w/v*) BSA, and 30% glycerol. About 0.1 g of leaf or root samples was ground with

1.6 mL of extraction solution. The tissue homogenate (about 1.7 mL) was centrifuged at 13,000× *g* for 5 min, and the supernatant was used for enzyme activity measurement.

The assay of the enzyme related to N metabolism in rice was carried out according to the method described by Hageman et al. [40]. The extraction solution of NR contained 10 mM cysteine, 1 mM EDTA-Na_2, 5% PVPP, and 25 mM phosphate buffer (pH = 8.7). Approximately 0.1 g of rice leaf or root samples was added to pre-cooled mortar containing a small amount of quartz sand and the above phosphate buffer for grinding and extraction. The homogenate was centrifuged at 4000× *g* for 15 min at 4 °C. The reaction mixture for NR activity contained 0.2 mL of extract supernatant, 100 mMKNO_3, and 2 mMNADH. The reaction mixture was incubated in a 30 °C water bath for 30 min. At the end of the reaction, 1% (*w/v*) sulfonamide was added immediately to terminate the reaction, and then 0.02% (*w/v*) naphthylamine solution was added to the mixture for reaction at 30 °C for 15 min. The total volume of the reaction mixture was 2 mL. Finally, the absorbance of the mixed solution was measured at 540 nm, and a blank control was set for each sample. $NaNO_2$ was used to make the standard curve, and the concentrations for the curve ranged from 0.01 μg/mL to 0.1 μg/mL of NO_2^-.

GS and GOGAT extraction solution contained 100 mM Tris HCl (pH = 7.5), 1 mM $MgCl_2$, 1 mM EDTA-Na_2, 1mM β-mercaptoethanol, 0.3% Triton X-100, and 5% PVPP. After grinding and centrifugation at 16,000× *g* 4 °C for 30 min, the supernatant was taken for enzyme determination. The GS activity measurement mixture contained a mixture of 80 mM Tris HCl, 40 mM $MgCl_2$, 40 mM sodium glutamate, 16 mM hydroxylamine hydrochloride, and 8 mM ATP. The mixed solution was measured to react for 30 minutes at 30 °C in a water bath. After the reaction, a chromogenic agent containing 2% TCA solution (*w/v*), 3.5% $FeCl_3$ (*w/v*), and 1.8% HCl (*v/v*) was added to the mixture. After the reaction for 5 min, the reaction mixture was centrifuged at 10,000× *g* for 5 min, and the supernatant was taken to measure its absorbance at 540 nm. A blank control was set for each sample. The reaction system for GOGAT determination contained 100 mM2-ketoglutarate, 10 mM KCl, 3 mM NADH, and 25 mM Tris HCl (pH = 7.6), and mixed well. Finally, 20 mM glutamate was added to start the reaction, and the enzymatic kinetic reaction was measured at 340 nm for 60 s.

4.5. Experimental Design and Statistic Analysis

In this experiment, three culture pots were set for each N treatment, and each culture pot contained 18 rice seedlings. Duncan's multiple range test was used for variance analysis between different treatments. Correlation plots were generated by Origin software (Version: Pro 2022b SR1). The experimental data and PCA were processed by Microsoft Excel 2010, analyzed by SPSS 19.0, and visualized by Sigmaplot 10.0 [41].

5. Conclusions

In this study, we found that low N and N deficiency reduced rice biomass, plant height, and N content, increased the content of main organic acids, and decreased the total free amino acids and leaf-soluble proteins in *O. sativa*. Although the dynamics of the contents of organic acids in rice leaves and roots were basically the same under low N and N deficiency, the dynamics of the activities of organic-acid-metabolism-related enzymes caused by reducing N-supplying level were different in rice leaves and roots.

With the decreased N-supplying level, the activities of N-metabolism-related enzymes, such as NR, GS, and GOGAT, decreased in rice leaves and roots, resulting in the decreased contents of TFAAs and total soluble protein, and finally resulting in the inhibition of plant growth. The signal of N deficiency might not directly act on the TCA cycle, but on the organic acid biosynthesis step in the cytoplasm or the upstream transcriptional regulation level.

The dynamics of organic acids metabolism caused by N deficiency were different in rice leaves and roots. Interestingly, our PCA result demonstrated that leaf organic acids metabolism had a closer relationship with leaf N metabolism than that between root organic acids metabolism and root N metabolism. In the future study, more effort should be focused on the specific molecular and physiological mechanism underlying how N deficiency regulates organic acids metabolism in rice plants.

Supplementary Materials: The following supporting information can be downloaded at: https://www.mdpi.com/article/10.3390/plants11192576/s1, Table S1: Factor loadings of variables of first two principal components.

Author Contributions: Conceptualization, L.-H.C. and L.-T.Y.; software, Z.-X.C.; methodology, M.X.; investigation, L.-H.C. and Z.-J.Y.; writing—original draft, L.-H.C.; writing—review and editing, L.-T.Y.; funding acquisition, L.-H.C. All authors have read and agreed to the published version of the manuscript.

Funding: This research was funded by the Leading Project of Department of Science and Technology of Fujian Province, China (2019N0004), Scientific Research Fund for Young Teachers of Jinshan College of Fujian Agriculture and Forestry University, China (kx211006), Collaborative Education Project of Industry-University Cooperation of Ministry of Education, China (202102549001), and National New Agricultural Science Research and Reform Practice Project of China (jx211001).

Institutional Review Board Statement: Not applicable.

Informed Consent Statement: Not applicable.

Data Availability Statement: Not applicable.

Conflicts of Interest: The authors declare no conflict of interest. The funders had no role in the design of the study; in the collection, analyses, or interpretation of data; in the writing of the manuscript, or in the decision to publish the results.

References

1. Sun, H.; Qian, Q.; Wu, K.; Luo, J.; Wang, S.; Zhang, C.; Ma, Y.; Liu, Q.; Huang, X.; Yuan, Q.; et al. Heterotrimeric G proteins regulate nitrogen-use efficiency in rice. *Nat. Genet* **2014**, *46*, 652. [CrossRef]
2. Shao, C.H.; Qiu, C.F.; Qian, Y.F.; Liu, G.R. Nitrate deficiency decreased photosynthesis and oxidation-reduction processes, but increased cellular transport, lignin biosynthesis and flavonoid metabolism revealed by RNA-Seq in *Oryza sativa* leaves. *PLoS ONE* **2020**, *15*, e0235975. [CrossRef]
3. Drincovich, M.F.; Voll, L.M.; Maurino, V.G. Editorial: On the diversity of roles of organic acids. *Front. Plant Sci.* **2016**, *7*, 1592. [CrossRef]
4. Hsieh, P.-H.; Kan, C.-C.; Wu, H.-Y.; Yang, H.-C.; Hsieh, M.-H. Early molecular events associated with nitrogen deficiency in rice seedling roots. *Sci. Rep.* **2018**, *8*, 12207. [CrossRef]
5. Palenchar, P.M.; Kouranov, A.; Lejay, L.V.; Coruzzi, G.M. Genome-wide patterns of carbon and nitrogen regulation of gene expression validate the combined carbon and nitrogen (CN)-signaling hypothesis in plants. *Genome Biol.* **2004**, *5*, R91. [CrossRef]
6. Yanagisawa, S.; Akiyama, A.; Kisaka, H.; Uchimiya, H.; Miwa, T. Metabolic engineering with Dof1 transcription factor in plants: Improved nitrogen assimilation and growth under low-nitrogen conditions. *Proc. Natl. Acad. Sci. USA* **2004**, *101*, 7833–7838. [CrossRef] [PubMed]
7. Ye, T.; Li, Y.; Zhang, J.; Hou, W.; Zhou, W.; Lu, J.; Xing, Y.; Li, X. Nitrogen, phosphorus, and potassium fertilization affects the flowering time of rice (*Oryza sativa* L.). *Glob. Ecol. Conserv.* **2019**, *20*, e00753. [CrossRef]
8. Agüera, E.; Cabello, P.; DeLaHaba, P. Induction of leaf senescence by low nitrogen nutrition in sunflower (*Helianthus annuus*) plants. *Physiol. Plantarum.* **2010**, *138*, 256–267. [CrossRef] [PubMed]
9. Lin, Z.H.; Chen, C.S.; Zhong, Q.S.; Ruan, Q.C.; Chen, Z.H.; You, X.M.; Shan, R.Y.; Li, X.L. The GC-TOF/MS-based metabolomic-analysis reveals altered metabolic profiles in nitrogen-deficient leaves and roots of tea plants (*Camellia sinensis*). *BMC Plant Biol.* **2021**, *21*, 506. [CrossRef] [PubMed]
10. Chen, H.; Jia, Y.; Xu, H.; Wang, Y.; Zhou, Y.; Huang, Z.; Yang, L.; Li, Y.; Chen, L.S.; Guo, J. Ammonium nutrition inhibits plant growth and nitrogen uptake in citrus seedlings. *Sci. Hortic-Amst.* **2020**, *272*, 109526. [CrossRef]
11. DeBona, F.D.; Fedoseyenko, D.; vonWirén, N.; Monteiro, F.A. Nitroge nutilization by sulfur-deficient barley plants depends on the nitrogen form. *Environ. Exp. Bot.* **2011**, *74*, 237–244. [CrossRef]
12. Wang, R.; Okamoto, M.; Xing, X.; Crawford, N.M. Microarray analysis of the nitrate response in *Arabidopsis* roots and shoots reveals over 1,000 rapidly responding genes and new linkages to glucose, trehalose-6-phosphate, iron, and sulfate metabolism. *Plant Physiol.* **2003**, *132*, 556–567. [CrossRef] [PubMed]
13. Wickert, E.; Marcondes, J.; Lemos, M.V.; Lemos, E.G.M. Nitrogen assimilation in *Citrus* based on CitEST data mining. *Genet. Mol. Biol.* **2007**, *30*, 810–818. [CrossRef]

14. Wang, Y.; Zhang, P.; Li, M.; Guo, Z.; Ullah, S.; Rui, Y.; Lynch, I. Alleviation of nitrogen stress in rice (*Oryza sativa*) by ceria nanoparticles. *Environ. Sci. Nano* **2020**, *7*, 2930–2940. [CrossRef]
15. Tschoep, H.; Gibon, Y.; Carillo, P.; Armengaud, P.; Szecowka, M.; Nunes-Nesi, A.; Fernie, A.R.; Koehl, K.; Stitt, M. Adjustment of growth and central metabolism to a mild but sustained nitrogen-limitation in *Arabidopsis*. *Plant Cell Environ.* **2009**, *32*, 300–318. [CrossRef]
16. Le, X.H.; Lee, C.-P.; Millar, A.H. The mitochondrial pyruvate carrier (MPC) complex mediates one of three pyruvate-supplying pathways that sustain *Arabidopsis* respiratory metabolism. *Plant Cell* **2021**, *33*, 2776–2793. [CrossRef] [PubMed]
17. Shi, J.H.; Yi, K.K.; Liu, Y.; Xie, L.; Zhou, Z.J.; Chen, Y.; Hu, Z.H.; Zheng, T.; Liu, R.H.; Chen, Y.L.; et al. Phosphoenolpyruvatecarboxylase in *Arabidopsis* leaves plays a crucial role in carbon and nitrogen metabolism. *Plant Physiol.* **2015**, *167*, 671–681. [CrossRef]
18. Masumoto, C.; Miyazawa, S.-I.; Ohkawa, H.; Fukuda, T.; Taniguchi, Y.; Murayama, S.; Kusano, M.; Saito, K.; Fukayama, H.; Miyao, M. Phosphoenolpyruvatecarboxylase intrinsically located in the chloroplast of rice plays a crucial role in ammonium assimilation. *Proc. Natl. Acad. Sci. USA* **2010**, *107*, 5226–5231. [CrossRef]
19. Chen, F.X.; Liu, X.H.; Chen, L.S. Developmental changes in pulp organic acid concentration and activities of acid-metabolising enzymes during the fruit development of two loquat (*Eriobotrya japonica* Lindl.) cultivars differing in fruit acidity. *Food Chem.* **2009**, *114*, 657–664. [CrossRef]
20. Zhang, X.M.; Du, L.Q.; Sun, G.M.; Gong, D.Q.; Chen, J.Y.; Li, W.C.; Xie, J.H. Changes in organic acid concentrations and the relative enzyme activities during the development of Cayenne pineapple fruit. *J. Fruit Sci.* **2007**, *3*, 381–384.
21. Yao, Y.X.; Li, M.; Liu, Z.; You, C.X.; Wang, D.M.; Zhai, H.; Hao, Y.J. Molecular cloning of three malic acid related genes *MdPEPC*, *MdVHA-A*, *MdcyME* and their expression analysis in apple fruits. *Sci. Hortic-Amst.* **2009**, *122*, 404–408. [CrossRef]
22. Famiani, F.; Battistelli, A.; Moscatello, S.; Cruz-Castillo, J.G.; Walker, R.P. The organic acids that are accumulated in the flesh of fruits: Occurrence, metabolism and factors affecting their contents—Areview. *Rev. Chapingo Ser. Hortic.* **2015**, *21*, 97–128. [CrossRef]
23. Degu, A.; Hatew, B.; Nunes-Nesi, A.; Shlizerman, L.; Zur, N.; Katz, E.; Fernie, A.R.; Blumwald, E.; Sadka, A. Inhibition of aconitase in citrus fruit callus results in a metabolic shift towards amino acid biosynthesis. *Planta* **2011**, *234*, 501–513. [CrossRef]
24. Sadka, A.; Dahan, E.; Or, E.; Cohen, L. $NADP^+$-isocitrate dehydrogenase gene expression and isozyme activity during citrus fruit development. *Plant Sci.* **2000**, *158*, 173–181. [CrossRef]
25. Huang, W.T.; Zheng, Z.C.; Hua, D.; Chen, X.F.; Zhang, J.; Chen, H.H.; Ye, X.; Guo, J.X.; Yang, L.T.; Chen, L.S. Adaptive responses of carbon and nitrogen metabolisms to nitrogen-deficiency in *Citrus sinensis* seedlings. *BMC Plant Biol.* **2022**, *22*, 370. [CrossRef] [PubMed]
26. Bao, A.; Liang, Z.; Zhao, Z.; Cai, H. Overexpressing of *OsAMT1-3*, a high affinity ammonium transporter gene, modifies rice growth and carbon–nitrogen metabolic status. *Int. J. Mol. Sci.* **2015**, *16*, 9037–9063. [CrossRef]
27. Lee, S. Recent advances on nitrogen use efficiency in rice. *Agronomy* **2021**, *11*, 753. [CrossRef]
28. Xiao, M.; Zang, H.; Ge, T.; Chen, A.; Zhu, Z.; Zhou, P.; Atere, C.T.; Wu, J.; Su, Y.; Kuzyakov, Y. Effect of nitrogen fertilizer on rice photosynthate allocation and carbon input in paddy soil. *Eur. J. Soil Sci.* **2019**, *70*, 786–795. [CrossRef]
29. FAOSTAT. The Fertilizers by Nutrient Dataset Contains Information on the Totals in Nutrients for Production, Trade and Agriculture Use of inorganic (Chemical or Mineral) Fertilizers, over the Time Series 1961-Present. Available online: http://www.fao.org/faostat/zh/#data/RFN (accessed on 1 September 2022).
30. Singh, B. Are nitrogen fertilizers deleterious to soil health? *Agronomy* **2018**, *8*, 48. [CrossRef]
31. He, L.; Jing, Y.; Shen, J.; Li, X.; Liu, H.; Geng, Z.; Wang, M.; Li, Y.; Chen, D.; Gao, J.; et al. Mitochondrial pyruvate carriers prevent cadmium toxicity by sustaining the TCA cycle and glutathione synthesis. *Plant Physiol.* **2019**, *180*, 198–211. [CrossRef] [PubMed]
32. Yuan, Y.; Ou, J.; Wang, Z.; Zhang, C.; Zhou, Z.; Lin, Q. Regulation of carbon and nitrogen metabolisms in rice roots by 2-oxoglutarate at the level of hexokinase. *Physiol. Plantarum.* **2007**, *129*, 296–306. [CrossRef]
33. Sung, J.; Sonn, Y.; Lee, Y.; Kang, S.; Ha, S.; Krishnan, H.B.; Oh, T.-K. Compositional changes of selected amino acids, organic acids, and soluble sugars in the xylem sap of N, P, or K-deficient tomato plants. *J. Plant Nutr. Soil Sci.* **2015**, *178*, 792–797. [CrossRef]
34. Wiren, N.V.; Lauter, F.R.; Ninnemann, O.; Gillissen, B.; Walch-Liu, P. Differential regulation of three functional ammonium transporter genes by nitrogen in root hair sand by light in leaves of tomato. *Plant J.* **2010**, *21*, 167–175. [CrossRef]
35. Xiong, Q.; Tang, G.; Zhong, L.; He, H.; Chen, X. Response to nitrogen deficiency and compensation on physiological characteristics, yield formation, and nitrogen utilization of rice. *Front. Plant Sci.* **2018**, *9*, 1075. [CrossRef]
36. Wang, Y.Y.; Cheng, Y.H.; Chen, K.E.; Tsay, Y.F. Nitrate transport, signaling, and use efficiency. *An. Rev. Plant Biol.* **2018**, *69*, 1–38. [CrossRef]
37. Lu, Y.B.; Yang, L.T.; Li, Y.; Xu, J.; Liao, T.T.; Chen, Y.B.; Chen, L.S. Effects of boron deficiency on major metabolites, key enzymes and gas exchange in leaves and roots of *Citrus sinensis* seedlings. *Tree Physiol.* **2014**, *34*, 608–618. [CrossRef]
38. Li, C.P.; Qi, Y.P.; Zhang, J.; Yang, L.T.; Wang, D.H.; Ye, X.; Lai, N.W.; Tan, L.L.; Lin, D.; Chen, L.S. Magnesium-deficiency-induced alterations of gas exchange, major metabolites and key enzymes differ among roots, and lower and upper leaves of *Citrus sinensis* seedlings. *Tree Physiol.* **2017**, *37*, 1564–1581. [CrossRef] [PubMed]
39. Bradford, M.M. A rapid and sensitive method for quantitation of microgram quantities of protein utilizing the principle of protein-dye binding. *Anal Biochem.* **1976**, *72*, 248–254.

40. Hageman, R.H.; Reed, A.J.; Femmer, R.A.; Sherrard, J.H.; Dalling, M.J.; Yoder, O.C.; Ferrari, T.F.; Filner, P. Some new aspect sof the in vivo assay for nitrate reductase in wheat (*Triticum aestivum* L.) leaves. 1: Re-evaluation of nitrate pool sizes. *Plant Physiol.* **1980**, *65*, 27–32. [CrossRef] [PubMed]
41. Yang, L.T.; Liu, J.W.; Wu, Y.M.; Qi, Y.P.; Wang, J.L.; Lai, N.W.; Ye, X.; Chen, L.S. Proteome profile analysis of boron-induced alleviation of aluminum-toxicity in *Citrus grandis* roots. *Ecotox. Environ. Saf.* **2018**, *162*, 488–498. [CrossRef] [PubMed]

plants

MDPI

Article

Modulation of Photosynthesis and ROS Scavenging Response by Beneficial Bacteria in *Olea europaea* Plantlets under Salt Stress Conditions

Estrella Galicia-Campos, Ana García-Villaraco Velasco, Mª Belén Montero-Palmero, F. Javier Gutiérrez-Mañero and Beatriz Ramos-Solano *

Faculty of Pharmacy, Universidad San Pablo-CEU Universities, 28668 Madrid, Spain
* Correspondence: bramsol@ceu.es; Tel.: +34-659-439-380

Abstract: Climate change consequences for agriculture involve an increase of saline soils which results in lower crop yields due to increased oxidative stress in plants. The present study reports the use of Plant Growth Promoting Bacteria (PGPB) as a tool to modulate plant innate mechanisms of adaptation to water stress (salinity and drought) in one year-old olive plantlets var. Arbosana and Arbequina. Integration of external changes in plants involve changes in Reactive Oxygen Species (ROS) that behave as signals to trigger plant adaptative mechanisms; however, they become toxic in high concentrations. For this reason, plants are endowed with antioxidant systems to keep ROS under control. So, the working hypothesis is that specific beneficial strains will induce a systemic response able to modulate oxidative stress and improve plant adaptation to water stress. Ten strains were assayed, evaluating changes in photosynthesis, pigments, ROS scavenging enzymes and antioxidant molecules, osmolytes and malondialdehyde, as oxidative stress marker. Photosynthesis and photosynthetic pigments were the most affected variables. Despite the specific response of each variety, the favorite targets of PGPBs to improve plant fitness were photosynthetic pigments and the antioxidant pools of glutathione and ascorbate. Our results show the potential of PGPBs to improve plant fitness modulating oxidative stress.

Keywords: PGPB; adaptation; salinity; abiotic stress; photosynthesis; hydric stress; oxidative stress; ROS scavenging; antioxidant; induced systemic tolerance (IST)

Citation: Galicia-Campos, E.; García-Villaraco Velasco, A.; Montero-Palmero, M.B.; Gutiérrez-Mañero, F.J.; Ramos-Solano, B. Modulation of Photosynthesis and ROS Scavenging Response by Beneficial Bacteria in *Olea europaea* Plantlets under Salt Stress Conditions. *Plants* **2022**, *11*, 2748. https://doi.org/10.3390/plants11202748

Academic Editor: Mingxun Chen

Received: 12 September 2022
Accepted: 7 October 2022
Published: 17 October 2022

1. Introduction

The global increase of temperature and CO_2 seen in the last decades has dramatically affected rainfall and therefore, water availability for crops. This limitation is not only due to lower water input through rainfall but also to the increase in salt concentration which has turned fertile soils into arid or saline soils, therefore limiting productive soils. In 2050, 50%reduction in productivity is expected due to drought, reaching up to 90% in 2100. Consequently, an increase in food prices, a decrease of world agronomic status and loss of world PIP around 0.3% is foreseen [1]. In view of this, the great challenge for the upcoming decades is to increase agronomic yields under water limiting conditions to feed the increasing world population.

Plants have survived to water stress due to salinity or lack of water along the years thanks to a series of mechanisms that allow a healthy energetic flux. Among these mechanisms are buffering ROS (reactive oxygen species) bursts with ROS-scavenging enzymes and antioxidant molecules, activating signaling cascades for ROS dependent regulatory genes, reversible regulation of proteins with disulfide bridges or phosphoproteins [2–6]. On the other hand, upon stress, ROS are able to activate nuclear gene transcription [7] or trigger systemic signals [8]. Therefore, as keeping ROS levels within a healthy concentration level is key for plant survival, activity of ROS scavenging systems appear as good markers of ROS homeostasis in cells.

Among the innate mechanisms to regulate water homeostasis in plants are (i) to increase internal solute concentration and (ii) to prevent water loss by transpiration. Osmolytes (i.e., proline), epidermal ion carriers in roots as well as ion carriers through xylem, phloem and leaf vacuoles are involved in the first approach, being all of them regulated by activating gene transcription. On the other hand, prevention of water loss affects plant hormonal balance, involving ethylene and abscisic acid levels which results in growth arrest. Either approach involve changes in ROS [9] that need to be brought back to low levels in order to be able to perform as second messengers in future stress events, controlling plant development and adaptation [10,11]. The main mechanisms to control ROS levels are antioxidant systems, both enzymatic and non-enzymatic. The enzymatic system consists of superoxide dismutase (SOD) and of those enzymes in the ascorbate-glutathione cycle (Ascorbate peroxidase (APX), Dehydroascorbate reductase (DHAR), Glutathione reductase (GR)) key to remove ROS and regulate H_2O_2 levels [12]. These enzymes are further supported by antioxidant molecules among which are ascorbate, glutathione, phenols and flavonols [13].

Interestingly, a certain overlap in adaptation to biotic and abiotic stress mechanisms has been described, sharing the ROS burst that triggers the systemic signal transduction leading to adaptation [14–16]. Therefore, similarly to the improvement of plant protection to biotic stress described for PGPB (Plant Growth Promoting Bacteria) and known as Induced Systemic Resistance (ISR), abiotic stress protection may also be enhanced by PGPB in a process known as Induced Systemic Tolerance (IST). Far from being alternative processes, specific PGPB may trigger simultaneously several plant mechanisms that result in a better adaptation [17]. Among many mechanisms by which PGPB increase IST are water and nutrient exchange, osmolyte accumulation, photosynthesis optimization, regulation of plant hormonal balance and stimulation of antioxidant mechanisms [18,19].

The aim of this study was to evaluate the ability of 10 putative PGPB to improve one-year old *Olea europaea* plantlets adaptation to saline stress, in two varieties, Arbosana (AS) and Arbequina (AQ), after 6 doses of PGPB along 6 months. To achieve our goal, the following parameters were evaluated: (i) changes in photosynthesis as physiological marker; (ii) changes in metabolic markers (photosynthetic pigments, proline, soluble sugars, antioxidant molecules and enzymes); and (iii) malondialdehyde (MDA) as oxidative stress marker, in order to identify the different adaptative mechanisms to salt stress induced by PGPB.

2. Results

This study reports effects of 10 putative PGPB on one year old olive plantlets of Arbosana (AS) and Arbequina (AQ), growing in high saline stress due to soil conditions (6.7 ds m^{-1}) and low rainfall, in a 6 month experiment open air in the Guadalquivir Marshes (Spain).

Since the recorded data was abundant, a multivariate analysis was performed initially in order to identify the most relevant factors in our experiment (variety, bacterial strain, marker). Data from all parametres measured on the two varieties treated with the 10 PGPB and controls were analyzed with a principal component analysis (PCA) and results are presented in Figure 1, where axis I absorbs 95.3% of the varianze while axis II absorbs 2.5%. Ordination of samples in the PCA revealed that the genetic variety was the most important factor, as AS samples group towards the positive values of axis I, separating from AQ samples which group towards the negative values of axis I. The factors that determine separation towards the positive values were osmolites (proline and soluble sugars), two photosynthetic parametres (the effective PSII quantum yield, PSR, and the maximal PSII quantum yield, Fv/Fm) and the gluthation pool, while non-enzymatic antioxidants (phenols and flavonols) were responsible for ordination towards the negative values of axis I. Separation along axis II was determined by ascorbic acid pull towards the positive values and photosynthetic pigments (chlorophyll a, b and carotenes) to the opposite end.

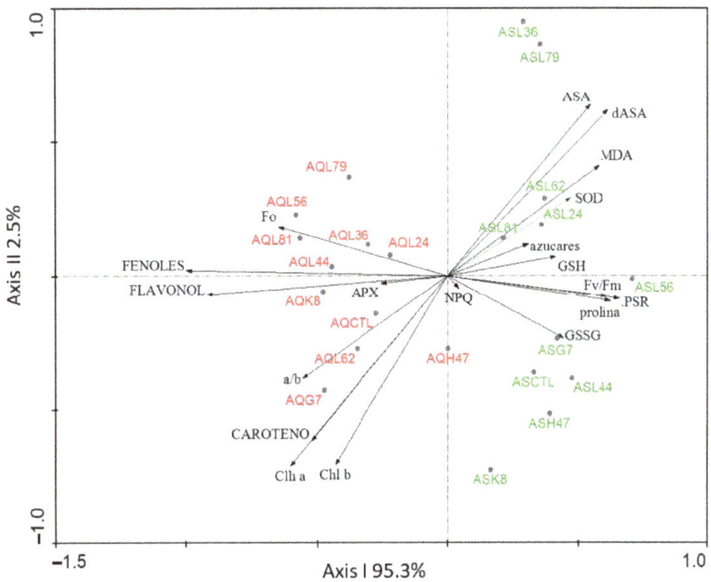

Figure 1. Principal component analysis (PCA) performed with data from physiological, metabolic and oxidative stress markers from AS and AQ inoculated with the 10 PGPB (G7, H47, K8, L24, L36, L44, L56, L62, L79, L81), and non-inoculated controls (CTL). Chl a, Chlorophyll a; Chl b, Chlorophyll b, a/b, ratio Chla/Chlb; CAROTENO, Carotenoids; azucares, Soluble sugars; prolina, proline; FENOLES: Phenols; FLAVONOL: flavonols; SOD, Superoxide Dismutase; APX, Ascorbate Peroxidase; GSH, reduced Glutathione; GSSG, oxidized Glutathione; AsA, reduced Ascorbic; dAsA, oxidized ascorbic acid; F_0, minimum fluorescence; Fv/Fm, maximal PSII quantum yield; Fv, maximal Fluorescence; NPQ, non-photochemical quenching; PSR: the effective PSII quantum yield. Varianze absorbed by each axis is represented on the figure.

In view of the differences of the two varieties, a comparation between them was carried out (Table 1). In short, AS had a higher photosynthetic capacity, with higher chlorophyll b content, higher SOD activity, higher antioxidant molecules content (ascorbate-glutathione cycle) and more osmolytes (proline and soluble sugars) associated to a higher oxidative stress based on MDA (malondialdehyde) values; interestingly, phenolics were in lower concentrations.

Photosynthetic parameters of both varieties inoculated with the 10 strains and the non-inoculated controls appear in Table 2. In AS, only F_0 and NPQ are affected by inoculation; only 4 strains increase F_0 (L79, L62, G7, K8) and one (L44) decreases it, while NPQ was increased by all strains. In AQ, all parameters are affected: F_0 decreases under the influence of 5 strains (L56, L24, L44, K8, H47), up to 10%, while all strains increased Fv/Fm and decreased photosynthetic quantum yield (PSR); finally, only G7 increased NPQ while all other strains decreased it.

While chlorophyll a and b concentration was similar in both varieties, carotene concentration was higher in AQ (Table 1). In AS, (Figure 2a) only K8 increases chlorophyll a level, while all other strains decrease it, being significant only with L79; in AQ (Figure 2b) values increase with L62 and G7, and decrease with L79 and L24, being significant with L79 only. As regards to chlorophyll b (Figure 2c), strains L56, L24, L62 and L36 significantly decrease values; in AQ (Figure 2d), L79 significantly decreases it, while L62, G7 and K8 increase chlorophyll b content. Finally, carotenoids significantly increased (Figure 2e) with K8 in AS while in AQ (Figure 2f) 5 the trend was to decrease them except for G7 which increased.

Table 1. Physiological and metabolic characterization of AS and AQ and differences between both expressed as % of variation (increase or decrease) of AS compared to AQ. Values are the average ± SE (n = 3). F_0, minimum fluorescence; Fv, Maximal Fluorescence; Fv/Fm, maximal PSII quantum yield; NPQ, non-photochemical quenching; PSR: the effective PSII quantum yield; Chl a, Chlorophyll a; Chl b, Chlorophyll b; Carotenes; SOD, Superoxide Dismutase; APX, Ascorbate Peroxidase; GSSG, oxidized Glutathione; GSH, reduced Glutathione; AsA, reduced Ascorbic acid; dAsA, oxidized Ascorbic acid; phenols; flavonols; proline; soluble sugars.

Parameters	Control AS	Control AQ	% AS vs. AQ
F_0	157.67 ± 8.76	216.78 ± 16.84	−27%
Fv/Fm	0.85 ± 0.003	0.77 ± 0.05	10%
PSR	0.77 ± 0.02	0.74 ± 0.02	4%
NPQ	0.10 ± 0.03	0.24 ± 0.02	−59%
Chl a (μg g^{-1})	125.54 ± 19.67	137.02 ± 13.65	−8%
Chl b (μg g^{-1})	60.25 ± 5.39	52.29 ± 5.15	15%
Carotenes (μg g^{-1})	92.93 ± 16.35	134.3 ± 7.91	−19%
SOD (% inhibition g^{-1} protein)	102.2 ± 3.86	90.65 ± 0.9	13%
APX (μmol g^{-1} protein)	0.03 ± 0.00	0.04 ± 0.01	−17%
GSSG (mg g^{-1} FW)	2.24 ± 0.087	2.55 ± 0.087	−12%
GSH (mg g^{-1} FW)	6.82 ± 1.05	1.82 ± 0.26	275%
AsA (mg g^{-1} FW)	61.56 ± 0.15	16.64 ± 1.39	270%
dAsA (mg g^{-1} FW)	92.08 ± 3.82	33.38 ± 0.82	176%
MDA	3.16 ± 0.48	2.06 ± 0.32	53%
Phenols (meq gallic acid per 100 g FW)	739.94 ± 70.42	1031.81 ± 89.63	−28%
Flavonols (meq catechin per 100 g FW)	6.73 ± 0.21	12.02 ± 1.45	−44%
Proline (nmol g^{-1} FW)	0.45 ± 0.02	0.39 ± 0.02	15%
Soluble sugars (mg g^{-1} FW)	4.93 ± 0.60	4.62 ± 0.74	7%

Table 2. Photosynthetic parameters of plants from AS and AQ, inoculated with the 10 strains and the non-inoculated controls. Minimum florescence (F_0). Maximum photosynthetic potential of PSII (Fv/Fm). Photosynthetic quantum yield (PSR) and non-photochemical quenching (NPQ). Values are the average ± SE (n = 3). Asterisks indicate significant differences according to *t*-Student test $p < 0.05$.

	ARBOSANA				**ARBEQUINA**			
	F_0	**Fv/Fm**	**PSR**	**NPQ**	**F_0**	**Fv/Fm**	**PSR**	**NPQ**
Control	157.67 ± 8.76	0.85 ± 0.003	0.77 ± 0.02	0.10 ± 0.03	216.78 ± 16.84	0.77 ± 0.05	0.74 ± 0.02	0.24 ± 0.02
L79	195.22 ± 17.96	0.83 ± 0.01	0.77 ± 0.03	0.15 ± 0.02	211.33 ± 8.97	0.84 ± 0.01	0.7 ± 0.02	0.23 ± 0.02
L81	150.67 ± 14.71	0.85 ± 0.01	0.8 ± 0.02	0.18 ± 0.03	213.67 ± 8.24	0.82 ± 0.01	0.7 ± 0.02	0.18 ± 0.01
L56	163 ± 0.01	0.85 ± 0.01	0.82 ± 0	0.61 ± 0.01	185 ± 7.62	0.84 ± 0.01	0.75 ± 0.02	0.15 ± 0.01 *
L24	166.67 ± 0.27	0.87 ± 0.01	0.78 ± 0.02	0.12 ± 0.02	190.33 ± 4.48	0.85 ± 0.01	0.72 ± 0.02	0.17 ± 0.01
L62	179.33 ± 5.46	0.85 ± 0.01	0.8 ± 0.01	0.1 ± 0.01	206.67 ± 8.45	0.83 ± 0.01	0.72 ± 0.01	0.23 ± 0.03
L36	170.67 ± 8.02	0.84 ± 0.01	0.74 ± 0.02	0.15 ± 0.02	219 ± 8.26	0.8 ± 0.02	0.67 ± 0.03	0.22 ± 0.03
G7	174 ± 9.81	0.87 ± 0.01	0.78 ± 0.02	0.13 ± 0.01	212.78 ± 13.39	0.82 ± 0.02	0.69 ± 0.02	0.25 ± 0.03
L44	133 ± 18.12	0.85 ± 0.01	0.77 ± 0.02	0.13 ± 0.03	160.67 ± 10.39	0.84 ± 0.02	0.82 ± 0.05	0.2 ± 0.04
K8	179.67 ± 11.47	0.85 ± 0.01	0.77 ± 0.01	0.15 ± 0.02	192.33 ± 9.82	0.82 ± 0.01	0.7 ± 0.04	0.21 ± 0.01
H47	148 ± 21.92	0.86 ± 0.01	0.8 ± 0.01	0.16 ± 0.05	195.75 ± 7.54	0.83 ± 0.01	0.75 ± 0.02	0.16 ± 0.02

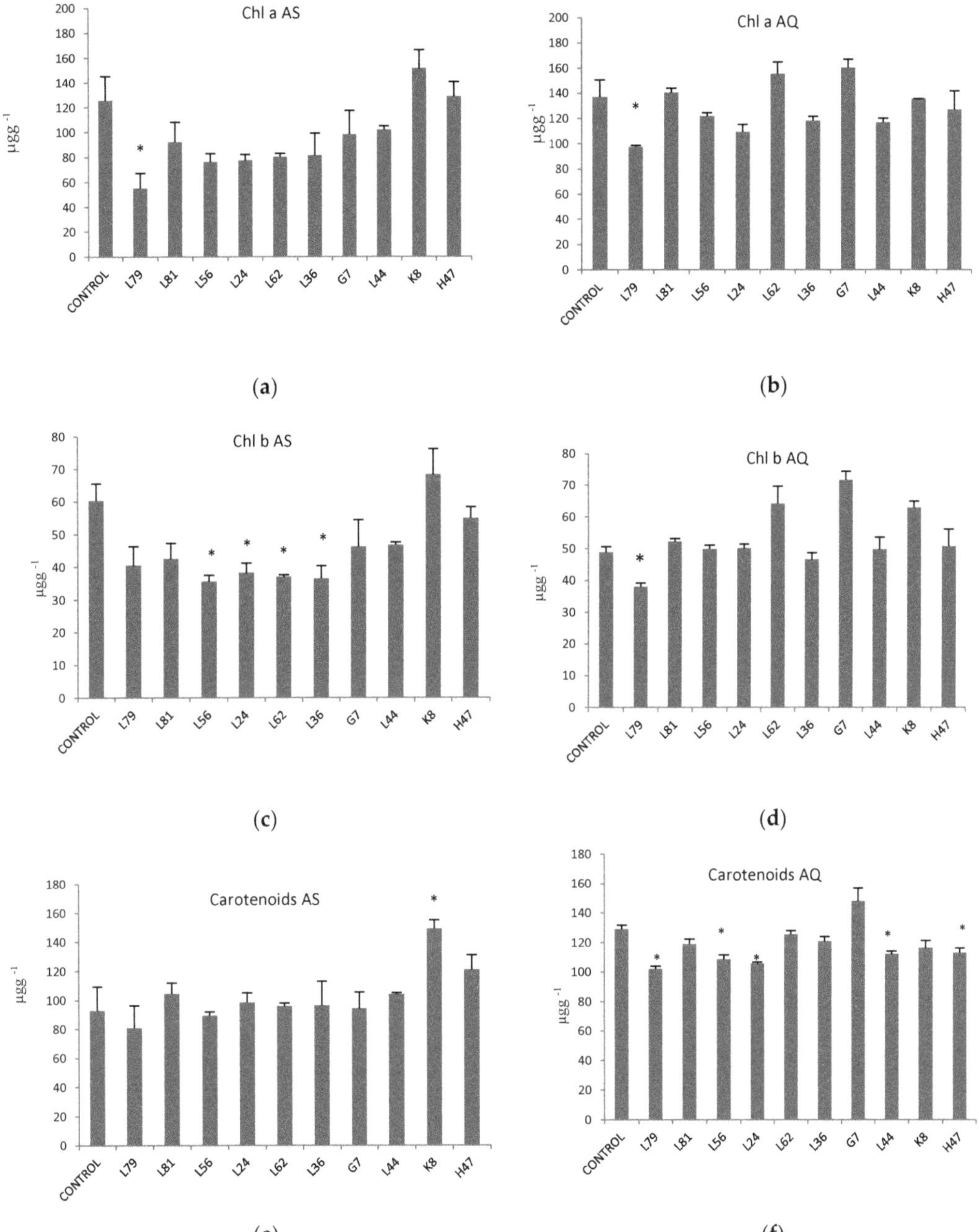

Figure 2. Photosynthetic pigments (µg g^{-1} FW) (**a**) chlorophyll a AS, (**b**) chlorophyll a AQ (**c**) chlorophyll b AS, (**d**) chlorophyll b AQ, (**e**) carotenoids AS and (**f**) carotenoids AQ in leaves of plants inoculated with the 10 strains and non-inoculated controls. Values are the average ± SE (n = 3). Asterisks (*) indicate significant differences with the controls according to T student ($p < 0.05$).

Effects of PGPB on osmolytes (proline and soluble sugars) was evaluated. Proline was not affected by any treatment in any of the two varieties (Figure S1) but soluble sugars did (Figure 3). In AS (Figure 3a), L79, L62 and L36 increased, values being significant only with L79, while L56 decreased them. In AQ, (Figure 3b) L62 significantly increased and H47 significantly decreased them.

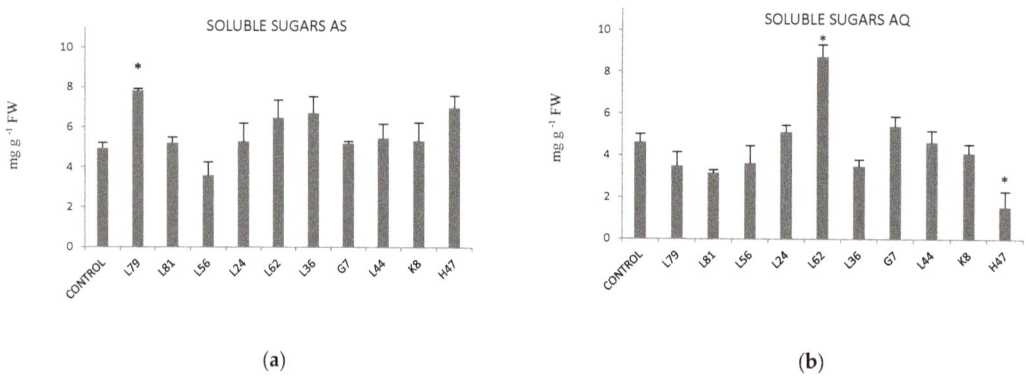

Figure 3. Soluble sugars (mg g^{-1} FW) in (**a**) var. AS and (**b**) var. AQ olive leaves. Values are the average ± SE (n = 3). Asterisks (*) indicate significant differences with the controls according to T student ($p < 0.05$).

As far as modification of antioxidant enzymes activity by bacterial strains, in AS no variation was detected in SOD (Figure 4), while in AQ it was significantly increased by L24, L62, L36 and G7.

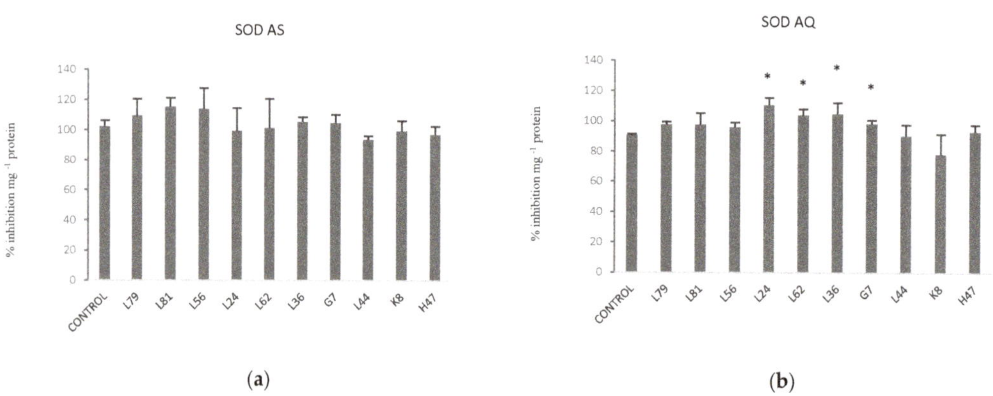

Figure 4. Superoxide dismutase activity (% inhibition mg^{-1} protein) in olive leaves (**a**) var. AS and (**b**) var. AQ, inoculated with the 10 PBPB and non-inoculated controls. Values are the average ± SE (n = 3). Asterisks (*) indicate significant differences with the controls according to t student ($p < 0.05$).

As regards to Ascorbate peroxidase activity (APX), in AS (Figure 5a) a significant decrease with L24 and a significant increase with L44 were registered. In AQ, a non-significant increase with L79 was registered.

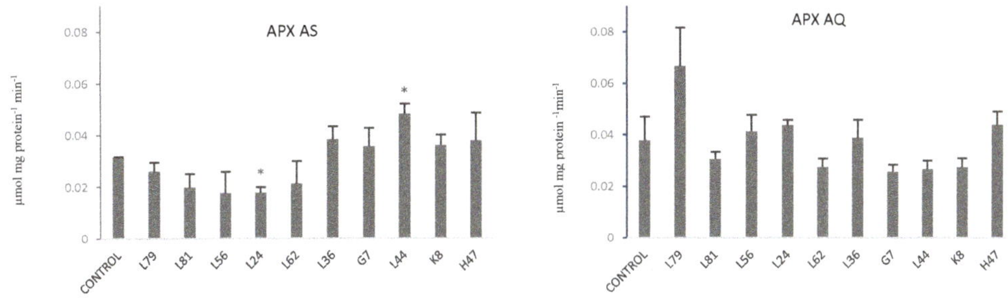

(a) (b)

Figure 5. Ascorbate peroxidase activity (μmol mg protein^{-1} min^{-1}) in (**a**) AS and (**b**) AQ leaves in plants inoculated with the 10 PGPB and non-inoculated controls. Values are the average ± SE (n = 3). Asterisks (*) indicate significant differences with the controls according to *t* student ($p < 0.05$).

Antioxidants concentration was affected by inoculation of PGPBs in both varieties (Figures 6–8), both in total amounts and in the oxidized/reduced ratio for AsA and glutathione. For the ascorbate pool (AsA/dAsA) is 30% lower in AQ tan in AS (Table 1). The trend in AS was to significantly reduce the total amount of ascorbate, reflected in both species except for L62, that caused a significant increase in both (Figure 6a). In AQ (Figure 6b), all strains except L62, induced a significant decrease in AsA, except K8 and L44 that also decreased dAsA.

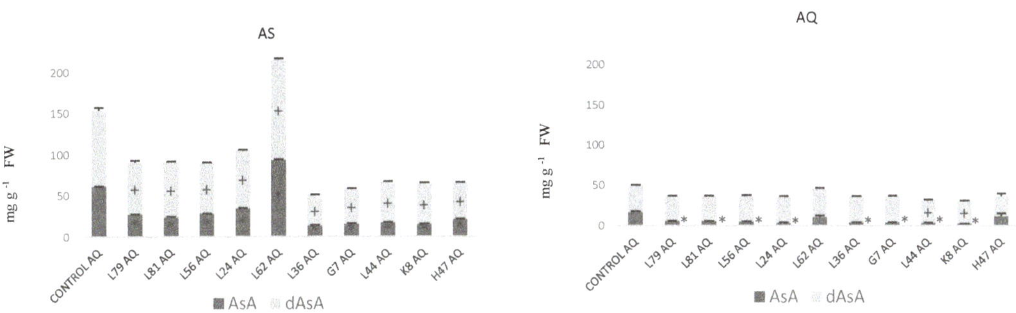

(a) (b)

Figure 6. Ascorbate concentration (mg g^{-1} FW) in the oxidized (dAsA) and reduced (AsA) forms in leaves of (**a**) AS and (**b**) AQ, in plants inoculated with the 10 PGPB and the non-inoculated controls. Values are the average ± SE (n = 3). Asterisks (*) or (+) indicate significant differences with the controls according to *t* student ($p < 0.05$) for AsA, and dAsA, respectively.

Glutathione (Figure 7) is less abundant in AQ than AS, mainly due to the lower concentration of reduced glutathione (GSH) in AQ (Table 1). In AS, the general trend was to significantly increase glutathione, mainly in the oxidized pool, being especially effective L62, G7 and L44; conversely, some strains (L79, L56, L24, L36, H47) significantly decreased GSH. In AQ (Figure 7b), the glutathione pool is significantly increased by L79, L24, L36 and K8, due to increases in GSH, except for L36 that increases both GSH and GSSG; however, strains L81 and G7 modified the balance GSH/GSSG, significantly decreasing GSH.

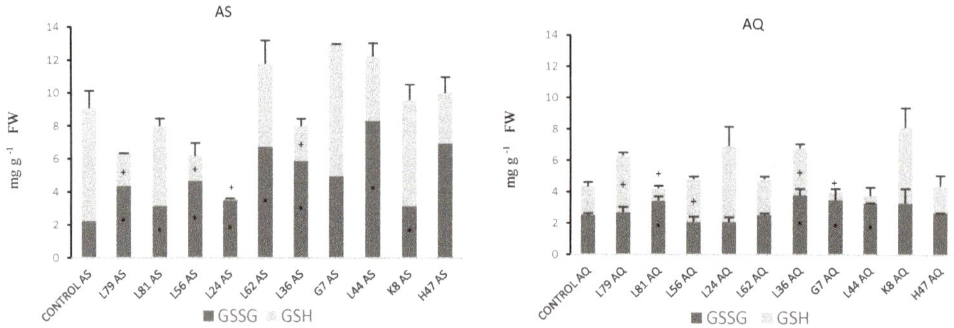

(**a**) (**b**)

Figure 7. Glutathione concentration (mg g^{-1} FW) of oxidized glutathione (GSSG) and reduced glutathione (GSH) (**a**) in AS and (**b**) AQ, in leaves of plants inoculated with the 10 PGPB and non-inoculated controls. Values are the average ± SE (n = 3). Asterisks (*) and (+) indicate significant differences with the controls according to *t* student ($p < 0.05$) for GSSG and GSH, respectively.

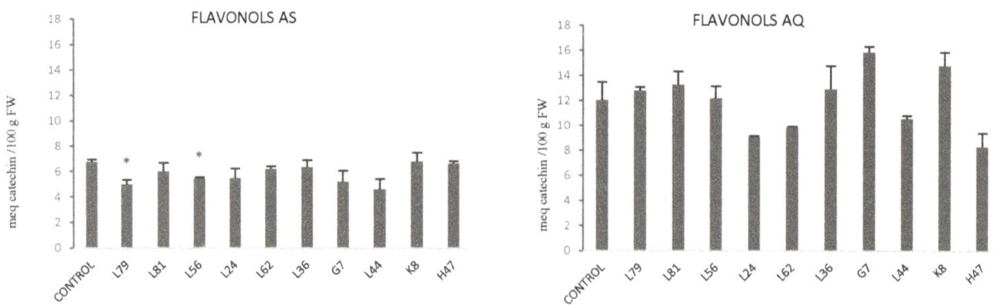

(**a**) (**b**)

Figure 8. Flavonol concentration (meq catechin/100 g fresh weight) in (**a**) AS and (**b**) AQ leaves of plants inoculated with the 10 PGPB and non-inoculated controls. Values are the average ± SE (n = 3). Asterisks (*) indicate significant differences with the controls according to *t* student ($p < 0.05$).

As regards to the antioxidant molecules phenols and flavonols, phenols were not affected by either treatment in AS or AQ (Figure S2) but flavonols did (Figure 8). Flavonols were almost two-fold higher in AQ than AS (Table 1). While in AS only two strains (L79, L56) triggered flavonoid metabolism, significantly lowering values (Figure 8a), in AQ L24, L62, L44 and H47 followed the same trend while G7 and K8 increased flavonol concentration.

Finally, concentration of MDA was analyzed being higher in AS than in AQ (Table 1). In AS, all treatments except L62, followed the same trend, tending to lower MDA, although only significant with L24 and L36 (Figure 9a). Interestingly in AQ, this parameter was not affected by any treatment (Figure 9b).

Ordination of samples in the PCA revealed that the genetic variety was the most important factor, so individual multivariate analyses were conducted (Figure 10), using only those parameters that were affected by bacterial treatments in both varieties: photosynthetic pigments, NPQ, Superoxide-dismutase, Ascorbate peroxidase, glutathione (oxidized and reduced), asocorbate (AsA/dAsa). In AS (Figure 10a), separation on axis I is driven by pigments towards the negative end and AsA/dAsA towards the positive end; in this figure, axis I accounts for 72.1% of the varianze while axis II accounts for 12%. In AQ (Figure 10b), separation along axis I is driven by pigments towards the positive end and APX and GSH towards the opposite end; in this analysis, axis I accounts for 81.1 % of the varianze and axis II, 10.1%.

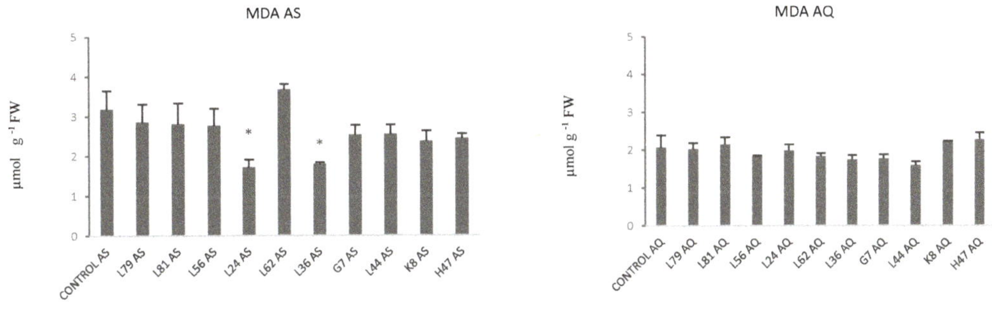

(a) (b)

Figure 9. Malondialdehyde concentration (μmol g^{-1} FW), in (**a**) AS and (**b**) AQ leaves of plants inoculated with the 10 PGPB and non-inoculated controls. Values are the average ± SE (n = 3). Asterisks (*) indicate significant differences with the controls according to *t* student ($p < 0.05$).

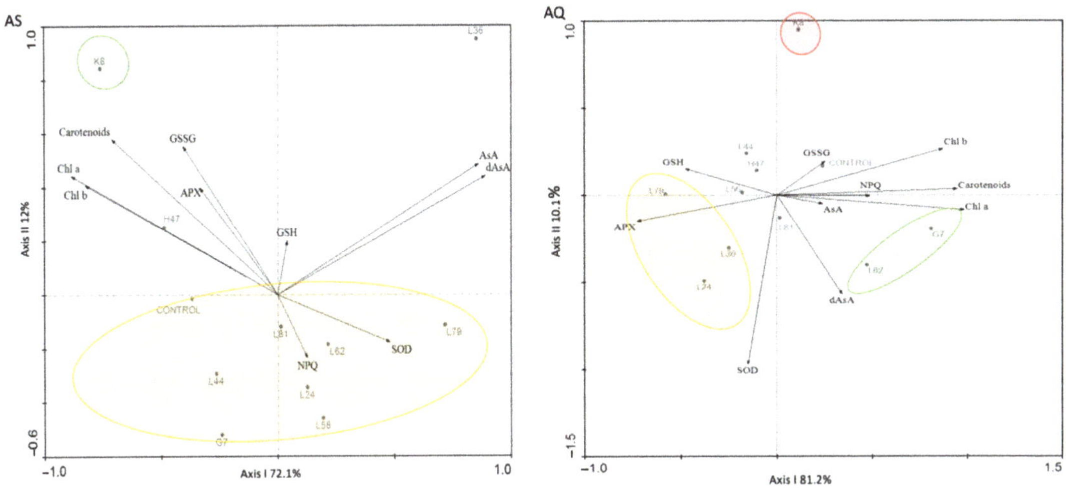

(a) (b)

Figure 10. Principal component analysis (PCA) performed with data from physiological, metabolic and oxidative stress markers from (**a**) AS and (**b**) AQ inoculated with the 10 PGPB (G7, H47, K8, L24, L36, L44, L56, L62, L79, L81), and non-inoculated controls (CONTROL). Chl a, Chlorophyll a; Chl b, Chlorophyll b; Carotenoids; SOD, Superoxide Dismutase; APX, Ascorbate Peroxidase; GSH, reduced Glutathione; GSSG, oxidized Glutathione; AsA, reduced Ascorbic; dAsA, oxidized ascorbic acid; NPQ, non- photochemical quenching. Varianze absorbed by each axis is represented on the figure.

3. Discussion

The present study shows that photosynthesis and its associated metabolic marker, photosynthetic pigments, are the most affected parameters by PGPB in one-year old olive plantlets. Secondly, ROS regulation mechanisms, both enzymatic and non-enzymatic, are the priority targets, and modification is not necessarily associated to increase their activity or concentration [20]. On the other hand, osmolytes are barely affected suggesting that modification of osmolyte concentration is not a priority mechanism to maintain cell homeostasis at this point of development, despite being priority at later stages [21].

The physiological differences between the two olive varieties were evidenced by the multivariate analysis in which all parameters under study were considered, as samples from each variety grouped together (Figure 1). Photosynthetic performance was different in AS and AQ (Table 1). AS was more efficient under stress conditions as revealed by higher values of maximal potential (Fv/Fm) and real (PSR) photosynthetic efficiency while AQ dissipated more energy (NPQ), confirming that AQ was more sensitive to stress [22,23]. Furthermore, adaptation to oxidative stress relies in different mechanisms in each variety: AS is genetically endowed with stronger SOD activity, a high concentration of ascorbate and glutathione as antioxidant molecules, and high proline content, consistent with its high photosynthetic activity; on the other hand, AQ´s endowment of antioxidant molecules is higher in phenols and flavonols. According to this statement, the potential targets of PGPB to improve adaptation to osmotic stress should be different for each variety [17].

Photosynthetic parameters have been used as indicators of plant responses to biotic stress. More precisely, increases in basal photosynthesis (F_0) have been reported to indicate an efficient inoculation [24,25], while decreases in F_0 after inoculation have been interpreted as plant reprogramming for a more relaxed status, that suggests strains protecting the photosynthetic apparatus [26]. Accordingly, fifty percent of strains decreased F_0 values, while all strains brought up the maximal potential photosynthetic energy (Fv/Fm) to regular healthy values (0.82–0.85), also increasing energy dissipation (NPQ) to release the excess of absorbed energy [22,26–28].

Modifications in the photosynthetic process involve photosynthetic pigments. As the role of chlorophylls is to absorb energy, an increase in pigments associated to growth increase upon PGPB treatment could be expected [29]. However, the effect was rather the opposite. Decreases in chlorophyll concentration induced by PGPB has been interpreted as a way to limit energy absorption, resulting in lower oxidative stress [30], a very frequent effect as reported here. As regards to carotenoids (Figure 2), they play several roles either as direct antioxidants [31,32], or directly involved in photosynthesis. In the latter process they play a dual role, either absorbing energy in antennas or dissipating part of the absorbed energy through the xanthophyll cycle [12]; the energy dissipation upon violaxanthin-zeaxanthin conversion is a photoprotective mechanism of PSII, as xanthophylls react with excited chlorophylls preventing $^1O_2$ formation [33] and releasing energy as heat [9,27]. As antioxidants, carotenoids quench lipid peroxidation products hence stopping oxidative cascade [34,35]. Connecting with the effects of PGPB, the increase in carotenoids could contribute (i) only to energy dissipation, or (ii) to adjust the energy flow increasing absorption in non-stressed conditions while increasing dissipation upon stress, in line to the plant needs to adjust to changing conditions. In summary, the general effect of PGPB in photosynthetic pigments are to decrease chlorophylls while maintaining carotenoid concentration, that is, lowering energy absorption rather than increasing dissipation, in order to slow down the system [30]. Interestingly, only two strains show a different pattern and still, it is true for the two varieties: K8 increases carotenoids supporting the dual role of carotenoids for this case, and L79 that lowers both chlorophylls (56%) and carotenes so, as the energy entrance is decreased, carotenoids do not need to dissipate energy. Finally, an interesting behavior is detected in AQ only, with strains G7 and L62, which increase chlorophyll concentration and therefore, energy absorption but also increase energy dissipation, allowing a higher energy flow through the system with a more relaxed status, preventing ROS formation [35].

Despite all efforts to prevent ROS formation due to their toxicity, non-toxic concentrations are still necessary [12] as they play a role as second messengers in growth and development [36,37], in adaptation to environmental changes [38], activating gene transcription in the nucleus [7], or triggering systemic processes [8]. Based on the different genetic endowment of each variety to keep oxidative stress under control, our data confirms the strong influence on the plant-bacteria interaction [21,26,39], as well as ROS involvement on the response to biotic stress. Each PGPB finds different targets to alter ROS levels, suggesting the existence of different isoenzymes for an optimal adaptation [40]. In AS, PGPBs modify the antioxidant profile targeting APX and ascorbate and glutathione pools as the main antioxidant molecules. Conversely, in AQ PGPBs modify the antioxidant pool targeting SOD, the glutathione and ascorbate pools as well as phenols and flavonols; interestingly, increases in flavonols represent an additional antioxidant mechanism as they reflect UV irradiation [25] and affect lipid peroxidation by increasing phospholipid packaging to prevent ROS diffusion [41]. Increases in SOD, together with decreases in APX and in AsA will result in keeping H_2O_2 levels high, as it is the systemic signal to activate IST in the plant, improving adaptation to stress [10]. Beyond the signaling effect, protection of the photosynthetic apparatus to oxidative stress by PGPB is partly achieved by modifying the redox status of AsA and GSH [4]. Modification of antioxidant profiles described above confirm the different strategies of each PGPB to ameliorate oxidative stress due to salinity [42,43], resulting in a better physiological status of the plant due to improved adaptative capacity [44].

In summary, PGPB use 3 general strategies to improve plant adaptation to salt stress: (i) Lowering energy absorption by decreasing photosynthetic pigments, which results in a lower oxidative stress and a concomitant decrease of non-enzymatic antioxidants, being this the most frequent option; (ii) Optimizing the energy absorption/dissipation system by increasing chlorophylls and carotenes, without modifying ROS scavenging mechanisms since carotenes are able to play a dual role to absorb and dissipate; unique option for K8 in AQ;. (iii) Optimizing the energy absorption/dissipation system by increasing chlorophylls and carotenes, which results in a higher oxidative stress to be controlled by enhancing antioxidant systems; only two strains in AQ (G7, L62) and K8 in AS. Among all the studied parameters, photosynthetic pigments appear as the most direct marker to detect PGPB effects. Despite the specific response of each variety, the favorite targets of PGPBs to improve plant fitness were the antioxidant pools of glutathione and ascorbate. Our results show the potential of PGPBs to improve plant fitness modulating oxidative stress.

4. Materials and Methods

4.1. Plant Material

Two olive varieties were selected Arbosana (AS) and Arbequina (AQ). One-year old plants were bought from a local provider Planta Continental (Rivero de Posada, Córdoba). Plants were transplanted to 5L pots filled with soil from Guadalquivir marshes and peat (3:1).

4.2. Bacterial Strains and Inoculum Preparation

The 10 beneficial strains (L79, L81, L56, L24, L62, L36, G7, L44, K8, and H47) assayed in this study were isolated from the rhizosphere of *Pinus pinea* and *P. pinaster* [45]. They were able to produce siderophores (L79, L81, G7, H47), auxins (L56, L24, L44), auxins and siderophores (L62, L36) or auxins and degrade 1-amincyclopropane-1-carboxylate (ACC) (K8). Except for L62, a Gram-positive non-sporulated rod, all other strains are Gram-positive sporulated bacilli [21].

Bacterial strains are kept at -80 °C in nutrient broth amended with 20% glycerol. To prepare inocula, strains were plated (PCA) and incubated for 24 h at 28 °C. Then, they were inoculated on liquid broth (Nutrient Broth (NB) for L62 and Luria Broth (LB) for all other) and incubated 24 h at 28 °C, under shaking. Cultures were diluted to 1×10^8 cfu (Colony forming units) mL^{-1} for inoculation.

4.3. Experimental Set Up

Seventy plants from each variety were transplanted to pots and placed on open air at the Guadalquivir marshes (37°11′25.9″ N 6°13′59.3″ W). Ten treatments plus a control were defined for each variety, with 6 plants per treatment. A total of eight inoculations were delivered in October, November, March and April, twice a month, by soil drench, with 400 mL of a bacterial solution (1 × 10^8 cfu mL^{-1}) per plant. Soil moisture was maintained with saline water (electric conductivity 8′2 ds m^{-1}), reaching a soil conductivity of 6′07 ds m^{-1}. In april 2018, photosynthesis was measured, and leaves were sampled and brought to the laboratory at 4 °C. Leaves from two plants were pulled and constituted an analytical replicate; leaves were powdered with liquid nitrogen to carry on the following determinations: the antioxidant enzymes superoxide dismutase (SOD) and ascorbate peroxidase (APX); the antioxidant molecules phenols, flavonols, Glutathione (oxidized and reduced) and ascorbate (oxidized and reduced); the osmoprotectants proline and soluble sugars; the photosynthetic pigments Chlorophyll a, Chlorophyll b and carotenoids; and malondialdehyde concentration, as oxidative stress marker.

4.4. *Photosynthesis (Chlorophyll Fluorescence)*

Photosynthetic efficiency was determined through the chlorophyll fluorescence emitted by photosystem II. A pulse amplitude modulated (PAM) fluorometer (Hansatech FM2, Hansatech, Inc., UK) was used to measure chlorophyll fluorescence. After dark-adaptation of leaves, a weak modulated irradiation (1 μmol m^{-2} s^{-1}) was applied to measure the minimal fluorescence (F_0; dark-adapted minimum fluorescence). Maximum fluorescence (Fm) was determined from the dark-adapted state delivering a 700 ms saturating flash (9000 μmol m^{-2} s^{-1}). The variable fluorescence (Fv) was calculated as the difference between the maximum fluorescence (Fm) and the minimum fluorescence (F_0). The maximum photosynthetic efficiency of photosystem II (maximal PSII quantum yield) was calculated as Fv/Fm. Immediately, the leaf was continuously irradiated with red-blue actinic beams (80 μmol m^{-2} s^{-1}) and after equilibrating for 15 s, Fs was recorded (steady-state fluorescence signal). Then, another saturation flash (9000 μmol m^{-2} s^{-1}) was applied to determine Fm′ (maximum fluorescence under light-adapted conditions). Other fluorescent parameters were calculated as follows: the effective PSII quantum yield PSR = (Fm′ − Fs)/Fm′ [46,47]; and the non-photochemical quenching coefficient NPQ = (Fm − Fm′)/Fm′. All measurements were carried out in the 6 plants of each treatment.

4.5. *Photosynthetic Pigments: Chlorophyll a, Chlorophyll b and Carotenoids*

Extraction was done according to [48], keeping tubes in dark throughout the process. One hundred mg of leaves powdered in liquid nitrogen was dissolved in 1 mL of acetone 80% (*v*/*v*), incubated overnight at 4 °C and then centrifuged 5 min at 10,000× *g* rpm in a Hermle Z233 M-2 centrifuge. One mL of acetone 80% was added to the supernantant and vortexed. Immediately, absorbance at 647, 663, and 470 nm was measured on a Biomate 5 spectrophotometer to calculate chlorophyll a, chlorophyll b, and carotenoids (xanthophylls + carotenes) using the formulas indicated below [47,48]

$$\text{Chl a } (\mu g\ g\ FW^{-1}) = [\ (12.25 \times \text{Abs}_{663}) - (2.55 \times \text{Abs}_{647})\ ] \times V(mL)/(g).$$

$$\text{Chl b } (\mu g\ g\ FW^{-1}) = [\ (20.31 \times \text{Abs}_{647}) - (4.91 \times \text{Abs}_{663})\ ] \times V(mL)/(g).$$

$$\text{Carotenoids } (\mu g\ g\ FW^{-1}) = [\ ((1000 \times \text{Abs}_{470}) - (1.82 \times \text{Chl a}) - (85.02 \times \text{Chl b}))/198] \times V(mL)/(g).$$

4.6. *Osmoprotectants: Proline and Soluble Sugars*

An ethanolic extract was prepared diluting 0.25 g of powder in 5 mL of 70% ethanol (*v*/*v*) incubated at 100 °C for 20 min. The extract was kept at 4 °C until analysis of proline and soluble sugars.

For proline determination 1 mL of ninhydrin reagent freshly prepared (1 g of ninhydrin dissolved in 60 mL of glacial acetic acid, 20 mL of ethanol and 20 mL of water) was mixed with 0.5 mL of the plant ethanol extract and heated at 95 °C for 20 min. Finally, absorbance at 520 nm was measured. Results are expressed as μmol g^{-1} [49].

Soluble sugars were determined according to Yemm and Willis [50]. Briefly, the following reaction was prepared: 3 mL of the reactive (200 mg of antrone + 100 mL of 72%

sulfuric acid) and 0.1 mL of the plant ethanol extract. The reaction was incubated in a bath at 100 °C for 10 min. Once it was cold, absorbance was measured at 620 nm. To calculate soluble sugar concentration the following equation was used.

$$\mu g\ g^{-1} = [(Abs620 - 0.016)/0.02]/(g)/1000$$

4.7. Enzymatic Antioxidants: Superoxide Dismutase (SOD) and Ascorbate Peroxidase (APX)

Prior to assessment of enzymatic activities, soluble proteins were extracted. One hundred mg of powder were suspended in 1 mL of 0.1 M potassium phosphate buffer, pH 7.0, containing 2 mM phenylmethylsulfonyl fluoride (PMSF). After sonication for 10 min followed by centrifugation for 10 min at 14,000× *g* rpm, the supernatant was aliquoted, frozen in liquid nitrogen and stored at −80 °C for further analysis of APX, SOD, and proteins. All the above operations were carried out at 0–4 °C.

To determine the amount of total protein in plant extracts, 250 μL of Bradford reagent, 5 μL samples and BSA (Bovine Serum Albumin) dilutions were inoculated in ELISA 96 well plates and incubated for 30 min at room temperature and then measured using a plate reader at 595 nm. Commercial BSA was used for a calibration curve. Total protein was expressed as mg μL^{-1}.

APX was measured as described in [51]. The reaction mixture consisted of 50 mM potassium phosphate buffer, pH 7.0, 0.25 mM sodium ascorbate, 5 mM H2O2 and 100 μL of enzyme extract in a final volume of 1.2 mL. H2O2 was used to start the reaction and ascorbate oxidation was determined by the decrease in A290. The extinction coefficient of 2.8 mM−1 cm−1 was used to calculate activity. One unit of APX activity is defined as the amount of enzyme that oxidizes 1 mmol min−1 of ascorbate under the above assay conditions.

SOD activity was determined as described in the detection kit (SOD Assay Kit-WST, Sigma-Aldrich, Darmstadt, Germany). With this method, xanthine is converted to superoxide radical ions, uric acid, and hydrogen peroxide by xanthine oxidase (XO). Superoxide reacts with WST1 to generate a product that absorbs at around 440 nm. SOD prevents the reduction of WST1 to WST-1formazan, thus reducing the absorption at 440 nm, which is proportional to SOD activity; the rate of the reduction of WST1 with O2 is linearly related to the xanthine oxidase (XO) activity. The unit used for this activity was: % inhibition of WST reduction per mg protein.

4.8. Non-Enzymatic Antioxidants: Ascorbate, Glutathione, Phenols and Flavonols

Ascorbate and glutathione were determined according to [52]. An extract was prepared by suspending 1 g leaf powder in 10 mL of 5% metaphosphoric acid and centrifuging for 15 min at 22,000× *g* at 4 °C.

Total ascorbate was determined by fully reducing dAsA to AsA with dithiothreitol (DTT), and then, dAsA was estimated calculating the difference between total ascorbate and AsA. The mixture to determine total ascorbate is as follows: 300 μL supernatant, 750 μL of 150 mM phosphate buffer (pH 7.4) with 5 mM EDTA and 150 μL DTT. After 10 min at room temperature, 150 μL of 0.5% N-ethylmaleimide were added to remove remaining DTT. To determine AsA, a similar mixture was used but DTT and N-ethylmaleimide were replaced by water (300 μL). The reaction was started by adding 600 μL of 10% TCA, 600 μL of 44%ortophosphoric acid, 600 μL of 4% α, α′-dipyridyl in 70% ethanol and 0.3% $FeCl_3$ (*w*/*v*). The mixture was vortexed and incubated for 40 min before measuring absorbance at 525 nm. A calibration curve between 0–100 $\mu g\ mL^{-1}$ AsA was done.

Total, oxidized (GSSG) and reduced glutathione (GSH) were determined in the supernatant. First, one mL of the supernatant was neutralized with 1.5 mL of 0.5M phosphate buffer (pH 7′5) and 50 μL water to determine total glutathione. Similarly, another mL of the supernatant was also neutralized and supplemented with 50 μL 2-vinylpyridine to mask GSH, by gently shaking to form an emulsion; then, a 60-minute incubation followed and GSSG was determined. GHS concentration was estimated from the difference between total and oxidized glutathione. Glutathione concentration was measured in a 3 mL final volumen reaction containing 0.2 mM NADPH, 100 mM (pH 7.5) phosphate buffer, 5 nM EDTA, DNTB 0.6 mM and 3 units of the enzyme Glutathione Reductase; changes in absorbance at 412 nm for 1 min were recorded. Concentration was calculated from a calibration curve from 0–50 μg mL^{-1}.

To determine phenols and flavonols, methanolic extracts were prepared from 0.25 g of leaves (powdered in liquid nitrogen) in 2.25 mL methanol 80%, sonicated for 10 min and centrifuged for 5 min at 5000 rpm.

Total phenols were quantitatively determined with Folin-Ciocalteu agent (Sigma. Aldrich, St. Louis, MO, USA) by a colorimetric method described by Xu and Chang [53], with some modifications; gallic acid was used as standard (Sigma-Aldrich, St. Louis, MO, USA). Twenty μL of extract were mixed with 250 μL of Folin-Ciocalteu 2 N and 750 μL of Na_2CO_3 20% solution. After 30 min at room temperature, absorbance was measured at 760 nm. A gallic acid calibration curve was made (r = 0.99). Results are expressed in mg of gallic acid equivalents per 100 g of fresh weight (FW).

Quantification of total flavonols was done as in [54], using catechin as standard (Sigma-Aldrich, St Louis, MO, USA). One milliliter of the extract was added to a flask of 10 mL with 4 mL of distilled water. Then, 300 μL of $NaNO_2$ 5%, and the same volume of $AlCl_3$ 10% were added after 5 min. One minute later, 2 mL of NaOH 1 M were added, and adjusted to a total volume of 10 mL with distilled water. The solution was mixed and measured at 510 nm. A catechin calibration curve was made (r = 0.99). Results are expressed as mg of catechin equivalents per 100 g of fresh weigh (FW).

4.9. *Malondialdehyde*

MDA content was determined as in [55]. One hundred mg of leaf powder as suspended in 2 mL trichloroacetic acid 10%. After 2–3 min vortex, it was centrifuged at 20,000× *g* for 30 min at 4 °C. One mL of the supernatant was added to 4 mL 0.5% (*v*/*v*) thiobarbituric acid (TBA) and 20% (*v*/*v*) trichloroacetic acid (TCA). The mixture was heated at 95 °C for 30 min, stopping the reaction on ice. After 10 min centrifugation, absorbance was determined 532 and 600 nm. The MDA content was calculated using the formula: MDA (nmol $g\ FW^{-1}$) = [(OD532-OD600)]/(ε × W), where FW is the fresh weight and ε the extinction coefficient (155 $mM^{-1}cm^{-1}$). Data were expressed as μmol g FW^{-1} (fresh weight).

4.10. *Statistics*

A Principal Components Analysis (PCA) with all the parameters measured for the ten strains was performed with CanocoTM for Windows v.4.5 software (Microcomputer power, Ithaca, NY, USA). To evaluate treatment effects, *t*-Student test were carried out for each variable (Statpgraphcis Centurion XVIII).

5. Conclusions

In view of the results presented here, it is evidenced that plant genotype is the most relevant factor that determine plant's response to stress. As regards to PGPB, all affected photosynthetic pigments, modulating energy flow through the system under saline stress. PGPB use 3 general strategies to improve plant adaptation to salt stress: (i) Lowering energy absorption by decreasing photosynthetic pigments, which results in a lower oxidative stress and a concomitant decrease of non-enzymatic antioxidants, (ii) Optimizing the energy absorption/dissipation system by increasing chlorophylls and carotenes, without modifying ROS scavenging mechanisms since carotenes are able to play a dual role to absorb and dissipate. (iii) Optimizing the energy absorption/dissipation system by increasing chlorophylls and carotenes, which results in a higher oxidative stress to be controlled by enhancing antioxidant systems. Our results show the potential of PGPBs to improve plant fitness modulating oxidative stress although further studies need to be carried out to confirm improvement of plant growth and/or protection to other stress conditions.

Supplementary Materials: The following are available online at https://www.mdpi.com/article/10.3390/plants11202748/s1, Figure S1: proline contents in As and AQ; Figure S2: total phenols concentration in As and AQ.

Author Contributions: Conceptualization, F.J.G.-M., A.G.-V.V. and B.R.-S.; formal analysis, E.G.-C. and M.B.M.-P.; resources, F.J.G.-M.; data curation, A.G.-V.V., M.B.M.-P. and B.R.-S.; writing—original draft preparation, E.G.-C.; writing—review and editing, B.R.-S. and A.G.-V.V.; supervision, F.J.G.-M.; funding acquisition, F.J.G.-M. and B.R.-S. All authors have read and agreed to the published version of the manuscript.

Funding: This research received no external funding.

Acknowledgments: The authors thank J.C. for providing plants and land for the experiment as well as supervision.

Conflicts of Interest: The authors declare no conflict of interest.

References

1. Malhi, G.S.; Kaur, M.; Kaushik, P. Impact of Climate Change on Agriculture and Its Mitigation Strategies: A Review. *Sustainability* **2021**, *13*, 1318. [CrossRef]
2. Mittler, R.; Vanderauwera, S.; Gollery, M.; Van Breusegem, F. Reactive oxygen gene network of plants. *Trends Plant Sci.* **2004**, *9*, 490–498. [CrossRef] [PubMed]
3. Noctor, G.; De Paepe, R.; Foyer, C.H. Mitochondrial redox biology and homeostasis in plants. *Trends Plant Sci.* **2007**, *12*, 125–134. [CrossRef] [PubMed]
4. Foyer, C.H.; Noctor, G. Redox regulation in photosynthetic organisms: Signaling, acclimation, and practical implications. *Antioxid. Redox Signal.* **2009**, *11*, 861–905. [CrossRef] [PubMed]
5. Foyer, C.H.; Bloom, A.J.; Queval, G.; Noctor, G. Photorespiratory metabolism: Genes, mutants, energetics, and redox signaling. *Annu. Rev. Plant Biol.* **2009**, *60*, 455–484. [CrossRef]
6. Pfannschmidt, T.; Bräutigam, K.; Wagner, R.; Dietzel, L.; Schröter, Y.; Steiner, S.; Nykytenko, A. Potential regulation of gene expression in photosynthetic cells by redox and energy state: Approaches towards better understanding. *Ann. Bot.* **2009**, *103*, 599–607. [CrossRef]
7. Woodson, J.D.; Chory, J. Coordination of gene expression between organellar and nuclear genomes. *Nat. Rev. Genet.* **2008**, *9*, 383–395. [CrossRef]
8. Karpinski, S.; Reynolds, H.; Karpinska, B.; Wingsle, G.; Creissen, G.; Mullineaux, P. Systemic signaling and acclimation in response to excess excitation energy in Arabidopsis. *Science* **1999**, *284*, 654–657. [CrossRef]
9. Suzuki, N.; Koussevitzky, S.; Mittler, R.; Miller, G. ROS and redox signalling in the response of plants to abiotic stress. *Plant Cell Environ.* **2012**, *35*, 259–270. [CrossRef]
10. Garg, N.; Manchanda, G. ROS generation in plants, boon or bane? *Plant Biosyst.* **2009**, *143*, 81–96. [CrossRef]
11. Huang, H.; Ullah, F.; Zhou, D.; Yi, M.; Zhao, Y. Mechanisms of ROS regulation of plant development and stress responses. *Front. Plant Sci.* **2019**, *10*, 800. [CrossRef] [PubMed]
12. Foyer, C.H.; Shigeoka, S. Understanding oxidative stress and antioxidant functions to enhance photosynthesis. *Plant Physiol.* **2011**, *155*, 93–100. [CrossRef]
13. Noctor, G.; Foyer, C.H. Intracellular redox compartmentation and ROS-related communication in regulation and signaling. *Plant Physiol.* **2016**, *171*, 1581–1592. [CrossRef] [PubMed]
14. Mullineaux, P.M.; Baker, N.R. Oxidative stress, antagonistic signaling for acclimation or cell death? *Plant Physiol.* **2010**, *154*, 521–525. [CrossRef] [PubMed]
15. Griebel, T.; Zeier, J. Light regulation and daytime dependency of inducible plant defenses in Arabidopsis: Phytochrome signaling controls systemic acquired resistance rather than local defense. *Plant Physiol.* **2008**, *147*, 790–801. [CrossRef]
16. Muühlenbock, P.; Szechynska-Hebda, M.; Płaszczyca, M.; Baudo, M.; Mateo, A.; Mullineaux, P.M.; Parker, J.E.; Karpińska, B.; Karpiński, S. Chloroplast signaling and LESION SIMULATING DISEASE1 regulate crosstalk between light acclimation and immunity in Arabidopsis. *Plant Cell* **2008**, *20*, 2339–2356. [CrossRef] [PubMed]
17. Ilangumaran, G.; Smith, D.L. Plant growth promoting rhizobacteria in amelioration of salinity stress: A systems biology perspective. *Front. Plant Sci.* **2017**, *8*, 1768. [CrossRef] [PubMed]
18. Kumar, A.; Singh, S.; Gaurav, A.K.; Srivastava, S.; Verma, J.P. Plant growth-promoting bacteria, biological tools for the mitigation of salinity stress in plants. *Front. Microbiol.* **2020**, *11*, 1216. [CrossRef]
19. Rojas-Tapias, D.; Moreno-Galván, A.; Pardo-Díaz, S.; Obando, M.; Rivera, D.; Bonilla, R. Effect of inoculation with plant growth-promoting bacteria (PGPB) on amelioration of saline stress in maize (*Zea mays*). *Appl. Soil Ecol.* **2012**, *61*, 264–272. [CrossRef]
20. Gutierrez Albanchez, E.; García-Villaraco, A.; Lucas, J.A.; Gutierrez, F.J.; Ramos-Solano, B. Priming fingerprint induced by *Bacillus amyloliquefaciens* QV15, a common pattern in *Arabidopsis thaliana* and in field-grown blackberry. *J. Plant Interact.* **2018**, *13*, 398–408. [CrossRef]

21. Galicia-Campos, E.; Ramos-Solano, B.; Montero-Palmero, M.; Gutierrez-Mañero, F.J.; García-Villaraco, A. Management of Plant Physiology with Beneficial Bacteria to Improve Leaf Bioactive Profiles and Plant Adaptation under Saline Stress in *Olea europea* L. *Foods* **2020**, *9*, 57. [CrossRef] [PubMed]
22. Yamane, K.; KawAsAki, M.; Taniguchi, M.; Miyake, H. Correlation between chloroplast ultrastructure and chlorophyll fluorescence characteristics in the leaves of rice (*Oryza sativa L.*) grown under salinity. *Plant Prod. Sci.* **2008**, *11*, 139–145. [CrossRef]
23. Lutts, S.; Kinet, J.; Bouharmont, J. NaCl-induced senescence in leaves of rice (*Oryza sativa* L.) cultivars differing in salinity resistance. *Ann. Bot.* **1996**, *78*, 389–398. [CrossRef]
24. Tsai, Y.; Chen, K.; Cheng, T.; Lee, C.; Lin, S.; Tung, C. Chlorophyll fluorescence analysis in diverse rice varieties reveals the positive correlation between the seedlings salt tolerance and photosynthetic efficiency. *BMC Plant Biol.* **2019**, *19*, 403. [CrossRef] [PubMed]
25. Bonilla, A.; Sarria, A.; Algar, E.; Ledesma, F.M.; Solano, B.R.; Fernandes, J.; Mañero, F.G. Microbe associated molecular patterns from rhizosphere bacteria trigger germination and *Papaver somniferum* metabolism under greenhouse conditions. *Plant Physiol. Biochem.* **2014**, *74*, 133–140. [CrossRef]
26. Guerfel, M.; Baccouri, O.; Boujnah, D.; Chaïbi, W.; Zarrouk, M. Impacts of water stress on gas exchange, water relations, chlorophyll content and leaf structure in the two main Tunisian olive (*Olea europaea* L.) cultivars. *Sci. Hortic.* **2009**, *119*, 257–263. [CrossRef]
27. Takahashi, S.; Badger, M.R. Photoprotection in plants, a new light on photosystem II damage. *Trends Plant Sci* **2011**, *16*, 53–60. [CrossRef]
28. Abdallah, M.B.; Trupiano, D.; Polzella, A.; De Zio, E.; Sassi, M.; Scaloni, A.; Zarrouk, M.; Youssef, N.B.; Scippa, G.S. Unraveling physiological, biochemical and molecular mechanisms involved in olive (*Olea europaea* L. cv. Chétoui) tolerance to drought and salt stresses. *J. Plant Physiol.* **2018**, *220*, 83–95. [CrossRef]
29. Gutiérrez-Albanchez, E.; Gradillas, A.; García, A.; García-Villaraco, A.; Gutierrez-Mañero, F.J.; Ramos-Solano, B. Elicitation with Bacillus QV15 reveals a pivotal role of F3H on flavonoid metabolism improving adaptation to biotic stress in blackberry. *PLoS ONE* **2020**, *15*, e0232626.
30. Trabelsi, L.; Gargouri, K.; Hassena, A.B.; Mbadra, C.; Ghrab, M.; Ncube, B.; Van Staden, J.; Gargouri, R. Impact of drought and salinity on olive water status and physiological performance in an arid climate. *Agric. Water Manag.* **2019**, *213*, 749–759. [CrossRef]
31. Sgherri, C.L.; Pinzino, C.; Navari-Izzo, F. Sunflower seedlings subjected to increasing stress by water deficit: Changes in O2—Production related to the composition of thylakoid membranes. *Physiol. Plant.* **1996**, *96*, 446–452. [CrossRef]
32. Boo, Y.C.; Jung, J. Water deficit—Induced oxidative stress and antioxidative defenses in rice plants. *J. Plant Physiol.* **1999**, *155*, 255–261.
33. Srivalli, B.; Sharma, G.; Khanna-Chopra, R. Antioxidative defense system in an upland rice cultivar subjected to increasing intensity of water stress followed by recovery. *Physiol. Plant.* **2003**, *119*, 503–512. [CrossRef]
34. Burton, G.W.; Ingold, K. β-Carotene, an unusual type of lipid antioxidant. *Science* **1984**, *224*, 569–573. [CrossRef] [PubMed]
35. Kchaou, H.; Larbi, A.; Chaieb, M.; Sagardoy, R.; Msallem, M.; Morales, F. Genotypic differentiation in the stomatal response to salinity and contrasting photosynthetic and photoprotection responses in five olive (*Olea europaea* L.) cultivars. *Sci. Hortic.* **2013**, *160*, 129–138. [CrossRef]
36. Foreman, J.; Demidchik, V.; Bothwell, J.H.; Mylona, P.; Miedema, H.; Torres, M.A.; Linstead, P.; Costa, S.; Brownlee, C.; Jones, J.; et al. Reactive oxygen species produced by NADPH oxidase regulate plant cell growth. *Nature* **2003**, *422*, 442–446. [CrossRef]
37. Foyer, C.H.; Noctor, G. Redox homeostasis and antioxidant signaling: A metabolic interface between stress perception and physiological responses. *Plant Cell* **2005**, *17*, 1866–1875. [CrossRef]
38. Doke, N.; Miura, Y.; Sanchez, L.M.; Kawakita, K. Involvement of superoxide in signal transduction: Responses to attack by pathogens, physical and chemical shocks, and UV irradiation. In *Causes of Photooxidative Stress and Amelioration of Defense Systems in Plants*; CRC Press: Boca Raton, FL, USA, 2019; pp. 177–198.
39. Ramos-Solano, B.; Algar, E.; Garcia-Villaraco, A.; Garcia-Cristobal, J.; Lucas Garcia, J.A.; Gutierrez-Mañero, F.J. Biotic elicitation of isoflavone metabolism with plant growth promoting rhizobacteria in early stages of development in Glycine max var. *Osumi. J. Agric. Food Chem.* **2010**, *58*, 1484–1492. [CrossRef]
40. Dietz, K.J. Thiol-Based Peroxidases and Ascorbate Peroxidases, Why Plants Rely on Multiple Peroxidase Systems in the Photosynthesizing Chloroplast? *Mol. Cells* **2016**, *39*, 20–25.
41. Arora, A.; Sairam, R.; Srivastava, G. Oxidative stress and antioxidative system in plants. *Curr. Sci.* **2002**, *82*, 1227–1238.
42. Asghari, B.; Khademian, R.; Sedaghati, B. Plant growth promoting rhizobacteria (PGPR) confer drought resistance and stimulate biosynthesis of secondary metabolites in pennyroyal (*Mentha pulegium* L.) under water shortage condition. *Sci. Hortic.* **2020**, *263*, 109132. [CrossRef]
43. Khademian, R.; Asghari, B.; Sedaghati, B.; Yaghoubian, Y. Plant beneficial rhizospheric microorganisms (PBRMs) mitigate deleterious effects of salinity in sesame (Sesamum indicum L.), physio-biochemical properties, fatty acids composition and secondary metabolites content. *Ind. Crops Prod.* **2019**, *136*, 129–139. [CrossRef]
44. Chin, D.; Kumar, R.S.; Suen, C.; Chien, C.; Hwang, M.; Hsu, C.; Xuhan, X.; Lai, Z.X.; Yeh, K.-W. Plant cytosolic ascorbate peroxidase with dual catalytic activity modulates abiotic stress tolerances. *iScience* **2019**, *16*, 31–49. [CrossRef] [PubMed]

45. Barriuso, J.; Ramos-Solano, B.; Santamaria, C.; Daza, A.; Gutierrez-Mañero, F. Effect of inoculation with putative plant growth-promoting rhizobacteria isolated from *Pinus* spp. on *Pinus pinea* growth, mycorrhization and rhizosphere microbial communities. *J. Appl. Microbiol.* **2008**, *105*, 1298–1309. [CrossRef]
46. Genty, B.; Briantais, J.; Baker, N.R. The relationship between the quantum yield of photosynthetic electron transport and quenching of chlorophyll fluorescence. *Biochim. Biophys. Acta Gen. Subj.* **1989**, *990*, 87–92. [CrossRef]
47. Lichtenthaler, H.K. Chlorophylls and carotenoids, pigments of photosynthetic biomembranes. *Meth. Enzymol.* **1987**, *148*, 350–382.
48. Porra, R.; Thompson, W.; Kriedemann, P. Determination of accurate extinction coefficients and simultaneous equations for assaying chlorophylls a and b extracted with four different solvents: Verification of the concentration of chlorophyll standards by atomic absorption spectroscopy. *Biochim. Biophys. Acta Bioenerg.* **1989**, *975*, 384–394. [CrossRef]
49. Carillo, P.; Gibon, Y. Protocol: Extraction and determination of proline. *Prometheuswiki* **2011**, 1–5.
50. Yemm, E.; Willis, A. The estimation of carbohydrates in plant extracts by anthrone. *Biochem. J.* **1954**, *57*, 508–514. [CrossRef]
51. García-Limones, C.; Hervás, A.; Navas-Cortés, J.A.; Jiménez-Díaz, R.M.; Tena, M. Induction of an antioxidant enzyme system and other oxidative stress markers associated with compatible and incompatible interactions between chickpea (*Cicer arietinum* L.) and *Fusarium oxysporum* f. sp. *ciceris*. *Physiol. Mol. Plant Pathol.* **2002**, *61*, 325–337. [CrossRef]
52. Zhang, J.; Kirkham, M. Antioxidant responses to drought in sunflower and sorghum seedlings. *New Phytol.* **1996**, *132*, 361–373. [CrossRef] [PubMed]
53. Xu, B.J.; Chang, S. A comparative study on phenolic profiles and antioxidant activities of legumes as affected by extraction solvents. *J. Food Sci.* **2007**, *72*, S159–S166. [CrossRef] [PubMed]
54. Zhishen, J.; Mengcheng, T.; Jianming, W. The determination of flavonoid contents in mulberry and their scavenging effects on superoxide radicals. *Food Chem.* **1999**, *64*, 555–559. [CrossRef]
55. Lucas, J.A.; Gutierrez-Albanchez, E.; Alfaya, T.; Feo-Brito, F.; Gutiérrez-Mañero, F.J. Oxidative stress in ryegrass growing under different air pollution levels and its likely effects on pollen allergenicity. *Plant Physiol. Biochem.* **2019**, *135*, 331–340. [CrossRef] [PubMed]

plants

Article

New Genomic Regions Identified for Resistance to Spot Blotch and Terminal Heat Stress in an Interspecific Population of *Triticum aestivum* and *T. spelta*

Sudhir Navathe [1,2], Ajeet Kumar Pandey [1], Sandeep Sharma [1], Ramesh Chand [1,*], Vinod Kumar Mishra [1], Dinesh Kumar [3,4], Sarika Jaiswal [3], Mir Asif Iquebal [3], Velu Govindan [5], Arun Kumar Joshi [6,7] and Pawan Kumar Singh [5,*]

1 Institute of Agricultural Sciences, Banaras Hindu University, Varanasi 221005, India
2 Agharkar Research Institute, G.G. Agharkar Road, Pune 411004, India
3 Centre for Agricultural Bioinformatics, ICAR-Indian Agricultural Statistics Research Institute, Library Avenue, PUSA, New Delhi 110012, India
4 Department of Biotechnology, School of Interdisciplinary and Applied Sciences, Central University of Haryana, Mahendergarh 123031, India
5 International Maize and Wheat Improvement Center (CIMMYT), Veracruz 56237, Mexico
6 Borlaug Institute for South Asia, NASC Complex, DPS Marg, New Delhi 110012, India
7 International Maize and Wheat Improvement Center (CIMMYT), G-2, B-Block, NASC Complex, DPS Marg, New Delhi 110012, India
* Correspondence: rc_vns@yahoo.co.in (R.C.); pk.singh@cgiar.org (P.K.S.)

Citation: Navathe, S.; Pandey, A.K.; Sharma, S.; Chand, R.; Mishra, V.K.; Kumar, D.; Jaiswal, S.; Iquebal, M.A.; Govindan, V.; Joshi, A.K.; et al. New Genomic Regions Identified for Resistance to Spot Blotch and Terminal Heat Stress in an Interspecific Population of *Triticum aestivum* and *T. spelta*. *Plants* **2022**, *11*, 2987. https://doi.org/10.3390/plants11212987

Academic Editor: Mingxun Chen

Received: 10 October 2022
Accepted: 3 November 2022
Published: 5 November 2022

Abstract: Wheat is one of the most widely grown and consumed food crops in the world. Spot blotch and terminal heat stress are the two significant constraints mainly in the Indo–Gangetic plains of South Asia. The study was undertaken using 185 recombinant lines (RILs) derived from the interspecific hybridization of '*Triticum aestivum* (HUW234) × *T. spelta* (H$^+$26)' to reveal genomic regions associated with tolerance to combined stress to spot blotch and terminal heat. Different physiological (NDVI, canopy temperature, leaf chlorophyll) and grain traits (TGW, grain size) were observed under stressed (spot blotch, terminal heat) and non-stressed environments. The mean maturity duration of RILs under combined stress was reduced by 12 days, whereas the normalized difference vegetation index (NDVI) was 46.03%. Similarly, the grain size was depleted under combined stress by 32.23% and thousand kernel weight (TKW) by 27.56% due to spot blotch and terminal heat stress, respectively. The genetic analysis using 6734 SNP markers identified 37 significant loci for the area under the disease progress curve (AUDPC) and NDVI. The genome-wide functional annotation of the SNP markers revealed gene functions such as plant chitinases, NB-ARC and NBS-LRR, and the peroxidase superfamily Cytochrome P450 have a positive role in the resistance through a hypersensitive response. Zinc finger domains, cysteine protease coding gene, F-box protein, ubiquitin, and associated proteins, play a substantial role in the combined stress of spot blotch and terminal heat in bread wheat, according to genomic domains ascribed to them. The study also highlights *T. speltoides* as a source of resistance to spot blotch and terminal heat tolerance.

Keywords: *Bipolaris sorokiniana*; Indo–Gangetic plain; resistance; terminal heat

1. Introduction

Wheat is one of the most widely grown and consumed food crops globally, having exceptionally high importance in the food system of South Asia. The eastern part of South Asia, which encompasses the eastern Gangetic Plains (EGP) of India, Nepal, and Bangladesh, is one of the most heavily populated parts of the world. In the EGP, where wheat is grown in about 10 m ha, the two major stresses to the wheat crop are spot blotch (SB) and terminal heat [1–3]. In EGP, SB caused by *Bipolaris sorokiniana* causes considerable yield loss between 15.5 and 19.6% annually [4]. However, when the disease is initiated at the flag leaf stage, losses of grains are estimated to be up to 24.2% [2]. (Singh et al., 2015). SB is normally a weak disease that takes advantage of heat stress (Rosyara et al., 2009), nutrient deficiency [5], and water stress [6] to induce significant grain damage. SB is favoured by cloudy and foggy days during the post-heading stage [7,8] and is expanding towards nontraditional cooler regions such as India's North West Plain Zone (NWPZ) [2,4]. Further, this disease is predicted to become more severe due to climate change, nutritional and water deficiencies, and increased heat stress [9].

Terminal heat and spot blotch lead to premature leaf senescence, reduced grain filling, low kernel weight, and yield reductions [1,3]. NDVI has been used as an indirect criterion for stay-green and higher grain yield under drought or heat conditions and spot blotch resistance [10]. NDVI and yield associations have been well recorded [10–12]. Grain yield, controlled by several component traits, is important for overall production and food security. However, component traits are equally important for production and the market value and milling yield of bread wheat [13,14].

Due to their importance in affecting wheat production, breeding for heat stress tolerance and spot blotch resistance are the two critical objectives of wheat improvement programs targeting the EGP of South Asia. Genetic evaluation for heat tolerance and spot blotch in cultivated wheat has been attempted separately, and resistance sources have been identified [3,15–18]. However, limited information is available for the genomic regions providing tolerance to the combined stress of spot blotch and terminal heat stress. Hence, this study was initiated to identify the genomic regions associated with the combined stress of spot blotch and terminal heat, wherein a population derived from the cross of *T. aestivum* and *T. spelta* was utilized.

2. Results

2.1. Descriptive Statistics for Quantitative Traits Indicate Reductions in Yield Contributing Traits Due to the Combined Stress of SB and Terminal Heat

The distribution of 185 RILs and the parents for mean values for the nine phenotypic traits under different sowing dates and treatments—control, spot blotch, terminal heat stress, and combined stress spot blotch and terminal heat, is presented in Figure 1. The mean and range of phenotypic traits in the RILs across the environments and treatments are given in Table S1. Under control conditions, the mean TGW was 34.17 ± 2.85, about 12% lower than the mean of 30.05 ± 3.33 g under SB infections. The decrease in grain area was observed from 11.57 ± 056 mm^2 to 7.84 ± 0.53 mm^2 (32.23%), which implicated in the reduction of thousand kernel weight from 34.17 ± 2.85 gm to 24.75 ± 2.46 gm (27.56%) (Table 1, Figure 1). The mean maturity duration (115.2 ± 1.19 days) under protected conditions decreased by >5% to 107.32 ± 4.06 days when infected to SB and further to 106.12 ± 1.22 and 103.76 ± 3.57 days under terminal heat stress and combined stress, respectively. The mean CT of the RILs under-protected was 23.11 ± 0.81 °C, while 25.73 ± 0.95 °C under spot blotch infection. A slight increase was noticed under terminal heat stress (30.23 ± 1.01 °C) and combined stress (31.75 ± 0.8 °C). The NDVI varied between 0.52–0.72, while after infection mean NDVI ranged between 0.39–0.59. It decreased to 0.42 ± 0.04 and 0.31 ± 0.03 units under terminal heat stress and combined SB + terminal heat stress (Figures S1 and S2).

The mean SPAD values were noted at 48.66 ± 2.4 units under protected conditions. However, this depleted to 43.43 ± 3.25 in response to SB infection. Under terminal heat stress, the SPAD mean was 48.63 ± 3.99, significantly decreasing to 26.1 under combined stress.

The AUDPC ranged between 299.31 and 689.35 with a mean of 504.69 7± 71.97. It was elevated to 731.14 ± 127.64 under combined stress. The decrease in grain area was 11.57 ± 056 mm^2 to 7.84 ± 0.53 mm^2 (32.23%) which was implicated in the reduction of TKW from 34.17 ± 2.85 gm to 24.75 ± 2.46 gm (27.56%) (Table 1).

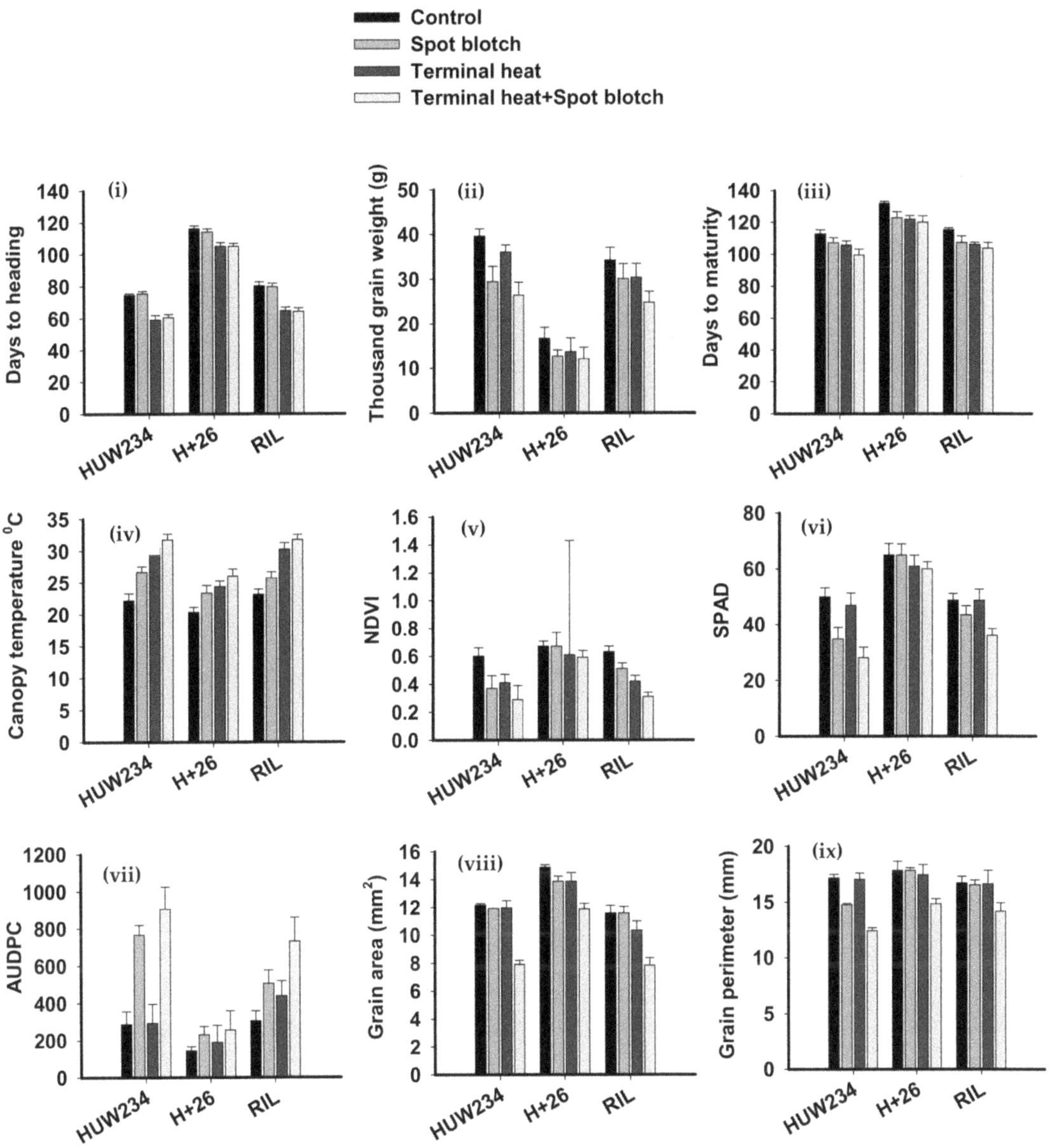

Figure 1. Summary of the effect of biotic and abiotic stresses on nine quantitative traits: (**i**) days to heading, (**ii**) thousand kernel weight, (**iii**) days to maturity, (**iv**) canopy temperature, (**v**) normalized distributed vegetation index (NDVI), (**vi**) soil plant analysis development (SPAD), (**vii**) area under disease progress curve (AUDPC), (**viii**) grain area, and (**ix**) grain perimeter.

Table 1. Effect of biotic and abiotic stress on the performance of nine traits in wheat.

Sr. No.	Phenotypic Trait	Control	Spot Blotch	Terminal Heat Stress	Spot Blotch + Terminal Heat Stress
1	Days to heading (days)	90.23 ± 1.79	89.71 ± 1.92	76.2 ± 2.46	76.57 ± 2.01
2	Thousand-grain weight (g)	30.15 ± 2.32	24.06 ± 2.71	26.68 ± 2.6	21.07 ± 2.65
3	Days to Maturity (days)	119.95 ± 1.66	112.49 ± 3.65	111.26 ± 2.06	107.75 ± 3.82
4	Canopy Temperature (°C)	21.88 ± 0.9	25.24 ± 1.01	28.36 ± 0.91	29.84 ± 0.94
5	NDVI	0.63 ± 0.05	0.52 ± 0.08	0.48 ± 0.09	0.4 ± 0.06
6	SPAD	54.5 ± 3.24	47.7 ± 3.82	52.14 ± 4.11	41.37 ± 2.91
7	AUDPC	244.69 ± 48.09	500.28 ± 56.39	306.91 ± 91.36	630.62 ± 116.58
8	Grain Area (mm^2)	12.87 ± 0.28	12.45 ± 0.42	12.05 ± 0.6	9.21 ± 0.41
9	Grain Perimeter (mm)	17.24 ± 0.57	16.37 ± 0.26	17.03 ± 0.89	13.8 ± 0.5

Data is mean ± SD.

The combined ANOVA for phenotypic traits indicated significant ($p \leq 0.0001$) differences for a year, sowing condition, treatment, RILs, and their interactions (Table 2). A highly significant and positive correlation was found for NDVI with DH, TGW, DM, SPAD, grain area, and grain perimeter (Table S2). A considerable grain area and grain perimeter correlation was obtained with DH, TGW, DM, SPAD, and NDVI, whereas AUDPC positively and significantly correlated with CT. Negative but highly significant associations were found between AUDPC with days to heading, TKW, DM, SPAD, NDVI, grain area, and grain perimeter.

Table 2. Analysis of variance for nine traits during the interaction of various treatments and environments (2015–2018).

Source	DF	Mean Sum of Squares								
		DH	TKW	DM	CT	NDVI	SPAD	AUDPC	Grain Area	Grain Perimeter
Year	2	36.71 *	9301.43 *	1515.34 *	5085.82 *	2.93 *	1275.91 *	19,890,523.95 *	38,070.98 *	32,605.08 *
Condition	1	266,584.51 *	23,293.19 *	44,301.94 *	47,917.63 *	47.16 *	15,318.28 *	35,978,424.99 *	6833.56 *	1701.23 *
Treatment	1	242.67 *	25,999.75 *	29,070.34 *	4757.77 *	15.16 *	88,172.20 *	67,495,554.29 *	1716.54 *	1928.05 *
RILs	184	81.23 *	116.31 *	52.74 *	6.082 *	0.0094 *	67.52 *	80,137.94 *	3.33 *	5.78 *
Rep	1	26.34	802.07 *	19.6	118.02 *	0.00011	962.57 *	1,885,835.47 *	1.29	40.88 *
Year × RILs	368	1.55	21.63 *	8.31 *	2.13 *	0.00345 *	8.86	44,840.46 *	0.632 *	0.52
Condition × RILs	184	8.58 *	18.05 *	44.91 *	4.26 *	0.0051 *	49.02 *	24,927.34 *	1.01 *	3.12 *
Treatment × RILs	184	36.72 *	56.38 *	48.92 *	4.92 *	0.00706 *	56.99 *	57,215.63 *	2.08 *	4.10 *
Year × Condition × Treatment × RILs	1295	2.04	19.51 *	21.52 *	6.50 *	0.0111 *	80.076 *	48,598.18 *	5.20 *	8.93 *

* significant at $p < 0.0001$. DF: degrees of freedom, DH: days to heading, TKW: thousand kernel weight DM: days to maturity, CT: canopy temperature, NDVI: normalized distributed vegetative index, AUDPC: area under disease progress curve, SPAD: soil plant analysis development.

2.2. *Diversity and Population Structure Analysis by SNP and DArT Markers*

The summary of minor allele frequency (MAF) and density of 5812 polymorphic SNP and DarT markers distributed on 21 chromosomes is given in Table S3. The population STRUCTURE analysis over 187 lines revealed the presence of three populations. The proportion of each population in the three clusters was 0.612, 0.056, and 0.332, respectively, indicating that the three clusters contained 114, 11, and 62 genotypes. Average distances (expected heterozygosity) between individuals within each cluster (K1–K3) were 0.2243, 0.1352, and 0.2207, respectively. The net nucleotide distance among structures, i.e., the average probability that a pair of alleles was different between K1 vs. K2, K1 vs. K3, and K2 vs. K3, was 0.2591, 0.1190, and 0.2834, respectively. The mean value of alpha was observed at 0.0485. Further, for each cluster (K1–K3), the mean value of Fst was e 0.5572, 0.7292, and 0.5630, respectively (Table S4).

The three-dimensional plot of the principal component analysis showing the genetic difference among RILs is shown in Figure 2a, while the heat map developed from 6734 SNP markers is in Figure 2b. The proportion and cumulative variances of the first three (3) PCAs were 16.80%, 6.10%, and 3.41%, respectively (Figures 2 and S3).

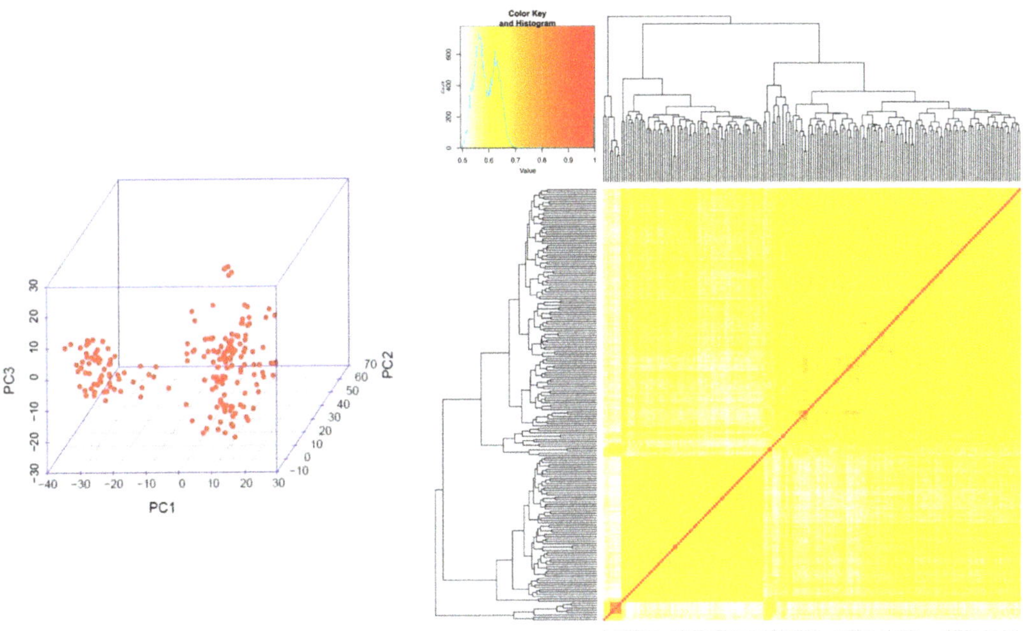

(a) (b)

Figure 2. (**a**) Three-dimensional plot of the first three principal components showing the genetic differences among 185 RILs and parents. (**b**) The heat map developed from 6734 SNP markers showed clustering of 185 RILs and parents.

Analysis of linkage disequilibrium: Out of 6734 markers used for the association mapping, 6369 markers were included in linkage disequilibrium analysis (LD). We filtered the markers with a minor allele threshold of 0.05, missing genotype 0.05, and removed individuals with a genotyping error of 0.1. (Table S5). We arrived at 2611 markers from the whole genome, which were further obtained for LD analysis. A total of 129,276 locus pairs were detected, and 32,221 locus pairs (24.92%) were found to be in LD at $p < 0.001$, of which 23,281 locus pairs (72.25%) were found at r2 > 0.1 and $p < 0.001$ (Figure S4).

2.3. Marker Trait Analysis Identifies the Unique SNP and Candidate Genes for Terminal Heat Stress and Spot Blotch Resistance

Eighty-five (85) significant marker–trait associations were identified for the nine different phenotypic traits over four different environments (Figure 3; Table S6). These marker–trait associations comprised thirty-seven (37) makers distributed majorly on nine chromosomes 1A, 1B, 2A, 3A, 5A, 5B, 6B, 7A, and 7B. The details of SNP makers and sequences are detailed in Table S6.

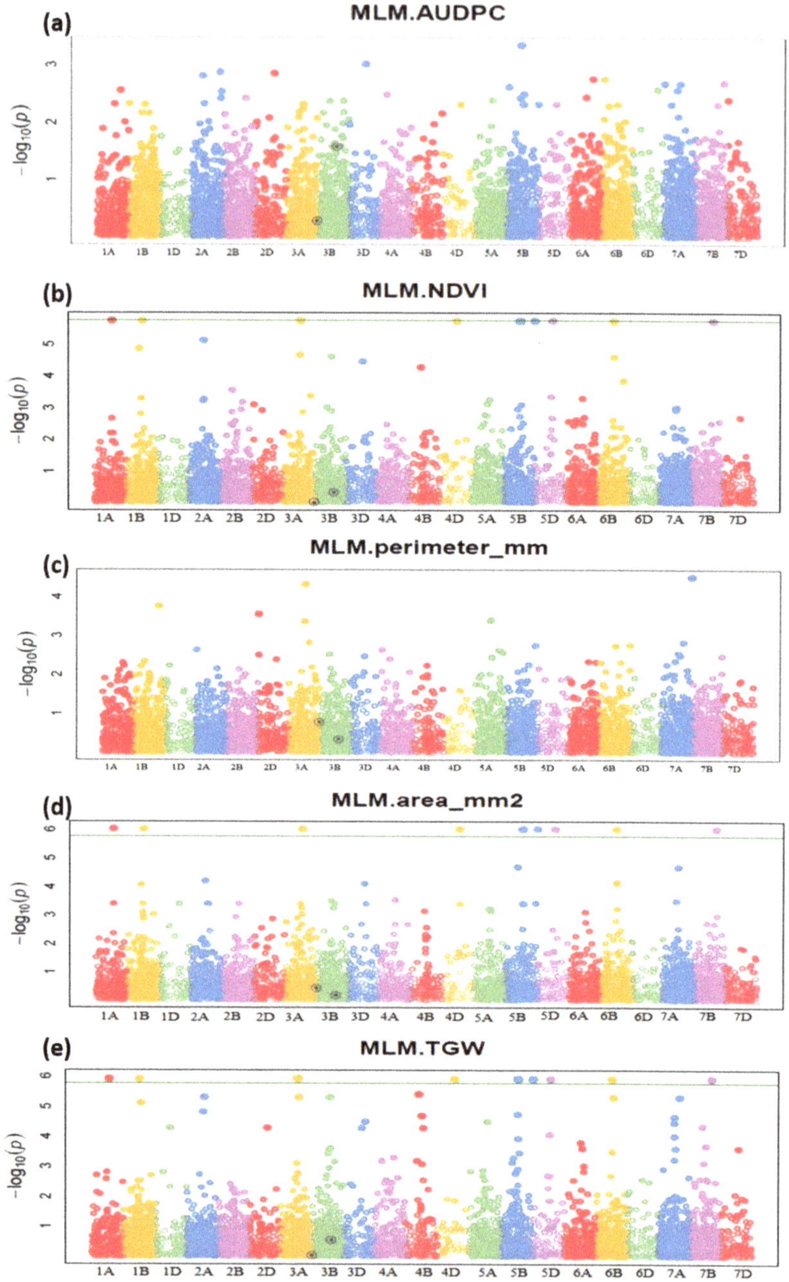

Figure 3. Genome-wide association scan for (**a**) area under disease progress (AUDPC), (**b**) normalized distributed vegetation index (NDVI), (**c**) grain perimeter, (**d**) grain area, (**e**) thousand-grain weight in RILs. The Manhattan plot was developed using a mixed linear model (MLM). The −log10 (p) values from a genome-wide scan are plotted against positions on each of the 21 wheat chromosomes. Horizontal lines indicate genome-wide significance thresholds.

The group of seven markers viz., 1125940 | F | 0 (1A), 1395486 | F | 0 (1B), 2256281 | F | 0, 980238 | F | 0 (3A), 1050819 | F | 0 (4D), 1029559 | F | 0, and 1020582 | F | 0 (5B) was commonly associated with the traits—grain area, days to heading, days to maturity, SPAD, and TKW. Seventeen identified markers are commonly associated with days to heading and days to maturity. For NDVI, nine unique marker–trait associations were identified, out of which two were on Chromosome 1A, three on 5A, and one each on 1B, 2A, 6B, and 7B. For AUDPC, five marker–trait associations were identified on chromosomes 2A, 5B, and 2D. The markers associated with SPAD were commonly associated with days to heading and days to maturity. Similarly, the markers associated with grain area, CT and TKW, were commonly linked with days to heading and maturity (Tables 3 and S6).

Table 3. Significant SNP and annotated proteins and transcripts on the high confidence genes based on wheat reference genome RefSeq1.1 (Ensembl Plant release 50, IWGSC RefSeq v1.1, October 2022).

Sr	Markers	Genomic Location	Trait	Transcript ID	Description
1	1058939 | F | 0	1A:530168043–530267926	NDVI	TraesCS1A02G340000	IPR027145: Periodic tryptophan protein 2, IPR020472: G-protein beta WD-40 repeat
				TraesCS1A02G339900	IPR003960: ATPase, AAA-type, conserved site, CDC48, IPR003338: N-terminal subdomain
				TraesCS1A02G339800	IPR006852: Glycosyltransferases, MUCI70
				TraesCS1A02G340100	IPR027417: P-loop containing nucleoside triphosphate hydrolase, IPR018368: ClpA/B, conserved site 1
2	1102573 | F | 0	1A:403359429–403388874	NDVI	TraesCS1A02G230600	Kinesin-like protein KIN-7N, IPR027417: P-loop containing nucleoside triphosphate hydrolase
3	1125940 | F | 0	1A: 499801953–499801988	Grain area, Days to heading, days to maturity, SPAD, Test weight (TGW), Canopy temperature	TraesCS1A02G309000	A0A1D5RW30 IPR001611: Leucine-rich repeat IPR013210: Leucine-rich repeat-containing N-terminal, plant-type IPR032675: Leucine-rich repeat domain superfamily
				TraesCS1A02G308800	A0A341NPS0 IPR002048: EF-hand domain IPR011992: EF-hand domain pair IPR018247: EF-Hand 1, calcium-binding site IPR039030: Calmodulin
				TraesCS1A02G308900	IPR032640: AMP-activated protein kinase, glycogen-binding domain
4	1241625 | F | 0	1B:687710001–687804688	Grain perimeter	TraesCS1B02G480300	IPR013210: Leucine-rich repeat-containing N-terminal, plant-type, IPR017441: Protein kinase, ATP binding site
				TraesCS1B02G480100	IPR012337: Ribonuclease H-like superfamily, IPR014811: Argonaute, linker 1 domain
				TraesCS1B02G480400	IPR045877: RNA-binding protein ZFP36-like, IPR000571: Zinc finger, CCCH-type
				TraesCS1B02G480200	IPR003388: Reticulon-like protein B12
5	1395486 | F | 0	1B:315459354–315462324	Grain Area, Days to heading, days to maturity, SPAD, Test weight (TGW), Canopy temperature	TraesCS1B02G174800	A0A341PBP3 IPR001210: Ribosomal protein S17e IPR036401: Ribosomal protein S17e-like superfamily
				TraesCS1B02G174900	A0A1D5SGD8 IPR001107: Band 7 domain IPR027705: Flotillin family IPR036013: Band 7/SPFH domain superfamily
16	3064765 | F | 0	1B:683455649–683585555	NDVI	TraesCS1B02G474800	IPR044798, Chromatin modification-related protein EAF1A/B
				TraesCS1B02G475200	IPR032675: Leucine-rich repeat domain
				TraesCS1B02G475400	IPR004907, ATPase, V1 complex, subunit C
7	1077356 | F | 0	2A:617631545–617741244	Days to heading, Days to maturity	TraesCS2A02G375100	IPR004182: GRAM domain, GEM-like protein
				TraesCS2A02G374800	IPR030847: Mitochondrial glycine transporter Hem25/SLC25A38

Table 3. *Cont.*

Sr	Markers	Genomic Location	Trait	Transcript ID	Description
8	2281188 | F | 0	2A:42147190–42231485	AUDPC	TraesCS2A02G088800	IPR000719: Protein kinase domain IPR001245: Serine-threonine/tyrosine-protein kinase, catalytic domain
				TraesCS2A02G088900	IPR042449: Ubiquitin-activating enzyme E1, inactive adenylation domain, subdomain 1 THIF-type NAD/FAD-binding (IPR000594)
9	1039495 | F | 0	2A:703378875–703411784	AUDPC	TraesCS2A02G454500	IPR013088, Zinc finger, NHR/GATA-type
				TraesCS2A02G454600	IPR044533, FCS-Like Zinc finger 1/2/3
10	2253029 | F | 0	2A:718816359–718988414	NDVI	TraesCS2A02G482000	CASP-like protein, IPR006702: Casparian strip membrane protein
				TraesCS2A02G482100	IPR001509, NAD-dependent epimerase/dehydratase
				TraesCS2A02G482400	IPR044661, Mediator of RNA polymerase II transcription subunit 15a/b/c-like
				TraesCS2A02G482700	IPR029063: S-adenosyl-L-methionine-dependent methyltransferase
11	3028841 | F | 0	2D:648337989–648380136	AUDPC	TraesCS2D02G594900	F-box domain-containing protein-related
				TraesCS2D02G594800	IPR000504: RNA recognition motif, IPR012677: Nucleotide-binding alpha-beta plait domain
				TraesCS2D02G594700	RING-type E3 ubiquitin transferase, IPR013083: Zinc finger
12	1019339 | F | 0	2D:625109256–625163527	Grain perimeter	TraesCS2D02G549600	IPR027417: P-loop containing nucleoside triphosphate hydrolase, IPR042197: Apoptotic protease-activating factors, helical domain
				TraesCS2D02G549700	IPR032675: Leucine-rich repeat domain superfamily, IPR044974: Disease resistance protein, plants
13	983670 | F | 0	3A:640746667–640773225	Days to heading, Days to maturity	TraesCS3A02G392900	IPR000109: Proton-dependent oligopeptide transporter, IPR018456: PTR2 family proton/oligopeptide symporter, conserved site
				TraesCS3A02G393000	IPR029058: Alpha/Beta hydrolase fold, IPR002168: Lipase, GDXG, putative histidine active site
14	3064641 | F | 0	3A:13351697–13418670	Grain perimeter	TraesCS3A02G025000	IPR023213: Chloramphenicol acetyltransferase-like domain
				TraesCS3A02G025300	Zinc finger FYVE domain-containing protein, IPR035669: GDSL lipase/esterase-like, plant
				TraesCS3A02G025200	E3 ubiquitin-protein ligase, IPR013083: Zinc finger, RING/FYVE/PHD-type
15	2256281 | F | 0	3A: 512307520–512555787	Grain area, Days to heading, days to maturity, SPAD, Test weight (TGW), Canopy temperature	TraesCS3A02G284100	Hexosyltransferase
				TraesCS3A02G283600	Peptidase S8 propeptide/proteinase inhibitor
				TraesCS3A02G283700	IPR007608: Senescence regulator S40
				TraesCS3A02G283900	IPR013857: NADH: ubiquinone oxidoreductase intermediate-associated protein 30
				TraesCS3A02G284100	Hexosyltransferase IPR002495: Glycosyl transferase, family 8 IPR029044: Nucleotide-diphospho-sugar transferases IPR029993: Plant galacturonosyltransferase GAUT
				TraesCS3A02G284200	Protodermal factor 1
16	980238 | F | 0	3A:638969536–639257641	Grain Area, Days to heading, days to maturity, SPAD, Test weight (TGW), Canopy temperature	TraesCS3A02G390800	tetratrico peptide repeat region (TPR)
				TraesCS3A02G390900	IPR007234:Vps53-like, N-terminal IPR039766: Vacuolar protein sorting-associated protein 53
				TraesCS3A02G391000	IPR025993: Ceramide glucosyltransferase IPR029044: Nucleotide-diphospho-sugar transferases
				TraesCS3A02G391100	IPR029768: Fructose-bisphosphate aldolase class-I active site
				TraesCS3A02G391400	IPR017907: Zinc finger, RING-type, conserved site

Table 3. *Cont.*

Sr	Markers	Genomic Location	Trait	Transcript ID	Description
17	2275693 | F | 0	3A:647604415–647821505	Days to heading, Days to maturity, Test weight	TraesCS3A02G402100	IPR011009: Protein kinase-like domain, IPR036426: Bulb-type lectin domain
				TraesCS3A02G402200	IPR001763: Rhodanese-like domain
				TraesCS3A02G402300	AS2, IPR004883: Lateral organ boundaries, LOB
18	976829 | F | 0	3B:672380580–672583235	Days to heading, Days to maturity	TraesCS3B02G433400	A0A077S6B7: BTB/POZ and TAZ domain-containing protein 3
				TraesCS3B02G433500	IPR018247: EF-Hand 1, calcium-binding site
				TraesCS3B02G433600	IPR001245: Serine-threonine/tyrosine-protein kinase, catalytic domain, IPR000719: Protein kinase domain, IPR008271: Serine/threonine-protein kinase, active site
				TraesCS3B02G433900	IPR013210: Leucine-rich repeat-containing N-terminal, plant-type
19	1088945 | F | 0	3D:436639761–436684217	AUDPC	TraesCS3D02G323600	IPR032675: Leucine-rich repeat domain, IPR044997 F-box protein, plant
				TraesCS3D02G323700	DNA-(apurinic or apyrimidinic site) lyase, IPR005135: Endonuclease/exonuclease/phosphatase
				TraesCS3D02G323800	FK506-binding-like protein (PTHR34567)
20	1034888 | F | 0	4A:597665575–598072618	Canopy temperature	TraesCS4A02G298500	IPR036410: Heat shock protein DnaJ, cysteine-rich domain
				TraesCS4A02G298600	IPR002068: Alpha crystallin/Hsp20 domain IPR008978:HSP20-like chaperone
				TraesCS4A02G299400	IPR002347: Short-chain dehydrogenase/reductase, IPR036291: NAD(P)-binding domain
				TraesCS4A02G298700	IPR013215: Cobalamin-independent methionine synthase MetE,
21	1050819 | F | 0	4D:46589710–46738670	Grain area, Days to heading, days to maturity, SPAD, Test weight	TraesCS4D02G071900	IPR044837 B3 domain-containing protein REM16-like, IPR015300 DNA-binding pseudo barrel domain superfamily
				TraesCS4D02G072000	IPR001461: Aspartic peptidase A1, IPR034161: Pepsin-like domain, plant
22	1088359 | F | 0	5A:503866360–503955851	Grain perimeter	TraesCS5A02G295400	IPR017736: Glycoside hydrolase, family 1, beta-glucosidase
				TraesCS5A02G295800	IPR003527: Mitogen-activated protein (MAP) kinase, conserved site IPR008271: Serine/threonine-protein kinase, active site
23	1029767 | F | 0	5A:615146872–615353569	NDVI	TraesCS5A02G431100	IPR007275: YTH domain
				TraesCS5A02G431600	Casein kinase I, photoperiodic control of flowering time, long-day repression, IPR008271: Serine/threonine-protein kinase, active site
				TraesCS5A02G431500	IPR032675: Leucine-rich repeat domain superfamily
				TraesCS5A02G431300	IPR001810: F-box domain IPR032675: Leucine-rich repeat domain
24	1045022 | F | 0	5A:691658614–691905149	NDVI	TraesCS5A02G534500	Flavin-containing monooxygenase, IPR036188: FAD/NAD(P)-binding domain superfamily
				TraesCS5A02G534800	IPR001810: F-box domain
				TraesCS5A02G534900	IPR042101: Signal recognition particle SRP54, IPR027417:P-loop containing nucleoside triphosphate hydrolase
				TraesCS5A02G534200	IPR039605: AT-hook motif nuclear-localized protein
25	3064380 | F | 0	5A:27509863–27509903	NDVI	TraesCS5A02G042600LC	NA
				TraesCS5A02G042700LC	NA

Table 3. *Cont.*

Sr	Markers	Genomic Location	Trait	Transcript ID	Description
26	1126383 I F I 0	5B:568398994–568517930	Days to heading, Days to maturity, SPAD, Test weight, Grain area	TraesCS5B02G389200	IPR002885: Pentatricopeptide repeat IPR011990: Tetratricopeptide-like helical domain
				TraesCS5B02G389300	EDA15, R022192: Mitochondrial degradosome RNA helicase subunit, C-terminal domain
				TraesCS5B02G389400	IPR044593, FCS-Like Zinc finger 8/MARD1
27	3064429 I F I 0	5B:596900954–596988713	AUDPC	TraesCS5B02G421900	IPR044991, Tetraspani, plant, auxin-activated signalling pathway
				TraesCS5B02G421100	IPR044659, Protein PELPK-like, Proline-rich protein 10, At5g09530
28	1029559 I F I 0	5B:281567207–281859354	Grain area, Days to heading, days to maturity, SPAD, Test weight (TGW)	TraesCS5B02G152400	IPR018247: EF-Hand 1, calcium-binding site IPR039647: EF-hand domain pair protein CML-like
				TraesCS5B02G152100	IPR029962 Trichome birefringence-like family
				TraesCS5B02G152300	IPR000547: Clathrin, heavy chain/VPS, 7-fold repeat IPR011990: Tetratricopeptide-like helical domain
				TraesCS5B02G152200	IPR014014: RNA helicase, DEAD-box type, Q motif IPR027417:P-loop containing nucleoside triphosphate hydrolase
				TraesCS5B02G152500	Ribosome assembly factor mrt4 IPR040637: 60S ribosomal protein L10P, insertion domain
				TraesCS5B02G152600	IPR017932 Glutamine amidotransferase type 2 domain
				TraesCS2D02G534800	IPR008271: Serine/threonine-protein kinase, active site
29	1020582 I F I 0	5B:609824667–609977091	Grain area, Days to heading, Days to maturity, SPAD, Test weight (TGW)	TraesCS5B02G435300	IPR002213: UDP-glucuronosyl/UDP-glucosyltransferase
				TraesCS5B02G435600	IPR043325: Alpha-Amylase Inhibitors (AAI), Lipid Transfer (LT) and Seed Storage (SS) Protein
30	987983 I F I 0	5D:104592141–104634865	Days to heading, Days to maturity, SPAD, Grain area	TraesCS5D02G095300	IPR001611: Leucine-rich repeat IPR008271: Serine/threonine-protein kinase, active site, IPR000719: Protein kinase domain
				TraesCS5D02G095400	IPR002171: Ribosomal protein L2 IPR008991: Translation protein SH3-like domain
31	2266275 I F I 0	6B:708055234–708286758	Days to heading, Days to maturity	TraesCS6B02G448700	IPR035896: AN1-like Zinc finger
				TraesCS6B02G447800	IPR044974, Disease resistance protein, plants IPR038005: Virus X resistance protein-like, coiled-coil domain
32	987210 I F I 0	6B:5683365–5845027	Days to heading, Days to maturity	TraesCS6B02G008700	IPR044814: Terpene cyclases, class 1, plant
				TraesCS6B02G008900	IPR008271: Serine/threonine-protein kinase, active site, IPR017441: Protein kinase IPR032675: Leucine-rich repeat domain
				TraesCS6B02G008800	IPR001232: S-phase kinase-associated protein 1-like
				TraesCS6B02G009105	IPR001881: EGF-like calcium-binding domain IPR008271: Serine/threonine-protein kinase, active site IPR011009: Protein kinase-like domain superfamily IPR018097: EGF-like calcium-binding, conserved site IPR025287: Wall-associated receptor kinase, galacturonan-binding domain
33	995480 I F I 0	6B:80769409–81048854	NDVI	TraesCS6B02G102800	IPR001810: F-box domain
				TraesCS6B02G102900	IPR008176: Defensin, plant, Amylase inhibitor-like protein
				TraesCS6B02G103200	IPR006813: Glycosyl transferase, family 17
34	1021511 I F I 0	7A:83081610–83137614	Days to heading, Days to maturity	TraesCS7A02G129000	IPR003311: AUX/IAA protein
35	2280866 I F I 0	7A:4249205–4264215	Grain perimeter	TraesCS7A02G009600	IPR023296, Glycosyl hydrolase, five-bladed beta-propellor domain

Table 3. *Cont.*

Sr	Markers	Genomic Location	Trait	Transcript ID	Description
36	2278379 | F | 0	7B:134493827–134645674	Days to heading, Days to maturity	TraesCS7B02G115900	IPR032799: Xylanase inhibitor, IPR001461: Aspartic peptidase A1 family
				TraesCS7B02G116200	PTHR31989: NAC domain-containing protein 82
37	1079395 | F | 0	7B:666498423–666698769	NDVI	TraesCS7B02G399800	IPR022991: Ribosomal protein L30e, conserved site
				TraesCS7B02G400300	IPR023213: Chloramphenicol acetyltransferase-like domain

The associated markers were linked to various important annotated gene families. The detailed annotation and their location in the whole genome sequence of wheat are presented in Table 4. The genome-wide functional annotation revealed that the gene functions such as plant chitinases, NB-ARC and NBS-LRR, are associated with many annotated SNP markers. A few other gene annotations—peroxidase superfamily and Cytochrome P450, appear to show a positive role in NAD(P) H-based regulation of oxidoreductase activity during the hypersensitive response (Table 3).

Table 4. Experimental layout for three consecutive cropping seasons (2015 to 2018).

Year	Environment	Treatment	Population
2015–2018 Cropping seasons November to April	EN1: Timely sown (last week of November)	Treatment 1: Control (no pathogen inoculation, protected using fungicide)	185 RILs + Parents (2 replications)
		Treatment 2: Spot blotch (inoculation by spot blotch pathogen, no protection by fungicide)	185 RILs + Parents (2 replications)
	EN2: Late sown (last week of December)	Treatment 3: Terminal Heat stress (no pathogen inoculated/protected using fungicide)	185 RILs + Parents (2 replications)
		Treatment 4: Spot blotch + terminal heat stress (inoculation by spot blotch pathogen no protection by the fungicide)	185 RILs + Parents (2 replications)

3. Discussion

Biotic stresses such as spot blotch and abiotic, which are mainly terminal heat, challenge field realities while cultivating wheat in South Asia. Spot blotch and heat stress at post-anthesis become critical during grain filling; hence this stage needs special protection [8,19]. The high temperature during grain filling stages affects photosynthesis and slashes the yield [20,21]. Recently, some wheat genotypes have been identified as being tolerant to abiotic and biotic stresses [2,9,22], and new varieties are being released to sustain wheat production. This has been mainly achieved through screening materials under heat stress and disease nurseries, which is costly and time-consuming. The multi-location shuttle breeding strategy has proven helpful and successfully selected the most favourable alleles contributing to resistance/tolerance toward important stresses [2,23]. The kernel size and grain yield are affected by heat stress and spot blotch events near anthesis [1,2]. The simulated reduction in kernel size of up to 3% per degree Celsius rise in temperature is well within the range of 2–7% from field experiments [24]. Likely, the loss in the green area due to spot blotch and terminal heat affects grain size due to the less remobilization of water-soluble carbohydrates stored in stem and leaf sheaths to developing grains under high temperature and disease [25].

We identified a group of seven SNP markers associated with six phenotypic traits that control the combined and individual stress of spot blotch and terminal heat. Additionally, several QTLs were identified for the grain attributes, such as higher TGW, grain weight/per spike, spikelet number/per spike, grain size and grain area. NDVI, which was first used to map spot blotch resistance by Kumar et al. [16], who mapped the resistance locus Sb2 and reported a negative correlation between the NDVI and AUDPC, which was also confirmed in the present study. Markers associated with NDVI can be effectively used to select resistant genotypes with most of the fitness traits. NDVI is influenced by the days to heading and days to maturity. Therefore, a marker–trait association for days to heading and maturity, TKW, and yield depend on healthy leaf area measured as NDVI. Markers associated with these traits can be essential in selecting promising spot blotch-resistant genotypes with higher yields under heat-stressed environments. In synthetic hexaploids derived from *Ae. tauschii*, Okamoto et al. [26] identified QTLs responsible for grain size and shape variation in the D genome.

Similarly, Williams and Sorrells [27] (2014) reported 31 QTLs for Seed size and shape in Synthetic W7984 × Opata M85 (SynOpDH) population. Additionally, environmentally stable QTLs on 1A and 2D and a pleiotropic QTL on 5A were also detected. Recently, Yan et al. [13] extensively studied the genetic factors in the 2D and 7D controlling grain size and shape variation. Similarly, Kumari et al. [14] identified seven markers associated with grain area, days to heading, days to maturity, SPAD, and test weight (TGW), indicating the important genomic regions associated with these traits.

Gene annotation of 21 SNP markers linked to the spot blotch and terminal heat-associated traits was also identified. The SNP 3026360 on chromosome 2D was associated with NBS-LRR and S/TPK protein; these are the most common R-gene. Another maker, 1125940 on chromosome 1A, was annotated to the potato virus X resistance protein (RX), and Peptidase S8, subtilisin, Asp-active site that took part in the resistance against potato virus X and belongs to an N-terminal coiled-coil domain, a nucleotide-binding domain, and leucine-rich repeats (CC-NB-LRR) [28,29]. One more SNP marker, 1079395 (chromosome 1A), was annotated to peroxidase superfamily protein. This protein plays a role in self-defence [30] by catalyzing oxide reduction of H_2O_2. Moreover, it has multiple tissue-specific functions during the hypersensitive response (HR).

The SNP 1122111 on chromosome 5A is annotated to plant phospholipase D (PLD), a calcium-dependent enzyme. This enzyme is linked with drought tolerance [31]. Similarly, another SNP marker, 1395486, on chromosome 2A, was annotated to cysteine peptidases belonging to the papain-like cysteine peptidase. This superfamily involved programmed cell death (PCD) based on disease resistance in various pathosystems [32]. Few markers were associated with the EF-hand motif, calcium-binding domains, and Cytochrome P450, which has a positive role in NAD (P) H based on the regulation of oxidoreductase activity during the hypersensitive response. A study by Ayana et al. [33] identified genomic regions on chromosomes 2D, 5A, and 7B linked to NBS-LRR, S/TPK, and many plants' defence-related protein families as Chitinase class I and peroxidases for spot blotch resistance.

Another gene with Zinc finger CCCH domain-containing protein pathogen-associated molecular pattern (PAMP)–that triggers immune responses was found in the genomic regions of SNP 1029559 | F | 0 and 1029767 | F | 0 in *Arabidopsis thaliana* [34].

In the genomic area of the SNP 995480 | F | 0 and wheat, two genes coding for Cytochrome P450 were also identified. The cysteine protease coding gene in the area is especially crucial since extracellular cysteine protease is required for pathogen recognition. Stress recognition causes an oxidative burst, followed by transcriptional reprogramming and HR, resulting in disease resistance [35]. Six F-box family proteins were also found in the region (SNPs 1034888 | F | 0, 1079395 | F | 0, 1045022 | F | 0, 1088945 | F | 0, and 3028841 | F | 0). F-box family protein controls various biological processes, including leaf senescence and responses to biotic [36] and abiotic stresses [37] independent of SAR via the ubiquitin–proteasome pathway. A ubiquitin family protein gene was discovered spanning the SNPs 2275693 | F | 0 and 1029767 | F | 0. Ubiquitin and associated proteins, which are components of the ubiquitin–proteasome system (UPS), regulate a variety of pathways, including responses to biotic and abiotic stimuli [38], and are one of the most important systems in plant defence [39].

In the backcross introgression lines produced from *T. durum* (cv. PDW274 susceptible) and *Ae. speltoides*, Kaur et al. [40] discovered five QTLs connected to SB resistance: Q.Sb.pau-2A, Q.Sb.pau-2B, Q.Sb.pau-3B, Q.Sb.pau-5B, and Q.Sb.pau-6A. The functional annotations for the previously published genomic regions are identical to those in the current work. At the same time, Tomar et al. [41] identified four new QTLs on Chr. 1A, 1D, 2B, and 6D that are associated with NBS-LRR, MADS-box transcription factors, and other disease-resistance protein families. Additionally, stable QTLs were detected on chromosomes 1B, 5A, 5B, 6A, 7A, and 7B in the CC population, explaining 2.89–10.32% of PV and collectively 39.91% of the total PV [42,43]. The quantitative genetic control of the spot blotch resistance, including markers linked to the *Lr46*, *Sb1*, *Sb2* and *Sb3* genes, has been reported recently [44]. The association of the 2NS translocation from *Ae. ventricosa* with spot blotch resistance and the spot blotch favourable alleles at the 2NS translocation, along with two markers on chromosome 3BS (3B_2280114 and 3B_5601689), has been reported first time from the multiple environment studies from Mexico and India. The findings of this study indicate the possibility of using the SNP linked for multiple stress regimes.

4. Materials and Methods

4.1. Plant Material, Experimental Design, and Layout of the Experiment

The experiments were conducted for three years (2014–2017) during the main wheat growing season (*Rabi/winter* season) at the Agricultural Research Farm of Banaras Hindu University, Varanasi (25.2° N and 83.0° E). One hundred eighty-five recombinant inbred lines (F_{10}) of '*T. aestivum* (HUW 234) × *T. spelta* ($H^{+}26$)' cross and their parents were evaluated for spot blotch, terminal heat stress, and their combined effect under field conditions. This is the same population that Pandey et al. [1] used from the same institution—Banaras Hindu University.

The experiment was conducted using an incomplete lattice design with four replications under two different environments—the third week of November was considered as timely sown (no terminal heat stress) but favourable for spot blotch only (EN1). The next sowing was carried out in the last week of December, considered late sown and favourable for both—spot blotch and terminal heat (EN2) [21,45]. The experiment was plated in plots of 1.2 m × 2 rows at a 22 cm distance between the rows. The plot area was considered to be 0.5 m^2. Approximately 50 seeds per row were sown. The detailed layout of the experiment is presented in Table 1. The crop was grown following prescribed agronomic practices (120 kg N: 60 kg P_2O_5: 40 kg K_2O per hectare) along with four irrigations. Two replications in each year/environment were protected with fungicide (Azoxystrobin 125 a.i. g/h), while two replications were inoculated with an aggressive isolate of *B. sorokiniana*. Fungicide was applied twice in GS 45 and GS 65 on Zadok's scale [46].

4.2. Pathogen Isolate and Inoculations

The *B. sorokiniana* isolate HD 3069 (MCC-1572) was multiplied by culturing on sorghum grain, following Chand et al. [47]. The spore suspensions were 10^4/mL in water containing 0.1 mL/L Tween 20. Plants were sprayed in the evening at growth stage ZGS 55 [46], and the field was irrigated the same day for optimal disease development.

4.3. Phenotyping for the Assessment of Spot Blotch and Terminal Heat Stress

4.3.1. Assessment of Disease Components

Scoring for disease reaction was initiated as soon as the first symptoms had appeared on all the accessions. The second scoring was conducted at ZGS 69, and the final was at ZGS 77. The scoring was conducted using a double-digit scale [48,49]. A disease severity

(DS) index was calculated from the ratio (D1/9) × (D2/9) × 100. AUDPCs were derived from the DS, as outlined by Shaner and Finney [50,51], based on the expression

$$AUDPC = \sum_{i=0}^{n=1} [\{(Y_i + Y(i+1)) \div 2\} \times (t(i+1) - t_i)] \quad (1)$$

where y_i is an assessment of disease at the ith observation, t_i is time (in days) at the ith observation, and n is the total number of observations.

4.3.2. Estimation of Chlorophyll Content by Soil Plant Analysis Development (SPAD)

A Minolta SPAD-502 m (Minolta Camera Ltd., Osaka, Japan) was used for the non-destructive assessment of leaf chlorophyll content described by Schlemmer et al. [52]. SPAD value was obtained as the mean of three measurements (base, middle, and apex) of the flag leaf (F). Three plants were recorded for each line in each replication. SPAD values were recorded 14 days after inoculation (dai), and at 21 dai, and an average was determined.

4.3.3. Canopy Temperature (CT)

The infrared gun LT 300 IRT was used to record CT; the readings were noted between 11:00 h to 14:00 h on cloudless, bright days within 0–4 days of disease assessment in the treated plots [53]. Canopy temperature was recorded at 14 dai and 21 dai and then averaged.

4.3.4. Normalized Difference Vegetative Index (NDVI)

A hand-held GreenSeeker crop device (Trimble Navigation Ltd., Sunnyvale, CA, USA) was used to measure NDVI [10]; the readings were obtained between 11.00 and 14.00 h. within 0–4 days of disease assessment in the treated plots.

4.3.5. Phenological Traits

Days to heading and physiological maturity (when the peduncle became yellow) were recorded from each RIL in each environment. The weight of 1000 kernels of individual RIL in each environment and each treatment was also recorded.

4.3.6. Grain Scan for Measurement of Grain Area and Perimeter

A grain scan tool was used to measure the grain size and area [54]. For further analysis, the grain scan generated data on grain area (mm^2) and perimeter (mm).

4.4. *Genetic Analysis of Spot Blotch and Heat Stress Associated with Phenotypic Traits*

4.4.1. Genotyping

The genomic DNA was extracted from 21-day-old seedlings of 185 RILs, and their parents using the Diversity Array Technology protocol described online http://www.diversityarrays.com/sites/default/files/pub/DArT_DNA_isolation.pdf (accessed on 10 November 2015). The resulting DNA was used for SNP and DArT array through Diversity Arrays Technology Pty. Ltd. University of Canberra, Australia. The 13,460 single nucleotide polymorphism (SNP) and 14,791 DArT loci obtained [55] were used for genome-wide association studies (GWAS) of various phenotypic traits associated with spot blotch and heat stress.

4.4.2. Population Structure Analysis

Population structure (Q) was analyzed using a model-based clustering method named STRUCTURE [56]. The number of subgroups (ΔK) in the panel was estimated following [57]. The fixation index (F_{ST}) of subpopulations was obtained through STRUCTURE run outputs. Population Matrix Q was also obtained for further analysis. Model-based cluster analysis implemented in STRUCTURE determines LnPD values for grouping 185 wheat genotypes into distinct groups. These values were used to determine the number of genetically distinct sub-populations implemented in the web-based tool Structure Harvester [58].

4.4.3. Genome-Wide Marker–Trait Association Analysis

The TASSEL 5.0 program [59] was used to calculate the population Kinship matrix based on the scaled identity by state (IBS) method using marker data that had passed quality filtering. Significant marker–trait associations (MTAs) were identified using a Mixed Linear Model (MLM) in TASSEL 5.0 (http://www.maizegenetics.net/; accessed on 20 May 2022) [59]. The analysis was carried out in PLINK [60], TASSEL [59], DARWIN [61,62] and GAPIT platforms in sequential order. The analysis was performed with a compressed mixed linear model [63] implemented in the GAPIT R package [64]. The MLM was run with the optimum compression level and previously determined population parameters [65]. To overcome the limitations of linkage mapping, LD mapping, a complementary strategy based on the correlation of genotype with phenotype in domesticated and natural populations, was used. This aided in shifting the emphasis from families to populations. The underlying principle of this approach is that LD between linked loci must be maintained over many generations. Linkage disequilibrium mapping exploits all historical recombination events in the population since the origin of the marker–trait association. However, to reduce the possibility of false positives in LD mapping. The population structure (Q) was estimated and then used in a mixed linear model to test for associations. The kinship relationships of the samples were also estimated for better control of type I error rates in association mapping, which accounts for population structure and relatedness.

4.4.4. In-Silico Analysis

The physical starting point of the marker preceded by the chromosome name was brought to Ensembl. A few thousand base pairs were added before and after (e.g., if the marker's position was 943389 on chromosome 2A, we used 2A: 942423–946423) to find the candidate genes linked to significant markers. The number of base pairs added varied for each marker depending on its proximity to the genes, but only the genes in the same genetic position were considered. The interval was then explored for predicted genes, and annotations from the IWGSC (https://www.wheatgenome.org/ accessed on 5 June 2022) were obtained. For several genes, the IWGSC annotations were not available. So, they evaluated based on orthologous genes in related species with known predicted functions using the comparative genomics tool in Plant Ensembl. In some cases, when the genes had a less similar disease resistance orthologue (<70%) in the annotated genomes of the related species in Ensembl, the sequence of the *T. aestivum* gene was brough to NCBI. The nucleotide basic local alignment search tool (BLAST) (http://blast.ncbi.nlm.nih.gov/Blast.cgi accessed on 5 June 2022) was used where only highly similar sequences (mega-blast) were considered. This search also included the gene predictions in different species available in GenBank but not in Ensembl. The *T. aestivum* gene transcripts and their available domains in Ensembl were also used (using the show transcript table link).

The blast (https://wheat.triticeaetoolbox.org/tools/blast/ accessed on 5 June 2022) in the Triticeae Toolbox website was used to perform a nucleotide BLAST (BLAST-n) of the significant marker sequences against the GBS markers in the Triticeae Toolbox (T3) database. Moreover, the JBrowse tool from T3 and GBrowse from URGI (https://urgi.versailles.inra.fr/gb2/gbrowse/wheat_survey_sequence_annotation; accessed on 25 May 2022) was also used to identify annotation to SNP markers.

4.5. Statistical Analysis

The statistical analysis was carried out using SAS software (version 9.2) [64]. The Sapiro and Wilnks test was first used to assess the normality of data, and the homogeneity of variance was determined using the Levene test. Field data from three consecutive years were subjected to variance analysis to determine significant differences among treatments using PROC GLM and the mixed model of SAS software. Correlation among the variables was established by PROC CORR using replicated data, and Bonferroni's adjustments at $p = 0.05$ were used to differentiate and group the genotype based on different variables.

5. Conclusions

Spot blotch and terminal heat tolerance are major constraints on wheat harvest, particularly in hot and humid climates prevailing in South Asia. Terminal heat and spot blotch lead to premature leaf senescence, reduced grain filling, low kernel weight, and reduced yield. The new sources of resistance must be continually identified and introgressed to counteract the restrictions posed by these stresses. The current work sheds light on the genetic regions that confer resistance to the combined stress of spot blotch and terminal heat stress. This research also specifies the possible use of NDVI, canopy temperature, and gain characteristics as indicator characteristics for high-throughput screening for these stresses during the vegetative and grain-filling stages. The genomic domains annotated to Zinc finger domains, cysteine protease coding gene, F-box family protein, ubiquitin and related proteins, and Cytochrome P450 reveal a significant role in the combined stress of spot blotch and terminal heat in bread wheat. The study also emphasizes *T. speltoides* as a source of resistance to spot blotch and terminal heat tolerance.

Supplementary Materials: The following supporting information can be downloaded at: https://www.mdpi.com/article/10.3390/plants11212987/s1, Figure S1: Frequency distribution of various phenotypic traits among RILs along with their parents under control (without inoculation) condition; Figure S2: Frequency distribution of various phenotypic traits among RILs along with its parents in response to spot blotch; Figure S3: Frequency distribution of various phenotypic traits among RILs along with its parents under terminal heat stress; Figure S4: Frequency distribution of various phenotypic traits among RILs along with its parents under combined stress of spot blotch and terminal heat stress; Figure S5: The plot of K versus Delta K showing variations (the steep change in slope indicates K = 3 as the best choice for the number of clusters); Figure S6: The plot showing the population structure of different recombinant inbred lines (RILs) along with parents in clusters for k = 3. (The numbers on the horizontal axis are the line numbers); Figure S7: Linkage disequilibrium (LD) plot based on Kinship matrix and SNP markers; Figure S8: Linkage disequilibrium decay plots are displaying r2 vs. genetic distance (cM) in 185 RILs along with parents. LD was calculated from intra-chromosomal pairs of the marker for the whole genome with 95 percentile confidence; Figure S9: Quantile–Quantile (Q-Q) plot showing the distribution of the recombinant inbred lines (RILs) analyzed in multiple linear models; Table S1: Mean performance for parent and recombinant inbred lines (RILs) across various environments and treatments; Table S2: Correlation coefficients between different nine phenotypic traits using pooled different sowing dates and treatment; Table S3: Summary of number, minor allele frequency (MAF) and density of single nucleotide polymorphism (SNP) markers used; Table S4: The Evanno table output at different values of K; Table S5: Linkage disequilibrium (LD) for the whole, A, B, and D genomes of wheat; Table S6: SNPs associated with spot blotch resistance identified through GWAS in the 185 RILs from the cross of *T. aestivum* (HUW 234) and *T. spelta* ($H^{+}26$).

Author Contributions: Conceptualization, R.C., V.K.M. and A.K.J.; data curation, A.K.P., D.K., S.J., M.A.I. and V.G.; formal analysis, D.K., S.J. and M.A.I.; funding acquisition, R.C. and A.K.J.; investigation, S.N. and A.K.P.; project administration, R.C. and A.K.J.; resources, V.G.; supervision, R.C., V.K.M., A.K.J. and P.K.S.; visualization, S.S.; writing—original draft, S.N.; writing—review and editing, S.S. and P.K.S. All authors have read and agreed to the published version of the manuscript.

Funding: S. Navathe received a Government of India Department of Science and Technology INSPIRE fellowship (Grant No: IF150037). The financial support received by the first and last authors from the Indian Council of Agriculture Research (ICAR), India, is also acknowledged.

Data Availability Statement: Raw phenotypic and genotypic data is submitted at "Mendeley Data" and available with the link, Mendeley Data, V1, https://doi.org/10.17632/k3ms7wmcjy.1 (accessed on 27 October 2022).

Acknowledgments: Authors acknowledge Banaras Hindu University for the infrastructure.

Conflicts of Interest: The authors declare no conflict of interest.

References

1. Pandey, A.K.; Mishra, V.K.; Chand, R.; Navathe, S.; Budhlakoti, N.; Srinivasa, J.; Sharma, S.; Joshi, A.K. Crosses with spelt improve tolerance of South Asian spring wheat to spot blotch, terminal heat stress, and their combination. *Sci. Rep.* **2021**, *11*, 6017. [CrossRef] [PubMed]
2. Singh, P.K.; Zhang, Y.; He, X.; Singh, R.P.; Chand, R.; Mishra, V.K.; Malaker, P.K.; Mostofa, A.R.; Mokhlesur, M.A.; Rahman, M.R.R.; et al. Development and characterization of the 4th CSISA-spot blotch nursery of bread wheat. *Eur. J. Plant Pathol.* **2015**, *143*, 595–605. [CrossRef]
3. Joshi, A.K.; Kumari, M.; Singh, V.P.; Reddy, C.M.; Kumar, S.; Rane, J.; Chand, R. Stay green trait: Variation, inheritance and its association with spot blotch resistance in spring wheat (*Triticum aestivum* L.). *Euphytica* **2007**, *153*, 59–71. [CrossRef]
4. Gupta, P.K.; Chand, R.; Vasistha, N.K.; Pandey, S.P.; Kumar, U.; Mishra, V.K.; Joshi, A.K. Spot blotch disease of wheat: The current status of research on genetics and breeding. *Plant Pathol.* **2018**, *67*, 508–531. [CrossRef]
5. Regmi, A.P.; Ladha, J.K.; Pasuquin, E.M.; Pathak, H.; Hobbs, P.R.; Shrestha, L.L.; Gharti, D.B.; Duveiller, E. The role of potassium in sustaining yields in a long-term rice-wheat experiment in the Indo-Gangetic plains of Nepal. *Biol. Fertil. Soils* **2002**, *36*, 240–247. [CrossRef]
6. Sharma, R.C.; Duveiller, E. Spot blotch continues to cause substantial grain yield reductions under resource-limited farming conditions. *J. Phytopathol.* **2006**, *154*, 482–488. [CrossRef]
7. Duveiller, E.; Kandel, Y.R.; Sharma, R.C.; Shrestha, S.M. Epidemiology of foliar blights (spot blotch and tan spot) of wheat in the plains bordering the Himalayas. *Phytopathology* **2005**, *95*, 248–256. [CrossRef]
8. Joshi, A.K.; Mishra, B.; Chatrath, R.; Ortiz-Ferrara, G.; Singh, R.P. Wheat improvement in India: Present status, emerging challenges and future prospects. *Euphytica* **2007**, *157*, 431–446. [CrossRef]
9. Crespo-Herrera, L.A.; Crossa, J.; Huerta-Espino, J.; Mondal, S.; Velu, G.; Juliana, P.; Vargas, M.; Pérez-Rodríguez, P.; Joshi, A.K.; Braun, H.J.; et al. Target Population of Environments for Wheat Breeding in India: Definition, Prediction and Genetic Gains. *Front. Plant Sci.* **2021**, *12*, 638520. [CrossRef]
10. Kumar, S.; Röder, M.S.; Singh, R.P.; Kumar, S.; Chand, R.; Joshi, A.K.; Kumar, U. Mapping of spot blotch disease resistance using NDVI as a substitute to visual observation in wheat (*Triticum aestivum* L.). *Mol. Breed.* **2016**, *36*, 95. [CrossRef]
11. Hazratkulova, S.; Sharma, R.C.; Alikulov, S.; Islomov, S.; Yuldashev, T.; Ziyaev, Z.; Khalikulov, Z.; Ziyadullaev, Z.; Turok, J. Analysis of genotypic variation for normalized difference vegetation index and its relationship with grain yield in winter wheat under terminal heat stress. *Plant Breed.* **2012**, *131*, 716–721. [CrossRef]
12. Liu, C.; Sukumaran, S.; Claverie, E.; Sansaloni, C.; Dreisigacker, S.; Reynolds, M. Genetic dissection of heat and drought stress QTLs in phenology-controlled synthetic derived recombinant inbred lines in spring wheat. *Mol. Breed.* **2019**, *39*, 34. [CrossRef]
13. Yan, L.; Liang, F.; Xu, H.; Zhang, X.; Zhai, H.; Sun, Q.; Ni, Z. Identification of QTL for Grain Size and Shape on the D Genome of Natural and Synthetic Allohexaploid Wheats with Near-Identical AABB Genomes. *Front. Plant Sci.* **2017**, *8*, 1705. [CrossRef]
14. Kumari, S.; Jaiswal, V.; Mishra, V.K.; Paliwal, R.; Balyan, H.S.; Gupta, P.K. QTL mapping for some grain traits in bread wheat (*Triticum aestivum* L.). *Physiol. Mol. Biol. Plants* **2018**, *24*, 909–920. [CrossRef]
15. Joshi, A.K.; Ortiz-Ferrara, G.; Crossa, J. Associations of environments in South Asia based on spot blotch disease of wheat caused by Bipolaris sorokiniana. *Crop Sci.* **2007**, *47*, 1071–1081. [CrossRef]
16. Rosyara, U.R.; Subedi, S.; Sharma, R.C.; Duveiller, E. Spot blotch and terminal heat stress tolerance in south Asian spring wheat genotypes. *Acta Agron. Hung.* **2009**, *57*, 425–435. [CrossRef]
17. Rosyara, U.R.; Vromman, D.; Duveiller, E. Canopy temperature depression as an indication of correlative measure of spot blotch resistance and heat stress tolerance in spring wheat. *J. Plant Pathol.* **2008**, *90*, 103–107.
18. Sukumaran, S.; Reynolds, M.P.; Sansaloni, C. Genome-wide association analyses identify QTL hotspots for yield and component traits in durum wheat grown under yield potential, drought, and heat stress environments. *Front. Plant Sci.* **2018**, *9*, 81. [CrossRef]
19. Chaurasia, S.; Joshi, A.K.; Dhari, R.; Chand, R. Resistance to foliar blight of wheat: A search. *Genet. Resour. Crop Evol.* **1999**, *46*, 469–475. [CrossRef]
20. Feng, B.; Liu, P.; Li, G.; Dong, S.T.; Wang, F.H.; Kong, L.A.; Zhang, J.W. Effect of heat stress on the photosynthetic characteristics in flag leaves at the grain-filling stage of different heat-resistant winter wheat varieties. *J. Agron. Crop Sci.* **2014**, *200*, 143–155. [CrossRef]
21. Paliwal, R.; Röder, M.S.; Kumar, U.; Srivastava, J.P.; Joshi, A.K. QTL mapping of terminal heat tolerance in hexaploid wheat (*T. aestivum* L.). *Theor. Appl. Genet.* **2012**, *125*, 561–575. [CrossRef] [PubMed]
22. Pask, A.; Reynolds, M.; Sharma, I.; Chatrath, R.; Singh, G.P.; Sohu, V.S.; Balasubramaniam, A. The CSISA wheat phenotyping network. In Proceedings of the International Workshop of the Wheat Yield Consortium, Cd. Obregon, Mexico, 5–7 March 2013; CIMMYT: Texcoco, Mexico, 2013; pp. 70–72.
23. Pask, A.; Joshi, A.K.; Manès, Y.; Sharma, I.; Chatrath, R.; Singh, G.P.; Mishra, V.K. A wheat phenotyping network to incorporate physiological traits for climate change in South Asia. *Field Crops Res.* **2014**, *168*, 156–167. [CrossRef]
24. Asseng, S.; Foster, I.A.N.; Turner, N.C. The impact of temperature variability on wheat yields. *Glob. Chang. Biol.* **2011**, *17*, 997–1012. [CrossRef]
25. Rehman, H.U.; Tariq, A.; Ashraf, I.; Ahmed, M.; Muscolo, A.; Basra, S.M.A.; Reynolds, M. Evaluation of physiological and morphological traits for improving spring wheat adaptation to terminal heat stress. *Plants* **2021**, *10*, 455. [CrossRef] [PubMed]

26. Okamoto, Y.; Nguyen, A.T.; Yoshioka, M. Identification of quantitative trait loci controlling grain size and shape in the D genome of synthetic hexaploid wheat lines. *Breed. Sci.* **2013**, *63*, 423–429. [CrossRef] [PubMed]
27. Williams, K.; Sorrells, M.E. Three-dimensional seed size and shape QTL in hexaploid wheat (*Triticum aestivum* L.) populations. *Crop Sci.* **2014**, *54*, 98–110. [CrossRef]
28. Martin, G.B.; Bogdanove, A.J.; Sessa, G. Understanding the functions of plant disease resistance proteins. *Annu. Rev. Plant Biol.* **2003**, *54*, 23–61. [CrossRef]
29. Rairdan, G.J.; Collier, S.M.; Sacco, M.A.; Baldwin, T.T.; Boettrich, T.; Moffett, P. The coiled-coil and nucleotide-binding domains of the potato Rx disease resistance protein function in pathogen recognition and signalling. *Plant Cell* **2008**, *20*, 739–751. [CrossRef]
30. Hiraga, S.; Sasaki, K.; Ito, H.; Ohashi, Y.; Matsui, H. A large family of class III plant peroxidases. *Plant Cell Physiol.* **2001**, *42*, 462–468. [CrossRef]
31. Peng, Y.; Zhang, J.; Cao, G.; Xie, Y.; Liu, X.; Lu, M.; Wang, G. Overexpression of a PLDα1 gene from *Setaria italica* enhances the sensitivity of Arabidopsis to abscisic acid and improves its drought tolerance. *Plant Cell Rep.* **2010**, *29*, 793–802. [CrossRef]
32. Niño, M.; Kim, J.; Lee, H.J.; Abdula, S.E.; Nou, I.S.; Cho, Y.G. Key roles of cysteine protease in different plant pathosystem. *Plant Breed. Biotech.* **2014**, *2*, 97–109. [CrossRef]
33. Ayana, G.T.; Ali, S.; Sidhu, J.S.; Gonzalez-Hernandez, J.L.; Turnipseed, B.; Sehgal, S.K. Genome-wide association study for spot blotch resistance in hard winter wheat. *Front. Plant Sci.* **2018**, *9*, 926. [CrossRef]
34. Maldonado-Bonilla, L.D.; Eschen-Lippold, L.; Gago-Zachert, S.; Tabassum, N.; Bauer, N.; Scheel, D.; Lee, J. The arabidopsis tandem zinc finger 9 protein binds RNA and mediates pathogen-associated molecular pattern triggered immune responses. *Plant Cell Physiol.* **2013**, *55*, 412–425. [CrossRef]
35. Thomas, E.L.; Van der Hoorn, R.A.L. Ten prominent host proteases in plant-pathogen interactions. *Int. J. Mol. Sci.* **2018**, *19*, 639–653. [CrossRef]
36. Kim, H.S.; Delaney, T.P. Arabidopsis SON1 is an F-box protein that regulates a novel induced defense response independent of both salicylic acid and systemic acquired resistance. *Plant Cell* **2002**, *14*, 1469–1482. [CrossRef]
37. Calderón-Villalobos, L.I.A.; Nill, C.; Marrocco, K.; Kretsch, T.; Schwechheimer, C. The evolutionarily conserved *Arabidopsis thaliana* F-box protein AtFBP7 is required for efficient translation during temperature stress. *Gene* **2007**, *392*, 106–116. [CrossRef]
38. Sadanandom, A.; Bailey, M.; Ewan, R.; Lee, J.; Nelis, S. The ubiquitin-proteasome system: Central modifier of plant signalling. *New Phytol.* **2012**, *196*, 13–28. [CrossRef]
39. Üstün, S.; Sheikh, A.; Gimenez-Ibanez, S.; Jones, A.; Ntoukakis, V.; Börnke, F. The proteasome acts as a hub for plant immunity and is targeted by Pseudomonas type III effectors. *Plant Physiol.* **2016**, *172*, 1941–1958. [CrossRef]
40. Kaur, J.; Kaur, J.; Dhillon, G.S.; Kaur, H.; Singh, J.; Bala, R.; Srivastava, P.; Kaur, S.; Sharma, A.; Chhuneja, P. Characterization and mapping of spot blotch in *Triticum durum-Aegilops speltoides* Introgression Lines Using SNP Markers. *Front. Plant Sci.* **2021**, *12*, 650400. [CrossRef]
41. Tomar, V.; Singh, D.; Singh, R.P.; Poland, J.; Joshi, A.K.; Dhillon, G.S.; Singh, P.K.; Kumar, S.; Rahman, M.M.; Tiwari, B.S.; et al. New QTLs for spot blotch disease resistance in wheat (*Triticum aestivum* L.) using genome-wide association mapping. *Front. Genet.* **2020**, *11*, 1740. [CrossRef]
42. Roy, C.; Gahtyari, N.C.; He, X.; Mishra, V.K.; Chand, R.; Joshi, A.K.; Singh, P.K. Dissecting quantitative trait loci for spot blotch resistance in South Asia using two wheat recombinant inbred line populations. *Front. Plant Sci.* **2021**, *12*, 641324. [CrossRef] [PubMed]
43. Gahtyari, N.C.; Roy, C.; He, X.; Roy, K.K.; Reza, M.M.A.; Hakim, M.A.; Malaker, P.K.; Joshi, A.K.; Singh, P.K. Identification of QTLs for spot blotch resistance in two bi-parental mapping populations of wheat. *Plants* **2021**, *10*, 973. [CrossRef] [PubMed]
44. Juliana, P.; He, X.; Poland, J.; Shrestha, S.; Joshi, A.K.; Huerta-Espino, J.; Govindan, V.; Crespo-Herrera, L.A.; Mondal, S.; Kumar, U.; et al. Genome-wide association mapping indicates quantitative genetic control of spot blotch resistance in bread wheat and the favorable effects of some spot blotch loci on grain yield. *Front. Plant Sci.* **2022**, *13*, 835095. [CrossRef] [PubMed]
45. Tiwari, C.; Wallwork, H.; Kumar, U.; Dhari, R.; Arun, B.; Mishra, V.K.; Joshi, A.K. Molecular mapping of high-temperature tolerance in bread wheat adapted to the Eastern Gangetic Plain region of India. *Field Crops Res.* **2013**, *154*, 201–210. [CrossRef]
46. Zadoks, J.C.; Chang, T.T.; Konzak, C.F. A Decimal Code for the Growth Stages of Cereals. *Weed Res.* **1974**, *14*, 415–421. [CrossRef]
47. Chand, R.; Yadav, O.P.; Bashyal, B.M.; Prasad, L.C.; Joshi, A.K. Technique for the maintenance of heterokaryotic isolates of Bipolaris sorokiniana. *Indian Phytopathol.* **2013**, *66*, 61–65.
48. Saari, E.E.; Prescott, J.M. A scale for appraising the foliar intensity of wheat diseases. *Plant Dis. Rep.* **1975**, *59*, 377–380.
49. Eyal, Z.; Scharen, A.L.; Prescott, J.M.; Van Ginkel, M. *The Septoria Disease of Wheat: Concepts and Methods of Disease Management*; CIMMYT: Texcoco, Mexico, 1987.
50. Shaner, G.; Finney, R.E. The effect of nitrogen fertilization on the expression of slow-mildewing resistance in Knox wheat. *Phytopathology* **1977**, *67*, 1051–1056. [CrossRef]
51. Madden, L.V.; Hughes, G.; Bosch, F. *The Study of Plant Disease Epidemics*; American Phytopathological Society (APS Press): St. Paul, MN, USA, 2007.
52. Schlemmer, M.R.; Francis, D.D.; Shanahan, J.F.; Schepers, J.S. Remotely measuring chlorophyll content in corn leaves with different nitrogen levels and relative water content. *Agron. J.* **2005**, *97*, 106–112. [CrossRef]

53. Reynolds, M.; Manes, Y.; Rebetzke, G. Application of physiology in breeding for heat and drought stress. In *Physiological Breeding I: Interdisciplinary Approaches to Improve Crop Adaptation*; Reynolds, M.P., Pask, A.J.D., Mullan, D.M., Eds.; CIMMYT: Texcoco, Mexico, 2012; pp. 18–32.
54. Whan, A.P.; Smith, A.B.; Cavanagh, C.R.; Ral, J.P.F.; Shaw, L.M.; Howitt, C.A.; Bischof, L. GrainScan: A low cost, fast method for grain size and colour measurements. *Plant Methods* **2014**, *10*, 23. [CrossRef]
55. Srinivasan, J. Molecular Mapping and Development of Zinc and Iron Rich Wheat Lines Using *Triticum aestivum* and *T. spelta* Crosses. Ph.D. Thesis, Banaras Hindu University, Varanasi, India, 2013.
56. Pritchard, J.K.; Stephens, M.; Rosenberg, N.A.; Donnelly, P. Association mapping in structured populations. *Am. J. Hum. Genet.* **2000**, *67*, 170181. [CrossRef]
57. Evanno, G.; Regnaut, S.; Goudet, J. Detecting the number of clusters of individuals using the software STRUCTURE: A simulation study. *Mol. Ecol.* **2005**, *14*, 26112620. [CrossRef]
58. Earl, D.A.; VonHoldt, B.M. STRUCTURE HARVESTER: A website and program for visualizing STRUCTURE output and implementing the Evanno method. *Conserv. Genet. Resour.* **2012**, *4*, 359–361. [CrossRef]
59. Bradbury, P.J.; Zhang, Z.; Kroon, D.E.; Casstevens, T.M.; Ramdoss, Y.; Buckler, E.S. TASSEL: Software for association mapping of complex traits in diverse samples. *Bioinformatics* **2007**, *23*, 2633–2635. [CrossRef]
60. Purcell, S.; Neale, B.; Todd-Brown, K.; Thomas, L.; Ferreira, M.A.; Bender, D.; Maller, J.; Sklar, P.; De Bakker, P.I.; Daly, M.J.; et al. PLINK: A tool set for whole-genome association and population-based linkage analyses. *Am. J. Hum. Genet.* **2007**, *81*, 559–575. [CrossRef]
61. Perrier, X.; Flori, A.; Bonnot, F. Data analysis methods. In *Genetic Diversity of Cultivated Tropical Plants*; Hamon, P., Seguin, M., Perrier, X., Glaszmann, J.C., Eds.; Enfield, Science Publishers: Montpellier, France, 2003; pp. 43–76.
62. Perrier, X.; Jacquemoud-Collet, J.P. DARwin Software. 2006. Available online: http://darwin.cirad.fr/ (accessed on 25 May 2022).
63. Zhang, Z.; Ersoz, E.; Lai, C.; Todhunter, R.J.; Tiwari, H.K.; Gore, M.A. Mixed linear model approach adapted for genome-wide association studies. *Nat. Genet.* **2010**, *42*, 355–360. [CrossRef]
64. Lipka, A.E.; Tian, F.; Wang, Q.; Peiffer, J.; Li, M.; Bradbury, P.J.; Gore, M.A.; Buckler, E.S.; Zhang, Z. GAPIT: Genome association and prediction integrated tool. *Bioinformatics* **2012**, *28*, 2397–2399. [CrossRef]
65. SAS Institute Inc. *SAS 9.1.3 Help and Documentation*; SAS Institute Inc.: Cary, NC, USA, 2002–2004.

plants

MDPI

Article

Evaluation of the Impacts of Potassium Bicarbonate, *Moringa oleifera* Seed Extract, and *Bacillus subtilis* on Sugar Beet Powdery Mildew

Baher A. El-Nogoumy [1], Mohamed A. Salem [2,3], Gabr A. El-Kot [4], Salem Hamden [4], Mohamed D. Sehsah [5], Abeer H. Makhlouf [6] and Yasser Nehela [7,*]

1 Microbiology Department, Faculty of Science, Kafrelsheikh University, Kafr El-Sheikh 33516, Egypt
2 Department of Chemistry, Faculty of Science & Arts, King Khalid University, Abha 62529, Saudi Arabia
3 Department of Chemistry, Faculty of Science, Al-Azhar University, Nasr City, Cairo 11284, Egypt
4 Department of Agricultural Botany, Faculty of Agriculture, Kafrelsheikh University, Kafr El-Sheikh 33516, Egypt
5 Plant Pathology Research Institute, Agricultural Research Center, Giza 12619, Egypt
6 Faculty of Agriculture, Minufiya University, Shibin El-Kom 32511, Egypt
7 Department of Agricultural Botany, Faculty of Agriculture, Tanta University, Tanta 31511, Egypt
* Correspondence: yasser.nehela@ufl.edu

Abstract: Powdery mildew disease, caused by *Erysiphe betae*, is one of the most threatening diseases on sugar beet plants worldwide. It causes a great loss in the root yield, sugar percentage, and quality of produced sugar. In the current study, we aimed to evaluate the susceptibility of 25 sugar beet cultivars to infection with powdery mildew disease under Egyptian conditions. Moreover, we evaluated the impacts of three eco-friendly materials, including potassium bicarbonate ($KHCO_3$; at 5 and 10 g L^{-1}), *Moringa oleifera* seed extract (25 and 50 g L^{-1}), and the biocontrol agent, *Bacillus subtilis* (10^8 cell suspension) against *E. betae* in two successive seasons 2020 and 2021. Our findings showed that there were significant differences between these 25 cultivars in their susceptibility to the disease under study. Using the detached leaves technique in vitro, *B. subtilis* showed strong antifungal activity against *E. betae*. Moreover, both concentrations of $KHCO_3$ and moringa seed extract significantly reduced the disease severity. Under field conditions, tested treatments significantly reduced the severity of powdery mildew disease and prevented *E. betae* from producing its conidiophores and conidia. Scanning electron microscope examination of treated leaves demonstrated the presence of the decomposition of fungal hyphae, conidiophores, conidia, and the occurrence of plasmolysis to fungal cells and spores on the surface of the leaves. Furthermore, these treatments greatly improved the percent of sucrose and soluble solids content, as well as the enzymatic activity of peroxidase, polyphenol oxidase, and phenylalanine ammonia-lyase. It is noteworthy that treatment with moringa seed extract gave the best results, followed by potassium bicarbonate, then *B. subtilis* cell suspension. Generally, it is recommended to use the substances used in this research to combat powdery mildew to minimize or prevent the use of chemical fungicides harmful to public health and the environment.

Keywords: antioxidant; *Bacillus subtilis*; *Erysiphe betae*; *Moringa oleifera*; PAL; potassium bicarbonate; salicylic acid; scanning electron microscope; sugar beet

Citation: El-Nogoumy, B.A.; Salem, M.A.; El-Kot, G.A.; Hamden, S.; Sehsah, M.D.; Makhlouf, A.H.; Nehela, Y. Evaluation of the Impacts of Potassium Bicarbonate, *Moringa oleifera* Seed Extract, and *Bacillus subtilis* on Sugar Beet Powdery Mildew. *Plants* **2022**, *11*, 3258. https://doi.org/10.3390/plants11233258

Academic Editors: Mingxun Chen, Lixi Jiang and Yuan Guo

Received: 21 October 2022
Accepted: 22 November 2022
Published: 27 November 2022

1. Introduction

Sugar beet (*Beta vulgaris* L.) is an important crop for sugar production worldwide. It ranks second in sugar production after sugar cane. In Egypt, about 269 thousand hectares are planted annually with the sugar beet crop with an average production of 18.7 tons per feddan [1]. The demand for sugar has increased due to the annual increase in the population. The Egyptian Ministry of Agriculture, represented by the Sugar Crops Council, aims to increase the cultivated areas of the sugar beet crop in the new season 2022–2021 to approximately 302 thousand hectares.

Powdery mildew of sugar beet, caused by *Erysiphe betae*, is a serious disease worldwide that causes a significant reduction in root yield [2]. The phytopathogenic fungus *E. betae* mainly attacks the leaves of sugar beet. Upon severe attack, up to a 22% reduction in root yield, as well as a 13% reduction in sucrose content in roots, has been recorded [3], thus reducing the yield and quality of the harvested crop [4]. It also reduces extractable root crops and sucrose and increases impurity concentrations, requiring higher processing and significant losses [5].

One of the best means of controlling plant diseases, including the powdery mildew of beets, is the production of resistant varieties [6]. However, the production of resistant varieties requires a lot of time, effort, and money and may not completely meet the required specifications [7]. The most traditional management of powdery mildew mainly relies on the use of fungicides [8]. However, the use of fungicides is instrumental in controlling plant diseases, and chemical control is economically costly and environmentally undesirable. In addition to their harmful effects on the environment and public health, the extensive use of fungicides leads to the development of fungicide-resistant strains of phytopathogenic fungi which has limited their use [9].

Accordingly, there is a necessity to search for new fungicides with different active ingredients. However, chemical fungicides cause serious problems in public health and the ecological system, where their ability to cause many diseases to animals and humans has been proven and they can kill beneficial organisms [10]. Several eco-friendly strategies have been developed to manage various diseases as alternative means for these chemical fungicides. Samples enclose environmentally pleasant chemicals such as potassium bicarbonate, commercially named Armicarb®, magnesium sulfate, copper sulfate, and potassium oxide [11,12], plant oils, and plant extracts, such as *Moringa oleifera* plant excerpt [12,13], or the use of microorganisms such as *Bacillus megaterium* and *Trichoderma album*, whether as cultural filtrates or as a suspension [12,14–17]. At least eighty years ago, Marloth demonstrated the fungal activity of bicarbonate salts [18].

Exogenous application of alternative disease management means may affect the biochemical and physiological traits within treated plants under abiotic stress [19–21] and biotic challenges such as viral [22], bacterial [23], and fungal [24–26] phytopathogens. The major biochemical changes are via the induction of enzymatic and nonenzymatic antioxidant defense machinery [24,25,27,28]. Enzymatic antioxidant defense machinery mainly relies on the activity of some enzymes such as polyphenol oxidase, phenylalanine, ammonia-lyase, peroxidase, and other enzymes [12,14,29]. The scanning electron microscope has been used by many researchers, including Jackowiak [30] and Sehsah [12], to examine the effect of the use of some materials on the growth of pathogens and their fruit structures.

Previously, bicarbonate salts, particularly sodium and potassium bicarbonate, have been proposed as alternative strategies to control several plant pathogens such as *Venturia inequalis* [11,31], *Pencillium italicum* and *P. digitatuim* [18], *Diplocarpon rosae, Sphaerotheca pannosa* var. *rosae* [32], and *Sphaerotheca fuliginea* [33]. Likewise, products of *M. oleifera* (crude extracts and essential oil) showed strong antifungal activity against some plant pathogenic fungi [13,34–37]. Previously, we tested the efficacy of *Bacillus subtilis*, *M. oleifera* extract, and potassium bicarbonate on the necrotrophic phytopathogenic fungus *Cercospora beticola*, the causal agent of Cercospora leaf spot on sugar beet [12]. Our previous findings showed that the three proposed treatments significantly enhanced the total soluble solids (TSS) contents and sucrose percentage of treated sugar beet plants, as well as the enzymatic activity of polyphenol oxidase, peroxidase, and phenylalanine ammonia-lyase [12]. However, to the best of our knowledge, the potential effect(s) of *B. subtilis*, *M. oleifera* extract, and potassium bicarbonate on the biotrophic fungus *Erysiphe betae*, the causal agent of sugar beet powdery mildew have been poorly studied.

Moreover, although the potential effects of Moringa extract on the physio-biochemical attributes of environmentally stressed treated plants have been well-studied [38–41], its physiological roles with plant response to biotic stress, such as pathogen infection, are poorly studied. For instance, exogenous applications of moringa leaf extract significantly increased the Chlorophyll a, chlorophyll b, carotenoids, and total photosynthetic pigments in stressed common bean leaves [38]. Likewise, it increased the chlorophyll fluorescence (F_v/F_m and PI), photosynthetic pigments, relative water content (RWC%), and membrane stability index (MSI%) of squash (*Cucurbita pepo*) plants exposed to salt and drought stresses [39]. Additionally, the application of Moringa extract increased the content of chlorophyll a, chlorophyll b, and carotenoids in maize (*Zea mays*) under heat stress [40] and rice (*Oryza sativa*) under drought stress [41]. In terms of plant diseases, soil amendment using Moringa extract significantly reduced the leaf litter, improved the chlorophyll content, and enhanced the response of tomato plants against the soil-borne pathogen, *Fusarium oxysporum f.* sp. *lycopersici,* the causal agent of Fusarium wilt disease [42].

In the current study, we aimed to find efficient, economic, and eco-friendly alternatives that could replace chemical fungicides partially or entirely. For this reason, we first evaluated the susceptibility of 25 sugar beet cultivars to infection with the powdery mildew disease under Egyptian conditions. Additionally, we evaluated the impacts of three eco-friendly materials, including potassium bicarbonate, *M. oleifera* seed extract, and the cell suspension of the biocontrol agent, *B. subtilis* on the disease severity, soluble solid contents (TSS), sucrose percentage, and enzymatic activity of peroxidase (POX), polyphenol oxidase (PPO), and phenylalanine ammonia-lyase (PAL) in powdery mildew-infected sugar beet plants during two successive seasons in 2020 and 2021. Moreover, to better understand the potential mechanism(s) by which the tested treatments affected the fungal pathogen, fungal morphological characteristics from the powdery mildew spots on treated plants were examined using scanning electron microscopy (SEM).

2. Results

2.1. Identification and Pathogenicity of E. betae

The phytopathogenic fungus used in this study was identified based on its morphological and microscopical characteristics and described as *E. betae*. To confirm the pathogenicity of this causal agent and to evaluate the susceptibility of twenty-five sugar beet cultivars, ninety-day old plants were artificially inoculated using conidia of *E. betae*. One-week post-inoculation, the disease severity of powdery mildew in the first season ranged between 2.33 and 70.00% and between 1.33 and 41.66% in the second season.

Two-way hierarchical cluster analysis (HCA) showed that the 25 cultivars were clustered into three main clusters. Cluster-I contained 10 cultivars that were least susceptible to powdery mildew infection including 9k887, Poseidon, Idira-KWS, Fantazja, MK 4200, Allanya-KWS, MK 4199, B 8141, Pintea, and FD17B4010. Cluster-II contained 10 cultivars that were susceptible to powdery mildew disease during 2020 but showed more tolerance during the 2021 season (Figure 1). These nine cultivars included Smart Djerba-KWS, Smart Jella-KWS, Melooia, Vangeus, Gregoria-KWS, LP 17B4011, SHRB21802, FD18B4018, and SI21801. On the other hand, Cluster-III contained only six cultivars that were susceptible and had higher disease severity during both seasons, including Carma, Dipendra-KWS, Zeppeun, Frappina-KWS, SV-2173, and Hammond (Figure 1).

It Is worth mentioning that both Allanya-KWS and FD17B4010 were the least sensitive cultivars (disease severity = 2.33%) during the 2020 season, whereas 9k887 was the least sensitive cultivar (disease severity = 1.33%) during the 2021 season. On the other hand, Smart Djerba-KWS was the most susceptible cultivar (disease severity = 70.00%) during the 2020 season and Zeppeun (disease severity = 41.22%) in the second season.

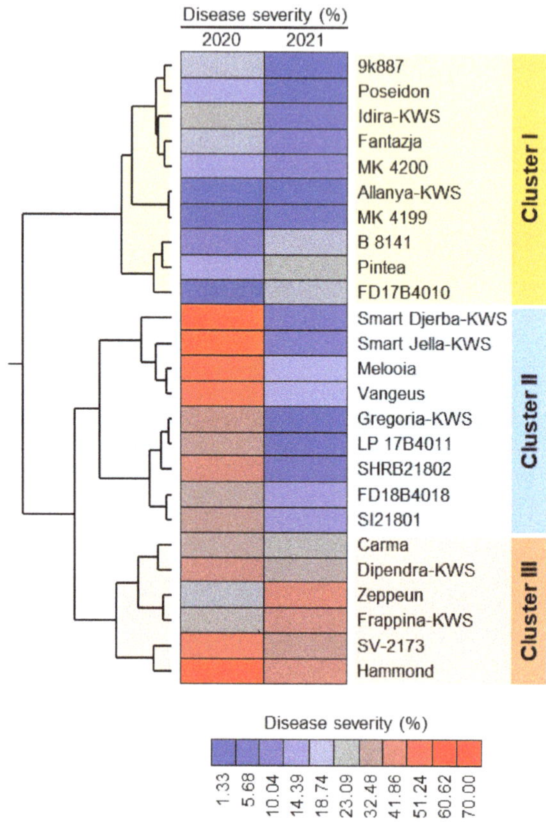

Figure 1. Two-way hierarchical cluster analysis (HCA) of the reaction of 25 sugar beet cultivars to powdery mildew disease caused by *Erysiphe betae* under greenhouse conditions during the 2020 and 2021 seasons. The data presented are the means of three biological replicates (n = 3). Rows represent different cultivars, whereas columns represent disease severity (%). Heatmap was colored based on the disease severity percentages where the red color represents a higher disease severity (%) and the blue color represents a lower disease severity (%). See the scale at the bottom of the heatmap. The experiment was repeated twice with similar results.

2.2. *Efficacy of Different Treatments against Sugar Beet Powdery Mildew Disease under Field Conditions*

The efficacy of foliar application of tested treatments included $KHCO_3$ (5 g L^{-1}), $KHCO_3$ (10 g L^{-1}), Moringa seed extract (25 mL L^{-1}), Moringa seed extract (50 mL L^{-1}), *B. subtilis* (1×10^8 cell mL^{-1}), Amisto fungicide (azoxystrobin; 1 mL L^{-1}), or just sprayed with water (mock control) against sugar beet powdery mildew disease was evaluated under field conditions. Interestingly, all tested treatments significantly decreased the severity of powdery mildew compared to mock-treated control during the 2020 and 2021 seasons (Table 1). The most efficient treatments were the foliar application of moringa seeds extract at 50 mL L^{-1} and *B. subtilis* cell suspension in both tested seasons (Table 1). $KHCO_3$ at 10 g L^{-1} followed these treatments in controlling powdery mildew on sugar beet and all of them were similar to Amisto fungicide treatment.

Table 1. Efficacy of different treatments against powdery mildew disease caused by *Erysiphe betae* under field conditions during the 2020 and 2021 seasons [a].

Treatments	2020 Season					
	Disease Severity (%)			Efficacy (%)		
	1st Spray	2nd Spray	3rd Spray	1st Spray	2nd Spray	3rd Spray
Control (Mock)	2.52 ± 0.11 a	3.95 ± 1.34 a	5.60 ± 1.13 a	0.00 ± 0.00 g	0.00 ± 0.00 d	0.00 ± 0.00 d
Amisto fungicide (Azoxystrobin)	0.72 ± 0.16 b	0.91 ± 0.16 b	1.50 ± 0.85 b	72.10 ± 1.56 a	76.90 ± 1.27 a	73.20 ± 0.99 a
Potassium bicarbonate (5 g L^{-1})	1.40 ± 0.14 ab	1.70 ± 0.14 b	2.30 ± 0.99 b	44.10 ± 1.27 f	56.40 ± 1.70 c	59.00 ± 0.71 c
Potassium bicarbonate (10 g L^{-1})	1.08 ± 0.18 ab	1.53 ± 0.18 b	2.50 ± 0.85 b	56.15 ± 1.63 d	61.50 ± 1.56 b	55.42 ± 0.96 c
Moringa seeds extract (25 mL L^{-1})	1.31 ± 0.16 ab	1.61 ± 0.16 b	2.00 ± 1.56 b	48.00 ± 1.41 e	58.90 ± 1.27 c	64.30 ± 1.13 b
Moringa seeds extract (50 mL L^{-1})	1.01 ± 0.15 ab	1.21 ± 0.13 b	1.90 ± 0.99 b	60.00 ± 2.83 c	69.20 ± 1.13 b	66.40 ± 0.85 b
Bacillus subtilis	0.91 ± 0.13 b	1.32 ± 0.12 b	1.90 ± 0.85 b	64.05 ± 1.48 b	66.60 ± 0.99 b	66.10 ± 1.56 b
p-value	<0.0001	<0.0001	=0.0002	<0.0001	<0.0001	<0.0001
	2021 season					
	Disease severity (%)			Efficacy (%)		
	1st spray	2nd spray	3rd spray	1st spray	2nd spray	3rd spray
Control (Mock)	3.60 ± 0.71 a	5.80 ± 1.27 a	8.70 ± 0.99 a	0.00 ± 0.00 e	0.00 ± 0.00 d	0.00 ± 0.00 d
Amisto fungicide (Azoxystrobin)	1.10 ± 0.57 c	1.90 ± 0.99 b	2.50 ± 0.85 b	69.40 ± 0.85 a	67.20 ± 0.85 b	71.30 ± 1.70 b
Potassium bicarbonate (5 g L^{-1})	2.20 ± 1.56 ab	2.70 ± 0.85 b	3.40 ± 1.27 b	38.30 ± 2.40 d	53.40 ± 0.71 c	60.10 ± 0.71 c
Potassium bicarbonate (10 g L^{-1})	1.90 ± 1.27 b	2.60 ± 0.71 b	2.80 ± 1.41 b	47.20 ± 1.56 c	55.20 ± 0.85 b	67.80 ± 0.99 b
Moringa seeds extract (25 mL L^{-1})	2.31 ± 1.26 ab	2.80 ± 1.13 b	3.20 ± 0.99 b	36.10 ± 1.27 d	51.70 ± 1.56 c	63.20 ± 0.85 c
Moringa seeds extract (50 mL L^{-1})	2.20 ± 1.70 ab	2.80 ± 1.41 b	3.00 ± 1.27 b	38.80 ± 1.56 d	60.30 ± 0.99 b	65.50 ± 1.27 b
Bacillus subtilis	1.60 ± 0.99 c	2.40 ± 0.85 b	2.80 ± 0.71 b	55.50 ± 1.70 b	58.60 ± 0.71 b	67.80 ± 0.99 b
p-value	=0.0080	=0.0005	<0.0001	<0.0001	<0.0001	<0.0001

[a] The data presented are the means ± standard deviation (mean ± SD) of three biological replicates (n = 3). Different letters signify statistically significant differences between treatments according to Tukey's HSD test ($p \leq 0.05$).

2.3. *Effect of the Used Treatments on Soluble Solids Contents and Sucrose Percentage of Powdery Mildew-Infected Sugar Beet Plants*

To better understand the potential effect(s) of tested treatments on the economic value and quality indices of treated sugar beet, soluble solids contents (SSC; %) and sucrose percentage were determined in the fresh root of sugar beet (Figure 2). Briefly, soluble solid content (%) was significantly increased in all tested treatments compared to the mock-control with the superiority of Moringa seed extract (50 mL L^{-1}) during the 2020 season (Figure 2A) and the application of $KHCO_3$ (10 g L^{-1}) during the 2021 (Figure 2B) season. Likewise, the sucrose percentage was significantly increased due to the foliar application of all tested treatments compared with both negative (mock-treated) and positive (fungicide-treated) controls. It is worth mentioning that during the 2020 season, application of *B. subtilis* cell suspension showed the highest sucrose percentage, followed by Moringa seed extract (50 mL L^{-1}), Moringa seed extract (25 mL L^{-1}), and $KHCO_3$ (10 g L^{-1}) which were similar to Amisto fungicide (Figure 2C). In the second season, $KHCO_3$ (10 g L^{-1}) had the highest sucrose percentage, followed by Moringa seed extract (50 mL L^{-1}), $KHCO_3$ (5 g L^{-1}), and *B. subtilis* (1×10^8 cell mL^{-1}) which were significantly higher than both controls (Figure 2D).

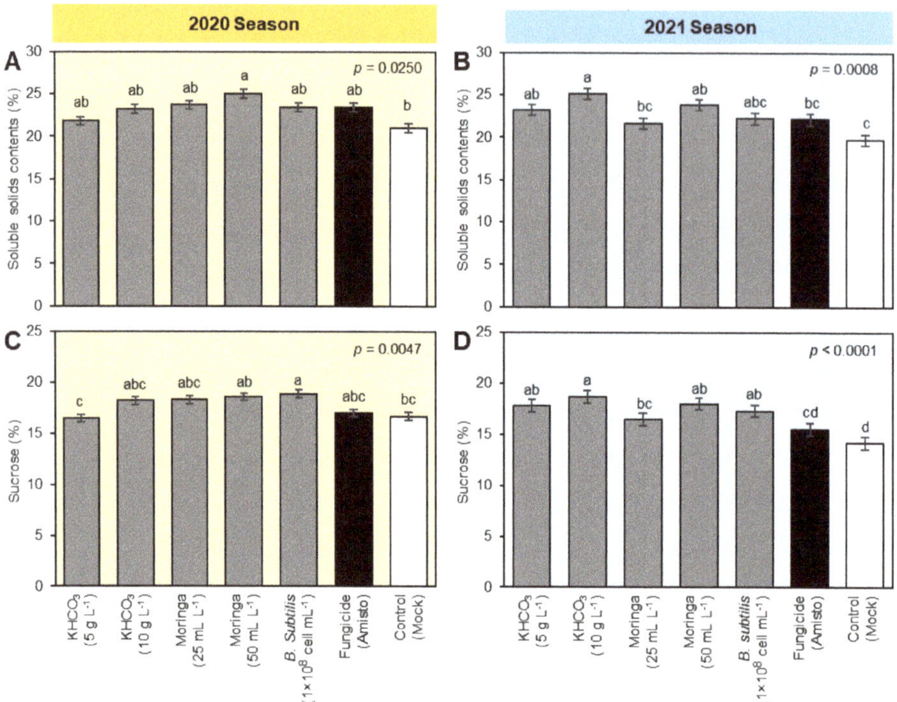

Figure 2. Effect of different treatments on the quality indices of sugar beet plants infected with powdery mildew disease caused by *Erysiphe betae* under field conditions during the 2020 and 2021 seasons. (**A**,**B**) Soluble solid contents (%) during the 2020 and 2021 seasons, respectively, and (**C**,**D**) Sucrose (%) during the 2020 and 2021 seasons, respectively. The data presented are the means ± standard deviation (mean ± SD) of three biological replicates (n = 3). Different letters signify statistically significant differences between treatments according to Tukey's HSD test ($p \leq 0.05$).

2.4. Effect of the Used Treatments on the Activity of Peroxidase (POX), Polyphenol Oxidase (PPO), and Phenylalanine Ammonia-Lyase (PAL) of Powdery Mildew-Infected Sugar Beet Plants

To better understand the biochemical mechanisms of tested treatments, their effect on two main components of the enzymatic antioxidant machinery including peroxidase (POX; Table 2) and polyphenol oxidase (PPO; Table 2) were evaluated. Generally, all tested treatments positively affected the enzymatic activity of POX and PPO in treated leaves of sugar beet with a notable peak at 5 days post first spray. Interestingly, foliar application of moringa seed extract (25 mL L^{-1}) or *B. subtilis* cell suspension had the highest POX (Table 2) and PPO (Table 2) activities at 5 days post first spray.

Likewise, the effect of tested treatments on the activity of phenylalanine ammonia-lyase (PAL; a key enzyme in the salicylic acid (SA) biosynthesis pathway) was also evaluated (Table 2). The enzymatic activity of PAL was incrementally enlarged over time with its highest activity at 15 days post first spray (5 days post second spray). Although there were no significant differences among all treatments at the beginning of the experiment (0 days post first spray), foliar application of moringa seed extract (50 mL L^{-1}), moringa seed extract (25 mL L^{-1}), or *B. subtilis* cell suspension had the highest PAL activity at 5- and 15-days post first spray (Table 2).

Table 2. Effect of different treatments on the enzymatic activity of peroxidase (POX; antioxidant enzyme), polyphenol oxidase (PPO; antioxidant enzyme), and phenylalanine ammonia-lyase (PAL; SA biosynthetic enzyme) of sugar beet plants infected with powdery mildew disease caused by *Erysiphe betae* under field conditions during the 2020 and 2021 seasons *.

	Peroxidase Activity (Δabs. Min^{-1} g^{-1})			**Polyphenol Oxidase Activity (Δabs. Min^{-1} g^{-1})**			**Phenylalanine Ammonia-Lyase Activity (µg *t*-Cinnamic Acid g^{-1} FW)**		
	0 dpt	**5 dpt**	**10 dpt**	**0 dpt**	**5 dpt**	**10 dpt**	**0 dpt**	**5 dpt**	**10 dpt**
Control (Mock)	0.034 ± 0.002 ns	0.113 ± 0.006 c	0.054 ± 0.003 d	0.032 ± 0.002 c	0.265 ± 0.013 b	0.215 ± 0.011 c	0.119 ± 0.006 ns	0.214 ± 0.011 c	0.304 ± 0.015 c
Amisto fungicide (Azoxystrobin)	0.034 ± 0.002 ns	0.135 ± 0.007 b	0.122 ± 0.006 cd	0.037 ± 0.002 c	0.263 ± 0.013 b	0.263 ± 0.013 b	0.130 ± 0.007 ns	0.276 ± 0.014 b	0.387 ± 0.019 b
Potassium bicarbonate (5 g L^{-1})	0.048 ± 0.002 ns	0.178 ± 0.009 a	0.189 ± 0.009 a	0.070 ± 0.004 b	0.241 ± 0.012 b	0.340 ± 0.017 a	0.122 ± 0.006 ns	0.203 ± 0.010 c	0.326 ± 0.016 c
Potassium bicarbonate (10 g L^{-1})	0.047 ± 0.002 ns	0.151 ± 0.008 b	0.149 ± 0.007 b	0.071 ± 0.004 b	0.234 ± 0.012 b	0.331 ± 0.017 a	0.123 ± 0.006 ns	0.206 ± 0.010 c	0.331 ± 0.017 c
Moringa seeds extract (25 mL L^{-1})	0.042 ± 0.002 ns	0.186 ± 0.009 a	0.184 ± 0.009 a	0.110 ± 0.006 a	0.368 ± 0.018 a	0.230 ± 0.011 c	0.132 ± 0.007 ns	0.304 ± 0.015 a	0.411 ± 0.021 a
Moringa seeds extract (50 mL L^{-1})	0.027 ± 0.001 ns	0.139 ± 0.007 b	0.145 ± 0.007 b	0.095 ± 0.005 a	0.247 ± 0.012 b	0.231 ± 0.012 c	0.134 ± 0.007 ns	0.309 ± 0.015 a	0.427 ± 0.021 a
Bacillus subtilis	0.032 ± 0.002 ns	0.184 ± 0.009 b	0.131 ± 0.007 bc	0.057 ± 0.003 b	0.378 ± 0.019 a	0.291 ± 0.015 b	0.099 ± 0.005 ns	0.337 ± 0.017 a	0.412 ± 0.021 a
p-value	=0.0594	<0.0001	<0.0001	<0.0001	<0.0001	<0.0001	<0.0601	<0.0001	<0.0001

* The data presented are the means ± standard deviation (mean ± SD) of three biological replicates (n = 3). Different letters signify statistically significant differences between treatments, whereas "ns" signify no significant differences between them according to Tukey's HSD test ($p \leq 0.05$). The experiment was repeated twice with similar results. dpt: Days post 1st treatment/spray.

2.5. Correlation Analysis between Disease Severity, Soluble Solids, Sucrose Contents, and Antioxidant Enzymes of Powdery Mildew-Infected Sugar Beet Plants

The correlation coefficient (r) between disease severity, quality traits (i.e., soluble solids content (SSC) and sucrose percentage), enzymatic activity (i.e., peroxidase activity (POX), polyphenol oxidase activity (PPO), and phenylalanine ammonia-lyase activity (PAL)) of powdery mildew-infected sugar beet plants were determined. In mock-treated plants, disease severity was negatively correlated with all other parameters (Figure 3A). However, treatment with Amisto fungicide significantly reversed this correlation (Figure 3A). In $KHCO_3$-treated plants, the disease severity was highly and negatively correlated with the quality traits (SSC and sucrose) but positively correlated with the enzymatic activities of POX, PPO, and PAL (Figure 3B). it is worth mentioning that the high concentration of $KHCO_3$ (10 g L^{-1}) significantly strengthened this relationship. Likewise, in Moringa-treated plants, the disease severity (%) was positively correlated with the enzymatic activities of POX, PPO, and PAL at 0, 5, and 15 dpt, but negatively correlated with SSC and sucrose percentage (Figure 3C). The foliar application of Moringa seed extracts (50 mL L^{-1}) noticeably enhanced this relationship. In *B. subtilis*-treated plants, disease severity was positively correlated with the enzymatic activities of POX, PPO, and PAL at all studied time points, except with PPO at 0 dpt (Figure 3D). On the other hand, disease severity (%) was negatively correlated with quality traits (i.e., SSC and sucrose percentage) and PPO activity at 0 dpt.

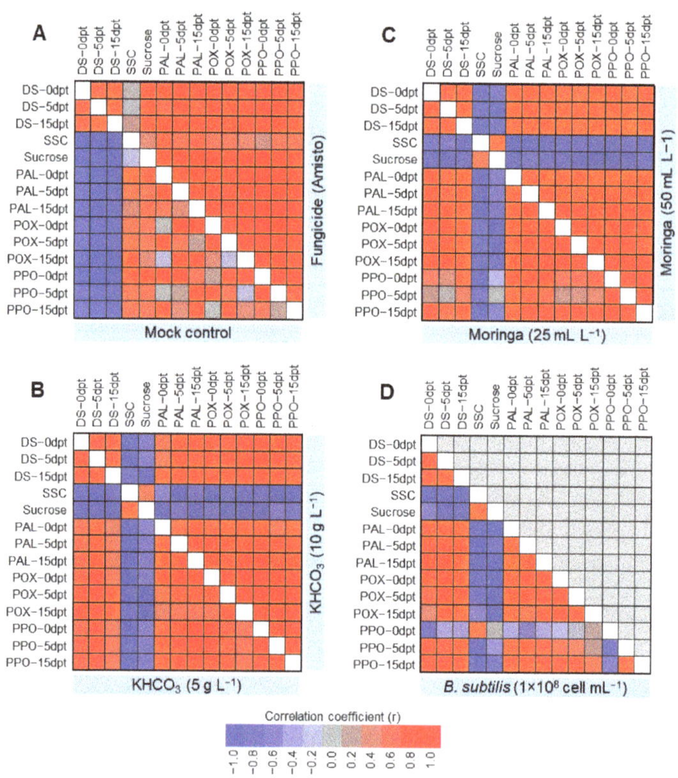

Figure 3. Correlation analysis between disease severity, soluble solids and sucrose contents, and antioxidant enzymes assessed in powdery mildew-infected sugar beet plants. (**A**) Mock and fungicide (Amisto) controls, (**B**) $KHCO_3$ (5 vs. 10 g L^{-1}), (**C**) Moringa seeds extract (25 vs. 50 mL L^{-1}), and (**D**) *B. subtilis* (1×10^8 cell mL^{-1}). Three biological replicates were used (n = 3). DS: Disease severity (%), SSC: Soluble solids contents (%), POX: Peroxidase activity, PPO: Polyphenol oxidase activity, PAL: Phenylalanine ammonia-lyase activity, and dpt: days post 1st treatment.

2.6. Scanning Electron Microscopy (SEM) Examination of the Interaction among the Most Promising Treatments and E. betae on Leaves of Sugar Beet

To gain greater knowledge about the potential mechanism(s) by which the tested treatments affected the fungal morphology, several fungal morphological characteristics were also examined from the powdery mildew spots on treated plants compared to those from mock-treated ones (control) using scanning electron microscopy (SEM). The examined fungal morphological characteristics included growth, conidiophores and conidia density, and mycelium and conidia decomposition. Our SEM findings showed that the foliar application of $KHCO_3$ (10 g L^{-1}; Figure 4B), Moringa seed extract (50 mL L^{-1}; Figure 4C), or *B. subtilis* (1×10^8 cell mL^{-1}; Figure 4D) significantly decreased the density of the fungal mycelium, especially on leaves treated with Moringa extract (Figure 4C) and potassium bicarbonate (Figure 4B), followed by the suspension of Bacillus cells (Figure 4D).

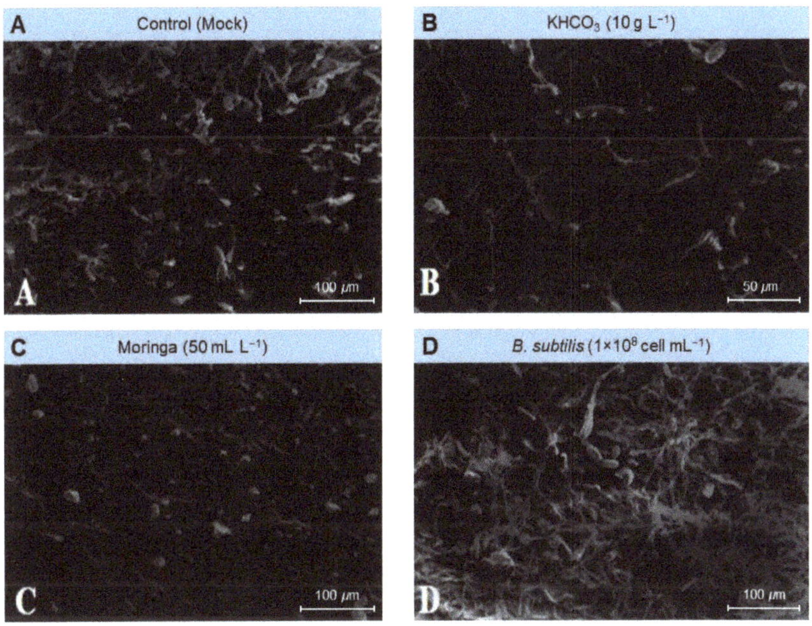

Figure 4. Effect of the most promising treatments on the mycelium density and conidia formation of *Erysiphe betae*, the causal agent of powdery mildew on sugar beet using scanning electron microscopy (SEM). (**A**) mock-treated plants (tap water), (**B**) plants treated with 10 g L^{-1} Potassium bicarbonate, (**C**) plants treated with 50 mL L^{-1} Moringa seeds extract, and (**D**) plants treated with a cell suspension of *Bacillus subtilis* (1×10^8 cell mL^{-1}).

Similarly, the above-mentioned treatments markedly dropped the number of conidia formed by the *E. betae* and diminished the capacity of the phytopathogenic fungus to form conidiophores and conidia (Figure 5). Moreover, it caused plasmolysis and decomposition of the mycelium of *E. betae* and its conidia. Interestingly, the conidia on treated leaves appeared in an incomplete state, and their construction appeared in a distorted form (Figures 5 and 6).

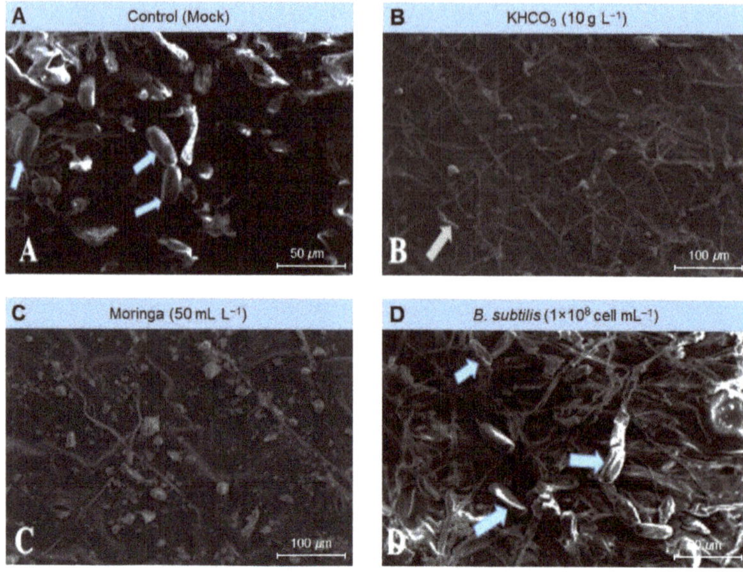

Figure 5. Effect of the most promising treatments on the morphology and number of conidia formed by *Erysiphe betae*, the causal agent of powdery mildew on sugar beet using scanning electron microscopy (SEM). (**A**) mock-treated plants (tap water), (**B**) plants treated with 10 g L^{-1} potassium bicarbonate, (**C**) plants treated with 50 mL L^{-1} Moringa seeds extract, and (**D**) plants treated with a cell suspension of *Bacillus subtilis* (1×10^8 cell mL^{-1}).

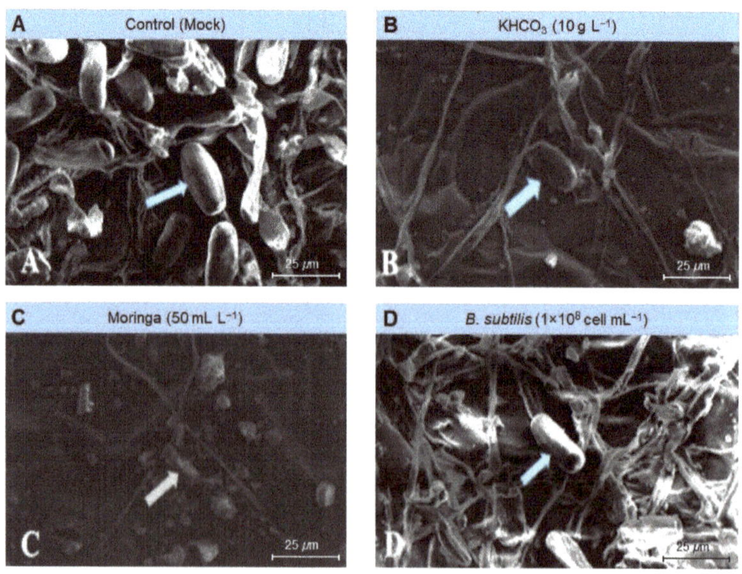

Figure 6. Effect of the most promising treatments on the decomposition of mycelium and conidia of *Erysiphe betae*, the causal agent of powdery mildew on sugar beet using scanning electron microscopy (SEM). (**A**) mock-treated plants (tap water), (**B**) plants treated with 10 g L^{-1} potassium bicarbonate, (**C**) plants treated with 50 mL L^{-1} Moringa seeds extract, and (**D**) plants treated with a cell suspension of *Bacillus subtilis* (1×10^8 cell mL^{-1}).

3. Discussion

In the current study, we evaluated the utilization of *B. subtilis* cell suspension as a biocontrol strategy against *E. betae*, the causal agent of powdery mildew of sugar beet along with another four eco-friendly strategies including two concentrations of potassium bicarbonate and two concentrations of *M. oleifera* seeds extract.

Initially, the susceptibility of 25 plant specimens was evaluated under the artificial inoculation of *E. betae*. These varieties showed great variation in the extent of their susceptibility to powdery mildew caused by the fungus *E. betae*, after artificially infecting them by dispersing the fungus spores from old leaves of infection on the leaves of resident varieties plants. These results are in agreement with the previous research that was performed on powdery mildew, as it indicated a difference in the susceptibility of plant varieties to powdery mildew due to the genetic differences between those varieties [16,17].

Allanya-KWS, 9k887, and FD17B4010 were the least susceptible (most resistant) cultivars to the infection with the fungus *E. betae* and could be generalized in future cultivation. On the other hand, Smart Djerba-KWS and Zeppeun were the most susceptible cultivars which should be eliminated and not be cultivated under Egyptian conditions. Previous studies assessed the resistance of some sugar beet varieties to powdery mildew using crossed immune-electrophoresis technique [43,44]. They explained that there are several *E. betae* resistance genes. The variety containing these genes appears to be more resistant to the disease than the variety lacking these genes.

Chemical control, despite containing substances that kill beneficial microbes and negatively affect public health and the environment, is still the first choice for combating plant diseases in general and powdery mildew in particular. Therefore, there will always be a need to search for safe alternatives to fungicides and at the same time give results that are close or similar and perhaps better than those agrochemicals and less harmful to public health and the environment. Biological control, whether by using bioagents, oils, plant extracts, or other eco-friendly materials, is considered one of the most promising options that attract great attention in this field.

Our findings showed that foliar application of *B. subtilis* cell suspension (1×10^8 cell mL^{-1}) significantly reduced the disease severity of sugar beet powdery mildew disease under field conditions. Moreover, it increased the sucrose percentage, and the enzymatic activity of POX, PPO, and PAL but decreased the density of the fungal mycelium on treated leaves. These results are in agreement with those of Bakeer et al. [45] who recently reported that the commercial biocide Biobac 50% WP (*B. subtilis*; 1×10^9 cell g^{-1}) showed significant antifungal activity against *E. heraclei*, the causal agent of powdery mildew of parsley *under* in vitro conditions. We suggest that the antifungal activity of biological agents, such as *B. subtilis*, might be due to their ability to biosynthesize and produce several metabolites such as antibiotics, hydrolytic enzymes, and toxic compounds, these metabolites could cause degradation and lyse to the cell wall of fungal mycelium and spores or inhibit some enzymes that are essential for spore germination [45]. However, further studies are required to deeply analyze the chemical composition of extracellular metabolites of *B. subtilis*.

Several studies have attributed the inhibition of the growth of various pathogens by *Bacillus subtilis* to its ability to produce some antibiotics such as mycoseptillin and trippilin [14,46,47]. Others have demonstrated that there are other ways for *Bacillus* sp. to inhibit the growth of phytopathogens or reduce the competitiveness with other microorganisms, such as the production of antibiotics or vital growth enzymes. Antibiotics, enzymes, and peptide antibiotics are substances with amphiphilic membrane activity and surfactants for pathogens. In the current study, and under field conditions, *B. subtilis* showed a significant reduction in the severity of powdery mildew disease. This is due to its high ability to spread rapidly on the surface of plant leaves and prevent pathogenic germs from reaching the natural openings, thus preventing infection.

Biocontrol agents also have the advantage of being able to compete intensely for oxygen and nutrients on the surface of leaves, prohibiting pathogens of access them, and starving them [48,49]. Examination using scanning electron microscopy (SEM) also demonstrated that *Bacillus subtilis* induced plasmolysis and lysis of conidiophores and conidia of *E. betae*. This finding is in line with the demonstration from several similar cases with phytopathogenic fungi, reported previously, which indicated that *B. subtilis* secretes fungal cell-degrading enzymes including b-1, 3-gluconase, and protease [12,50].

Other eco-friendly alternatives to chemical fungicides use beneficial chemicals and or plant extracts. For instance, potassium bicarbonate ($KHCO_3$) is a beneficial eco-friendly chemical for plants; however, our knowledge about its role in plant response to biotic stressors is still limited. Likewise, antifungal activities of plant extracts, such as extracts derived from *M. oleifera*, have been reported against several phytopathogens [34,45,51]. Nevertheless, these fungistatic activities were due to the utilization of *M. oleifera* essential oil and seed extract, but not crude leaf extract [35]. Herein, we tested the effectiveness of two different concentrations of $KHCO_3$ (5 and 10 g L^{-1}) and two concentrations of moringa seed extract to control of powdery mildew of sugar beet. Our finding showed that both Moringa seed extract and $KHCO_3$ caused a significant decrease in the growth of *E. betae* to varying degrees. Furthermore, they showed a great ability to combat the powdery mildew of sugar beet under field conditions. Additionally, certain concentrations of Moringa seed extract and $KHCO_3$ significantly increased SSC and sucrose percentage, and the enzymatic activity of POX, PPO, and PAL.

Scanning electron microscope (SEM) examination of samples treated with potassium bicarbonate showed that the fungal structures were decomposed and their inability to produce both conidia and conidiophores. Previous research suggests that potassium bicarbonate can be used to combat certain plant diseases such as powdery mildew, apple scorch spots, and scab [11,31], grapevine powdery mildew [52], powdery mildew on cucumber [33], Cercospora leaf spot of sugar beet [12] and black spot and powdery mildew of roses [32]. We suggest that the action of potassium bicarbonate is associated with the occurrence of a disturbance in the osmotic pressure and the bio-carbon/bicarbonate ion balance of sensible fungi, and the pH. Bicarbonate prevents the development of fungi and their ability to sporulate through contact with fungi in an aqueous solution. It is therefore believed that the possibility of the emergence of resistant strains of bicarbonate is low due to the multiplicity of its modes of action [31,32].

Similarly, in this study, whether in the laboratory or the field, the two concentrations used of Moringa seed extract (25 and 50 g L^{-1}) significantly reduced the linear growth of *E. betae* and the severity of powdery mildew disease. The antimicrobial activity of *M. oleifera* extract was evaluated previously against *Fusarium solani*, *Pasturella multocida*, *Staphylococcus aureus*, *Escherichia coli*, and *Rhizopus solani* strains [36]. Previous reports on the antibiotic principle results of *M. oleifera* seeds through their purification, clarification, and antimicrobial properties, as well as on the antibiotic substance of *M. oleifera* roots have been demonstrated [13,53]. Several studies have proven that certain plant extracts contain many toxins and fungi inhibitors that negatively affect the growth of pathogens [54–56]. Phytochemical analysis revealed the presence of glycosides, alkaloids, triterpenoids, flavonoids, steroids, and tannins [37]. Seed extracts of moringa also contain organic compounds and pigments such as flavonoids, carotenoids, niacin, isothiocyanates, minerals, sterols, and glucosinolates, all of which are accountable for the fashioning of antioxidants [36,57].

To better understand the positive results of tested treatments to control powdery mildew of sugar beet, the effect of the studied treatments on the activity of some oxidation enzymes, such as POX and PPO, and the key enzyme in SA biosynthesis pathway, PAL, was evaluated, and its action was related to stimulating plants to resist pathogens [12,58,59]. Our findings from the current study proved that foliar application of sugar beet plants with $KHCO_3$, Moringa seed extract, or *B. subtilis* (1×10^8 cell mL^{-1}) led to a significant improvement in the activity of these enzymes. This improvement in the activity of defense enzymes helps to reveal additional explanations for the ability of the treatments to reduce the severity of the pathogen under study, in addition to the possibility of improving some of the good qualities of sugar beet, such as the content of soluble solids and sucrose percentage.

4. Materials and Methods

4.1. Pathogen Identification, Pathogenicity, and Susceptibility of Sugar Beet Cultivars

In the current study, the phytopathogenic fungus was identified according to Barnet and Hunter [60]. Subsequently, a pathogenicity test was carried out in 35 cm diameter pots under greenhouse conditions. Sandy-loam soil (1:2 *w*/*w*; pH = 6.7, available P_2O_5 = 284 mg Kg^{-1}, available K_2O = 252 mg Kg^{-1}, N = 15 g Kg^{-1}, and humus = 17.8 g Kg^{-1}) was used throughout the study. Twenty-five sugar beet varieties (namely 9k887, Allanya-KWS, B 8141, Carma, Dipendra-KWS, Fantazja, FD17B4010, FD18B4018, Frappina-KWS, Gregoria-KWS, Hammond, Idira-KWS, LP 17b4011, Melooia, MK 4199, MK 4200, Pintea, Poseidon, SHRB21802, SI21801, Smart Djerba KWS, Smart Jella-KWS, SV-2173, Vangeus, and Zeppeun) were utilized in this investigation to test the pathogenicity of the causal agent and to evaluate the susceptibility of these varieties to infection with powdery mildew. To test the sensitivity of these varieties to powdery mildew, they were artificially infected by applying conidia for *E. betae* on the leaves of those plants from aged, infected leaves. The percent of disease severity was documented according to Shane and Teng [61] after 100 days from sowing.

4.2. Treatments

In addition to the negative control (mock-treated plants), Amisto (Azoxystrobin) 25% SC, at its recommended rate (1 mL L^{-1}), was used as positive control throughout the study. Potassium bicarbonate ($KHCO_3$; Al-Gomhoria Company for Chemicals and Glasses, Cairo, Egypt) was used at the rate of 5 or 10 g L^{-1}. *Bacillus subtilis* was isolated from nourishing sugar beet leaves and identified according to Bergey's Manual of Systematic Bacteriology [62] and used as a bioagent at 1×10^8 cell mL^{-1} in this study. The seed extract of *Moringa oleifera* was applied at 25- or 50-mL L^{-1}. All experiments were carried out using a completely randomized design (CRD) with three biological replicates per treatment.

4.3. Evaluation of the Tested Treatments against Erysiphe betae under Field Conditions

The susceptible cultivar 'Pleno' was used throughout the field experiments. All experiments were carried out at the Research Experimental Farm of Plant Pathology Research Institute, Sakha Station, Kafrelsheikh, Egypt using a completely randomized design with three biological replicates during two successive seasons (2020 and 2021). Each biological replicate was composed of six rows with 900 cm length and 60 cm width. Each row contained 45 hills 20 cm apart. All recommended cultural practices were performed at the proper time. Plants were sprayed three times, with 10 day intervals between them, with one of the following treatments, $KHCO_3$ (5 g L^{-1}), $KHCO_3$ (10 g L^{-1}), Moringa seed extract (25 mL L^{-1}), Moringa seed extract (50 mL L^{-1}), *B. subtilis* (1×10^8 cell mL^{-1}), Amisto fungicide (1 mL L^{-1}), or just sprayed with water (Mock control). The first spray was after approximately 90 days of cultivation when disease symptoms were detected.

4.3.1. Disease Severity

Disease severity was estimated after 10 days of the first treatment (initial disease severity) and twice more with ten days intervals. In other words, the last appraisal took place 30 days after the first application (the final disease severity). The powdery mildew disease severity was estimated according to McGrath and others [63], by calculating observable sporulating mildew colonies on both adaxial and abaxial textures per leaf. Briefly, five old leaves per plant were examined for five plants in each plot (i.e., 25 leaves/treatment). Assessments were also made on fully expanded leaves from the middle and upper thirds of a plant. Data from all three age classes of leaves were averaged concurrently. The initial disease severity ranged widely under field conditions, due to the presence of powdery mildew colonies on sugar beet leaves before treatment. The experiment was carried out using a completely randomized design (CRD) with three biological replicates per treatment.

The corrected final disease severity (A) was calculated as the result of corrected disease severity (*CDS*) according to Equation (1) as described by Kamel [64] as follows:

$$CDS = \frac{L}{M} \times N \quad (1)$$

where *CDS*: Corrected disease severity, *L*: the initial disease severity of treatment, *M*: the initial disease severity of the check (control), and *N*: the final disease severity of the control.

4.3.2. The Soluble Solid Content and Sucrose Percentage

Soluble solid content (SSC; %) was determined in the fresh root of sugar beet using a hand refractometer according to McGinnis [65]. Moreover, the sucrose percentage was estimated according to AOAC [66].

4.3.3. Enzymatic Activity

Generally, leaf fresh samples were collected for the enzymatic activity assays at 0, 5, and 15 days post the first treatment (dpt). Briefly, one gram of leaf tissue was ground in 2 mL of 0.1 M sodium phosphate buffer (pH 7.1) using a porcelain mortar. Subsequently, samples were centrifuged at 3000 rpm for 20 min at 6 °C and the supernatants were collected and considered as a crude enzyme extract. Peroxidase (POX) activity was determined according to the method of Allam et al. [67] by estimating the oxidation of pyrogallol to pyrogalline in the presence of hydrogen peroxide. The enzymatic activity of POX was calculated by following the differences in absorbance at 425 nm every 1 min for five minutes using Beckman Spectrophotometer Du® 7400(Beckman Coulter Inc., Fullerton, CA, USA). Likewise, the enzymatic activity of polyphenol oxidase (PPO) was spectrophotometrically determined according to Maxwell and Batman [68]. Briefly, PPO activity was calculated by following the changes in absorbance at 495 nm every 1 min for five minutes using the same spectrophotometer mentioned above. Furthermore, the enzymatic activity of phenylalanine ammonia-lyase (PAL) was determined according to the method of Zucker [69]. Briefly, PAL activity was estimated in acetone powder prepared from leaves, using 0.75 gm acetone powder suspended in sodium borate buffer (pH 8.8). The reaction mixture included 0.5 mL enzyme preparation, 1.5 mL borate buffer 0.2 M (pH 8.8), 1 mL of 1% phenylalanine, and 2.5 mL deionized water. One mL of deionized water was added instead of phenylalanine as a blank. The mixture was incubated at 40 °C for one hour. The reaction was stopped by adding 0.5 mL of 5N HCl to each tube. The enzyme activity was measured at 290 nm and expressed as μg *t*-cinnamic acid g^{-1} FW. The experiment was carried out using a completely randomized design (CRD) with three biological replicates per treatment.

4.4. Scanning Electron Microscopy (SEM) Examination

The effectiveness of the used treatments on the formation of conidia and spores, as well as the growth of *E. betae* on sugar beet leaves, was investigated using scanning electron microscopy (SEM) according to Manzali et al. [70]. Interaction sites (spots) were marked and disc blocks of 1 cm^2 were taken for SEM using Jeol Scanning Electron Microscope model JSM-5500lv (JEOL Ltd., Akishima, Tokyo, Japan) at Electron Microscope Unit, Nanotechnology Institute, Kafrelsheikh University. Samples, illustrating the interaction region, were appointed with osmium oxide, and then dehydrated, using a serial dilution of ethyl alcohol and then finally acetone. Processed samples were then dried, using a critical point drier (EMS 850; Electron Microscopy Sciences [EMS], Hatfield, PA, USA), coated with gold using a sputter coater (EMS 550; EMS, Hatfield, PA, USA), then the samples were investigated using an SEM (Jeol 100cx–11 ASID—4D; JEOL Ltd., Akishima, Tokyo, Japan).

4.5. Statistical Analysis

All experiments were carried out using a completely randomized design with three biological replicates during two successive seasons (2020 and 2021). All data were statistically analyzed using the analysis of variance (ANOVA), followed by the Tukey–Kramer honestly significant difference test (Tukey HSD, $p \leq 0.05$) as a post hoc analysis for pairwise comparisons. Moreover, hierarchical cluster analysis (HCA) was used to better understand the variations in susceptibility of different cultivars. Finally, correlation analysis was conducted to evaluate the relationships between disease severity, quality traits (i.e., SSC and sucrose percentage), and enzymatic activity (i.e., POX, PPO, and PAL) of powdery mildew-infected sugar beet plants. Correlation coefficients (r) are presented as a heatmap.

5. Conclusions

Collectively, our findings demonstrated that foliar application of $KHCO_3$, Moringa seed extract, or *B. subtilis* cell suspension significantly enhanced sugar beet resilience to the infection with *E. betae*, the causal agent of powdery mildew disease. These treatments are eco-friendly, less expensive, and may replace commercial fungicides totally or partially. Application of $KHCO_3$, Moringa seed extract, or *B. subtilis* cell suspension significantly reduced the severity of powdery mildew disease but induced the SSC and sucrose percentage. The protective role(s) of these compounds might be due to the activation of enzymatic antioxidant machinery (as expressed by higher enzymatic activities of POX and PPO) or the enhancement of SA biosynthesis (as expressed by higher enzymatic activities of PAL). Nevertheless, further investigations are required to determine the chemical composition of Moringa seed extract and *B. subtilis* extracellular metabolites and their molecular mechanisms during the *E. betae*–beet interactions.

Author Contributions: Conceptualization, B.A.E.-N., G.A.E.-K. and Y.N.; methodology, B.A.E.-N., G.A.E.-K., S.H. and M.D.S.; software, Y.N.; validation, B.A.E.-N., G.A.E.-K., S.H. and M.D.S.; formal analysis, B.A.E.-N., G.A.E.-K., S.H., M.D.S., Y.N.; investigation, B.A.E.-N., G.A.E.-K., S.H. and M.D.S.; resources, B.A.E.-N., G.A.E.-K., M.A.S., S.H. and M.D.S.; data curation, G.A.E.-K. and S.H.; writing—original draft preparation, B.A.E.-N., M.A.S., G.A.E.-K., S.H., A.H.M. and M.D.S.; writing—review and editing, G.A.E.-K. and Y.N.; visualization, Y.N.; supervision, G.A.E.-K.; project administration, S.H., M.D.S. and Y.N; funding acquisition, M.A.S. All authors have read and agreed to the published version of the manuscript.

Funding: The ACP of this manuscript was funded by the Deanship of Scientific Research at King Khalid University for funding the APC through the Program of Research Groups under grant number (RGP. 2/106/43).

Data Availability Statement: The data collected and analyzed throughout the present research are available upon request.

Acknowledgments: Y.N., G.A.E.-K., and S.H. would like to extend their appreciation to the Graduate Student and Research Affairs Sector of Tanta University and Kafrelsheikh University, Egypt.

Conflicts of Interest: The authors declare no conflict of interest.

References

1. Statista. Agriculture in Egypt—Statistics & Facts. Available online: https://www.statista.com/topics/5674/agriculture-in-egypt/ (accessed on 2 November 2021).
2. Kontradowitz, L.; Verreet, J.A. Assessment of Resistance and Virulence in the Pathosystem Sugar Beet (*Beta vulgaris*)/Powdery Mildew (*Erysiphe betae*)—Development of Basics for an Effective Powdery Mildew Resistance Breeding. *J. Plant Dis. Prot.* **2010**, *117*, 49–54. [CrossRef]
3. Karaoglanidis, G.S.; Karadimos, D.A. Efficacy of Strobilurins and Mixtures with DMI Fungicides in Controlling Powdery Mildew in Field-Grown Sugar Beet. *Crop Prot.* **2006**, *25*, 977–983. [CrossRef]
4. El-Fahar, S.A.; Abou El-Magd, B.M. Effect of Infected Sugar Plants with Powdery Mildew Disease on Invertase Activity and Certain Quality Characteristic. *J. Agric. Res. Kafr El-Sheikh Univ.* **2008**, *34*, 940–949.
5. Lamey, H.A.; Cattanach, A.W.; Bugbee, W.M. *Cercospora Leaf Spot of Sugarbeet*; North Dakota State University: Fargo, ND, USA, 1987; pp. 1–4.
6. Miller, J.; Rekoske, M.; Quinn, A. Genetic Resistance, Fungicide Protection and Variety Approval Policies Controlling Yield Losses from Cercospora Leaf Spot Infections. *J. Sugar Beet Res.* **1994**, *31*, 7–12. [CrossRef]
7. Smith, G.A.; Campbell, L.G. Association between Resistance to Cercospora and Yield in Commercial Sugarbeet Hybrids. *Plant Breed.* **1996**, *115*, 28–32. [CrossRef]

8. Carlson, A.L.; Luecke, J.L.; Khan, M.F.R. Survey of Fungicide Use in Sugarbeet in Minnesota and Eastern North Dakota–2008. *Sugarbeet Res. Ext. Rep* **2009**, *39*, 195–199.
9. Weiland, J.J.; Smith, G.A. A Survey for the Prevalence and Distribution of *Cercospora beticola* Tolerant to Triphenyltin Hydroxide and Mancozeb and Resistant to Thiophanate Methyl in 1998. *Sugarbeet Res. Ext. Rep.* **1999**, *29*, 289–291.
10. Abdel-Monaim, M.F. Induced Systemic Resistance in Tomato Plants against Fusarium Wilt Diseases. *Int. Res. J. Microbiol.* **2012**, *3*, 14–23.
11. Marku, L.; Vrapi, H.; Hasani, M. Effect of Potassium Bicarbonate (Armicarb) on the Control of Apple Scab (*Venturia inequalis*) in the Region of Puka in Albania. *Int. Refereed J. Eng. Sci.* **2014**, *3*, 25–30.
12. Sehsah, M.D.; El-Kot, G.A.; El-Nogoumy, B.A.; Alorabi, M.; El-Shehawi, A.M.; Salama, N.H.; El-Tahan, A.M. Efficacy of *Bacillus subtilis*, *Moringa oleifera* Seeds Extract and Potassium Bicarbonate on Cercospora Leaf Spot on Sugar Beet. *Saudi J. Biol. Sci.* **2022**, *29*, 2219–2229. [CrossRef]
13. El–Mohamedy, R.S.; Abdallah, A.M. Antifungal Activity of *Moringa oleifera* Oil and Seed Extract against Some Plant Pathogenic Fungi. *Middle East J. Agric. Res.* **2014**, *3*, 242–249.
14. El-Kot, G.A.N.; Hegazi, M.A. Non-Chemical Control of Powdery Mildew Disease on Zinnia (*Zinnia elegans*, L.). *Alex. J. Agric. Res.* **2008**, *53*, 219–230.
15. Hegazi, M.A.; el Kot, G. Influences of Some Essential Oils on Vase-Life of *Gladiolus hybrida*, L. Spikes. *Int. J. Agro Vet. Med. Sci.* **2009**, *3*, 19–24. [CrossRef]
16. Derbalah, A.S.; el Kot, G.A.; Hamza, A.M. Control of Powdery Mildew in Okra Using Cultural Filtrates of Certain Bio-Agents Alone and Mixed with Penconazole. *Arch. Phytopathol. Plant Prot.* **2011**, *44*, 2012–2023. [CrossRef]
17. El-Moghazy, S.M.; El-kot, G.A.; Hamza, A.M. Control of Sugar Beet LEAF Spot Disease Caused by by the Fungus *Cercospora beticola* (Sacc). *J. Plant Prot. Pathol.* **2011**, *2*, 1037–1047. [CrossRef]
18. Marloth, R.H. The Influence of Hygrogen-Ion Concentration and of Sodium Bicarbonate and Related Substances on *Pencillium italicum* and *P. digitatuim*. *Phytopathology* **1931**, *21*, 169–198.
19. Hafez, E.M.; Gowayed, S.M.; Nehela, Y.; Sakran, R.M.; Rady, A.M.S.; Awadalla, A.; Omara, A.E.D.; Alowaiesh, B.F. Incorporated Biochar-Based Soil Amendment and Exogenous Glycine Betaine Foliar Application Ameliorate Rice (*Oryza sativa* L.) Tolerance and Resilience to Osmotic Stress. *Plants* **2021**, *10*, 1930. [CrossRef] [PubMed]
20. Nehela, Y.; Mazrou, Y.S.A.; Alshaal, T.; Rady, A.M.S.; El-Sherif, A.M.A.; Omara, A.E.D.; El-Monem, A.M.A.; Hafez, E.M. The Integrated Amendment of Sodic-Saline Soils Using Biochar and Plant Growth-Promoting Rhizobacteria Enhances Maize (*Zea mays* L.) Resilience to Water Salinity. *Plants* **2021**, *10*, 1960. [CrossRef]
21. Abdelrasheed, K.G.; Mazrou, Y.; Omara, A.E.D.; Osman, H.S.; Nehela, Y.; Hafez, E.M.; Rady, A.M.S.; El-Moneim, D.A.; Alowaiesh, B.F.; Gowayed, S.M. Soil Amendment Using Biochar and Application of K-Humate Enhance the Growth, Productivity, and Nutritional Value of Onion (*Allium cepa* L.) under Deficit Irrigation Conditions. *Plants* **2021**, *10*, 2598. [CrossRef]
22. Abdelkhalek, A.; Qari, S.H.; Abu-Saied, M.A.A.R.; Khalil, A.M.; Younes, H.A.; Nehela, Y.; Behiry, S.I. Chitosan Nanoparticles Inactivate *Alfalfa mosaic virus* Replication and Boost Innate Immunity in *Nicotiana glutinosa* Plants. *Plants* **2021**, *10*, 2701. [CrossRef]
23. Nehela, Y.; Killiny, N. Melatonin Is Involved in Citrus Response to the Pathogen Huanglongbing via Modulation of Phytohormonal Biosynthesis. *Plant Physiol.* **2020**, *184*, 2216–2239. [CrossRef] [PubMed]
24. Nehela, Y.; Taha, N.A.; Elzaawely, A.A.; Xuan, T.D.; Amin, M.A.; Ahmed, M.E.; El-Nagar, A. Benzoic Acid and Its Hydroxylated Derivatives Suppress Early Blight of Tomato (*Alternaria solani*) via the Induction of Salicylic Acid Biosynthesis and Enzymatic and Nonenzymatic Antioxidant Defense Machinery. *J. Fungi* **2021**, *7*, 663. [CrossRef]
25. Abdelrhim, A.S.; Mazrou, Y.S.A.; Nehela, Y.; Atallah, O.O.; El-Ashmony, R.M.; Dawood, M.F.A. Silicon Dioxide Nanoparticles Induce Innate Immune Responses and Activate Antioxidant Machinery in Wheat Against *Rhizoctonia solani*. *Plants* **2021**, *10*, 2758. [CrossRef] [PubMed]
26. Atallah, O.O.; Mazrou, Y.S.A.; Atia, M.M.; Nehela, Y.; Abdelrhim, A.S.; Nader, M.M. Polyphasic Characterization of Four *Aspergillus* Species as Potential Biocontrol Agents for White Mold Disease of Bean. *J. Fungi* **2022**, *8*, 626. [CrossRef]
27. El-Nagar, A.; Elzaawely, A.A.; Xuan, T.D.; Gaber, M.; El-Wakeil, N.; El-Sayed, Y.; Nehela, Y. Metal Complexation of Bis-Chalcone Derivatives Enhances Their Efficacy against Fusarium Wilt Disease, Caused by *Fusarium equiseti*, via Induction of Antioxidant Defense Machinery. *Plants* **2022**, *11*, 2418. [CrossRef] [PubMed]
28. El-Nagar, A.; Elzaawely, A.A.; Taha, N.A.; Nehela, Y. The Antifungal Activity of Gallic Acid and Its Derivatives against *Alternaria solani*, the Causal Agent of Tomato Early Blight. *Agronomy* **2020**, *10*, 1402. [CrossRef]
29. Bayoumi, Y.A.; El-Kot, G.A.N. Potential Impacts of Different Aqueous Compost Extracts on Growth, Yield, Enzyme Activities and Controlling Powdery Mildew of Cucumber Plants Grown under Plastic Houses. *Minufiya J. Agric. Res.* **2010**, *35*, 649–661.
30. Jackowiak, H.; Packa, D.; Wiwart, M.; Perkowski, J. Scanning Electron Microscopy of Fusarium Damaged Kernels of Spring Wheat. *Int. J. Food Microbiol.* **2005**, *98*, 113–123. [CrossRef]
31. Jamar, L.; Lefrancq, B.; Lateur, M. Control of Apple Scab (*Venturia inaequalis*) with Bicarbonate Salts under Controlled Environment. *J. Plant Dis. Prot.* **2007**, *114*, 221–227. [CrossRef]
32. Horst, R.K.; Kawamoto, S.O.; Porter, L.L. Effect of Sodium Bicarbonate and Oils on the Control of Powdery Mildew and Black Spot of Roses. *Plant Dis.* **1992**, *76*, 247. [CrossRef]
33. Homma, Y.; Arimoto, Y.; Misato, T. Effect of Sodium Bicarbonate on Each Growth Stage of Cucumber Powdery Mildew Fungus (*Sphaerotheca fuliginea*) in Its Life Cycle. *J. Pestic. Sci.* **1981**, *6*, 201–209. [CrossRef]

34. Goss, M. The Efficacy of *Moringa oleifera* Plant Extracts against Selected Fungal and Bacterial Plant Pathogens Infecting Selected Vegetable Crops in Zimbabwe. Doctoral Dissertation, College of Agriculture, Engineering and Science School of Agricultural, Earth and Environmental Sciences, Pietermaritzburg, South Africa, 2018.
35. Chuang, P.H.; Lee, C.W.; Chou, J.Y.; Murugan, M.; Shieh, B.J.; Chen, H.M. Anti-Fungal Activity of Crude Extracts and Essential Oil of *Moringa oleifera* Lam. *Bioresour. Technol.* **2007**, *98*, 232–236. [CrossRef]
36. Jabeen, R.; Shahid, M.; Jamil, A.; Ashraf, M. Microscopic Evaluation of the Antimicrobial Activity of Seed Extracts of *Moringa oleifera*. *Pak. J. Bot.* **2008**, *40*, 1349–1358.
37. Zaffer, M.; Ganie, S.A.; Gulia, S.S.; Yadav, S.S.; Singh, R.; Ganguly, S. Antifungal Efficacy of *Moringa oleifera* Lam. *Am. J. Phytomedicine Clin. Ther.* **2015**, *3*, 28–33.
38. Latif, H.H.; Mohamed, H.I. Exogenous Applications of Moringa Leaf Extract Effect on Retrotransposon, Ultrastructural and Biochemical Contents of Common Bean Plants under Environmental Stresses. *S. Afr. J. Bot.* **2016**, *106*, 221–231. [CrossRef]
39. Abd El-Mageed, T.A.; Semida, W.M.; Rady, M.M. Moringa Leaf Extract as Biostimulant Improves Water Use Efficiency, Physio-Biochemical Attributes of Squash Plants under Deficit Irrigation. *Agric. Water Manag.* **2017**, *193*, 46–54. [CrossRef]
40. Batool, A.; Wahid, A.; Abbas, G.; Shah, S.H. Application of *Moringa oleifera* Plant Extracts for Enhancing the Concentration of Photosynthetic Pigments Leading to Stable Photosynthesis under Heat Stress in Maize (*Zea mays* L.). *Pak. J. Bot.* **2019**, *51*, 2031–2036. [CrossRef] [PubMed]
41. Khan, S.; Basit, A.; Hafeez, M.B.; Irshad, S.; Bashir, S.; Bashir, S.; Maqbool, M.M.; Saddiq, M.S.; Hasnain, Z.; Aljuaid, B.S.; et al. Moringa Leaf Extract Improves Biochemical Attributes, Yield and Grain Quality of Rice (*Oryza sativa* L.) under Drought Stress. *PLoS ONE* **2021**, *16*, e0254452. [CrossRef] [PubMed]
42. Hlokwe, M.T.P.; Kena, M.A.; Mamphiswana, N.D. Evaluating Crude Extracts of Monsonia Burkeana and *Moringa oleifera* against Fusarium Wilt of Tomato. *Acta Agric. Scand.* **2018**, *68*, 757–764. [CrossRef]
43. Byford, W.J. A Survey of Foliar Diseases of Sugar Beet and Their Control in Europe. In Proceedings of the Comptes-Rendus des Congres de l'Institut International de Recherches Betteravieres (The 59th Meeting of International Institute of Sugar Beet Research), Brussels, Belgium, February 1996; pp. 1–10.
44. El-Sayed, A.E.N. Assessment of Resistance in Some Sugar Beet Varieties to Powdery Mildew Using Crossed Immuno Electrophoresis Technique. *Egypt. J. Phytopathol.* **2015**, *43*, 151–158. [CrossRef]
45. Bakeer, A.-R.T.; Ahmed, H.M.H.; Baiuomy, M.A.M.; Fatoh, E.F.M. Efficacy Evaluation of Inhibitory Activity of Some Biological Products, Essential Oils, Plant Extracts and Induced Resistance Agents against *Erysiphe heraclei* Dc, the Pathogenic Powdery Mildew of Parsley (*Petroselinum sativum* L.). *Egypt. J. Appl. Sci.* **2019**, *34*, 111–134. [CrossRef]
46. McGrath, M.T. Diseases of Cucurbits and Their Management. In *Diseases of Fruits and Vegetables Volume I*; Naqvi, S.A.M.H., Ed.; Springer: Dordrecht, The Nederlands, 2004; Volume I, pp. 455–510.
47. Elkot, G.A.N.; Belal, E.B.A. Biocontrol of Fusarium Damping-off of Pea by Certain Bacterial Antagonists. *J. Agric. Res. Tanta Univ.* **2006**, *32*, 225–241.
48. Romero, A.M.; Zapata, R.; Montecchia, M.S. First Report of Black Rot on Arugula Caused by *Xanthomonas campestris* Pv. *Campestris* in Argentina. *Plant Dis.* **2008**, *92*, 980. [CrossRef] [PubMed]
49. Hamza, A.M.; El-Kot, G.A.; El-Moghazy, S. Non-Traditional Methods for Controlling Maize Late Wilt Disease Caused by *Cephalosporium maydis*. *Egypt. J. Biol. Pest Control* **2013**, *23*, 87–93.
50. Antal, Z.; Manczinger, L.; Szakacs, G.; Tengerdy, R.P.; Ferenczy, L. Colony Growth, in Vitro Antagonism and Secretion of Extracellular Enzymes in Cold-Tolerant Strains of *Trichoderma* Species. *Mycol. Res.* **2000**, *104*, 545–549. [CrossRef]
51. Taha, M.A.; Abd El-All, A.M.; El-Shennawy, M.Z. Effect of Some Plant Aqueous Extracts on Lettuce Growth, Chemical Constituents, Yield and Downy Mildew Disease. *J. Plant Prod.* **2020**, *11*, 933–938. [CrossRef]
52. Sawant, S.D.; Sawant, I.S. Use of Potassium Bi-Carbonates for the Control of Powdery Mildew in Table Grapes. *Acta Hortic.* **2008**, *785*, 285–292. [CrossRef]
53. Bowers, J.H.; Locke, J.C. Effect of Botanical Extracts on the Population Density of *Fusarium oxysporum* in Soil and Control of Fusarium Wilt in the Greenhouse. *Plant Dis.* **2007**, *84*, 300–305. [CrossRef]
54. Abdel-Kader, M.M.; Abdel-Kareem, F.; El-Mougy, N.S.; El-Mohamady, R.S. Integration between Compost, *Trichoderma harzianum* and Essential Oils for Controlling Peanut Crown Rot under Field Conditions. *J. Mycol.* **2013**, *2013*, 262130. [CrossRef]
55. Tabassum, N.; Vidyasagar, G.M. Antifungal Investigations on Plant Essential Oils. A Review. *Int. J. Pharm.* **2013**, *5*, 19–28.
56. Hadi, M.; Kashefi, B.; Sobhanipur, A.; Rezaarabsorkhi, M. Study on Effect of Some Medicinal Plant Extracts on Growth and Spore Germination of *Fusarium oxysporum* Schlecht. In vitro. *Am. Eurasian J. Agric. Environ. Sci.* **2013**, *13*, 581–588.
57. Dwivedi, S.K. Enespa Effectiveness of Extract of Some Medicinal Plants against Soil-Borne Fusaria Causing Diseases on *Lycopersicon esculantum* and *Solanum melongena* Plants. *Int. J. Pharma Bio Sci.* **2012**, *3*, B-1171–B-1180.
58. Radjacommare, R.; Kandan, A.; Nandakumar, R.; Samiyappan, R. Association of the Hydrolytic Enzyme Chitinase against *Rhizoctonia solani* in Rhizobacteria-Treated Rice Plants. *J. Phytopathol.* **2004**, *152*, 365–370. [CrossRef]
59. El-kazzaz, M.K.; Salem, E.A.; Ghoneim, K.E.; Elsharkawy, M.M.; El-Kot, G.A.E.-w.N.; Kalboush, Z.A.E. Integrated Control of Rice Kernel Smut Disease Using Plant Extracts and Salicylic Acid. *Arch. Phytopathol. Plant Prot.* **2015**, *48*, 664–675. [CrossRef]
60. Barnett, H.L.; Hunter, B.B. *Illustrated Genera of Imperfect Fungi*, 3rd ed.; Burgess Publishing Company: Minneapolis, MN, USA, 1972.

61. Shane, W.W.; Teng, P.S. Impact of Cercospora Leaf Spot on Root Weight, Sugar Yield, and Purity of *Beta vulgaris*. *Plant Dis.* **1992**, *76*, 820. [CrossRef]
62. Bergey, D.H.; Krieg, N.R. *Bergey's Manual of Systematic Bacteriology*; Williams and Willkins: Baltimore, MD, USA, 1984.
63. McGrath, M.T.; Staniszewska, H.; Shishkoff, N.; Casella, G. Fungicide Sensitivity of *Sphaerotheca fuliginea* Populations in the United States. *Plant Dis.* **1996**, *80*, 697–703. [CrossRef]
64. Kamel, S.M.H. Antagonistic Effects of Some Microbial Inhabitants on Phylloplane of Squash Plants towards *Sphaerotheca fuliginea*. Master's Thesis, Faculty of Agriculture of Kafrelsheikh, Tanta University, Kafrelsheikh, Egypt, 2003.
65. McGinnis, R.A. *Beet-Sugar Technology*, 3rd ed.; Beet Sugar Development Foundation: Denver, CO, USA, 1982.
66. AOAC. *Official Methods of Analysis of Association of Official Agriculture Chemists*, 17th ed.; AOAC: Gaithersburg, MD, USA, 2000.
67. Allam, A.; Pitts, G.; Hollis, J. Sulfide Determination in Submerged Solls with AN Ion-Selective Electrode. *Soil Sci.* **1972**, *114*, 456–467. [CrossRef]
68. Maxwell, D.P.; Bateman, D.F. Changes in the Activities of Some Oxidases in Extracts of *Rhizoctonia*-Infected Bean Hypocotyls in Relation to Lesion Maturation. *Phytopathology* **1967**, *57*, 132–136.
69. Zucker, M. Sequential Induction of Phenylalanine Ammonia-Lyase and a Lyase-Inactivating System in Potato Tuber Disks. *Plant Physiol.* **1968**, *43*, 365. [CrossRef]
70. Manzali, D.; Nipoti, P.; Pisi, A.; Filippini, G.; D'Ercole, N. Scanning Electron Microscopy Study of in vitro Antagonism of *Trichoderma* Spp. Strains against Rhizoctonia Solani Kühn. *Phytopathol. Mediterr.* **1993**, *32*, 1–6.

plants

MDPI

Article

Induction of Male Sterility by Targeted Mutation of a Restorer-of-Fertility Gene with CRISPR/Cas9-Mediated Genome Editing in *Brassica napus* L.

Zunaira Farooq [1,2], Muhammad Nouman Riaz [1], Muhammad Shoaib Farooq [1], Yifan Li [1], Huadong Wang [1], Mayra Ahmad [1], Jinxing Tu [1], Chaozhi Ma [1], Cheng Dai [1], Jing Wen [1], Jinxiong Shen [1], Tingdong Fu [1], Shouping Yang [2], Benqi Wang [3,*] and Bin Yi [1,*]

[1] National Key Laboratory of Crop Genetic Improvement, National Center of Rapeseed Improvement, Huazhong Agricultural University, Wuhan 430070, China

[2] Soybean Research Institute, National Center for Soybean Improvement, Key Laboratory of Biology and Genetic Improvement of Soybean (General Ministry of Agriculture), Jiangsu Collaborative Innovation Center for Modern Crop Production, Nanjing Agricultural University, Nanjing 210095, China

[3] Wuhan Vegetable Research Institute, Wuhan Academy of Agricultural Science and Technology, Wuhan 430065, China

* Correspondence: wangbenqi@wuhanagri.com (B.W.); yibin@mail.hzau.edu.cn (B.Y.)

Citation: Farooq, Z.; Nouman Riaz, M.; Farooq, M.S.; Li, Y.; Wang, H.; Ahmad, M.; Tu, J.; Ma, C.; Dai, C.; Wen, J.; et al. Induction of Male Sterility by Targeted Mutation of a Restorer-of-Fertility Gene with CRISPR/Cas9-Mediated Genome Editing in *Brassica napus* L.. *Plants* **2022**, *11*, 3501. https://doi.org/10.3390/plants11243501

Academic Editors: Mingxun Chen, Lixi Jiang, Yuan Guo and Othmane Merah

Received: 13 September 2022
Accepted: 29 November 2022
Published: 13 December 2022

Abstract: *Brassica napus* L. (canola, oil seed rape) is one of the world's most important oil seed crops. In the last four decades, the discovery of cytoplasmic male-sterility (CMS) systems and the restoration of fertility (*Rf*) genes in *B. napus* has improved the crop traits by heterosis. The homologs of *Rf* genes, known as the restoration of fertility-like (*RFL*) genes, have also gained importance because of their similarities with *Rf* genes. Such as a high non-synonymous/synonymous codon replacement ratio (dN/dS), autonomous gene duplications, and a possible engrossment in fertility restoration. *B. napus* contains 53 *RFL* genes on chromosomes A9 and C8. Our research aims to study the function of *BnaRFL11* in fertility restoration using the CRISPR/Cas9 genome editing technique. A total of 88/108 (81.48%) T_0 lines, and for T_1, 110/145 (75%) lines carried T-DNA insertions. Stable mutations were detected in the T_0 and T_1 generations, with an average allelic mutation transmission rate of 81%. We used CRISPR-P software to detect off-target 50 plants sequenced from the T_0 generation that showed no off-target mutation, signifying that if the designed sgRNA is specific for the target, the off-target effects are negligible. We also concluded that the mutagenic competence of the designed sgRNAs mediated by U6-26 and U6-29 ranged widely from 31% to 96%. The phenotypic analysis of *bnarfl11* revealed defects in the floral structure, leaf size, branch number, and seed production. We discovered a significant difference between the sterile line and fertile line flower development after using a stereomicroscope and scanning electron microscope. The pollen visibility test showed that the pollen grain had utterly degenerated. The cytological observations of homozygous mutant plants showed an anther abortion stage similar to nap-CMS, with a *Orf222*, *Orf139*, *Ap3*, and *nad5c* gene upregulation. The *bnarfl11* shows vegetative defects, including fewer branches and a reduced leaf size, suggesting that PPR-encoding genes are essential for the plants' vegetative and reproductive growth. Our results demonstrated that *BnaRFL11* has a possible role in fertility restoration. The current study's findings suggest that CRISPR/Cas9 mutations may divulge the functions of genes in polyploid species and provide agronomically desirable traits through a targeted mutation.

Keywords: rapeseed CMS; *Rf*-like (*RFL*); CRISPR/Cas9; sgRNA; cytological study; genome editing

1. Introduction

Rapeseed is a globally important crop that provides ingestion products for humans and animal feed, such as oil products, proteins, and raw materials for industrialized products worldwide [1,2]. Male sterility is one of the most effective heterosis techniques for increasing crop production. Maternal inheritance is essential for hybrid breeding programs and mitochondrial and chloroplast genomes. CMS is a maternally inherited trait in flowering plants that cannot produce functional pollen [3]. Several cellular metabolic processes mandatory for higher plants, such as ATP production by oxidative phosphorylation, occur in mitochondria. Due to a mitochondrial impairment, cytoplasmic male sterility offers unique mechanisms for revealing the genetic association between nuclear genomes and mitochondria in plants [4]. The most common male sterility induction mechanisms in rapeseed are GMS, CMS, and chemical gametocide [5]. In many plants, CMS traits are determined by *ORFs* encoded by the mitochondrial genome. Typically, rapeseed has two CMS mitotypes, nap and Pol. According to previous studies, the CMS in rapeseed is of the following types: *nap* CMS [6], *ogu* CMS [7], *pol* CMS, Moricandia arvensis [8], *shan2A* [9], Tour CMS [10], *Nca* CMS [11], *oxa* CMS [12], *Nsa* [13], *inap* CMS, and *hau* CMS [14].

The effects of mitochondrial genes controlling cytoplasmic male sterility (CMS) can be suppressed by fertility restoration genes (*Rf*). Several *Rf* genes in rapeseed and other crops encode the pentatricopeptide repeat protein (PPR) family [15]. PPR proteins are a group of pattern-specific single-stranded RNA-associated proteins found in terrestrial plants that regulate post-transcriptional mechanisms, such as RNA degradation, cleavage, stability, splicing, and genome editing in mitochondrial and chloroplast genomes [16].

According to previous research, the *Rf* genes found in petunia [17], rice [18,19], radish [20,21], and sorghum [22] are members of the pentatricopeptide repeat gene family. In *Brassicaceae*, crops have multiple fertility restorer genes (*Rf*) encoding PPR proteins such as Nap-*orf222-Rfn* [23], Pol-*orf224-Rfp* [23], Ogu-*orf138-Rfo* [24], Tour-*orf263-Rf1_Rf2_* [25,26], Kos-*orf125-Rfk* [24,27], Nsa-*orf224,orf309*, *orf346*-unknown [28], and Shaan2a-*orf224-1-Shaan2b* [29]. Previous studies have revealed that both *Rf*-CMS genes interact throughout evolution, which may explain why all cloned *Rf*-PPR genes share a common ancestor [30,31].

According to several studies, *Rf-PPR* genes are also recruited to mitochondria, which block CMS gene products' accretion [32]. Additionally, many *Rf* genes encoding PPR proteins are invariably present in clusters of linked *Rf*-PPR-like genes (*RFL*) [33,34]. *RFL*-rich areas contain many *Rf*-PPR genes, and it is hypothesized that most *RFL* genes in the same chromosomal region are functional restorer genes. For instance, the rice chromosome 10 *RFL*-rich area contains *Rf1* and *Rf4* [35,36]. The *Rf8* locus in maize is located in an *RFL* cluster on chromosome 2 [37].

Fortunately, recent research has revealed that 53 *RFL* genes have been discovered in *Brassica napus*. It has been demonstrated that 10 and 18 *BnaRFLs* assemble into highly dense clusters on chromosomes A9 and C8, respectively. These nominated restorer *BnaRFLs* on chromosomes A9 and C8 likely play a conserved role in mitochondrial RNA processing. It can be concluded that six *BnaRFLs* are candidates for fertility restorer genes (*RFL3, RFL4, RFL5, RFL8, RFL15,* and *RFL41*), four genes are similar to restorer genes (*RFL2, RFL10, RFL11,* and *RFL42*), and two restore genes (*Rfn-RFL6* and *Rfp-RFL13*) clustered together in the phylogenetic tree, indicating that these genes were the most likely restorer gene members in the CMS rapeseed model. When comparing *BnaRFL* to other known restorer genes, it was discovered that all the identified *BnaRFLs* had 80 amino acid motifs, *suggesting* that *BnaRFL* could help in a rapeseed fertility restoration. It indicates that the *RFL* genes revealed in rapeseed have a possible role in fertility restoration on the A9 and C8 chromosomes [29].

Furthermore, they discovered *BnRFL13* (*Rfp*) and *BnRFL6* (*Rfn*), which have already been proven to be restorer genes in Pol and nap CMS in rapeseed [37–39]. *RFL* genes, grouped on chromosomes A9 and C8, have been identified as restorer genes in the rapeseed CMS system. These genes have a significant impact on mitochondrial RNA processing [29].

Consequently, we can learn how putative *RFL* genes restore rapeseed fertility. The *CRISPR/Cas9* genome editing method has recently been extensively used in rapeseed to develop *BnTT8* plants with a double mutant phenotype, displaying increased oil and protein levels with a changed fatty acid (F.A.) content but no major yield defects [40]. In the current study, we used this technique to develop a *BnaRFL11* mutant to induce mutations by knocking out the *RFL* gene, a candidate for fertility restoration in rapeseed CMS.

2. Material and Methods

2.1. Plant Materials

In the present study, a pure line of *B. napus* Westar (nap-cytoplasm) was used as the source of transformation. The seeds were obtained from the National Engineering Research Center of Rapeseed (Wuhan, China). Wild-type plants and transgenic lines were grown in a greenhouse under (16/8 h (h) light/dark cycle at 22 °C).

2.2. Phylogenetic and Bioinformatics Analysis

All *Brassica napus* selected gene sequences, and other species' homologs, were subjected to Mega 7 (https://www.megasoftware.net/ accessed on 20 May 2019) for phylogenetic analysis using the neighbor-joining method. The sequences for the trees that only contained *Brassica* and *Arabidopsis* were downloaded from BRAD and TAIR, respectively, and conserved motifs were identified using the MEME online suit (http://meme-suite.org/tools/meme, accessed on 25 May 2019) with the default settings [41].

2.3. Computational sgRNA Design and Selection

The sgRNAs were designed using the CRISPR-P web tool (http://cbi.hzau.edu.cn/cgi-bin/CRISPR accessed on 25 May 2019) [42]. First, we found all possible sgRNA sequences in specific target genes with a G.C. content (45~60%) and an on-target score (score value > 0.6). The off-target scores were then calculated based on a previous study's scoring system of off-target sites [43].

2.4. Construction of Binary Vectors and Genetic Transformation of Plants

To construct Cas9/sgRNA-expressing binary vectors, variant sgRNAs with a sequence specificity were designed for the gene of interest using the web-based tool CRISPR-P (http://cbi.hzau.edu.cn/cgi-bin/CRISPR, accessed on 25 May 2019). For constructing vectors using a multiplex genome targeting system, the binary pHSE401-2gR/Cas9 was arranged by Prof. Wang Zhiping (Chen Qijun, College of Biology, China Agricultural University China) and used to construct a library according to the findings of [44]. The desired constructs were monitored by sequencing and finally transformed into *B. napus* using the Agrobacterium-mediated hypocotyl method, as reported previously [45]. The transgenic lines were confirmed using a polymerase chain reaction (PCR) and antibiotic selection markers. The oligo primers used for preparing the sgRNA vectors are listed in (Table S1).

2.5. Identification of Transgenic Mutant Plants and Potential Off-Targets

The presence of the T-DNA construct was assessed by PCR using CPA F/R gene-specific primers. A PCR was performed to amplify the genomic region surrounding the CRISPR target sites using specific primers (Table S1), and the mutations were screened using the PAGE method previously described [46]. To confirm the PAGE-based genotyping results, we used the high-throughput tracking of mutations via the T.A. cloning of the PCR products. The pMD18-T vector (Takara) was used for T.A. cloning and sequencing. The potential off-target sites were identified using CRISPR-P (http://cbi.hzau.edu.cn/cgi-bin/CRISPR, accessed on 5 June 2020). The PCR amplified an approximately 250-bp DNA sequence covering each off-target site. The primers used are listed in (Table S1). For each target gene, the mixed genomic DNA from T_0-edited plants was used as the template, and wild-type DNA was included as a control. All the PCR products were purified and mixed in equal amounts (50 ng each) as a single sample. The DNA library construction and sequencing were performed using the T.A. cloning method.

2.6. RNA Extraction, RT-PCR

The total RNA was extracted by Trizol reagent (Invitrogen, Waltham, MA, USA). The total RNA (5 μg) was treated with DNase (Thermo Fermentas, Waltham, MA, USA), purified, and precipitated using ethanol. The cDNA was obtained by reverse transcription using Superscript III according to the manufacturer's instructions. The SYBR Green I master PCR kit was used for real-time PCR with gene-specific primers on a Light Cycler 480.

2.7. Histological Analyses

2.7.1. Scanning Electron Microscopy Analysis

Fresh WT (Westar) and mutant plant pollen grains were air-dried for approximately 30 min before viewing with a Hitachi S 4700 scanning electron microscope at an accelerating voltage of 5 kV.

2.7.2. Pollen Fertility Examination

The pollen sampled from the W.T. and mutant plants' rapeseed buds immediately before flowering were stained with 1% (*w*/*v*) acetocarmine solution to analyze the pollen viability. Three biological replicates were used for this study. The stained pollen grains were visualized, and the images were recorded using a Leica DMIRB fluorescence microscope.

2.7.3. Cytological Analysis

The sections of flower buds from the CMS and fertile plants were obtained following the method described in [47]. The sample was embedded and sectioned at a 6 to 10 μm thickness using a Leica 2035 Biocut. The sections were stained with hematoxylin and eosin (0.5%) for the anthers. The images were captured using a Nikon Eclipse 80i microscope.

2.8. Field Experiments and Phenotyping

Wild-type plants and T_0 and T_1 transgenic lines were kept in a greenhouse under a photoperiod (16/8 h light/dark at 22 °C) in 2019 and 2020. After screening, the homozygous mutant lines were grown during the rapeseed crop season (2020–2021) in the experimental field of Huazhong Agriculture University, Wuhan, China. The experiment was performed on field plants, followed by three replicates with a complete block design. In one row, approximately 11–12 plants were planted; each row followed the same procedure, with a distance observed in one row of 21 cm, followed by a 30 cm distance in each row. The field management of each line was performed using a standard breeding method.

3. Results

3.1. Bioinformatic Analysis of BnaRFL11 Gene in Westar

According to the genome information, *B. napus* contains one copy of *BnaRFL11*. (http://www.genoscope.cns.fr/brassicanapus/, accessed on 20 May 2019). The DNA and protein sequence alignments were similar for a single copy of *the BnaA09g45590D*, At*1g12300*, and *At1g12620* genes (Figures S1 and S2). *BnaRFL11 and Arabidopsis* homologs DNA sequence were given in to MEME "(http://meme-suite.org/, accessed on 25 May 2019)" to search for the conserved motifs between the selected gene protein sequences (Figure 1b). According to the output, all the motifs were highly conserved in these DNA sequences, suggesting a similarity in function (Figure 1b). The phylogenetic tree showed that a single copy of *BnaRFL11* was highly similar to its homologs in *A. thaliana* and *Rfn* (*RFL6*) (Figure 1a).

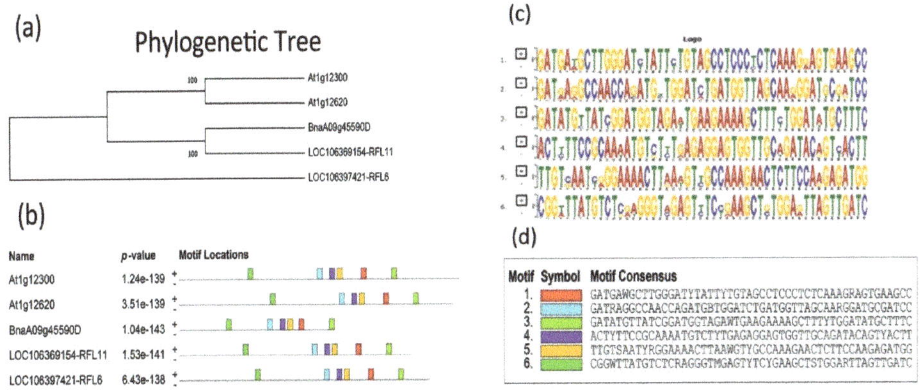

Figure 1. Conserved motif analysis and phylogenetic relationship of *BnaRFL11* with other gene species. (**a**) Phylogenetic tree showing an association between *BnaRFL11* and its sister genes in *Arabidopsis*. Protein sequences were obtained from the GenBank with the following gene ID: *BnaA09g45590D*, *AT1G12300*, and *AT1G12620*. (**b**) Presence of similar motifs between sequences. (**c**,**d**) The conserved motifs among *BnaRFL11*, known fertility restore genes *Rfn* and *Rfp*, and *Arabidopsis thaliana* sister genes.

3.2. CRISPR/Cas9 Vector Construction to Knock out the BnaRFL11 Gene in Rapeseed

Many fertility restorer genes have been reported to encode the PPR protein family essential for a fertility restoration in petunia, radish, rice, *Arabidopsis*, and *Brassica* species [48,49]. Thus, the modification of *PPR-encoding* genes may induce male sterility in rapeseed. We used the pure Westar line of *B. napus*, which is amenable to an Agrobacterium-mediated transformation. A single copy of *BnaRFL11* (BnaA09.PPR.AA09) has six exons, similar to the *Arabidopsis* sister genes. Two sgRNA (S1–S2) were designated, employing the "CRISPR-P" [44] to generate point mutations in a single copy of *BnaRFL11* using a Cas9 gene-editing technique (Figures 2 and S3). S1 and S2 are located on the different exons of the *BnaRFL11* gene.

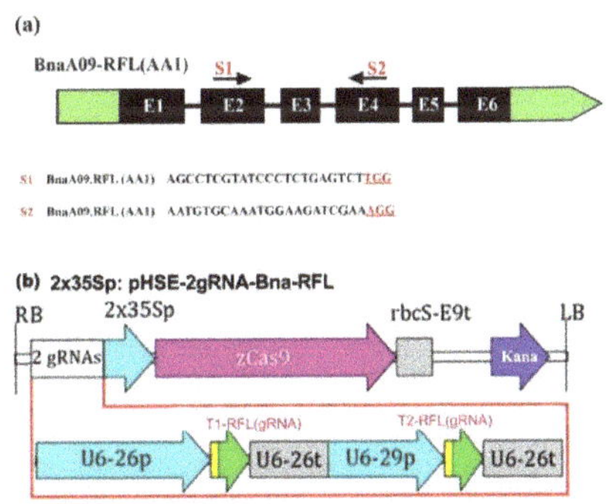

Figure 2. Model of *BnaRFL11* with target sequences and the binary plasmid Cas9 vector. (**a**) The *BnaRFL11* gene structure is shown in black. The vertical line indicates the gene model, and the arrow shows the sgRNA direction. The target sequences of sgRNAs are represented by PAM sites highlighted in red. (**b**) The kanamycin resistance cassette is driven by the CaMV35S promoter of the cauliflower mosaic virus. Two sgRNAs are driven by the U6-26p and U6-29p promoters and the U6-26t terminator.

We used a multiplex genome-editing technique with a single CRISPR/Cas9 vector. The CRISPR vector based on pCAMBIA was arranged with these two sgRNAs with Cas9 driven by U6-26p as the promoter of the *Arabidopsis* U6 gene; U6-26t is the terminator of the U6-26 gene with the downstream sequence. The CaMV35S promoter was used for the Cas9 protein expression analysis. In addition, the U6-26p and U6-29p promoters were used to drive S1 and S2 from *Arabidopsis* [44] (Figure 2b). Such a design will determine which promoter is best for generating an adequate mutation in the desired gene sequence with the Cas9 protein expression in *B. napus*.

3.3. Identification of Mutation Patterns in the Transgenic Plants

The *BnaRFL11* CRISPR/Cas9 vector was positively transferred into Westar by an Agrobacterium transformation, and 108 independent lines were developed for *bnarfl11* (Figure 3). To check the positive transgenic plants through a PCR using the construct-specific primers, 88/108 (81.48%) of the T_0 lines carried T-DNA insertions, and for T_1, 110/145 (75%) lines carried T-DNA insertions (Table 1). Polyacrylamide gel electrophoresis (PAGE) was used to identify the edited lines. Following the standard procedure, the PCR products from each target site were denatured, renatured, and separated. This method is based on the slower migration of heteroduplex DNA (mutation) than of homoduplex DNA (non-mutation) on native PAGE [46]. The genotyping results of many plants displayed band profiles that differed from those on the non-denaturing PAGE gels in wild-type plants (W.T.), showing the presence of mutations at the target sites (Figure S4).

Figure 3. Different stages of the tissue culture process for Agrobacterium-mediated transformation in rapeseed. (**a**) Sowing of wild-type Westar seeds on the seedling medium (M0). (**b**) Growth of plants on seedling medium after 6–7 days. (**c**) Spreading of hypocotyls for 48 h in the dark on (M1) media. (**d**) After 48 h, transfer explants were transferred to (M2) media and placed in a growth room for 15–20 days. (**e**) After 15–20 days, explants were on (M3) media for callus induction until the plants were regenerated. (**f**) Regenerated plants were transferred into (M4) rooting media to obtain better roots. (**g**) After obtaining roots, transgenic plants were transferred into pots under controlled conditions for their survival. (**h**) Transgenic plants were transferred to the field under normal conditions after survival. Bar = 4 cm.

Table 1. Statistical results for *the bnarfl11* transgenic generations.

Target Gene	Number of Targets	Generation	Transgene Rate (%)	Positive Rate (%)
BnaRFL11	2	T_0	81% (88/108)	89% (97/108)
	2	T_1	75% (110/145)	90% (131/145)

Numbers in brackets indicate the number of mutated plants divided by the total number of plants developed.

We used a high-throughput tracking system of mutations, the T.A. cloning of PCR products (gel recovery and ligation of DNA), to detect T_0 and T_1 positive transgenic plant mutations to authenticate the PAGE results. The site-specific primers for each target were designed based on the SNP differences. For each target, forward and reverse primers covered the target site, and both primers were designed within 20–80 nucleotides from the target site (Figure S5).

Transgenic plants were sequenced for each targeted area following T.A. cloning to calculate the mutation rate correctly. The sequencing chromatograms of each line were manually examined. When the sequencing chromatograms demonstrated a nucleotide substitution (deletion, insertion, or substitution) or several indications (overlapping peaks) at the sgRNA target locations, we concluded that mutations were successfully induced (Table S2), indicating that T. A. clone-based sequencing is an effective and valuable technique for identifying plant mutations. CRISPR/Cas9 induces mutations in the designed S1 and S2 targets in a single copy of the *BnaRFL11*. Of the 88 transgenic plants, a total of 70 T_0 mutant lines were examined; 25/70 (35.71%) loci of *bnarfl11* showed putative heterozygous (Hetro) alterations, 7/70 (10%) loci had chimeric changes, and more than half 38/70 (54.28%) of the loci were homozygous (Homo) mutations. Mutant lines with single and quadruple mutations were selected for the genotyping and phenotypic analysis. Seventeen mutant lines were selected from the 70 for the T_1 generation for a further confirmation (Table S2).

3.4. Isolation of Mutants with Transgenic Elements in T_0 and T_1 Generations

To develop stable mutant lines, 17 T_0 lines carrying mutations and a male-sterile phenotype in *bnarfl11* were crossed with the wild type to obtain the seeds, and an individual T_1 generation was genotyped through the T.A. clone. The observed transmission rate of the allelic mutations from T_0 mutant plants to T_1 progenies ranged from 35% to 100% in *bnarfl11*, but the average transmission rate from T_0 to T_1 plants with allelic mutations was 81% (Table S3). For example, the detected mutations in the T_1 generation matched the observed mutations in the related T_0 lines (Table S2). Homozygous mutations were detected in the T_1 plants generated at more than two loci, indicating that the next-generation mutant was stably inherited and permanently exhibited homozygous genotypes (Table S3). Interestingly, one and two new mutations were identified in the T_1 lines.

In contrast, these mutations were not present in T_0 lines at the S2 site of *bnarfl11* (Table S2), indicating that wild-type sites were modified during the development of independent lines with a low efficiency. Consistently, the observed mutation and male sterility phenotypes in the corresponding *bnarfl11-1*, *bnarfl11-5*, and *bnarfl11-8* T_0 plants matched the detected changes and detected phenotypes in the next T_1 (*bnarfl11-1-5*, *bnarfl11-5-8*, and *bnarfl11-8-3*) progeny lines (Table S3 and Figure 4). Thus, the sequencing data of the heritable changes at both target sites of *bnarfl11* (*BnaA09g45590D*) after a T_0 to T_1 generation provides strong evidence that the constant Cas9-induced germline mutation transfer of mutations in *B. napus* is achievable (Figure 4). To screen the mutants with a targeted modification without integrating foreign DNA into the *B. napus* genome, we performed a PCR analysis of the T_0 and T_1 generations using Cas9-888 and U6-26p primers to confirm the positive transgenic plants. The Cas9 vector with T-DNA was not detected in 20 of 108 (18.51%) T_0 plants and 35/145 (24.1%) T_1 plants originating from 17 independent T_0 lines (Table 1). Altogether, various *bnarfl11* single, double, triple, and quadruple homozygous transgene-free T-DNA mutants were obtained in the T_1 generation. Thus, transgene-free T-DNA plants carry out the desired gene modifications that might be obtained through heritable segregation in *B. napus*.

Line	S1		Line	S2	
bnarfl11-1	AGCCTCGTATCCCTCTGAGTCTTGG	WT	*bnarfl11*-1	AGCCTCGTATCCCTCTGAGTCTTGG	WT
aa09 T_0	AGCCTCGTATCCCATCTGTAGTCTTGG	+2bp	aa09 T_0	AGCCTCGT-TCCCTCTTGAGTCTTGG	-1bp/+1bp
bnarfl11-1-5	AGCCTCGTATCCCTCTGAGTCTTGG	WT	*bnarfl11*-1-5	AGCCTCGTATCCCTCTGAGTCTTGG	WT
aa09 T_1	AGCCTCGTATCCCATCTGTAGTCTTGG	+2bp	aa09 T_1	AGCCTCGT-TCCCTCTTGAGTCTTGG	-1bp/+1bp
bnarfl11-5	AGCCTCGTATCCCTCTGAGTCTTGG	WT	*bnarfl11*-5	AGCCTCGTATCCCTCTGAGTCTTGG	WT
aa09 T_0	AGCCTCGTATA-------------------------	+A/-31BP	aa09 T_0	AGCCTCGTTATCCCT-----TCTTGG	+T/-5bp
bnarfl11-5-8	AGCCTCGTATCCCTCTGAGTCTTGG	WT	*bnarfl11*-5-8	AGCCTCGTATCCCTCTGAGTCTTGG	WT
aa09 T_1	AGCCTCGTATA-------------------------	+A/-31BP	aa09 T_1	AGCCTCGTTATCCCT----TCTTGG	+T/-5bp
bnarfl11-15	AGCCTCGTATCCCTCTGAGTCTTGG	WT	*bnarfl11*-15	AGCCTCGTATCCCTCTGAGTCTTGG	WT
aa09 T_0	AGCCTCGTATCCCTCTGAGTCTTGG	Hetero	aa09 T_0	AGCCTCGTATCCCTCTGAGTCTTGG	Hetero
bnarfl11-15-3	AGCCTCGTATCCCTCTGAGTCTTGG	WT	*bnarfl11*-15-3	AGCCTCGTATCCCTCTGAGTCTTGG	WT
aa09 T_1	AGCCTCGTATCCCTCTGAGTCTTGG	Hetero	aa09 T_1	AGCCTCGTATCCCTCTGAGTCTTGG	Hetero
bnarfl11-23	AGCCTCGTATCCCTCTGAGTCTTGG	WT	*bnarfl11*-23	AGCCTCGTATCCCTCTGAGTCTTGG	WT
aa09 T_0	AGCCTCGTATCCCTACTGATGTCTTGG	+2bp	aa09 T_0	AGCCTCGTATCCCTCTGAGTCTT GG	-18bp
bnarfl11-23-4	AGCCTCGTATCCCTCTGAGTCTTGG	WT	*bnarfl11*-24	AGCCTCGTATCCCTCGAGTCTTGG	WT
aa09 T_1	AGCCTCGTATCCCTACTGATGTCTTGG	+2bp	aa09 T_1	AGCCTCGTATCCCTCTGAGTCTTGG	-18bp
bnarfl11-34	AGCCTCGTATCCCTCTGAGTCTTGG	WT	*bnarfl1*-34	AGCCTCGTATCCCTCTGAGTCTTGG	WT
aa09 T_0	AGCCTCGTATC--TCTGAGTCTTGG	-2bp	aa09 T_0	AGCCTCGTATCCCTCTG AGTCTTGG	Hetero
bnarfl11-34-8	AGCCTCGTATCCCTCTGAGTCTTGG	WT	*bnar11*-34-8	AGCCTCGTATCCCTCTGAGTCTTGG	WT
aa09 T_1	AGCCTCGTATC--TCTGAGTCTTGG	-2bp	aa0 T_1	AGCCTCGTATCCCTCTGAGTCTTGG	Hetero
bnarfl11-44	AGCCTCGTATCCCTCTGAGTCTTGG	WT	*barfl11*-44	AGCCTCGTATCCCTCTGAGTCTTGG	WT
aa09 T_0	AGCCTCGATATCCCGTCTGAGTCTTG	+2bp	a09 T_0	AGCCTCTGTATCCCTCTGTAGTCTTGG	+2bp
bnarfl11-44	AGCCTCGTATCCCTCTGAGTCTTGG	WT	*bnarfl11*-44	AGCCTCGTATCCCTCTGAGTCTTGG	WT
aa09 T_1	AGCCTCGATATCCCGTCTGAGTCTTGG	+2bp	aa09 T_1	AGCCTCTGTATCCCTCTGTAGTCTTGG	+2bp
bnarfl11-75	AGCCTCGTATCCCTCTGAGTCTTGG	WT	*bnarfl11*-75	AGCCTCGTATCCCTCTGAGTCTTGG	WT
aa09 T_0	AGCCTCGTGTATCCCATCTG------GG	+3bp/-6bp	aa09 T_0	AGCCT--TATCCCTCTGTAGTCTTGG	+1bp/-2bp
bnarfl11-75-6	AGCCTCGTATCCCTCTGAGTCTTGG	WT	*bnarfl11*-75-6	AGCCTCGTATCCCTCTGAGTCTTGG	WT
aa09 T_1	AGCCTCGTGTATCCCATCTG------GG	+3bp/-6b	aa09 T_1	AGCCT--TATCCCTCTGTAGTCTTGG	+1bp/-2bp
bnarfl11-83	AGCCTCGTATCCCTCTGAGTCTTGG	WT	*bnarfl11*-83	AGCCTCGTATCCCTCTGAGTCTTGG	WT
aa09 T_0	AGCCTCGTATCCCTCTGAGTCTTGG	Hetero	aa09 T_0	AGCCTCGTAATTGTCCCTC-GAGTCTTGG	+4bp/-1bp
bnarfl11-83-4	AGCCTCGTATCCCTCTGAGTCTTGG	WT	*bnarfl11*-83-4	AGCCTCGTATCCCTCTGAGTCTTGG	WT
aa09 T_1	AGCCTCGTATCCCTCTGAGTCTTGG	Hetero	aa09 T_1	AGCCTCGTAATTGTCCCTC-GAGTCTTGG	+4bp/-1bp

Figure 4. Mutation transmission from T_0 to T_1 generations. At both target sites in *bnarfl11* mutant plants. Red font and red hyphens signify CRISPR-based alterations, whereas PAM is underlined and highlighted in red. The altered alleles of *RFL11* on chromosome A were expressed as aa09. "−" and "+" signify losses.

3.5. CRISPR's Off-Target Effects on Mutant Lines (T_0) in *B. napus*

In the current study, we used CRISPR-P software to detect off-targets. We searched for putative off-targets with similar identities to the two sgRNAs used for an on-target mutation in *Brassica napus* with the help of "CRISPR-P" [50]. Some off-target genes were identified for the selected sgRNAs (Table S4). There were four off-targets for S1 with a maximum of four mismatches, whereas S2 had three off-targets with a maximum of four mismatches in *BnaRFL11*. We sequenced 50 plants from the T_0 generation that showed no off-target mutations, signifying that if the designed sgRNA is specific for the target, the off-target effects are negligible. It also emphasized that the male-sterile phenotype detected in *bnarfl11* was due to an induced mutation in *BnaA09g45590D* and not by off-target effects. Therefore, the CRISPR/Cas9 system can induce stable and specific mutations in the *B. napus* genome.

3.6. Editing Efficiency of sgRNAs in BnaRFL11

Different sgRNAs have different mutation rates. Therefore, the two sgRNAs for the gene ensured a high rate of mutation (Table S5). Interestingly, these sgRNAs created a targeted mutation in a single copy of *BnaRFL11* and developed single, double, triple, and quadrable mutations in the T_0 generation (Table S2, Figure 4). The highest sgRNA editing efficiency was recorded at S1 (64.81%) in *bnarfl11* and S2 (46.29%) in *bnarfl11* (Table S5). This suggests that the practical selection of sgRNAs is crucial for effectively generating mutations in the target sequence. The efficiency of sgRNA depends on the promoters expressing the Cas9 protein in *B. napus.* The mutagenic competence of the designed sgRNAs mediated by U6-26 and U6-29 ranged from 31% to 96% (Table S5), indicating that not every promoter can effectively mediate genome editing in *B. napus*. This is the first time a highly conserved *bnarfl11* gene has been knocked out in *B. napus* to generate quadrable mutations. The generation of quadrable mutations in a single copy of *bnarfl11* was due to the highly efficient design of sgRNAs.

3.7. Morphological Analysis of bnarfl11 Mutant Plants and Pollen Viability Test

For *bnarfl*11 gene-phenotype analysis, all homozygous mutant T_1 lines were grown in the experimental field of Huazhong Agricultural University, Wuhan, China.

There were no morphological differences between sterile and fertile plants during the shoot and root growth cycles. The morphology of fertile flower buds and flowers of *bnarfl11* differed significantly during the later stages of development. We discovered a significant difference between the sterile line and fertile line anther development after using a stereomicroscope and scanning electron microscopy. The petals shrank and contracted; there was no pollen and much less nectary in the sterile plants, whereas the fertile plant was stable, with wild-type stamen and petal growth and regular nectary. The fertile flowers were larger than the *bnarfl11* flowers. During the growth process, the filaments and anthers of sterile flowers are often smaller than those of fertile flowers. (Figure 5a,b) shows the wild-type side view and upper view with normal anther development (Figure 5c,d). The *bnarfl11* side view and upper view mutant anther development (Figure 5a–d).

Figure 5. Phenotypic analysis of CRISPR/Cas9 mutated *bnarfl11* plants. (**a**,**b**) Shows wild type with normal flower development. (**c**,**d**) Shows *bnarfl11* flower.

Furthermore, sterile anthers produced very diminutive but degenerated pollen and did not affect the pistil growth. For the pollen viability test, staining with a 1% (w/v) acetocarmine solution showed that *bnarfl11* had no pollen if some flowers succeeded in producing anthers. The pollen visibility test showed that the pollen grains were degenerated entirely (Figure 6a,b). As mentioned in many research articles, the PPR mutant showed many morphological defects; therefore, *bnarfl11* showed vegetative defects, including fewer branches and a reduced leaf size. (Figure 7a–f) shows various stages from bud development to seed production, and sterile plants produced significantly less seed than W.T. while crossing with the wild type to produce pure F1 hybrid seeds with a stable Cas9 vector transmission.

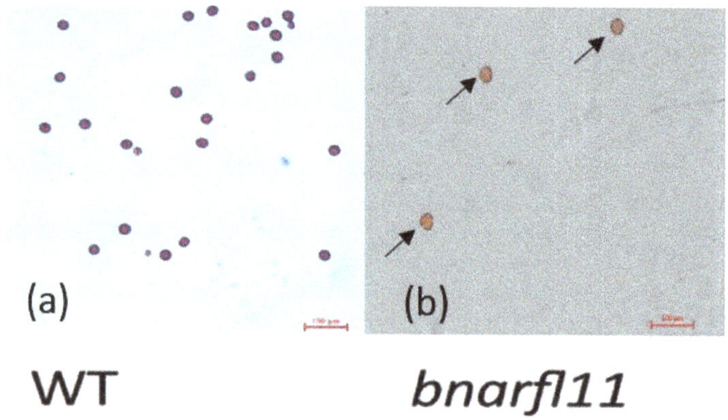

Figure 6. Pollen viability test staining with 1% (w/v) acetocarmine solution. (**a**) Shows WT pollens. (**b**) Arrows indicating *bnarfl11* degenerated pollens.

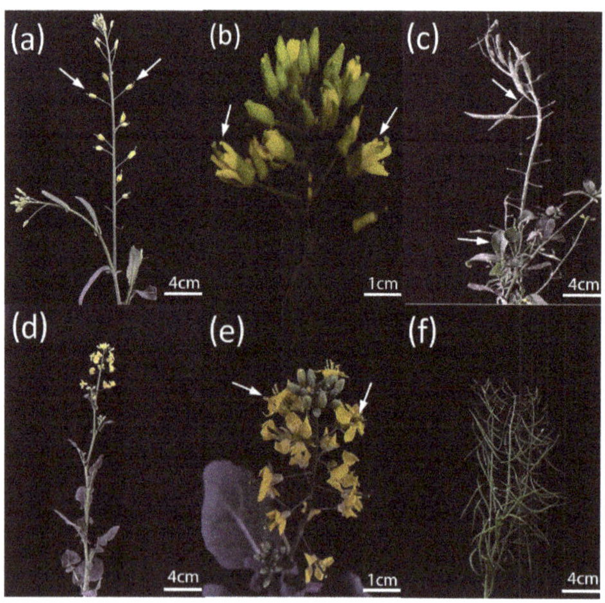

Figure 7. Plant growth stages of *bnarfl11* compared to W.T. (**a**) Male-sterile plant at the close flower stage. (**b**) Male-sterile plant at the open-flower stage. (**c**) Male-sterile plants after crossing with the wild type with significantly fewer seeds. (**d**–**f**) Wild-type growth stages compared with those of *bnarfl11*.

3.8. Cytological Analysis of Anther Abortion in *bnarfl11*

Theis and Robbelen (1990) described all the fertile anther stages of development; according to them, the results were as follows.

Development of *bnarfl11* Anthers

There was no structural distinction between fertile and non-fertile individuals in the 1st stage named primordial. Distinctions emerge when primordial tissue is divided into sporogenous, vascular, and parenchymal tissues. During the early stages, the sterile buds lose one to three locks per anther than fertile buds, producing four horizontal, angular locules. In sterile flowers, the number of locules varies from one to another.

Furthermore, in sterile buds, the development of locules inside an anther has polarity: adaxial locks often grow, while other locks are mostly sterile. The tapetum is frequently dense and disconnected from the microspore mother cells (MMC) of growing CMS locules. Surprisingly, this has little effect on their inconsistency: sexual cell division occurs in locules, and Tds segregate into immature pollen in the same way as fertile lines do. Both locules inside and between the fertile anthers mature simultaneously, going through various stages of development simultaneously. The fusion of two adaxial locks and the early displacement of a tapetum from its locule layer are typical defects in CMS anthers. Rather than dividing fibrous tissue throughout the MMC stage, the tapetum in sterile anthers is frequently distinguished during the MMC process. The pollen grains that grew within the formed locules grew as fertile pollen but became degenerated, considering the numerous structural variations detected in the CMS anthers (Figures 8a–p and 9a–d).

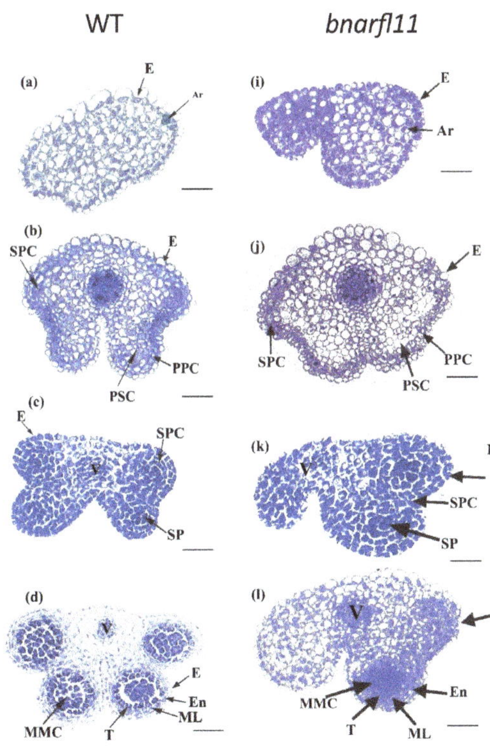

Figure 8. *Cont.*

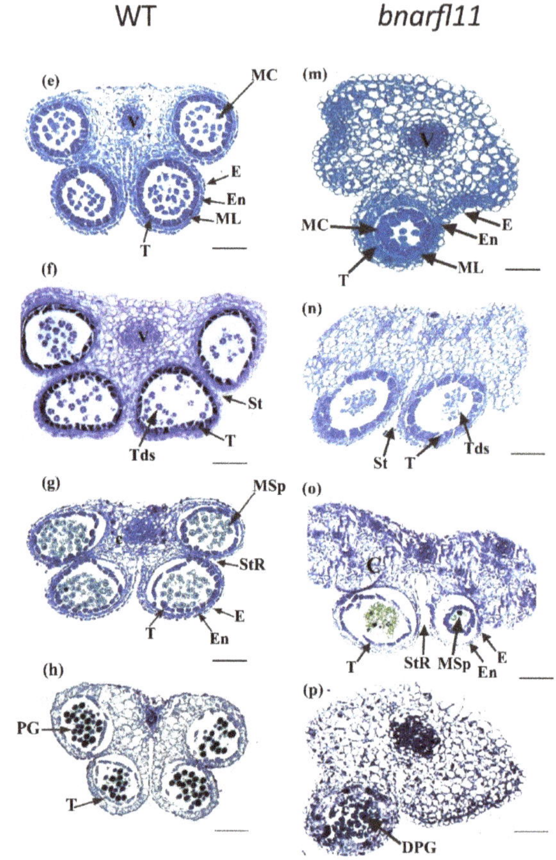

Figure 8. Cytological analysis of *bnarfl11* (**i–p**) compared to the wild type (**a–h**). There is no significant difference at early stages (**a**,**b**,**i**,**j**), but at later stages (**c**,**d**,**k**,**l**), the loss of one or two locules per anther is evident. In (**l**), MMCs are intermittently formed with an intact tapetum in the locks compared to W.T. **Ar**, archesporial cell; **E**, epidermis; **PSC**, primary sporogenous cell; **MMC**, microspore mother cells; **S.P.**, sporogenous cell; **En**, endothecium; **SPC**, secondary parietal cell; **ML**, middle layer; **T**, tapetum; V, vascular region; **PPC**, primary parietal cell Bar = 10 µm. Cytological analysis of *bnarfl11* compared to the wild type. (**e–h**) Stages of the wild type from **5–8**. (**m–p**) displays the stages of the sterile flower from **5–8**. (**m**) The Meiocyte stage with a highly dense callose wall around them and vacuolation in the tapetum was more confirmed than W.T. (**e**,**n**) At the tetrad stage, clumps of dense tissue in CMS locules, while in W.T. (**f**) locules and MC are typical. (**o**) At the pre-dehiscent stage, the locule increases in size, microspores form a cell clump, and the tapetum starts degrading; however, in W.T. (**g**), they have four large locules and normal microspores. (**p**) Degenerated pollen grains in the CMS flowers. (**h**) Normal pollen grains. **T**, tapetum; **MC**, meiotic cell; **E**, epidermis; **St**, stomium; **Msp**, microspores; **S.P.**, sporogenous cell; **Tds**, tetrads; **V**, vascular region; **P.G.**, pollen grain; **DGP**, degenerated pollen grains Bar = 10 µm.

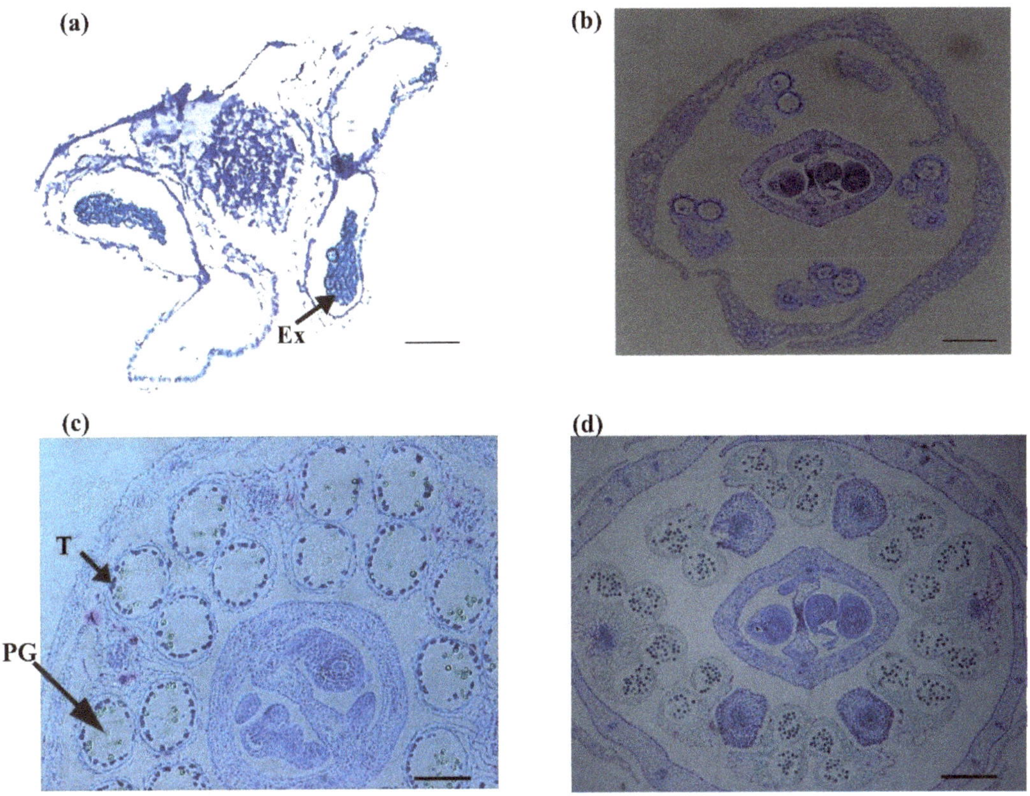

Figure 9. Dehiscence stage of *bnarfl11* compared with W.T. (**a**,**b**) *bnarfl11* dehiscence stage. (**a**) Heap of scarring in exine (Ex) if four locks succeed in development. (**b**) The loss of adaxial or abaxial locules is evident in *the bnarfl11*. (**c**,**d**) Wild-type flowers at dehiscence, with four locules and normal mature pollen grains. Bar = 10 μm.

3.9. The Behavior of nap-CMS Causing Gene in bnarfl11

First, we checked the cytoplasm type using the specific primers designed by [23] (Table S1) and found that the cytoplasm type was nap-cytoplasm (Figure 10a). Therefore, we can say that the sterility type observed was nap-CMS. To determine the expression of the nap-CMS-causing gene, we performed qRT-PCR. Because *AP3* is essential for determining the individuality and behavioral symmetry of petals and stamens, its expression was significantly high in buds and petals. Furthermore, the expression of *orf222*, *orf139*, and *nad5c* was significantly upregulated in the stamens and petals in the sterile bud compared to other organs of plants and the wild type (Figure 10b–f).

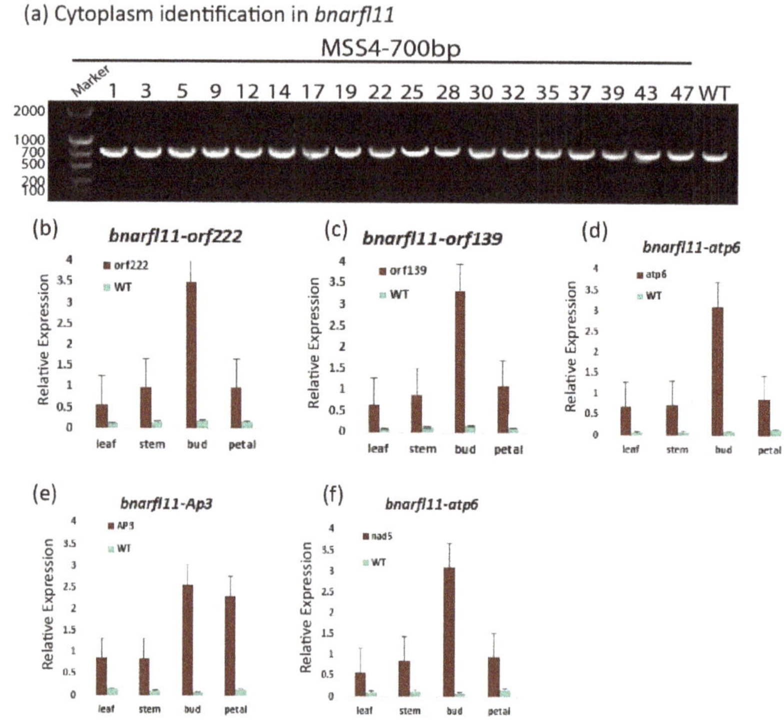

Figure 10. RT-PCR and qRT-PCR of *bnarfl11* plants. (**a**) RT-PCR analysis of nap-CMS cytoplasm in *bnarfl11* plants. For RT-PCR, total DNA was extracted from the Cas9 transgenic line, and the W.T. leaves. (**b–f**) qRT-PCR analysis of CMS-causing genes in *bnarfl11* plants. For qRT-PCR analysis, total RNA was extracted from leaves, stems, buds, and petals of the Cas9 transgenic line. Values represent the mean and standard deviation. Error bars indicate the standard deviation among triplicate experiments.

4. Discussion

4.1. In B. napus, the CRISPR/Cas9 System Is Highly Effective in Producing Targeted Mutations

Since it is easy to produce mutations in single genes or various genetic sequences with unknown functions, the CRISPR/Cas9 system has enormous potential for facilitating plant functional studies [51]. However, the challenge of rapidly producing and detecting high-efficiency stable homozygous mutations and the ability to mutate multiple targeted genes simultaneously is a real benefit of CRISPR for functional genomics in plants [52]. In the present study, we used Cas9 as a single guided RNA to generate targeted rapeseed mutations and to support a stable mutation transmission across the generations (Tables S2 and S3). In comparison to Pubi, we used the U6-26p promoter, which showed a higher Cas9 epigenetic modification and editing ability in rapeseed, which contradicted a previous study [44] that suggested the Pubi promoter in dicot plants for Cas9 proteins than U6-26p [53]. The efficient knockout of *bnarfl11* rapeseed homologs via the CRISPR/Cas9 system induced male sterility and similar observations in *Arabidopsis* and demonstrated the use of sgRNA via the CRISPR/Cas9 system to induce targeted mutations in the most critical *Brassica* crop cultivation traits based on the knowledge of the model plant gene function. This study evaluated the editing accuracy of a single copy of BnaRFL11. We examined 15 possible off-target loci from T_0 mutant plants, none of which showed induced CRISPR/Cas9 mutations, suggesting that well-designed specific sgRNA does not target any other site, and there is a marginal risk of off-target effects (Table S4).

Several main factors affect the sgRNA potency in plants, including Cas9 and sgRNA expression values, G.C. content %, targeting circumstances, and the key targeted structure of sgRNAs [53]. In the present study, targeted deletions in most transgenic positives, S1 and S2, resulted in high levels of single-target DSBs. We also discovered that the G.C. content of the current study sgRNAs ranged from 45% to 65%, optimal for sgRNAs, and the mutagenesis performance ranged from 0% to 65.10% at T_0 for the two sgRNAs (Table S5), suggesting that almost all the promoters were involved in rapeseed genome editing. The effectiveness of the target site, the guiding RNA design, the adequacy of the Cas9 expression, and the inter-analysis of the transformants for more numerous sgRNA-based plant editing are all factors that limit the editing performance. The editing efficacy of sgRNA (S1–S2) has been evaluated using a single copy of *the bnarfl11*. These sgRNAs demonstrated a dramatic efficiency in genome editing for targeting the same gene, such as 64.81% at S1 and 46.29% at S2. We also compared the editing efficiency of sgRNA in our present study with that of the previous study. [54] targeted three genes, *BnCLV1*, *BnCLV2*, and *BnCLV3*, with ten different sgRNA driven by the same promoter. [40] also targeted three genes, *BnIND*, *BnALC*, and *BnTT8*, using 12 different sgRNAs driven by the same promoter. The data in (Table S5) represent that the editing efficiency of the U6-26p promoter was higher than that of the other promoters. In short, variations in the efficacy of the genome editing of several sgRNAs most likely indicate changes in the nucleotide composition of the sgRNAs.

4.2. Precise Identification of Allelic Variation with T.A. Clone following Sequencing

A PAGE-based genotyping method for analyzing CRISPR/Cas9-related mutagenesis at the targets of *bnarfl11* was used in the current study. This helped develop a highly efficient and straightforward mutational detection method (Figure S4). Although the time efficiency of the PAGE-based genotyping method is evident, there is still a limited use, such as small nucleotide indels or substitution replacements in homozygous mutant plants, which are very difficult to detect [46]. T_0 validated the results of the PAGE screening, we used the high-throughput tracking of the T. A. clone. A total of 88 targeted mutant lines were identified in the T_0 generation using Sanger sequencing. In both mutant lines, various mutations were detected, including inserting and deleting different nucleotides at the S1 and S2 target sites (Figure 4). Thus, the TA clone-based sequencing method for mutation detection is an efficient and straightforward method with a high mutation frequency. Seventeen mutant lines were selected out of 88 for further generation mutation analyses (Figure S4, Table S3).

For the successful use of *B. napus*, it is essential to recognize desirable allelic variations and candidate genes for male sterility and proper leaf growth for photosynthesis in China's tri-annual crop rotation systems. Male sterility is a complex quantitative trait. Male sterility is induced only by mitochondrial ORFs when *Rf*, a nuclear gene, fails to diminish the *ORF* function. In the current study, *BnaRFL11* triple and quadrable homozygous mutants had severe male sterility and an abnormal leaf phenotype (Figures 4, 5a–f, 8a–p and 9a–d,). CRISPR/Cas9 has not yet developed impulsive or induced male sterility mutants in *B. napus*.

4.3. BnaRFL11 Gene Plays an Essential Role in Fertility Restoration in B. napus

Rf genes are well known in *Arabidopsis* and other crops, as they restore male fertility. *RFL* is a class of *PPR*-encoding genes; *RFL* genes have functions similar to PPR in RNA modification, stabilization, and cleavage. Thus, *RFL* is a successful *Rf* candidate that could play a key role in fertility restoration. However, we still have limited knowledge of the *RFL* genes in other species, such as rapeseed. In a previous study, 53 *BnaRFLs* were identified, of which *BnRFL13* (*Rfp*) and *BnaRFL6* (*Rfn*) were found on chromosome A9 inside the gene cluster [29]. The first study on the loss of the *RFL* mutant in *Arabidopsis* showed that the *AtRFL2* mutant had an abnormal phenotype [55].

RPF8 is an *RFL*-PPR protein [21,56]. A linkage study validated the involvement of *the RFP8* protein in producing the -141 $5'$ ends of the *nad3-rps12* mRNA in Van-0 [57]. The endonucleolytic cleavage of the mitochondrial gene *orf291*, which prevents transcript accumulation, is mediated by the *Arabidopsis* mitochondrial protein *RFL2*. *PRORP1*, a proteinaceous protein involved in mitochondrial RNA processing, facilitates this cleavage [48,58]. *RPF8* and *RFL2* from *A. thaliana* were similar to a single copy of *BnaRFL11* (Figures 1 and 2).

In the current study, all homozygous variations of the T_0 and T_1 lines were grown in the Hubei province to characterize *the BnaRFL11* mutant phenotypes. Compared to wild-type plants, all transgenic lines showed a dramatic change in the floral structure with stable male sterility, as explained previously [59]. Microsporocytes and other locule features are uncommon, and all sites within CMS anthers rarely fail to form, as shown by nap CMS in rapeseed. Because of the disturbance of the anther development symmetry and the polarity of the developmental anomalies, the locules proximal to the carpel have a much greater risk of forming than locules distal to the carpel; as a result, they are among the additional and hyper sexualized features of the morphology of *Brassica* CMS. The cytological analysis (Figure 8) and expression analysis by qRT-PCR (Figure 10) of *orf222/orf139/nad5c* in sterile transgenic lines confirmed that the anther abortion stage is closely similar to the nap-CMS type. The dramatic rise in mitochondrial transcripts, particularly in anther tissues in CMS plants, also includes copies of the CMS-related gene *orf222*. The N-terminal encoding segment of *orf222* is provided by atp8, which encodes a cellular ATP synthase moiety. As recommended by [60], by interfering with the function of its regular counterpart, the expression of a CMS-related gene product that mimics a regular mitochondrial product can cause mitochondrial damage. ATP8 is essential for ATP synthase aggregation. It is conceivable that in rapeseed CMS, a dysfunctional synthase array throughout phases of the successful mitochondrial function in sporogenous tissue causes oxidative damage that is severe enough to disrupt the average anther development [61]. *AP3* and other B function genes regulate the identification and functional symmetry of the stamen and petals. According to our findings, an *AP3* predictor array's CMS activity timing differs from that of fertile stamens and petals.

As a result, the specific elements of CMS morphology that we examined here, comprising both stamen and petal modifications, may be caused by changes in the *AP3* function. Our findings indicate that early in stamen growth, the cell-specific expression of *orf222* prevents sporogenesis from starting at specific sites, resulting in sterile anthers. Since the *Rfn* gene failed to restore fertility, our findings suggest that this consequence could be regulated by *orf222* activity on the output of the nuclear genes involved in anther processing, such as *AP3*.

Flowering plants have also developed a unique method for controlling organelle gene expression, in which post-transcriptional courses play a significant role in determining gene stoichiometry. The pentatricopeptide repeat function (PPR) proteins in plant RNA research have been among the most significant breakthroughs in the last few decades. Restorer of fertility (*Rf*) genes are nuclear loci suppressing CMS development. *The Rf* genes also encode pentatricopeptide repeat proteins. *RFL* genes have been implicated in the post-transcriptional alteration of mito-RNA in many studies. In *Arabidopsis*, pentatricopeptide repeat (*PPR*) protein RNA processing factor8 *(RPF8)* and *RFL2* are involved in the 5′ processing of various mitochondrial mRNAs. Both factors are similar to *Rf*, found in various plant species' CMS/restoration systems. These results indicate that *Rf*-like *PPR* proteins play a crucial role in post-transcriptional 5′-processing. Collectively, the *RFL* gene family provides a new way to explore restorer genes in other CMS systems to complement traditional genetic mapping to locate the candidate genes. The mutants of *BnaRFL11* developed in the current study might help researchers further investigate the molecular mechanism of *RFL* genes in fertility restoration and Mendelian inheritance in hybrid breeding (Figure 11).

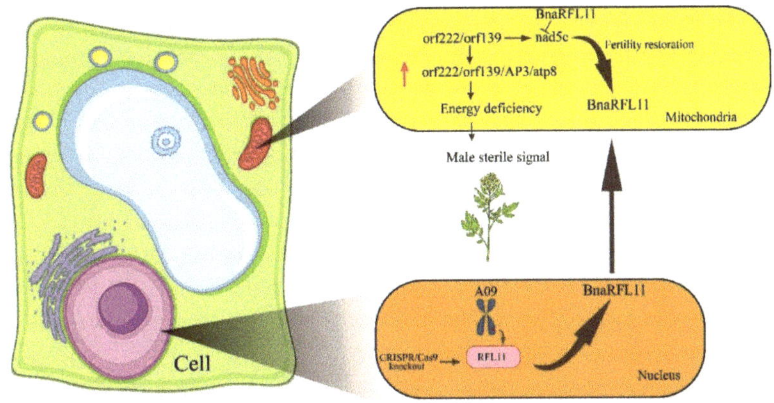

Figure 11. Putative model for male sterility in the current study on rapeseed. Red arrow indicating upregulation.

5. Conclusions

The present study utilized the CRISPR/Cas9 system to knock out a single copy of the *BnaRFL11* gene (*BnaA09g45590D*) to develop plants with male sterility in *B. napus*.

(1) In a single copy of *bnarfl11*, heritable mutant plants develop different mutant lines with a constant mutation transmission in the T_0 and T_1 generations. Free T-DNA mutant lines were generated across two successive generations.
(2) The phenotypic analysis of *bnarfl11* revealed defects in the floral structure, leaf size, branch number, and seed production. The phenotypic analysis showed that the mutant plant lines had male sterility with multiple branches and a reduced leaf size compared to the W.T. plant lines.
(3) To determine whether the off-target effect occurred in this study, we used the "CRISPR-P" software to identify a supposed off-target site in the rapeseed genome that is homologous to the two designed sgRNAs. The locations of potential off-target sites in the genome were noted. According to the sequencing (TA-Clone) of the PCR products, seven possible off-target sites from several T_0 mutated plants showed no mutations.
(4) The editing efficiency of the two sgRNAs of the *bnarfl11* gene ranged from 0% to 65.10% in T_0. We analyzed the editing efficiency of each sgRNA (S1–S2) that targets the *bnarfl11* gene.

The current study's findings suggest that CRISPR/Cas9 mutations may divulge the functions of genes in polyploid species and provide agronomically desirable traits through a targeted mutation.

6. Future Prospective

Future work can be conducted in the following ways.

(1) To further identify the signaling pathway and molecular mechanism of *BnaRFL11* in controlling other related genes involved in male sterility.
(2) The development of markers for the polymorphic identification of variations in different mutation sites for CMS breeding in *B. napus*.

Supplementary Materials: The following supporting information can be downloaded at: https://www.mdpi.com/article/10.3390/plants11243501/s1, Figure S1: Sequence alignment of BnaRFL11 with Arabidopsis homologs genes; Figure S2: Protein sequence alignment from *B. napus* (*BnaA09g45590D*) and *A. thaliana* (*AT1G12300, AT1G12620*); Figure S3: Sequence alignment of single functional *RFL11* gene copy in Westar; Figure S4: Positive, Transgenic and mutation screening in *bnarfl11*; Figure S5: The *BnaRFL11* gene contains a single copy with two target sites S1 and S2; Figure S6: Sequence alignment of *BnaRFL11* with *BnaRFL6*; Table S1: The primer used in all studies; Table S2: Genotypical study of *bnarfl11* transgenic plants and next-generation transmission analysis; Table S3: Molecular investigation of CRISPR/Cas9-tempted transmutations in *bnarfl11* and it has spread to from T0 to T1 generations; Table S4: Exposure of off-effects for designed guiding-RNA in mutant lines; Table S5: Comparison of editing efficiency for *BnaRFL11*-promoters with other published gene-editing studies.

Author Contributions: Conceptualization, Z.F. and B.W.; methodology, Z.F. and M.N.R.; software, Z.F.; validation, H.W. and Y.L.; formal analysis, Z.F. and B.W.; investigation, Z.F. and M.S.F.; resources, J.T., B.Y., T.F., J.S., S.Y., C.M., J.W. and C.D.; data curation, Z.F.; writing—original draft preparation, Z.F. and M.N.R.; writing—review and editing, Z.F., M.S.F. and M.A.; visualization, Z.F. and B.W.; supervision, B.Y.; project administration, B.Y. and T.F.; funding acquisition, B.Y. and B.W. All authors have read and agreed to the published version of the manuscript.

Funding: This project was supported by the National Natural Science Foundation of China supported this project (Grant No. 32201883) and the Hubei Key Research Project of CRISPR Gene-Editing Technology in Fruit Vegetable Crops (2022BBA0062). Funding agencies had no role in the study design, data collection, analysis, or interpretation.

Institutional Review Board Statement: Not applicable.

Informed Consent Statement: Not applicable.

Data Availability Statement: All data generated or analyzed during this study are included in this published article (and its Supplementary Materials).

Conflicts of Interest: The authors declare no conflict of interest.

References

1. Xiang, N.T.; Barbetti, M.J.; Jacqueline, B. Current Status and Challenges in Identifying Disease Resistance Genes in Brassica napus. *Front. Plant Sci.* **2017**, *8*, 1788.
2. Van de Wouw, A.P.; Idnurm, A.; Davidson, J.A.; Sprague, S.J.; Khangura, R.K.; Ware, A.H.; Lindbeck, K.D.; Marcroft, S.J. Fungal diseases of canola in Australia: Identification of trends, threats and potential therapies. *Australas. Plant Pathol.* **2016**, *45*, 415–423. [CrossRef]
3. Bharti, S.; Pedro, M.J.; Chrungu, K.B.; Aparajita, M. Chloroplast DNA Variations in Wild Brassicas and Their Implication in Breeding and Population Genetics Studies. *Scientifica* **2015**, *2015*, 952395.
4. Yamagishi, H.; Bhat, S.R. Cytoplasmic male sterility in Brassicaceae crops. *Breed. Sci.* **2014**, *64*, 38. [CrossRef]
5. Yang, L.Y.; Liu, P.W.; Yang, G.S. Development of Polima temperature-sensitive cytoplasmic male sterile lines of *Brassica napus* through isolated microspore culture. *Plant Breed.* **2010**, *125*, 368–371. [CrossRef]
6. Hompson, K.F. Cytoplasmic male-sterility in oil-seed rape. *Heredity* **1972**, *29*, 253–257. [CrossRef]
7. Ogura, H. Studies on the new male sterility in Japanese radish, with special references on the utilization of this sterility towards the practical raising of hybrid seeds. *Mem. Fac. Agric. Kagoshima Univ.* **1968**, *6*, 40–75.
8. Hu, Q.; Hua, W.; Yin, Y.; Zhang, X.; Liu, L.; Shi, J.; Zhao, Y.; Qin, L.; Chen, C.; Wang, H. Rapeseed research and production in China. *Crop J.* **2017**, *5*, 127–135. [CrossRef]
9. Liu, J.; Li, M.; Wang, H.; Yu, L.; Li, D. Sequence analysis and expression of orf224 gene associated with two types of cytoplasmic male sterility in *Brassica napus* L. *Z. Nat. C J. Biosci.* **2010**, *65*, 395–402. [CrossRef]
10. Banga, S.S.; Labana, K.S.; Banga, S.K. Male sterility in Indian mustard (*Brassica juncea* (L.) Coss.)—A biochemical characterization. *Theor. Appl. Genet.* **1984**, *67*, 515–519. [CrossRef]
11. Liu, J.; Xiang, R.; Wang, W.; Mei, D.; Li, Y.; Mason, A.S.; Fu, L.; Hu, Q. Cytological and molecular analysis of Nsa CMS in *Brassica napus* L. *Euphytica* **2015**, *206*, 279–286. [CrossRef]
12. Heng, S.; Wei, C.; Jing, B.; Wan, Z.; Wen, J.; Yi, B.; Ma, C.; Tu, J.; Fu, T.; Shen, J. Comparative analysis of mitochondrial genomes between the hau cytoplasmic male sterility (CMS) line and its iso-nuclear maintainer line in Brassica juncea to reveal the origin of the CMS-associated gene orf288. *BMC Genom.* **2014**, *15*, 322. [CrossRef] [PubMed]
13. Hu, Q.; Andersen, S.; Dixelius, C.; Hansen, L. Production of fertile intergeneric somatic hybrids between *Brassica napus* and *Sinapis arvensis* for the enrichment of the rapeseed gene pool. *Plant Cell Rep.* **2002**, *21*, 147–152.
14. Li, P.; Kang, L.; Wang, A.; Cui, C.; Li, Z. Development of a Fertility Restorer for *inap* CMS (*Isatis indigotica*) *Brassica napus* through Genetic Introgression of One Alien Addition. *Front. Plant Sci.* **2019**, *10*, 257. [CrossRef] [PubMed]
15. Castandet, B.; Araya, A. The nucleocytoplasmic conflict, a driving force for the emergence of plant organellar RNA editing. *IUBMB Life* **2012**, *64*, 120–125. [CrossRef] [PubMed]
16. Barkan, A.; Small, I. Pentatricopeptide Repeat Proteins in Plants. *Annu. Rev. Plant Biol.* **2014**, *65*, 415–442. [CrossRef] [PubMed]
17. Bentolila, S.; Alfonso, A.A.; Hanson, M.R. A pentatricopeptide repeat-containing gene restores fertility to cytoplasmic male-sterile plants. *Proc. Natl. Acad. Sci. USA* **2002**, *99*, 10887–10892. [CrossRef]

18. Akagi, H.; Nakamura, A.; Yokozeki-Misono, Y.; Inagaki, A.; Takahashi, H.; Mori, K.; Fujimura, T. Positional cloning of the rice Rf-1 gene, a restorer of BT-type cytoplasmic male sterility that encodes a mitochondria-targeting PPR protein. *Theor. Appl. Genet.* **2004**, *108*, 1449–1457. [CrossRef]
19. Fujii, S.; Kazama, T.; Ito, Y.; Kojima, S.; Toriyama, K. A candidate factor that interacts with RF2, a restorer of fertility of Lead rice-type cytoplasmic male sterility in rice. *Rice* **2014**, *7*, 21. [CrossRef]
20. Brown, G.G.; Formanová, N.; Jin, H.; Wargachuk, R.; Dendy, C.; Patil, P.; Laforest, M.; Zhang, J.; Cheung, W.Y.; Landry, B.S. The radish Rfo restorer gene of *Ogura cytoplasmic* male sterility encodes a protein with multiple pentatricopeptide repeats. *Plant J.* **2003**, *35*, 262–272. [CrossRef]
21. Desloire, S.; Gherbi, H.; Laloui, W.; Marhadour, S.; Clouet, V.; Cattolico, L.; Falentin, C.; Giancola, S.; Renard, M.; Budar, F. Identification of the fertility restoration locus, Rfo, in radish, as a member of the pentatricopeptide-repeat protein family. *EMBO Rep.* **2003**, *4*, 588–594. [CrossRef] [PubMed]
22. Klein, R.; Klein, P.; Mullet, J.; Minx, P.; Rooney, W.; Schertz, K. Fertility restorer locus Rf1 of sorghum (*Sorghum bicolor* L.) encodes a pentatricopeptide repeat protein not present in the colinear region of rice chromosome 12. *Theor. Appl. Genet.* **2005**, *111*, 994–1012. [CrossRef] [PubMed]
23. Heng, S.; Chen, F.; Wei, C.; Hu, K.; Yang, Z.; Wen, J.; Yi, B.; Ma, C.; Tu, J.; Si, P.; et al. identification of different cytoplasms based on newly developed mitotype-specific markers for marker-assisted selection breeding in *Brassica napus* L. *Plant Cell Rep.* **2017**, *36*, 901–909. [CrossRef] [PubMed]
24. Kim, Y.J.; Zhang, D. Molecular Control of Male Fertility for Crop Hybrid Breeding. *Trends Plant Sci.* **2018**, *23*, 53–65. [CrossRef]
25. Pahwa, R.; Banga, S.; Gogna, K.; Banga, S. Tournefortii male sterility system in *Brassica napus*. Identification, expression and genetic characterization of male fertility restorers. *Plant Breed.* **2004**, *123*, 444–448. [CrossRef]
26. Kang, L.; Li, P.; Wang, A.; Ge, X.; Li, Z. A Novel Cytoplasmic Male Sterility in *Brassica napus* (inap CMS) with Carpelloid Stamens via Protoplast Fusion with Chinese Woad. *Front. Plant Sci.* **2017**, *8*, 529. [CrossRef]
27. Iwabuchi, M.; Koizuka, N.; Fujimoto, H.; Sakai, T.; Imamura, J. Identification and expression of the kosena radish (*Raphanus sativus* cv. Kosena) homologue of the ogura radish CMS-associated gene, orf138. *Plant Mol. Biol.* **1999**, *39*, 183–188. [CrossRef]
28. Sang, S.-F.; Mei, D.-S.; Liu, J.; Zaman, Q.U.; Zhang, H.-Y.; Hao, M.-Y.; Fu, L.; Wang, H.; Cheng, H.-T.; Hu, Q. Organelle genome composition and candidate gene Identification for Nsa cytoplasmic male sterility in *Brassica napus*. *BMC Genom.* **2019**, *20*, 813. [CrossRef]
29. Ning, L.; Wang, H.; Li, D.; Li, Y.; Chen, K.; Chao, H.; Li, H.; He, J.; Li, M. Genome-wide identification of the restorer-of-fertility-like (RFL) gene family in Brassica napus and expression analysis in Shaan2A cytoplasmic male sterility. *BMC Genom.* **2020**, *21*, 765. [CrossRef]
30. Melonek, J.; Stone, J.D.; Small, I. Evolutionary plasticity of restorer-of-fertility-like proteins in rice. *Sci. Rep.* **2016**, *6*, 35152. [CrossRef]
31. Sykes, T.; Yates, S.; Nagy, I.; Asp, T.; Small, I.; Studer, B. In silico identification of candidate genes for fertility restoration in cytoplasmic male sterile perennial ryegrass (*Lolium perenne* L.). *Genome Biol. Evol.* **2017**, *9*, 351–362. [PubMed]
32. Kazama, T.; Nakamura, T.; Watanabe, M.; Sugita, M.; Toriyama, K. Suppression mechanism of mitochondrial ORF79 accumulation by Rf1 protein in BT-type cytoplasmic male sterile rice. *Plant J.* **2008**, *55*, 619–628. [CrossRef] [PubMed]
33. Uyttewaal, M.; Arnal, N.; Quadrado, M.; Martin-Canadell, A.; Vrielynck, N.; Hiard, S.; Gherbi, H.; Bendahmane, A.; Budar, F.; Mireau, H. Characterization of *Raphanus sativus* pentatricopeptide repeat proteins encoded by the fertility restorer locus for Ogura cytoplasmic male sterility. *Plant Cell* **2008**, *20*, 3331–3345. [CrossRef] [PubMed]
34. Barr, C.M.; Fishman, L. The nuclear component of a cytonuclear hybrid incompatibility in Mimulus maps to a cluster of pentatricopeptide repeat genes. *Genetics* **2010**, *184*, 455–465. [CrossRef]
35. Fujii, S.; Yamada, M.; Fujita, M.; Itabashi, E.; Hamada, K.; Yano, K.; Kurata, N.; Toriyama, K. Cytoplasmic–nuclear genomic barriers in rice pollen development revealed by comparison of global gene expression profiles among five independent cytoplasmic male sterile lines. *Plant Cell Physiol.* **2010**, *51*, 610–620. [CrossRef]
36. Luo, D.; Xu, H.; Liu, Z.; Guo, J.; Li, H.; Chen, L.; Fang, C.; Zhang, Q.; Bai, M.; Yao, N. A detrimental mitochondrial-nuclear interaction causes cytoplasmic male sterility in rice. *Nat. Genet.* **2013**, *45*, 573–577. [CrossRef]
37. Meyer, J.; Pei, D.; Wise, R.P. Rf8-mediated T-urf13 transcript accumulation coincides with a pentatricopeptide repeat cluster on maize chromosome 2L. *Plant Genome* **2011**, *4*. [CrossRef]
38. Liu, Z.; Yang, Z.; Wang, X.; Li, K.; An, H.; Liu, J.; Yang, G.; Fu, T.; Yi, B.; Hong, D. A Mitochondria-Targeted PPR Protein Restores pol Cytoplasmic Male Sterility by Reducing orf224 Transcript Levels in Oilseed Rape. *Mol. Plant* **2016**, *9*, 1082–1084. [CrossRef]
39. Liu, X.-Q.; Liu, Z.-Q.; Yu, C.-Y.; Dong, J.-G.; Hu, S.-W.; Xu, A.-X. TGMS in rapeseed (*Brassica napus*) resulted in aberrant transcriptional regulation, asynchronous microsporocyte meiosis, defective tapetum, and fused sexine. *Front. Plant Sci.* **2017**, *8*, 1268. [CrossRef]
40. Zhai, Y.; Yu, K.; Cai, S.; Hu, L.; Amoo, O.; Xu, L.; Yang, Y.; Ma, B.; Jiao, Y.; Zhang, C. Targeted mutagenesis of BnTT8 homologs controls yellow seed coat development for effective oil production in *Brassica napus* L. *Plant Biotechnol. J.* **2020**, *18*, 1153–1168. [CrossRef]
41. Bailey, T.L.; Boden, M.; Buske, F.A.; Frith, M.; Grant, C.E.; Clementi, L.; Ren, J.; Li, W.W.; Noble, W.S. MEME SUITE: Tools for motif discovery and searching. *Nucleic Acids Res.* **2009**, *37*, W202–W208. [CrossRef] [PubMed]
42. Liu, H.; Ding, Y.; Zhou, Y.; Jin, W.; Xie, K.; Chen, L.L. CRISPR-P 2.0: An improved CRISPR-Cas9 tool for genome editing in plants. *Mol. Plant* **2017**, *10*, 530–532. [CrossRef] [PubMed]

43. Hsu, P.D.; Scott, D.A.; Weinstein, J.A.; Ran, F.; Konermann, S.; Agarwala, V.; Li, Y.; Fine, E.J.; Wu, X.; Shalem, O. DNA targeting specificity of RNA-guided Cas9 nucleases. *Nat. Biotechnol.* **2013**, *31*, 827–832. [CrossRef] [PubMed]
44. Xing, H.-L.; Dong, L.; Wang, Z.-P.; Zhang, H.-Y.; Han, C.-Y.; Liu, B.; Wang, X.-C.; Chen, Q.-J. A CRISPR/Cas9 toolkit for multiplex genome editing in plants. *BMC Plant Biol.* **2014**, *14*, 327. [CrossRef] [PubMed]
45. Zhou, Y.; Wang, H.; Gilmer, S.; Whitwill, S.; Keller, W.; Fowke, L. Control of petal and pollen development by the plant cyclin-dependent kinase inhibitor ICK1 in transgenic *Brassica* plants. *Planta* **2002**, *215*, 248–257. [CrossRef]
46. Zhu, X.; Xu, Y.; Yu, S.; Lu, L.; Ding, M.; Cheng, J.; Song, G.; Gao, X.; Yao, L.; Fan, D. An efficient genotyping method for genome-modified animals and human cells generated with CRISPR/Cas9 system. *Sci. Rep.* **2014**, *4*, 6420. [CrossRef]
47. Peng, Y.; Shi, D.; Zhang, T.; Li, X.; Fu, T.; Xu, Y.; Wan, Z. Development and utilization of an efficient cytoplasmic male sterile system for Cai-xin (*Brassica rapa* L.). *Sci. Hortic.* **2015**, *190*, 36–42. [CrossRef]
48. Fujii, S.; Suzuki, T.; Giegé, P.; Higashiyama, T.; Koizuka, N.; Shikanai, T. The Restorer-of-fertility-like 2 pentatricopeptide repeat protein and R.N. ase P are required for the processing of mitochondrial orf291 RNA in Arabidopsis. *Plant J.* **2016**, *86*, 504–513. [CrossRef]
49. Bohra, A.; Jha, U.C.; Adhimoolam, P.; Bisht, D.; Singh, N.P. Cytoplasmic male sterility (CMS) in hybrid breeding in field crops. *Plant Cell Rep.* **2016**, *35*, 967–993. [CrossRef]
50. Lei, Y.; Lu, L.; Liu, H.-Y.; Li, S.; Xing, F.; Chen, L.-L. CRISPR-P: A web tool for synthetic single-guide RNA design of CRISPR-system in plants. *Mol. Plant* **2014**, *7*, 1494–1496. [CrossRef]
51. Hussain, B.; Lucas, S.J.; Budak, H. CRISPR/Cas9 in plants: At play in the genome and at work for crop improvement. *Brief. Funct. Genom.* **2018**, *17*, 319–328. [CrossRef] [PubMed]
52. Sander, J.D.; Joung, J.K. CRISPR-Cas systems for editing, regulating and targeting genomes. *Nat. Biotechnol.* **2014**, *32*, 347–355. [CrossRef] [PubMed]
53. Ma, X.; Zhang, Q.; Zhu, Q.; Liu, W.; Chen, Y.; Qiu, R.; Wang, B.; Yang, Z.; Li, H.; Lin, Y. A robust CRISPR/Cas9 system for convenient, high-efficiency multiplex genome editing in monocot and dicot plants. *Mol. Plant* **2015**, *8*, 1274–1284. [CrossRef] [PubMed]
54. Yang, Y.; Zhu, K.; Li, H.; Han, S.; Meng, Q.; Khan, S.U.; Fan, C.; Xie, K.; Zhou, Y. Precise editing of CLAVATA genes in *Brassica napus* L. regulates multilocular silique development. *Plant Biotechnol. J.* **2018**, *16*, 1322–1335. [CrossRef] [PubMed]
55. Jonietz, C.; Forner, J.; Hildebrandt, T.; Binder, S. RNA PROCESSING FACTOR3 is crucial for the accumulation of mature ccmC transcripts in mitochondria of Arabidopsis accession Columbia. *Plant Physiol.* **2011**, *157*, 1430–1439. [CrossRef]
56. Fujii, S.; Bond, C.S.; Small, I.D. Selection patterns on restorer-like genes reveal a conflict between nuclear and mitochondrial genomes throughout angiosperm evolution. *Proc. Natl. Acad. Sci. USA* **2011**, *108*, 1723–1728. [CrossRef]
57. Schleicher, S.; Binder, S. In Arabidopsis thaliana mitochondria 5′ end polymorphisms of nad4L-atp4 and nad3-rps12 transcripts are linked to RNA PROCESSING FACTORs 1 and 8. *Plant Mol. Biol.* **2021**, *106*, 335–348. [CrossRef]
58. Gutmann, B.; Gobert, A.; Giegé, P. PRORP proteins support RNase P activity in both organelles and the nucleus in Arabidopsis. *Genes Dev.* **2012**, *26*, 1022–1027. [CrossRef]
59. Geddy, R.; Mahé, L.; Brown, G.G. Cell-specific regulation of a Brassica napus CMS-associated gene by a nuclear restorer with related effects on a floral homeotic gene promoter. *Plant J.* **2005**, *41*, 333–345. [CrossRef]
60. Sabar, M.; Gagliardi, D.; Balk, J.; Leaver, C.J. ORFB is a subunit of F1 FO-ATP synthase: Insight into the basis of cytoplasmic male sterility in sunflower. *EMBO Rep.* **2003**, *4*, 381–386. [CrossRef]
61. Devenish, R.J.; Prescott, M.; Roucou, X.; Nagley, P. Insights into ATP synthase assembly and function through the molecular genetic manipulation of subunits of the yeast mitochondrial enzyme complex. *Biochim. Et Biophys. Acta (BBA) Bioenerg.* **2000**, *1458*, 428–442. [CrossRef]

plants

MDPI

Article

Exogenous Phytohormones and Fertilizers Enhance *Jatropha curcas* L. Growth through the Regulation of Physiological, Morphological, and Biochemical Parameters

Rahmatullah Jan [1,2,†], Murtaza Khan [3,†], Muhammad Adnan [4], Sajjad Asaf [5], Saleem Asif [1], Kyung-Min Kim [1,2,*] and Waheed Murad [6,*]

1 Department of Applied Biosciences, Graduate School, Kyungpook National University, Daegu 41566, Republic of Korea
2 Coastal Agriculture Research Institute, Kyungpook National University, Daegu 41566, Republic of Korea
3 Department of Horticulture and Life Science, Yeungnam University, Gyeongsan 38541, Republic of Korea
4 Department of Botanical and Environmental Sciences, Kohat University of Science and Technology, Kohat 26000, Pakistan
5 Natural and Medical Sciences Research Center, University of Nizwa, Nizwa 616, Oman
6 Department of Botany, Abdul Wali Khan University, Mardan 23200, Pakistan
* Correspondence: kkm@knu.ac.kr (K.-M.K.); waheedmurad@awkum.edu.pk (W.M.)
† These authors contributed equally to this work.

Abstract: *Jatropha curcas* L. is a perennial plant, that emerged as a biodiesel crop attracting the great interest of researchers. However, it is considered a semi-wild plant and needed to apply crop-improving practices to enhance its full yield potential. This study was conducted to improve the growth and development of the *J. curcas* plant by exogenous application of Gibberellic acid (GA), indole acetic acid (IAA), and fertilizer (nitrogen, phosphorus, potassium (NPK)). The experiment was conducted in pots in triplicate and 100 ppm and 250 ppm of GA and IAA were applied separately while NPK was applied in two levels (30 and 60 g/pot). The results revealed a significant difference in growth parameters with the application of hormones and fertilizer. The highest shoot length (47%), root length (63%), root fresh weight (72%), and root dry weight (172%) were shown by plants treated with GA 250 ppm. While plants treated with NPK 60 g showed the highest increases in shoot fresh weight and shoot dry weight compared to control plants. The highest increase in leaves number (274%) and branches number (266%) were shown by the plants treated with GA 100 ppm and GA 250 ppm, respectively, while GA 250 ppm and IAA 250 ppm highly increased stem diameter (123%) and stem diameter was also shown by GA 250 ppm-treated plants. NPK 60 g highly increased proximate composition (protein content, carbohydrate, fat, moisture content, and ash content) compare with hormones and control plants. Our results concluded the optimized concentration of IAA, GA, and NPK significantly increases *J. curcas* growth vigor.

Keywords: biodiesel; gibberellin; growth parameters; indole acetic acid; *Jatropha curcas*; proximate composition

Citation: Jan, R.; Khan, M.; Adnan, M.; Asaf, S.; Asif, S.; Kim, K.-M.; Murad, W. Exogenous Phytohormones and Fertilizers Enhance *Jatropha curcas* L. Growth through the Regulation of Physiological, Morphological, and Biochemical Parameters. *Plants* **2022**, *11*, 3584. https://doi.org/10.3390/plants11243584

Academic Editors: Andrzej Bajguz and Othmane Merah

Received: 27 October 2022
Accepted: 13 December 2022
Published: 19 December 2022

1. Introduction

Jatropha curcas L. is a perennial shrub, belonging to the family Euphorbiaceae. It is about 5 m tall and has smooth grey bark, leaves are large and usually pale green. Fruits are produced in winter or throughout the year depending on temperature and soil humidity. The indications show that it originated from South and Central America and some other parts of the tropical and subtropical regions of Africa and Asia [1]. Specifically, it is grown in Benin, Brazil, China, Egypt, Ethiopia, Ghana, Guinea, India, Madagascar, Mali, Mexico, Mozambique, Namibia, Senegal, South Africa, Sudan, Tanzania, Uganda, Zambia, and Zimbabwe [2]. Its name indicates that it is used as a medicinal plant in the Portuguese in the 16th century, as its name is derived from the Greek word "iatros" which means doctor, and "trophe" which means food [3]. Literature shows that the Portuguese established the plantation of *J. curcas* for the first time to make soap, lamp oil, and other medicine [1].

However, *J. curcas* is exotic to Pakistan and introduced in 2007 to Karachi. It is an oilseed plant that does not compete with food crops and its seed contains 34–60% oil contents [3,4]. It has the ability to adapt to a high range of agro-climatic conditions. It grows in gravelly, sandy, degraded, acidic, and poor stony soils. It is easy to establish, grows quickly, drought tolerant, grow well in low rainfall condition, and can be used to reclaim eroded areas [5]. Therefore, it is suggested for cultivation on poor degraded soil. However, it shows stunt growth under heavy metals stress [6]. It can survive in long dry periods and it is considered to be well adapted to arid and semi-arid conditions. In the initial stage, it requires a large amount of water but after maturity, it may survive without water for sixty days. In the rainfed area, it required 250 up to 3000 mm/year of rainfall for best growth [7]. Gadallah and Sayed (2001) reported that the exogenous application of hormones increases *J. curcas* resistance against environmental constraints and promotes plant growth and development, however, the effectiveness of growth regulators depends on their concentration and the method of application [8].

To promote plant growth and development in non-suitable conditions, phytohormones or macronutrients are important candidates. Trail studies indicated that phytohormones and fertilizers increase plant growth and development under stress conditions. Gibberellin (GA) increases plant growth, leaves number, bud formation, cell division and elongation, and flowering [9,10]. Indol acetic acid (IAA) is also involved in several developmental processes such as the development of vascular tissue, cell elongation, and apical dominance [11]. The growth regulator enhances the growth of *J. curcas* by regulating its morpho-physiological and biochemical processes [8]. Besides hormones, many other parameters also affect the growth of *J. curcas*. Fertilizer is one of the essential parameters which promote plant development in a harsh environment. Nitrogen, phosphorous, and potassium (NPK) are the main nutrients needed for plant growth and development. Without providing adequate NPK, plants cannot reach maximum growth [12]. The NPK supplementation increases the biomass and oil contents in *J. curcas,* which shows the importance of fertilizers for the growth and development of *J. curcas*. Being a biofuel plant, *J. curcas* attracted the researcher's attention. As it grows well in drought and high temperatures; therefore, it is suitable for arid regions where it is not competing with the crops for water and land.

Due to the potential demand and better opportunities, cultivation of *J. curcas* appears viable. It adapts well to marginal lands and the large-scale cultivation on wasteland with low water and rainfall could generate employment and increase the income of the locality. However, the ability of *J. curcas* to grow in marginal land and dry soil has not been properly explored [13]. Several studies have been conducted to evaluate the *J. curcas* performance under low water availability in marginal land [13]. In order to understand the plant growth performance in the marginal land of Kohat, Khyber Pakhtun Khwa, Pakistan, region. Kohat is an arid region containing a large quantity of marginal land, which is suitable for *J. curcas* cultivation. Based on *J. curcas* plant adaptation to marginal land, we conducted a trail base studyin the Kohat region (arid region) district in Khyber Pakhtunkhwa, Pakistan. However, the result of plant growth was not satisfactory under normal growth conditions (data not published). To further investigate the root cause of our previous project failure, we performed the current study to investigate the possible way to promote the growth of *J. curcas* in the same region. In the current study, we focus on exogenous hormonal application and fertilizer supplementation. We hypothesized that GA, IAA and NPK promote *J. curcas* growth and development by regulation of morphological, physiochemical and biochemical characteristics of *J. curcas*. Therefore, the current study aimed to investigate that the exogenous hormones and fertilizer induces morphological, physiological, and biochemical parameters of the *J. curcas* plant and enable it to grow and develop in arid regions. Based on our study, we can further suggest the cultivation of the *J. curcas* plant in the barren land after phytohormones and fertilizers supplementation. This study is of great importance and thereby suggests that *J. curcas* growth and development increases with the exogenous application of phytohormones and fertilizers. *J. curcas* is an economically important plant and unlike crop plants, it does not need significant attention and does not compete with other crops for land and water. The current study added a new insight to the field that application of phytohormones and fertilizers can promote *J. curcas* growth and development in the marginal land under limited water availability.

2. Results

2.1. Phytohormones and Fertilizer Enhances J. curcas Growth Vigor

In the present study, it is investigated that IAA, GA, and NPK increased *J. curcas* shoot length (Figure 1). IAA 100 ppm and IAA 250 ppm increased significantly the shoot length up to 25% and 29%, respectively, as compared to control plants after four months of growth. The GA 100 ppm, GA 250 ppm, and NPK 30 g significantly increased the shoot length up to 17, 47, and 41%, respectively, after four months of growth, compared with control plants (Figure 1A). The higher concentration of IAA, GA, and NPK also increased shoot fresh and dry weight (Figure 1B,C). The results indicated that IAA 250 ppm significantly increased fresh and dry weight up to 141 and 183%, respectively. GA 250 ppm increased shoot fresh and dry weight up to 141 and 179%, respectively, while NPK 60 g increased by 185 and 267%, respectively, compared to control plants. NPK-applied plants showed higher fresh and dry weight than the GA and IAA. On the other hand, hormones and fertilizer also induced root biomass. The IAA 250 ppm, GA 100 ppm, GA 250 ppm, and NPK 60 g significantly increased root length up to 31, 50, 63, and 40%, respectively (Figure 1D). Root fresh weight was significantly increased by 21% by IAA 250 ppm, 72% by GA 250 ppm, and 30% by NPK 60 g compared with control plants (Figure 1E). The root dry weight followed the same pattern of the root fresh weight. The IAA 250 ppm increased the root dry weight to 81%, GA 250 ppm increased to 172%, and NPK 30 g and 60 g increased to 63 and 118%, respectively (Figure 1F). We investigated that, among the hormones and fertilizer, the highest shoot length was shown by the GA 250 ppm-treated plants, while the highest shoot fresh and dry weight was shown by NPK 60 g-treated plants. However, the GA 250 ppm-treated plant showed the highest increase in root length, root fresh, and dry weight which shows that GA can induce *J. curcas* root growth and development.

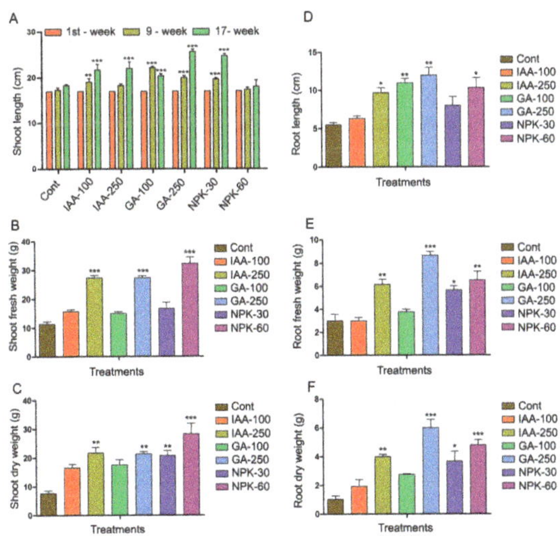

Figure 1. IAA, GA, and NPK regulate *J. curcas* plant soot, root length, shoot, root fresh and dry weight. (**A**) represents shoot length, (**B**) represent shoot fresh weight, (**C**) represent shoot dry weight. (**D–F**) represent foot length, root fresh weight, and root dry weight, respectively. Graphs show mean ± standard deviation, and asterisks show significant differences (* $p \leq 0.05$, ** $p \leq 0.01$, and *** $p \leq 0.001$) according to two-way ANOVA and Bonferroni post hoc tests.

2.2. *Phytohormones and Fertilizer Treatment Promotes Leaf and Branch Number and Stem Diameter*

In the current study, we found that IAA, GA, and NPK efficiently stimulated *J. curcas* number of leaves and branches, and stem diameter in four months (Figure 2). Both low and high concentrations of hormones and NPK application significantly increased the number of leaves, branches, and stem diameter as compared to control plants. IAA 100 ppm and 250 ppm increased leaves number by 262 and 211%, respectively, while branches number increased by 200 and 160%, respectively, and stem diameter increased by 100 and 123%, respectively. GA 100 ppm and 250 ppm increased leaves number up to 274 and 211%, branches number 229 and 260%, and stem diameter 109 and 123%, respectively, as compared to control plants after four months. The 30 g and 60 g of NPK application increased leaves number by 199 and 211%, branches number by 129 and 229%, and stem diameter by 76 and 80%, respectively, after four months of treatment. Among the IAA, GA, and NPK, IAA 100 ppm and GA 250 ppm showed the highest number of leaves increase while GA 250 and NPK 60 g showed the highest induction of branches. Whereas the highest increase in stem diameter was found in IAA 250 ppm and GA 250 ppm treated plants followed by GA 100 ppm. However, the lowest stem diameter and branches number were found in the NPK 30 g treated plants, whereas the lowest leaves number were found in the IAA 100 ppm, GA 250 ppm and NPK 60 g treated plants (equally reduced 211%). These results show that the optimized concentration of IAA, GA, and NPK is needed to increase *J. curcas* leaves, branches, and stem diameter.

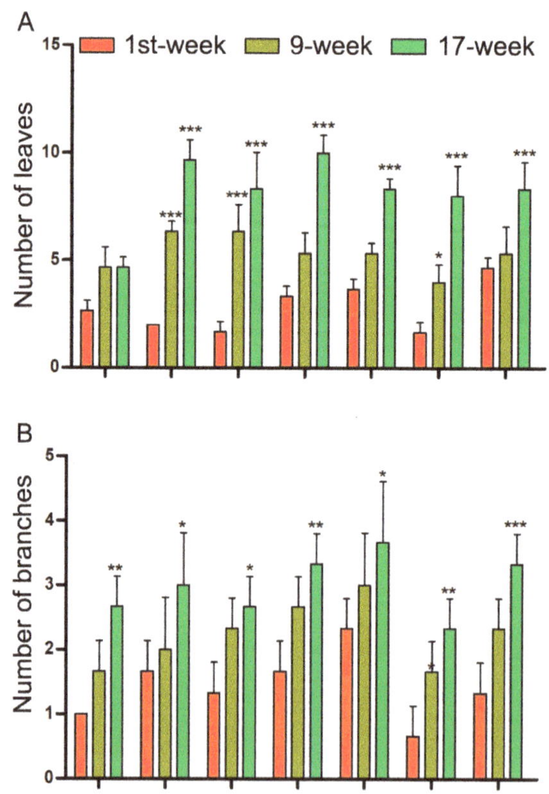

Figure 2. *Cont.*

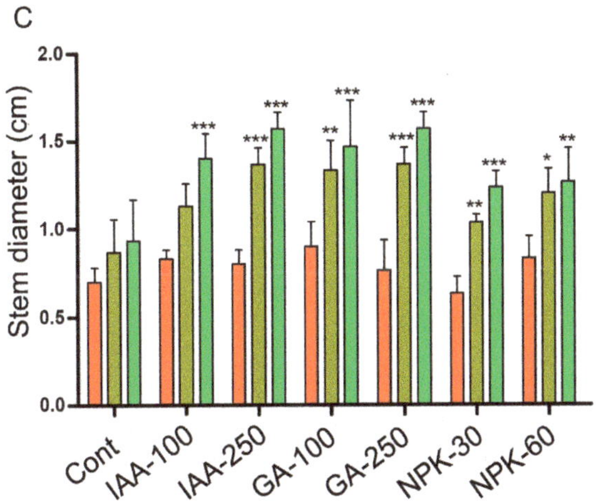

Figure 2. IAA, GA, and NPK regulate *J. curcas* leaves number, branches number, and stem diameter. (**A**) represents leaves number, (**B**) represents branches number, (**C**) represents stem diameter. Graphs show mean ± standard deviation, and asterisks show significant differences (* $p \leq 0.05$, ** $p \leq 0.01$, and *** $p \leq 0.001$) according to two-way ANOVA and Bonferroni post hoc tests.

2.3. Phytohormones and Fertilizers Promote J. curcas Plant Growth by Regulating Proximate Compositions

In the current study, we determined the protein, carbohydrate, and fat contents after four months of treatment of IAA, GA, and NPK in different concentrations. The higher concentration of hormones and NPK such as IAA 250 ppm, GA 250 ppm, and NPK 60 g showed a significant increase of protein, carbohydrates, and fat contents compared with control plants (Figure 3). The NPK 60 g treated plants showed the highest increase (34%) in protein contents followed by IAA 250 ppm (27%), and GA 250 ppm (25%) compared with control plants (Figure 3A). IAA 100 ppm also increased significantly the protein content up to 20%, which indicates that increasing the IAA concentration can increase protein contents. Carbohydrates were significantly increased by 61% by NPK 60 g followed by GA 250 ppm 42% and IAA 250 ppm 30%, compared with control plants (Figure 3B). Similarly, the highest increase (46%) in fat contents was found in NPK 60 g treated plants followed by IAA 250 ppm (39%) and GA 250 ppm (32%), compared with control plants (Figure 3C). The highest moisture and ash contents were found in NPK 60 g treated plants followed by IAA 250 ppm and GA 250 ppm treated plants compared with control plants (Figure 3E,F). These results suggested that phytohormones and NPK treatment enhance the growth of *J. curcas* by induction of proximate compositions. Among the phytohormones and NPK, it was found that NPK application highly increased the overall proximate composition compared with IAA and GA. These results show that fertilizers can better increase the *J. curcas* succession in temperate and barren land compared with hormones.

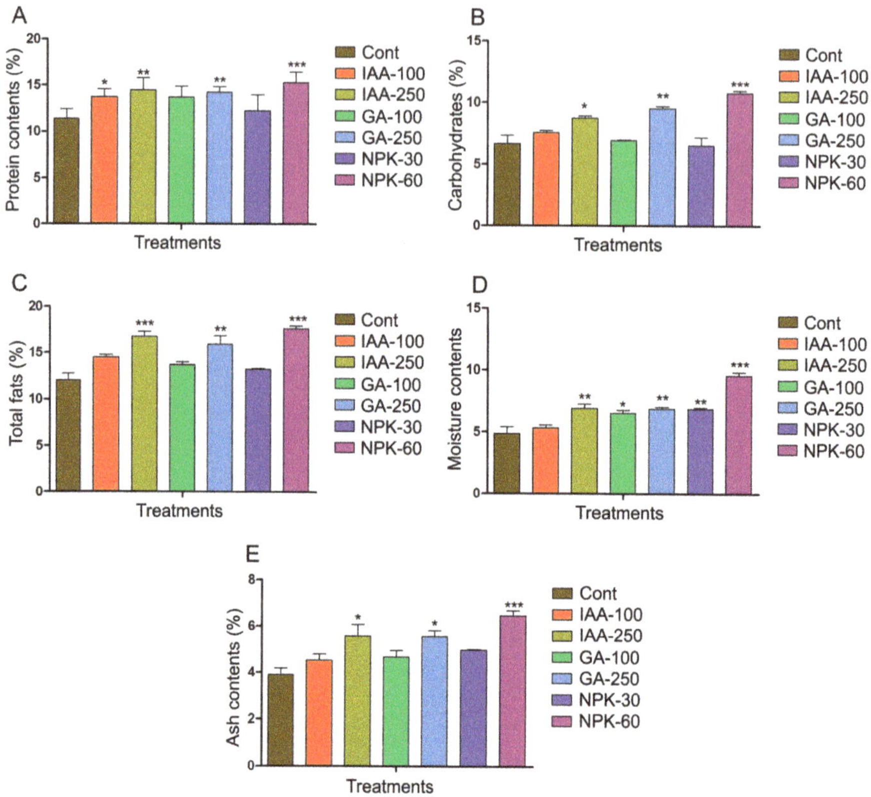

Figure 3. IAA, GA, and NPK induces biosynthesis of *J. curcas* plant proximate composition contents. (**A**) represents protein contents, (**B**) represents carbohydrate contents, (**C**) represents fat contents, (**D**) represents moisture contents, and (**E**) represents ash contents. Graphs show mean ± standard deviation, and asterisks show significant differences (* $p \leq 0.05$, ** $p \leq 0.01$, and *** $p \leq 0.001$) according to two-way ANOVA and Bonferroni post hoc tests.

3. Discussion

In the present study, we reported that IAA, GA, nitrogen, phosphorus, and potassium (NPK) act as growth enhancers of the *J. curcas* plant. We provided physiological, morphological, and biochemical evidence that *J. curcas* growth and development were enhanced by the application of hormones and NPK. Generally, the *J. curcas* plant grows in almost all kinds of soil and drought conditions; however, hormones and fertilizers increase their growth and development. The applied hormones (GA, IAA) and NPK greatly influenced the plant's morphological, physiological, and biochemical traits.

IAA is an essential auxin with significant in vivo roles such as stem growth, root growth, stem cambium cell activation, and lateral bud formation [14,15]. In most plants, IAA is the supreme active form of auxin. Recent studies revealed that IAA application has positive effects on growth parameters such as plant root and shoot length, fresh and dry weight, number of leaves, chlorophyll contents, carbohydrates, amino acids, and phenolic contents [16,17]. Our results also showed that IAA increased plant growth parameters (shoot length, root length, fresh and dry weight, branches and leaves number, and stem diameter) in *J. curcas* plants compared with control plants. In a recent study, it is predicted that phytohormones including IAA alter the sugar metabolism which is responsible for the modulation of biological processes that are involved in plant growth promotion [18,19]. IAA interacts with sucrose and alters plant morphogenesis which regulates leaf morphology [20]. In addition, IAA increases endogenous GA accumulation, which is a prominent growth regulator [18]. In our study, we found that IAA increased leaves and branches number. The possible reason for the increased number of leaves and branches number might be the suppression of the ABA hormone. ABA generally inhibits growth or keeps the apical tissue dormant while IAA reduces ABA and increases GA which breaks dormancy and results in plant growth and increasing branches number [18]. Besides other growth parameters, we also found that the IAA-treated plant's stem diameter was increased significantly than the control plants, which predicts that IAA is efficiently involved in vascular cell division and differentiation. It is reported that IAA increases *Persea americana* cell differentiation which increases vascular vessel density [21]. In *Glycin max*, IAA stimulated plant height, leaves number and area, number of branches, and seed per plant [22]. However, different plants show different reactions to IAA in various concentrations. In the *J. curcas* plant, between the IAA 100 ppm and 250 ppm, the IAA 250 ppm showed a higher increase in morphological, physiological, and biochemical parameters. However, IAA 100 ppm increased the *G. max* height more than that of 200 pmm [22]. Foliar application of IAA on cowpea plants also increased plant height, fresh and dry weight, number of branches, number of leaves, and yield components [23]. As far as IAA is concerned to plant height, some researchers provided evidence that IAA promoted GA synthesis. For instance, it is reported that exogenous IAA increased GA1 and GA_3 biosynthesis through the activation of GA_1 and GA_3 synthesis enzymes [24,25]. IAA not only induced morphological and physiological parameters of the *J. curcas* plant, but also induced their proximate composition (protein, carbohydrate, fat, moisture, and ash contents) compared to control plants. To the best of our knowledge, there is no data published on the effect of IAA and GA on the proximate composition of *J. curcas*; however, researchers investigated the effects of IAA and GA on the proximate composition of various plants. A recent study investigated that, exogenous application of IAA enhanced the photosynthesis rate in *Gossypium hirsutum* which resulted in increased biomass including fresh and dry weight [18]. In the *Balanites aegyptiaca* plant, different concentrations of IAA increased total protein and carbohydrates when compared to non-treated plants [26]. IAA also increased fresh and dry weight, relative water contents (moisture contents), and chlorophyll contents in white clover plants [24]. These results are in line with our findings, which show that IAA is involved in the biosynthesis of protein and carbohydrates and increases moisture contents in *J. curcas*.

GA is a key hormone that induces many parameters of plant growth and development such as seed germination, shoot elongation, leaf expansion, and flower and fruit development [27]. In our study, we found that GA exogenous application increased shoot and root length, fresh and dry weight, and proximate composition percentage in *J. curcas* plants compared with control plants. Plant biomass accumulation is associated with endogenous GA accumulation. GA enhances carbonic anhydrase activity which promotes CO_2 fixation in photosynthesis as this enzyme takes part in the hydration of CO_2 and is strictly related to chloroplast [28]. This phenomenon provides enough CO_2 to the site of fixation and increases the photosynthesis rate which as a result increases biomass accumulation [29,30]. In comparison to IAA and NPK, GA greatly influenced the shoot length, root length, and root fresh and dry weight (Figure 1). Researchers evaluated the basic mechanism of GA that regulates plant growth and development. Previous studies investigated that GA induces the transcription of genes that are involved in cell division and cell elongation occurring during plant growth [31]. GA also stimulates the induction of hydrolytic enzymes which are involved in the conversion of starch to sugar and the controlling of starch and sugar accumulation by GA can significantly influence plant growth and development [32]. Mostly the GA induces plant growth by alteration of certain genes expression. GA also plays a key role in several metabolic pathways that effecting plant growth such as chlorophyll biosynthesis, nitrogen metabolism and redistribution, and translocation of assimilates [28]. GA influences the differentiation of phloem fiber and enhances the length of bast fibers by inducing internode length. Reports show that a high level of GA results in primary phloem fiber elongation in the *Coleus blumei* plant and the length of differentiating internode is associated with the length of the primary phloem [33]. The increase in phloem fiber is associated with an increase in plant height and an increase in plant intermodal length [34]. Previous studies show that GA efficiently stimulates the lateral branches outgrowth in the *J. curcas* plant and it was investigated that increasing concentration increases shot branching [35]. In *Arabidopsis thaliana*, GA insensitive mutant shows a reduction in apical dominance and an increased number of axillary shoots [36]. GA sometimes shows an inhibitory role in lateral branching depending on plant species. In *Pisum sativum* plants, GA plays an inhibitory role in lateral bud formation [37]. The overexpression of the GA biosynthesis gene increased tillers or branches in *Paspalum distichum* and *Populus tremula*, suggesting that GA is significantly involved in the branching of these species [38–40]. However, GA-induced bud formation in citrus, sweet cherry, and rose [41–43]. Our results were consistent with the previously reported research; therefore, it is evident that GA is a positive regulator of stem elongation and buds and branches development. Our study further showed that GA-treated plants showed increased stem diameter, which suggested that GA is significantly involved in the secondary growth of *J. curcas*. Melanie Mauriat et al., (2011) reported that the expression of the GA biosynthesis gene in *Populus tremula* and *Nicotiana tabacum* enhanced the internode length and stem diameter [39]. Reports show that GA enhances the elongation and division of xylem and fiber cells in the vascular bundle region and increases cambium activity which increases stem diameter [44–46]. Compared to IAA and NPK, GA showed an increased number of leaves, branches, and stem diameter (Figure 2). Leaf number and growth are major determining factors contributing to shoot biomass and yield production. Leaves and branches number are regulated and controlled genetically and depend on developmental stage and species [47]. However, studies reported that growth regulators regulate leaves and branches development. In *Gladiolus grandifloras,* GA enhanced the leaves number and sprouting emergence compared with control plants [48]. Although there is a lack of data about GA induction of proximate composition in the *J. curcas* plant; however, it is reported that GA application increases carbohydrate contents in *Phalaenopsis* apex [49]. Reports show that GA accumulation enhances certain enzyme activity and increases cell wall plasticity which enhances membrane permeability and facilitates the uptake of mineral nutrients and transport of photosynthates [50–53]. The GA-facilitated uptake of nutrients and photosynthates transportation promotes plant growth and development. GA accelerates the cell cycle and starch hydrolysis to provide energy and a carbon skeleton for the synthesis of soluble sugar (carbohydrates) and other metabolites [54]. Our results showed that GA application increased protein contents in *J. curcas* which is in line with the results of Satendra Singh et al., (2014), they determined that GA_3 significantly increased the total amino acid (protein contents) of *Phaseolus vulgaris* L. plant [55]. Taken together with these results, it is revealed that the exogenous application of GA enhances *J. curcas* growth and development mediated through the regulation of morphological, physiological, and biochemical parameters.

J. curcas is a biodiesel-producing plant and possesses great characteristics of growing on barren land, low water, and harsh climatic condition. It is a nutrient-reactive plant and its requirement for nutrients varies with the soil fertility [56]. N, P, K, Zn, and B are the main fertilizers needed for the full-size growth of the plant to produce seeds [57,58]. N and P are the main nutrients that affect significantly the seed yield of *J. curcas* [59]. Researchers have suggested that, in the developmental stage of *J. curcas*, NPK is needed to build up the plant architecture such as root, stem, leaves, flowers, and seeds [57]. Analysis of our results revealed that NPK application to *J. curcas* showed significant differences in root shoot length, fresh dry weight, number of leaves, and branches. Nitrogen has a key role in chlorophyll biosynthesis and increases the rate of photosynthesis which results in increases in the dry biomass of the plant [60]. Similarly, phosphorus is involved in energy metabolism and photosynthesis while potassium plays a key role in carbohydrates and protein metabolism during plant growth [61,62]. In general, NPK fertilizer is easily absorbed by plants which play an important role in growth by supporting vegetative development including leaf, stem, and root development. Nitrogen is a building factor of protein, which greatly affect plant biomass [63]. Phosphorus is also used in protein and fat biosynthesis and transform adenosine diphosphate into adenosine triphosphate to generate energy [64]. Potassium enable CO2 during photosynthesis to enter through stomata, photosinthate transport, water, sugar, and protein and sugar synthesis [63]. Potassium availability increases energy that results into growth and development regulation [65]. Compared to GA and IAA, NPK 60 g showed the highest significant increase in total protein, carbohydrate, fat, moisture, and ash content percentage (Figure 3), which indicates that NPK higher concentration increases the proximate composition of *J. curcas* plant. Ali Sher Chandio et al., (2016) investigated that the highest amount of NPK application per hector showed maximum growth, fruit, and seed oil yield in the *J. curcas* plant [56]. Another research also revealed that the highest value of dry matter and seed fatty acid were obtained from NPK-treated *J. curcas* plants compared with non-treated plants [12]. Nitrogen has a key importance in plant growth than the other material as it plays a central role in many physiological and biochemical processes in plants [66]. It is a basic part of the structure of chlorophyll, protein, fats, and nucleic acid [66]. After nitrogen, phosphorus is an important macronutrient that affects plant growth. Without enough phosphorus, it is difficult for a plant to attain maximum growth and development as it has a key role in the storage and transfer of energy in plants [67,68]. Whereas potassium also plays a key role in metabolism and its application influence *J. curcas* seed oil and fatty acids [12]. All three macronutrients (NPK) are important growth inducers of plants but instead of using them separately, they are more efficient when used in combination [34]. Among the NPK, N plays a more important role in enhancing agriculture production by promoting chlorophyll, soluble protein, proline contents, and promoting fiber yield [69,70]. Compared to previous research, our study also determined that NPK application promoted *J. curcas* growth and development by enhancing morphological, physiological, and biochemical regime.

4. Materials and Methods

4.1. Experimental Design and Material Used

In the present study, six months old seedlings of equal size were collected from a common dealer in Multan, Pakistan. The seedling weight, stem width, and root length were different; therefore, the initial weight, stem width, and root length were measured and recorded. The whole experiment was conducted in pots and single seedlings were grown in each pot filled with an equal amount of soil and treated with hormones and fertilizer separately. The hormones Gibberellic acid GA3 (GA) and Indole acetic acid (IAA) were applied as 100 ppm and 250 ppm each to every single pot and NPK was applied 30 g and 60 g to each pot separately (mixed in the soil). Control plants were treated with only water. The plants were watered after two weeks with an equal amount of water. The data were collected after each week for four months and the whole experiment was conducted in triplicate. In fertilizer, nitrogen was 28%, phosphorus was 18% and potassium was 16%.

4.2. Parameters Studied

In the present study, three main parameters were studied, i.e., physiological, morphological, and biochemical parameters. In physiological parameters, root, shoot length, and fresh and dry weight of root and stem were measured. In morphological parameters, the number of leaves, number of branches, and stem diameter were studied. In biochemical parameters, carbohydrates, proteins, lipids, moisture, and ash contents were studied. The data were taken every week until four months.

4.3. Proximate Composition Analysis

To determine the effect of hormones and fertilizers on leaf moisture contents, the fresh leaves were randomly collected in triplicate from each treatment. The collected leaves were washed, weighed, and recorded as fresh weight (FW). The selected leaves were at 105 °C in the oven for 3 h. The fully dried leaves were again weighed and recorded as dried weight (Dw). The moisture content was calculated as the following:

$$\text{Moisture content (\%)} = (\text{Fw} - \text{Dw}) \times 100$$

To measure the ash content percentage, we collected 1 g of leaves and washed them to remove the contamination and the tissue dried. The leaves were burned in the muffle furnace at 450–550 °C to remove water and other volatile substances. The leaves samples were weighed before and after burning in the muffle furnace. The ash content was calculated as follows:

$$\text{Ash (\%)} = (M_{ash}/M_{dry}) \times 100$$

where M_{ash} is the mass of fresh ash and M_{dry} is the mass of dry ash.

To determine fat contents, we followed the method used by Aurea M. Almazan and Samuel O. Adeyeye (1998) [71]. About 1 g of fresh leaves were collected from *J. curcas* and dried in the oven at 60 °C for 15 h and then ground into fine powder. The powder was homogenized with hexane and extracted fat by using the Soxhlet method (AOAC, 1990). The extract was treated with methanolic sodium and methanolic boron trifluoride to convert the fatty acid into methyl ester, using the method followed by Paquot and Hautfenne in 1987 [72]. The fatty acid methyl ester in the hexane layer was dried at 90 °C by passing the nitrogen gas on its surface. The solution of hexane and methyl ester was filtered by a 45 μm filter and injected 1 μL into a Shimadzu gas chromatograph with a Perkin-Elmer PE-WAX capillary column (30 m × 025 mm) and flame ionization detector. The initial column temperature was 80 °C and increased to 260 °C kept for 7 min with the flow rate of gas (H_2) being 30 cm/s. The percentage of fat was calculated based on the total area of fatty acids. All these determinations were carried out according to the Association of Official Analytical Chemists (AOAC, 1990). For crude protein determination, the Micro Kjeldahl method was followed according to the AOAC international [73]. About 1 g of the leaf samples was collected randomly then ground into a fine powder and digested with 15 mL H_2SO_4 by heating in the presence of K_2SO_4 and selenium using a heating block at 420 °C for 2 h. The digested sample was then neutralized by adding NaOH, to convert ammonium sulfate into ammonia, which is further distilled off and collected in a flask, and added boric acid was to form ammonium borate. The residual boric acid was further titrated with H_2SO_4 with the use of an endpoint indicator to determine the total nitrogen contents. The amount of total nitrogen in the raw material was multiplied by the traditional conversion factor of 6.25 and the specific conversion factor [74]. Carbohydrate content was determined by calculating the difference between the sum of all the proximate compositions from 100% [75].

4.4. Soil Analysis

To find the nutrient deficiency in the soil, the soil was analyzed at the Barani Agriculture Research Center (BARS), Kohat. Two soil samples were randomly selected from the different sites of the field. The BARS report of soil is presented in Table 1.

Table 1. Parameters studied in soil analysis.

Sample No	pH	Electric Conductivity ds/m	$CaCO_3$ %	Organic Manure %	Nitrogen %	Texture
1	7.51	0.55	14	0.69	0.0345	Clay loam
2	7.31	0.67	16	0.759	0.0379	Clay loam

4.5. Statistical Analysis

All experiments were performed in triplicate, and the data from each replicate were pooled. Data were analyzed using one-way ANOVA with Bonferroni post hoc tests (* shows $p < 0.05$, ** shows $p < 0.01$, and *** shows $p < 0.001$ significant difference). A completely randomized design was used to compare the mean values of different treatments. Data were graphically plotted, and statistical analyses were performed using the GraphPad Prism software (version 5.01, GraphPad, San Diego, CA, USA).

5. Conclusions

This study demonstrated that the exogenous application of phytohormones and fertilization of the *J. curcas* plant promoted growth and development. The results confirmed that different concentrations of GA, IAA, and NPK induced various parameters differentially. GA 250 ppm increased shoot root length, root fresh and dry weight, branches number, and stem diameter while GA 100 ppm increased leaves number as compared to IAA and NPK. While IAA 250 ppm increased stem diameter, and NPK increased proximate compositions compared to hormones. Our study concluded that optimum hormones and NPK level is essential for the efficient promotion of morphological, physiological, and biochemical aspects of the *J. curcas* plant. Furthermore, the hormones were used for scientific validation, however on commercial basis application of phytohormones are less economical than the fertilizers. Our current study opens an important research area for the future study to investigate, how to improve the indigenous phytohormones of *J. curcas* to improve its growth and development in the barren land.

Author Contributions: Conceptualization, K.-M.K., W.M. and R.J.; methodology, R.J., S.A. (Sajjad Asaf) and M.A.; Formal analysis and software, R.J., S.A. (Saleem Asif) and M.K.; writing—original draft, R.J., K.-M.K. and W.M.; review, visualization and supervision, K.-M.K. and W.M. All authors have read and agreed to the published version of the manuscript.

Funding: This research received no external funding.

Data Availability Statement: The data presented in this study are available on request from the corresponding author.

Acknowledgments: This work was supported by a grant from the New Breeding Technologies Development Program (Project No. PJ016531012022), Rural Development Administration, Korea.

Conflicts of Interest: The authors declare no conflict of interest.

References

1. Lustosa Sobrinho, R.; Zoz, T.; Finato, T.; Oliveira, C.E.d.S.; Neto, S.S.d.O.; Zoz, A.; Alaraidh, I.A.; Okla, M.K.; Alwasel, Y.A.; Beemster, G. *Jatropha curcas* L. as a Plant Model for Studies on Vegetative Propagation of Native Forest Plants. *Plants* **2022**, *11*, 2457. [CrossRef] [PubMed]
2. Holl, M.; Gush, M.; Hallowes, J.; Versfeld, D. *Jatropha curcas in South Africa: An Assessment of Its Water Use and Bio-Physical Potential*; Pretoria South Africa Water Research Commission: Pretoria, South Africa, 2007.
3. Neupane, D.; Bhattarai, D.; Ahmed, Z.; Das, B.; Pandey, S.; Solomon, J.K.; Qin, R.; Adhikari, P. Growing Jatropha (*Jatropha curcas* L.) as a potential second-generation biodiesel feedstock. *Inventions* **2021**, *6*, 60. [CrossRef]
4. Salimon, J.; Abdullah, R. Physicochemical properties of Malaysian *Jatropha curcas* seed oil. *Sains Malays.* **2008**, *37*, 379–382.
5. Abou Kheira, A.A.; Atta, N.M. Response of *Jatropha curcas* L. to water deficits: Yield, water use efficiency and oilseed characteristics. *Biomass Bioenergy* **2009**, *33*, 1343–1350. [CrossRef]
6. Kumar, G.; Yadav, S.; Thawale, P.; Singh, S.; Juwarkar, A. Growth of *Jatropha curcas* on heavy metal contaminated soil amended with industrial wastes and Azotobacter–A greenhouse study. *Bioresour. Technol.* **2008**, *99*, 2078–2082. [CrossRef]
7. Achten, W.M.; Verchot, L.; Franken, Y.J.; Mathijs, E.; Singh, V.P.; Aerts, R.; Muys, B. Jatropha bio-diesel production and use. *Biomass Bioenergy* **2008**, *32*, 1063–1084. [CrossRef]
8. Joshi, G.; Shukla, A.; Shukla, A. Synergistic response of auxin and ethylene on physiology of *Jatropha curcas* L. *Braz. J. Plant Physiol.* **2011**, *23*, 66–77. [CrossRef]
9. Srivastava, L.M. *Plant Growth and Development: Hormones and Environment*; Elsevier: Amsterdam, The Netherlands, 2002.
10. Leite, V.M.; Rosolem, C.A.; Rodrigues, J.D. Gibberellin and cytokinin effects on soybean growth. *Sci. Agric.* **2003**, *60*, 537–541. [CrossRef]
11. Wang, Y.; Mopper, S.; Hasenstein, K.H. Effects of salinity on endogenous ABA, IAA, JA, and SA in Iris hexagona. *J. Chem. Ecol.* **2001**, *27*, 327–342. [CrossRef]
12. Akbarian, M.; Modafebehzadi, N.; Bagheripour, M. Study of fertilizer (NPK) effects on yield and triglycerids in Jatropha (*Jatropha curcas*). *Plant Ecophysiol.* **2010**, *2*, 169–172.
13. Sapeta, H.; Costa, J.M.; Lourenco, T.; Maroco, J.; van der Linde, P.; Oliveira, M.M. Drought stress response in *Jatropha curcas*: Growth and physiology. *Environ. Exp. Bot.* **2013**, *85*, 76–84. [CrossRef]

14. Zolman, B.K.; Yoder, A.; Bartel, B. Genetic analysis of indole-3-butyric acid responses in *Arabidopsis thaliana* reveals four mutant classes. *Genetics* **2000**, *156*, 1323–1337. [CrossRef]
15. El-Eslamboly, A.A. Effect of watermelon propagation by cuttings on vegetative growth, yield and fruit quality. *Egypt. J. Agric. Res.* **2014**, *92*, 553–579. [CrossRef]
16. Sadak, M.S.; Dawood, M.G.; Bakry, B.; El-Karamany, M. Synergistic effect of indole acetic acid and kinetin on performance, some biochemical constituents and yield of faba bean plant grown under newly reclaimed sandy soil. *World J. Agric. Sci.* **2013**, *9*, 335–344.
17. Ogwu, M.C.; Osas, A.R.; Edwin, O.M. Effects of Indole-3-Acetic Acid on Germination in Lead Polluted Petri Dish of *Citrullus lanatus* (Thunberg) Matsumura and Nakai, Cucurbitaceae. *Aceh Int. J. Sci. Technol.* **2015**, *4*, 107–113. [CrossRef]
18. Zhao, T.; Deng, X.; Xiao, Q.; Han, Y.; Zhu, S.; Chen, J. IAA priming improves the germination and seedling growth in cotton (*Gossypium hirsutum* L.) via regulating the endogenous phytohormones and enhancing the sucrose metabolism. *Ind. Crops Prod.* **2020**, *155*, 112788. [CrossRef]
19. Farooq, M.; Jan, R.; Kim, K.-M. Gravistimulation effects on *Oryza sativa* amino acid profile, growth pattern and expression of OsPIN genes. *Sci. Rep.* **2020**, *10*, 17303. [CrossRef] [PubMed]
20. Goren, S.; Lugassi, N.; Stein, O.; Yeselson, Y.; Schaffer, A.A.; David-Schwartz, R.; Granot, D. Suppression of sucrose synthase affects auxin signaling and leaf morphology in tomato. *PLoS ONE* **2017**, *12*, e0182334. [CrossRef]
21. Martens, D.; Luck, S.; Frankenberger, W., Jr. Role of plant growth regulators in vegetative spring flush, flowering, and fruit drop in avocado (*Persea americana*, Mill.). *Circular* **1994**, *CAS-94*, 1–14.
22. Sarkar, P.K.; Haque, M.S.; Abdul Karim, M. Effects of GA and IAA and their frequency of application on morphology, yield 3. *Pak. J. Agron.* **2002**, *1*, 119–122.
23. El-Bassiouny, H.M.; Shukry, W.M. Cowpea growth pattern, metabolism and yield in response to IAA and biofertilizers under drought conditions. *Egypt. J. Biol.* **2001**, *3*, 117–129.
24. Zhang, Y.; Li, Y.; Hassan, M.J.; Li, Z.; Peng, Y. Indole-3-acetic acid improves drought tolerance of white clover via activating auxin, abscisic acid and jasmonic acid related genes and inhibiting senescence genes. *BMC Plant Biol.* **2020**, *20*, 150. [CrossRef] [PubMed]
25. Pan, X.; Welti, R.; Wang, X. Quantitative analysis of major plant hormones in crude plant extracts by high-performance liquid chromatography–mass spectrometry. *Nat. Protoc.* **2010**, *5*, 986–992. [CrossRef] [PubMed]
26. Mostafa, G.G.; Alhamd, M. Effect of gibberellic acid and indole 3-acetic acid on improving growth and accumulation of phytochemical composition in *Balanites aegyptiaca* plants. *Am. J. Plant Physiol.* **2011**, *6*, 36–43. [CrossRef]
27. Hu, Y.-X.; Tao, Y.-B.; Xu, Z.-F. Overexpression of *Jatropha Gibberellin 2-oxidase 6* (*JcGA2ox6*) induces dwarfism and smaller leaves, flowers and fruits in *Arabidopsis* and *Jatropha*. *Front. Plant Sci.* **2017**, *8*, 2103. [CrossRef]
28. Miceli, A.; Moncada, A.; Sabatino, L.; Vetrano, F. Effect of gibberellic acid on growth, yield, and quality of leaf lettuce and rocket grown in a floating system. *Agronomy* **2019**, *9*, 382. [CrossRef]
29. Khan, N. Effect of gibberellic acid on carbonic anhydrase, photosynthesis, growth and yield of mustard. *Biol. Plant.* **1996**, *38*, 145–147. [CrossRef]
30. Khan, N. Variation in carbonic anhydrase activity and its relationship with photosynthesis and dry mass of mustard. *Photosynthetica* **1994**, *30*, 317.
31. Devies, P.; Sun, T. Gibberellin signal transduction in stem elongation and leaf growth. In *Plant Hormones*; Springer: Dordrecht, The Netherlands, 2004.
32. Basra, A. *Plant Growth Regulators in Agriculture and Horticulture: Their Role and Commercial Uses*; CRC Press: Boca Raton, FL, USA, 2000.
33. Aloni, R. Role of auxin and gibberellin in differentiation of primary phloem fibers. *Plant Physiol.* **1979**, *63*, 609–614. [CrossRef]
34. Ullah, S.; Anwar, S.; Rehman, M.; Khan, S.; Zafar, S.; Liu, L.; Peng, D. Interactive effect of gibberellic acid and NPK fertilizer combinations on ramie yield and bast fibre quality. *Sci. Rep.* **2017**, *7*, 10647. [CrossRef]
35. Ni, J.; Gao, C.; Chen, M.-S.; Pan, B.-Z.; Ye, K.; Xu, Z.-F. Gibberellin promotes shoot branching in the perennial woody plant *Jatropha curcas*. *Plant Cell Physiol.* **2015**, *56*, 1655–1666. [CrossRef] [PubMed]
36. Koorneef, M.; Elgersma, A.; Hanhart, C.v.; van Loenen-Martinet, E.; Van Rijn, L.; Zeevaart, J. A gibberellin insensitive mutant of *Arabidopsis thaliana*. *Physiol. Plant.* **1985**, *65*, 33–39. [CrossRef]
37. Scott, T.K.; Case, D.B.; Jacobs, W.P. Auxin-gibberellin interaction in apical dominance. *Plant Physiol.* **1967**, *42*, 1329–1333. [CrossRef] [PubMed]
38. Agharkar, M.; Lomba, P.; Altpeter, F.; Zhang, H.; Kenworthy, K.; Lange, T. Stable expression of AtGA2ox1 in a low-input turfgrass (*Paspalum notatum* Flugge) reduces bioactive gibberellin levels and improves turf quality under field conditions. *Plant Biotechnol. J.* **2007**, *5*, 791–801. [CrossRef] [PubMed]
39. Mauriat, M.; Sandberg, L.G.; Moritz, T. Proper gibberellin localization in vascular tissue is required to control auxin-dependent leaf development and bud outgrowth in hybrid aspen. *Plant J.* **2011**, *67*, 805–816. [CrossRef]
40. Zawaski, C.; Busov, V.B. Roles of gibberellin catabolism and signaling in growth and physiological response to drought and short-day photoperiods in Populus trees. *PLoS ONE* **2014**, *9*, e86217. [CrossRef]
41. Marth, P.C.; Audia, W.V.; Mitchell, J.W. Effects of gibberellic acid on growth and development of plants of various genera and species. *Bot. Gaz.* **1956**, *118*, 106–111. [CrossRef]
42. Elfving, D.C.; Visser, D.B.; Henry, J.L. Gibberellins stimulate lateral branch development in young sweet cherry trees in the orchard. *Int. J. Fruit Sci.* **2011**, *11*, 41–54. [CrossRef]

43. Choubane, D.; Rabot, A.; Mortreau, E.; Legourrierec, J.; Péron, T.; Foucher, F.; Ahcène, Y.; Pelleschi-Travier, S.; Leduc, N.; Hamama, L. Photocontrol of bud burst involves gibberellin biosynthesis in *Rosa* sp. *J. Plant Physiol.* **2012**, *169*, 1271–1280. [CrossRef]
44. Björklund, S.; Antti, H.; Uddestrand, I.; Moritz, T.; Sundberg, B. Cross-talk between gibberellin and auxin in development of Populus wood: Gibberellin stimulates polar auxin transport and has a common transcriptome with auxin. *Plant J.* **2007**, *52*, 499–511. [CrossRef]
45. Mauriat, M.; Moritz, T. Analyses of GA20ox-and GID1-over-expressing aspen suggest that gibberellins play two distinct roles in wood formation. *Plant J.* **2009**, *58*, 989–1003. [CrossRef] [PubMed]
46. Denis, E.; Kbiri, N.; Mary, V.; Claisse, G.; Conde e Silva, N.; Kreis, M.; Deveaux, Y. WOX 14 promotes bioactive gibberellin synthesis and vascular cell differentiation in Arabidopsis. *Plant J.* **2017**, *90*, 560–572. [CrossRef] [PubMed]
47. Xu, Q.; Krishnan, S.; Merewitz, E.; Xu, J.; Huang, B. Gibberellin-regulation and genetic variations in leaf elongation for tall fescue in association with differential gene expression controlling cell expansion. *Sci. Rep.* **2016**, *6*, 30258. [CrossRef] [PubMed]
48. Rahman, A.; Hussain, I.; Nabi, G. Exogenous gibberellic acid application influences on vegetative and reproductive aspects in gladiolus. *Ornam. Hortic.* **2020**, *26*, 244–250. [CrossRef]
49. Chen, W.-S.; Liu, H.-Y.; Liu, Z.-H.; Yang, L.; Chen, W.-H. Gibberellin and temperature influence carbohydrate content and flowering in *Phalaenopsis* [sucrose feeding]. *Physiol. Plant.* **1994**, *90*, 391–395. [CrossRef]
50. Crozier, A.; Turnbull, C. Gibberellins: Biochemistry and action in extension growth. *What New Plant Physiol.* **1984**, *15*, 9–12.
51. Wood, A.; Paleg, L. Alteration of liposomal membrane fluidity by gibberellic acid. *Funct. Plant Biol.* **1974**, *1*, 31–40. [CrossRef]
52. Ansari, H. Effect of Some Phytohormones and NPK on Growth and Metabolism of Mustard. Ph.D. Thesis, Aligarh Muslim University, Aligarh, India, 1996.
53. Estruch, J.J.; Peretó, J.G.; Vercher, Y.; Beltrán, J.P. Sucrose loading in isolated veins of *Pisum sativum*: Regulation by abscisic acid, gibberellic acid, and cell turgor. *Plant Physiol.* **1989**, *91*, 259–265. [CrossRef]
54. Yuxi, Z.; Yanchao, Y.; Zejun, L.; Tao, Z.; Feng, L.; Chunying, L.; Shupeng, G. GA3 is superior to GA4 in promoting bud endodormancy release in tree peony (*Paeonia suffruticosa*) and their potential working mechanism. *BMC Plant Biol.* **2021**, *21*, 323. [CrossRef]
55. Singh, S.; Kulkarni, M.G.; Van Staden, J. Biochemical changes associated with gibberellic acid-like activity of smoke-water, karrikinolide and vermicompost leachate during seedling development of *Phaseolus vulgaris* L. *Seed Sci. Res.* **2014**, *24*, 63–70. [CrossRef]
56. Chandio, A.S.; Depar, M.S.; Meghwar, B.L.; Waseem, M.; Kalroo, A.K.P.; Solangi, A.H.; Rind, P.A.; Ahmed, N. Nutrient Requirement of *Jatropha cultivars* of Diversified Origin. *Int. J. Biol. Biotechnol.* **2016**, *13*, 385–392.
57. Franken, Y.; Nielsen, F. Plantation establishment and management. In *The Jatropha Handbook-From Cultivation to Application (9–29)*; FACT Foundation: Eindhoven, The Netherlands, 2010.
58. Mohapatra, S.; Panda, P.K. Effects of fertilizer application on growth and yield of *Jatropha curcas* L. in an aeric tropaquept of Eastern India. *Not. Sci. Biol.* **2011**, *3*, 95–100. [CrossRef]
59. Patolia, J.; Ghosh, A.; Chikara, J.; Chaudhary, D.; Parmar, D.; Bhuva, H. Response of *Jatropha curcas* grown on wasteland to N and P fertilization. In Proceedings of the Expert seminar on *Jatropha curcas* L. Agronomy and Genetics, Wageningen, The Netherlands, 26–28 March 2007; pp. 26–28.
60. Duarah, I.; Deka, M.; Saikia, N.; Deka Boruah, H. Phosphate solubilizers enhance NPK fertilizer use efficiency in rice and legume cultivation. *3 Biotech* **2011**, *1*, 227–238. [CrossRef]
61. Yan, L.; Sunoj, V.J.; Short, A.W.; Lambers, H.; Elsheery, N.I.; Kajita, T.; Wee, A.K.; Cao, K.F. Correlations between allocation to foliar phosphorus fractions and maintenance of photosynthetic integrity in six mangrove populations as affected by chilling. *New Phytol.* **2021**, *232*, 2267–2282. [CrossRef] [PubMed]
62. Hassanein, R.A.; Hussein, O.S.; Abdelkader, A.F.; Farag, I.A.; Hassan, Y.E.; Ibrahim, M. Metabolic activities and molecular investigations of the ameliorative impact of some growth biostimulators on chilling-stressed coriander (*Coriandrum sativum* L.) plant. *BMC Plant Biol.* **2021**, *21*, 361. [CrossRef]
63. Umami, N.; Abdiyansah, A.; Agus, A. Effects of different doses of NPK fertilization on growth and productivity of *Cichorium intybus*. In Proceedings of the IOP Conference Series: Earth and Environmental Science, Moscow, Russia, 27 May–6 June 2019; p. 012097.
64. Goenadi, D. *Pupuk dan Teknologi Pemupukan Berbasis Hayati Dari Cawan Petri ke Lahan Petani*; Yayasan John Hi-Tech Idetama Jkt.: Jakarta, Indonesia, 2006.
65. Suprihati, N.D. Soil chemical and mineralogical characteristics and its relationship with the fertilizers requirement for rice (*Oriza sativa*), maize (*Zea mays*), and soybean (*Glycine max*). *Bul. Agron* **2005**, *33*, 40–47.
66. Hillel, D.; Hatfield, J.L. *Encyclopedia of Soils in the Environment*; Elsevier: Amsterdam, The Netherlands, 2005; Volume 3.
67. Fixen, P. Nutrient management following conservation reserve program. *Better Crops* **1996**, *80*, 16–19.
68. Lynch, J.; Läuchli, A.; Epstein, E. Vegetative growth of the common bean in response to phosphorus nutrition. *Crop Sci.* **1991**, *31*, 380–387. [CrossRef]
69. Khan, S.; Anwar, S.; Kuai, J.; Ullah, S.; Fahad, S.; Zhou, G. Optimization of nitrogen rate and planting density for improving yield, nitrogen use efficiency, and lodging resistance in oilseed rape. *Front. Plant Sci.* **2017**, *8*, 532. [CrossRef]
70. Huang, C.; Wei, G.; Luo, Z.; Xu, J.; Zhao, S.; Wang, L.; Jie, Y. Effects of nitrogen on ramie (*Boehmeria nivea*) hybrid and its parents grown under field conditions. *J. Agric. Sci.* **2014**, *6*, 230. [CrossRef]

71. Almazan, A.M.; Adeyeye, S.O. Fat and fatty acid concentrations in some green vegetables. *J. Food Compos. Anal.* **1998**, *11*, 375–380. [CrossRef]
72. Paquot, C.; Hautfenne, A. *Standard Methods for the Analysis of Oils, Fats and Derivatives*; Blackwell Scientific Publications Oxford: Hoboken, NJ, USA, 1987; Volume 73.
73. Latimer, G. *Official Methods of Analysis of AOAC International*; AOAC International: Rockville, MD, USA, 2016.
74. Mæhre, H.K.; Dalheim, L.; Edvinsen, G.K.; Elvevoll, E.O.; Jensen, I.-J. Protein determination—Method matters. *Foods* **2018**, *7*, 5. [CrossRef] [PubMed]
75. Marcel, A.; Bievenu, M.J. Proximate, mineral and phytochemical analysis of the leaves of *H. myriantha* and *Urera trinervis*. *Pak. J. Biol. Sci. PJBS* **2012**, *15*, 536–541. [CrossRef] [PubMed]

plants

MDPI

Article

Brassica napus Transcription Factor *Bna.A07.WRKY70* Negatively Regulates Leaf Senescence in *Arabidopsis thaliana*

Tiantian Liu, Yuxin Li, Chang Wang, Da Zhang, Jiajia Liu, Mingyuan He, Mingxun Chen * and Yuan Guo *

National Yangling Agricultural Biotechnology & Breeding Center, Shaanxi Key Laboratory of Crop Heterosis, and College of Agronomy, Northwest A&F University, Yangling 712100, China
* Correspondence: cmx786@nwafu.edu.cn (M.C.); guoyuan2109@163.com (Y.G.)

Abstract: Leaf senescence is the final stage of leaf development and is essential for storage properties and crop productivity. WRKY transcription factors have been revealed to play crucial roles in several biological processes during plant growth and development, especially in leaf senescence. However, the functions of *Brassica napus* WRKY transcription factors in leaf senescence remain unclear. In the present study, *Bna.A07.WRKY70*, one paralogue of *Brassica napus WRKY70*, was cloned from the *B. napus* cultivar "Zhongshuang11 (ZS11)". We found that Bna.A07.WRKY70 contains a highly conserved WRKY domain and is most closely related to *Arabidopsis thaliana* WRKY70. The subcellular localization and transcriptional self-activation assays indicated that Bna.A07.WRKY70 functions as a transcription factor. Meanwhile, RT-qPCR and promoter-GUS analysis showed that *Bna.A07.WRKY70* is predominantly expressed in the leaves of *B. napus* and rosette leaves of *A. thaliana*. In addition, our results demonstrated that ectopic expression of *Bna.A07.WRKY70* in *A. thaliana wrky70* mutants could restore the senescence phenotypes to wild-type levels. Consistently, the expression levels of three senescence-related marker genes of *wrky70* mutants were restored to wild-type levels by ectopic expression of *Bna.A07.WRKY70*. These findings improve our understanding of the function of *Bna.A07.WRKY70* in *B. napus* and provide a novel strategy for breeding the new stay-green cultivars in rapeseed through genetic manipulation.

Keywords: *Bna.A07.WRKY70*; leaf senescence; *Arabidopsis thaliana*; *Brassica napus*

Citation: Liu, T.; Li, Y.; Wang, C.; Zhang, D.; Liu, J.; He, M.; Chen, M.; Guo, Y. *Brassica napus* Transcription Factor *Bna.A07.WRKY70* Negatively Regulates Leaf Senescence in *Arabidopsis thaliana*. *Plants* **2023**, *12*, 347. https://doi.org/10.3390/plants12020347

Academic Editor: Vagner A. Benedito

Received: 14 December 2022
Revised: 3 January 2023
Accepted: 5 January 2023
Published: 11 January 2023

1. Introduction

Rapeseed (*Brassica napus* L., AACC, 2n = 38), an allotetraploid species, originated from spontaneous hybridization between two diploid *Brassica* species: *Brassica rapa* (AA, 2n = 20) and *Brassica oleracea* (CC, 2n = 18) [1]. It is a major oilseed crop grown worldwide for the production of edible oil in the human diet, livestock feed, and industrial materials [2]. Therefore, there is important social and economic significance for studying its associated biological processes, including leaf development. The leaf is the primary organ of photosynthesis and can produce nutrition and gather energy during plants' growth and maturation stages. Leaf senescence, as a type of programmed cell death (PCD), is the terminal stage of leaf development. During leaf senescence, the chloroplast first starts disassembling, and is followed by a loss of chlorophyll together with the catabolism of macromolecules such as protein, lipids, nucleic acids, and RNA [3]. By general catabolism, cellular materials are converted into easily exportable nutrients, which from senescing leaves were subsequently transported to reproductive and developing structures [4]. Consequently, leaf senescence is a critical process for crop fitness and is particularly essential for the optimization of crop productivity. Generally, leaf senescence is influenced by various external environmental and endogenous factors. The environmental cues that affect leaf senescence include high temperature, light signals, drought, and biotic stress [5–7]. The endogenous factors include the accumulation of reactive oxygen species (ROS), variation of plant hormones, and, most importantly, regulation of multiple senescence-associated genes [8–11]. Therefore, mining the key genes regulating the leaf senescence process is of great importance in rapeseed.

The WRKY proteins are one of the largest and most important superfamilies of transcription factors (TFs) in plants. WRKY transcription factors encompass a core motif WRKYGQK (a highly conserved WRKY domain) at the N-terminus and an atypical Zinc finger motif at the C-terminus [12]. On the basis of both the number of WRKY domains and the features of the Zn-finger motif in their evolutionary history, the WRKY TFs can be divided into three different groups (I, II, and III). Group I contains two WRKY domains and a finger motif whose pattern is conserved zinc ligands ($C–X_{4-5}–C–X_{22-23}–H–X_1–H$), which is the same as the Zinc finger motif of group II, but there is only one WRKY domain in group II. Instead of a C2–H2 pattern in group I and II, group III contains a pattern of C2–HC ($C–X_7–C–X_{23}–H–X_1–C$) zinc finger-like motif and have one WRKY domain [13,14]. All three groups' members of WRKY TFs have been demonstrated to interact with the specific DNA *cis*-acting element W-box (C/TTGACT/C) in the promoter regions of downstream genes and further regulate their expression [15]. In recent decades, experimental evidence has shown that WRKY proteins act as key regulators widely involved in various plant growth and development processes, such as leaf senescence, growth of roots [16], stem elongation [17], and multiple biotic and abiotic stressors [18,19]. In *Arabidopsis thaliana*, *WRKY53* acts in a complex transcription factor signaling network regulating leaf senescence-specific gene expression [20], WRKY45 interacts with the DELLA protein RGL1 to positively regulate age-triggered leaf senescence [21], and *WRKY71* mediates ethylene (ET) signaling and synthesis to hasten leaf senescence [22].

WRKY70 belongs to WRKY III TFs and has been reported in response to several developmental and physiological processes in diversified species [23]. In *Arabidopsis*, *WRKY70* acts as a negative regulator of leaf senescence, with gradually increasing expression during leaf development [3], and *WRKY70* is also crucial in plant defense against pathogens, controlling the cross-talk of salicylic acid (SA) and jasmonic acid (JA) signaling in plant defense [23,24]. Moreover, *WRKY70* is an important signaling component that is positively involved in brassinolide (BR)-regulated growth and negatively involved in drought responses by inhibiting drought-responsive genes [25]. In chickpeas, *WRKY70* was reported to regulate the expression of a chickpea HD-Zip transcription factor *CaHDZ12*, which improved tolerance to osmotic stresses under drought and salinity stress, and increased sensitivity to abscisic acid (ABA) in transgenic tobacco and chickpea [26]. In addition, it was suggested that GhWRKY70D13 negatively regulates cotton's resistance to *Verticillium dahliae* mainly through its effect on ET and JA biosynthesis and signaling pathways [27]. A recent study demonstrated that *TaWRKY70* positively regulates *TaCAT5* by directly binding to the *TaCAT5* promoter to enhance Cd tolerance in transgenic *Arabidopsis* [28]. In *B. napus*, the *BnWRKY70* knockout plants by CRISPR/Cas9 system enhanced *Sclerotinia sclerotiorum* resistances, while overexpression of *BnWRKY70* reduced resistance to *S. sclerotiorum* [29]. However, the roles of WRKY proteins in *B. napus* in the regulation of leaf senescence remain unclear.

In the current study, *Bna.A07.WRKY70*, one of the *AtWRKY70* orthologues in *B. napus*, was isolated and functionally characterized. We found that *Bna.A07.WRKY70* functioned as a TF and was specifically expressed in the leaves in *A. thaliana* and *B. napus*. We also demonstrated that ectopic expression of *Bna.A07.WRKY70* in the *A. thaliana wrky70* mutant restored the leaf senescence rate and chlorophyll content and greatly altered the expression of three senescence-related genes in this mutant. Our results may indicate that *Bna.A07.WRKY70* functions as a negative regulator of leaf senescence in *Arabidopsis*, which might reveal a conserved role of WRKY70 proteins in regulating leaf senescence between *A. thaliana* and *B. napus*.

2. Results

2.1. Sequence Analysis of BnaWRKY70 Paralogs

In the *B. napus* cultivar "Zhongshuang11 (ZS11)", six paralogs of BnaWRKY70 were predicted in BnPIR (http://cbi.hzau.edu.cn/bnapus/index.php, accessed on 9 September 2022) and were designated Bna.A07.WRKY70 (BnaA07G0195100ZS), Bna.C06.WRKY70 (BnaC06-

G0198900ZS), Bna.A04.WRKY70 (BnaA04G0035900ZS), Bna.C08.WRKY70 (BnaC08G0362900ZS), Bna.A09.WRKY70 (BnaA09G0519800ZS), and Bna.C04.WRKY70 (BnaC04G0308100ZS). With the multiple sequence alignment, we found that the WRKY70 protein from *B. napus* and *A. thaliana* possessed highly conserved WRKY domains, including WRKYGQ/KK core motif and a pattern of C2–HC zinc finger-like motif at the C-terminus (Figure 1A). Among them, Bna.A07.WRKY70 was predicted to share the highest identity in the amino acid sequence with the AtWRKY70 protein (66.01%) (Figure S1). A phylogenetic analysis was performed to investigate the evolutionary relationships between Bna.A07.WRKY70 and 20 WRKY70 proteins from seven plant species, including *A. thaliana, B. napus, B. rapa, Glycine max, Zea mays, Oryza sativa*, and *Setaria italic*. As illustrated in Figure 1B, Bna.A07.WRKY70 is most closely related to the WRKY70 protein from *B. rapa* (NP_001288847.1) and *A. thaliana* (AtWRKY70). These results suggested preliminarily that Bna.A07.WRKY70 might have similar functions as AtWRKY70.

2.2. Subcellular Localization and Transcriptional Activity of Bna.A07.WRKY70

For subcellular localization, Bna.A07.WRKY70 was expressed in tobacco (*Nicotiana benthamiana*) leaf cells as a recombinant protein fused to a green fluorescent protein marker. The fluorescence signal was detected in the nucleus by laser scanning confocal microscopy (Figure 2A), suggesting that Bna.A07.WRKY70 might function as a transcription factor.

To further characterize Bna.A07.WRKY70 function, we investigated whether Bna.A07. WRKY70 has transcription activation activity in yeast cells. The empty vector pGBKT7 as negative control and fusion construct (*pBD-Bna.A07.WRKY70*) were transformed separately into Y2HGold yeast cells, which were cultured on SDO (SD/-Trp) and TDO (SD/-Trp/-His/-Ade) medium. As shown in Figure 2B, on SDO (SD/-Trp) medium, all yeast transformants could grow normally, indicating that the constructs were transformed successfully into the Y2HGold yeast cells. Instead, on TDO (SD/-Trp/-His/-Ade) medium, the empty vector pGBKT7 did not grow, but yeast cells with Bna.A07.WRKY70 fusion constructs grew well, which demonstrated that Bna.A07.WRKY70 could activate the expression of the reporter genes. Given these findings, the Bna.A07.WRKY70 was testified to function as a transcription activator.

2.3. Analysis of Bna.A07.WRKY70 Expression Pattern

We further investigated the spatiotemporal expression pattern of *Bna.A07.WRKY70* by analyzing the relative abundance of the mRNA in various tissues of *B. napus* cultivar "ZS11" using quantitative reverse transcription PCR (RT-qPCR). The results showed that *Bna.A07.WRKY70* was widely expressed in different organs of *B. napus*, with higher expression in leaves, moderate in stems, flowers, and roots but low in developing seeds (Figure 3A). To comprehensively investigate the spatiotemporal expression pattern of *Bna.A07.WRKY70* in *A. thaliana*, we obtained 16 *pBna.A07.WRKY70:GUS* in wild-type background independent lines of *A. thaliana* and the one representative line was used for promoter-GUS analysis because of similar GUS staining patterns in most lines. Consistent with the RT-qPCR data in *B. napus*, promoter-GUS activity staining was predominantly detected in rosette leaves of *A. thaliana* (Figure 3D) and then was also slightly detected in other organs of *A. thaliana*, including stems (Figure 3C), roots (Figure 3B), and flower abscission zones (Figure 3E). Conversely, it was not detected in the embryo (Figure 3G) and siliques (Figure 3F). In summary, these observations suggested that *Bna.A07.WRKY70* might regulate a significant function during leaf development.

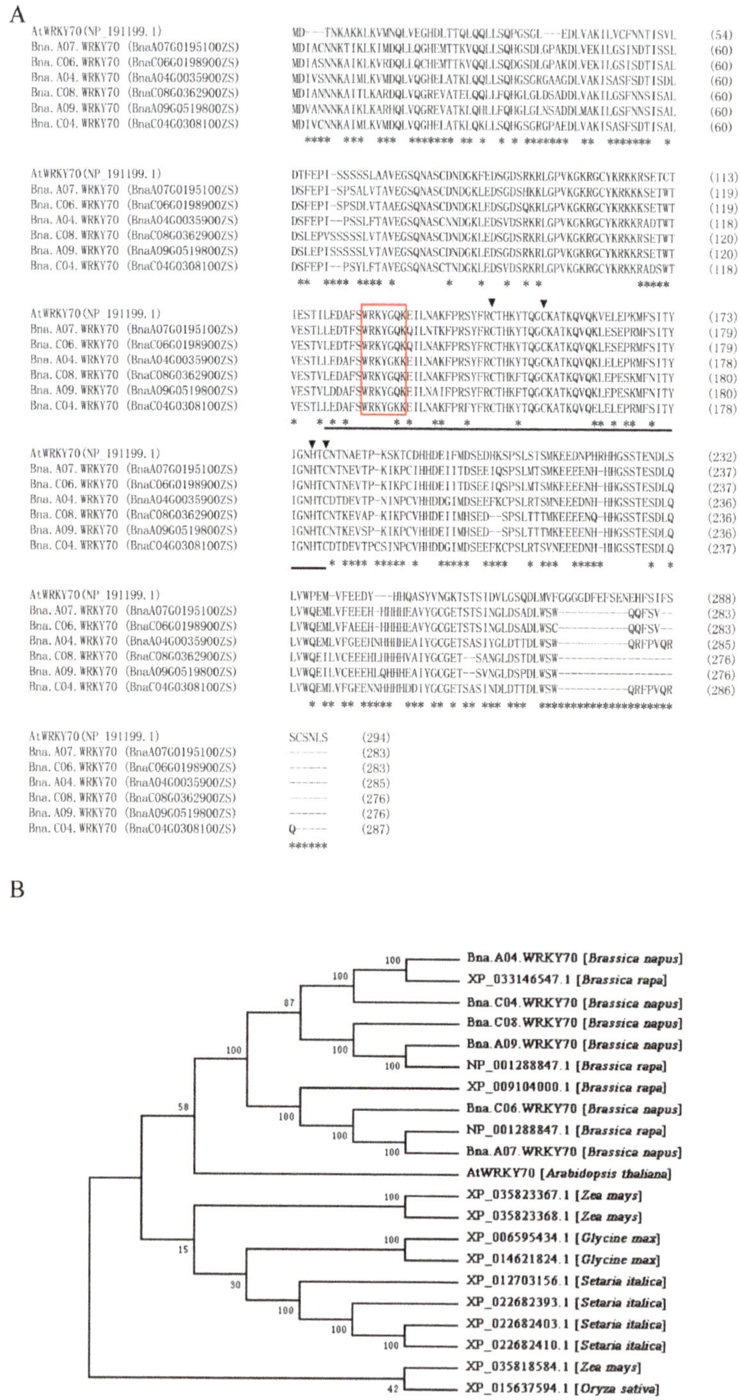

Figure 1. Protein sequence and phylogenetic analyses of WRKY70 proteins. (**A**) Protein sequence

alignment of WRKY70 from *A. thaliana* and *B. napus* was carried out using the MUSCLE program (http://www.ebi.ac.uk/Tools/msa/muscle/, accessed on 12 September 2022). Asterisks indicate non-conservative differences. The WRKY domain 125–185, as indicated by the Conserved Domain Search program (http://www.ncbi.nlm.nih.gov/Structure/cdd/wrpsb.cgi, accessed on 12 September 2022), is underlined. The highly conserved core sequence WRKYGQK in the WRKY domain is represented by a red box, together with the C and H residues in the CCHC zinc-finger-like motif indicated by a downward black triangle. (**B**) Phylogenetic analysis of Bna.A07.WRKY70 with 20 other WRKY70 proteins from seven plant species, including AtWRKY70 (*Arabidopsis thaliana*); Bna.A07.WRKY70 (BnaA07G0195100ZS), Bna.C06.WRKY70 (BnaC06G0198900ZS), Bna.A04.WRKY70 (BnaA04G0035900ZS), Bna.C08.WRKY70 (BnaC08G0362900ZS), Bna.A09.WRKY70 (BnaA09G0519800ZS), and Bna.C04.WRKY70 (BnaC04G0308100ZS (*Brassica napus*); NP_001288821.1, XP_033146547.1, XP_009104000.1, and NP_001288847.1 (*Brassica rapa*); XP_015637594.1 (*Oryza sativa*); XP_035823367.1, XP_035823368.1, and XP_035818584.1 (*Zea mays*); XP_012703156.1, XP_022682393.1, XP_022682403.1, and XP_022682410.1 (*Setaria italica*) and XP_006595434.1 and XP_014621824.1 (*Glycine max*). A neighbor-joining tree (Jones–Taylor–Thornton model) with 1000 replicates of bootstrap analysis was generated by MEGA7. Bootstrap values are indicated at the nodes, and the accession numbers of the species are labeled on the phylogenetic tree.

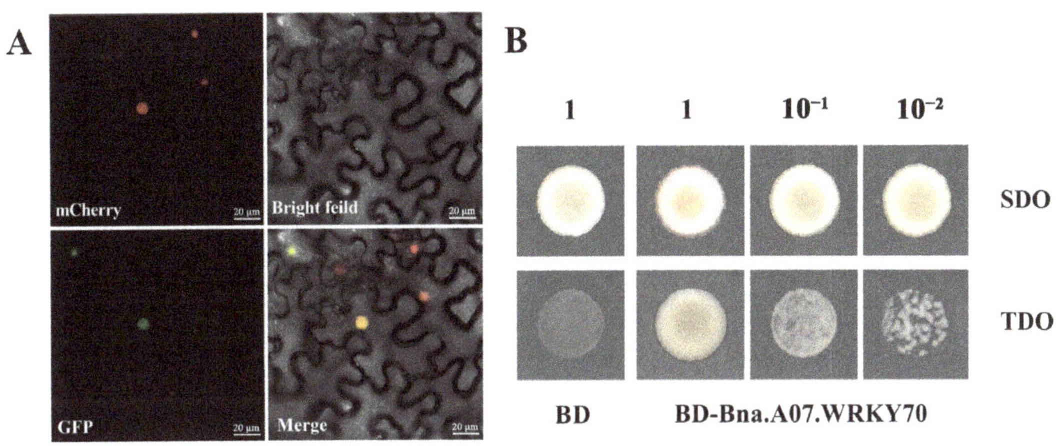

Figure 2. Transcription factor characterization of Bna.A07.WRKY70. (**A**) Subcellular localization of Bna.A07.WRKY70 protein fused with GFP (*35S:GFP-Bna.A07.WRKY70*) in tobacco leaves (*Nicotiana benthamiana*). mCherry, a nuclear-localized protein fused with a red fluorescent protein; merge, merge of mCherry, GFP, and bright field images. (**B**) Transcriptional activation assays of Bna.A07.WRKY70 in yeast. BD: empty vector that contains GAL4 DNA-binding domain, *BD-Bna.A07.WRKY70*: cDNAs encoding of *Bna.A07.WRKY70* transcripts were separately cloned into the pGBKT7/BD vector containing the GAL4 DNA binding domain, which transformed into the yeast strain Y2HGold, SDO: ability of yeast transformants to grow on medium lacking Trp, TDO: ability of yeast transformants to grow on medium lacking Trp, His and Ade indicates transcriptional activation. 1, 10^{-1}, 10^{-2}: the transformed strains were spotted on plates by diluting to different concentrations. The images show representative results from more than four independent yeast transformants.

2.4. *Bna.A07.WRKY70 Negatively Regulates Leaf Senescence in A. thaliana*

To further explore the function of *Bna.A07.WRKY70* on leaf development, we introduced the construct *35S:Bna.A07.WRKY70-GFP* into *A. thaliana wrky70* mutant (Figure 4A). Twelve independent T_1 transgenic plants were generated using hygromycin selection, and five independent T_3 homozygous transgenic lines *wrky70 35S:Bna.A07.WRKY70-GFP* were selected and confirmed by PCR amplification with the specific primers 35S-F/Bna.A07.WRKY70-GFP-BamHI-R (Figure 4B; Supplementary Table S1). Of these lines, three representative ones, *wrky70 35S:Bna.A07.WRKY70-GFP* #4, #6, and #12, with a relatively high expression level (Figure 4C), were selected for the follow-up experiment. As illustrated in Figure 5A,B, the loss-of-function mutants *wrky70* exhibited markedly yellowing of leaves at 35 DAG (days after germination) and indicated earlier senescence compared to wild-type plants, which is in line with previous findings [3]. Interestingly, we found that the *A. thaliana wrky70* mutant leaves were smaller than wild-type plants. Ectopic expression of *Bna.A07.WRKY70* fully restored the rate of leaf senescence to wild-type levels in *Arabidopsis wrky70* mutants (Figure 5A). Furthermore, by arranging the rosette leaves of 35-day-old Col-0, *wrky70* mutant, and transgenic plants (#4, #6, #12) according to their age from older to younger, we found that three *Bna.A07.WRKY70* transgenic lines in the *wrky70* background delayed the premature senescence of leaves and rescued the phenotype of leaves relatively smaller in size compared to wild-type plants (Figure 5B). The results of the chlorophyll content indicated that the chlorophyll content of the *wrky70* mutant intensified degradation from the fifth week, but the chlorophyll content of *Bna.A07.WRKY70* transgenic lines were in keeping with that of Col-0 and clearly higher than that of the *wrky70* mutant (Figure 5C).

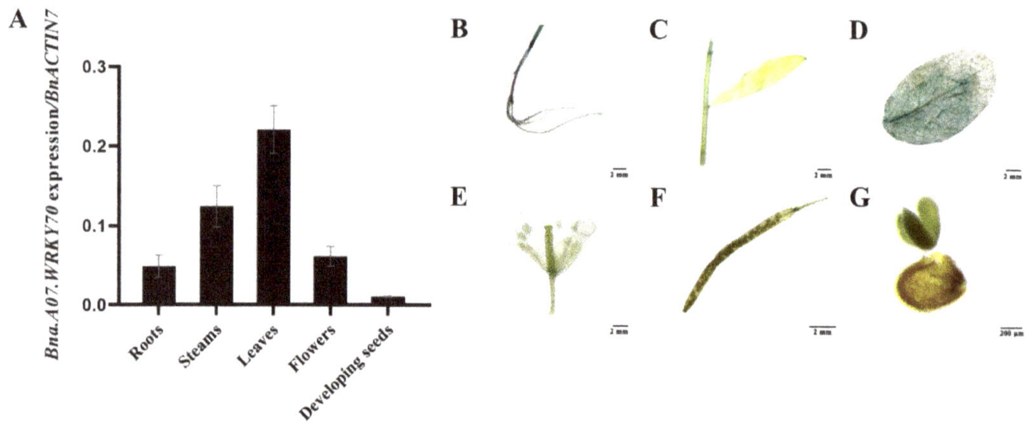

Figure 3. Analysis of the *Bna.A07.WRKY70* expression pattern. (**A**) RT-qPCR analysis of the *Bna.A07.WRKY70* expression in various tissues of *B. napus* cultivar "ZS11". The RT-qPCR result was normalized against the expression of *BnACTIN7* as an internal control. Values are means ± SD (n = 3). Error bars denote standard deviations. (**B**) to (**G**), Histochemical GUS staining in 35-day-old *ProBna.A07.WRKY70:GUS* transgenic *Arabidopsis* plants. (**B**) Roots (bar = 2 mM); (**C**) Stems and leaves (bar = 2 mM); (**D**) rosette leaves (bar = 2 mM); (**E**) Flowers (bar = 2 mM); (**F**) siliques 12 days after pollination (bar = 2 mM); (**G**) Developing seeds 12 days after pollination (bar = 200 μM).

In order to further confirm that the *Bna.A07.WRKY70* regulates the progress of leaf senescence in *A. thaliana*, we assessed the transcript levels of representative genes relating to senescence in the fifth and sixth rosette leaves of *A. thaliana* wild-type, the *wrky70* mutant, *wrky70 35S:Bna.A07.WRKY70-GFP* plants at 35 DAG. Compared to the wild type, the results showed that the expression of the senescence-related gene *AtSAG13* (*senescence-associated gene 13*) and *AtSEN1* (*senescence-associated gene 1*) were significantly increased, while the expression of photosynthesis-related *AtCAB1* gene (*chlorophyll a/b-binding protein*) was significantly decreased in 35-day-old *wrky70* mutant plants (Figure 6). However, when the *35S:Bna.A07.WRKY70-GFP* was introduced into the *wrky70* mutant, we found that the transcript abundance of these three senescence-related marker genes, including *AtSEN1*, *AtCAB1*, and *AtSAG13*, was restored to wild type levels. In brief, all results containing the premature senescence phenotype, chlorophyll content, and the expression of senescence-associated marker genes together revealed that *Bna.A07.WRKY70* may negatively regulate the leaf senescence by adjusting the expression of senescence genes in *A. thaliana* and play a similar role with *AtWRKY70* in *A. thaliana*.

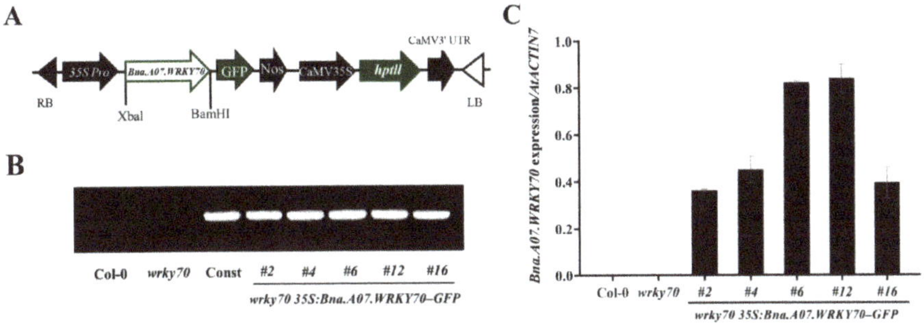

Figure 4. Molecular characterization of *wrky70 35S:Bna.A07.WRKY70-GFP* transgenic plants. (**A**) Schematic diagram of constitutive expression cassette of the *Bna.A07.WRKY70* gene in the binary vector pCAMBIA-1300 used for plant transformation. RB, right border; LB, left border; 35S Pro, CaMV 35S promoter; Nos, nopaline synthase terminator; CaMV35S, CaMV 35S promoter; *hptII*, hygromycin resistance gene. (**B**) PCR-based DNA genotyping of *wrky70 35S:Bna.A07.WRKY70-GFP* transgenic plants using specific primers of 35S_P/Bna.A07.WRKY70-GFP-BamHI-R. Const, *35S:Bna.A07.WRKY70-GFP* construct. Col-0 and *wrky70* indicate *A. thaliana* wild type and mutant plants, respectively. (**C**) Expression analysis of *Bna.A07.WRKY70* in *wrky70 35S:Bna.A07.WRKY70-GFP* transgenic plants using RT-qPCR. The expression level was normalized against the expression of *AtACTIN7*, which was used as an internal control. Values are the means ± SD (n = 3). Error bars indicate standard deviation. # indicates the transgenic lines.

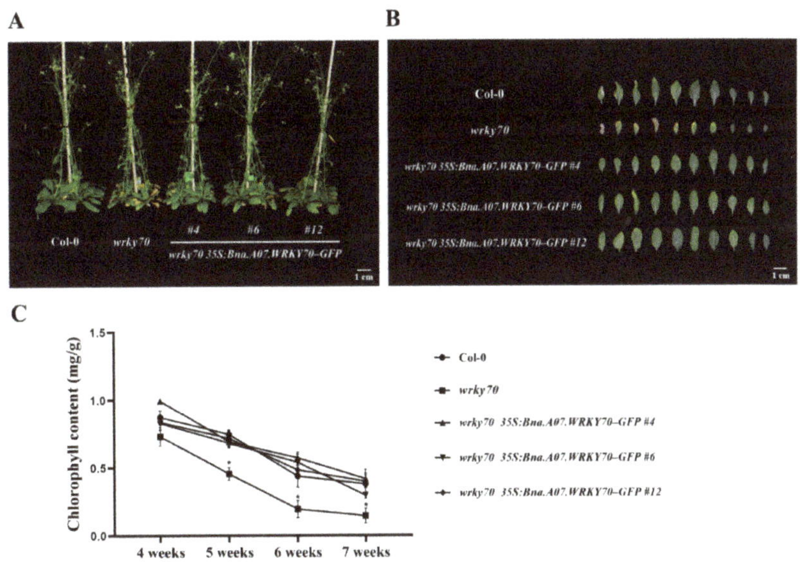

Figure 5. Effects of *Bna.A07.WRKY70* overexpression in the *wrky70* mutant background on leaf senescence in *A. thaliana*. (**A**) The whole plant phenotypes of leaf senescence in the wild type (Col-0), *wrky70* mutant, and *wrky70 35S:Bna.A07.WRKY70-GFP* transgenic plants. The images were taken 35 days after germination (DAG). Bar = 1 cM. (**B**) Phenotype of rosette leaves in 35-day-old plants, excised leaves are arranged according to age, from older to younger. Bar = 1 cM. (**C**) Comparisons of chlorophyll content of the fifth to sixth rosette leaves among wild-type (Col-0), *wrky70* mutant, and *wrky70 35S:Bna.A07.WRKY70* transgenic plants at the indicated ages. Values are means ± SD (n = 3). Asterisks indicate significant differences from wild-type (two-tailed paired Student's *t*-test, $p \leq 0.05$). Error bars indicate standard deviation. # indicates the transgenic lines.

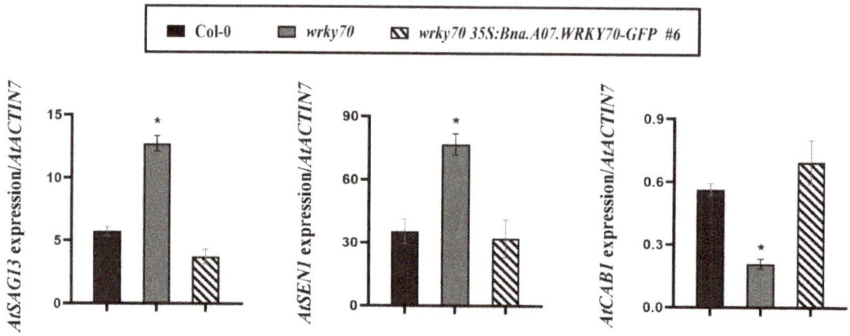

Figure 6. Expression analysis of leaf senescence marker genes in the rosette leaves among wild-type (Col-0), *wrky70* mutant, and *wrky70 35S:Bna.A07.WRKY70-GFP* transgenic plants at the 35 DAG, as measured by RT-qPCR. Expression levels were normalized to the expression of the internal reference gene, *AtACTIN7*. Values are means ± SD (n = 3). Error bars indicate standard deviations. Asterisks indicate statistically significant differences from wild type plants (two-tailed paired Student's *t*-test, $p \leq 0.05$). # indicates the transgenic lines.

3. Discussion

Leaf senescence is an indispensable portion and spans the latter half of leaf development. It is a highly intricate process regulated by multiple pathways [30]. As previously reported, the three largest groups of transcription factors, WRKY, NAC, and MYB superfamilies, are responsible for modulating transcriptional changes during leaf senescence [31], in which the *AtWRKY70* has already been confirmed with a high level of expression in the late stage of leaf development and functions as an essential repressor during leaf senescence in *A. thaliana* [32]. However, the roles of *WRKY70* transcription factors during leaf development in *B. napus* were lacking.

It has been widely known that *B. napus* was formed 7500 years ago by natural hybridization between *B. rapa* and *B. oleracea* [33]. *B. napus*, and diploid parental species *B. rapa* and *B. oleracea*, are believed to share a common ancestor with *A. thaliana*, a fact that has favored the transfer of knowledge from *Arabidopsis* to *B. napus*. As an allopolyploid plant, a large number and a high frequency of chromosome variation activities were identified, such as duplication, rearrangement, fusion, and deletion in the evolution processes of *B. napus*, which makes the genomics of *B. napus* more complicated. Generally, a single *Arabidopsis* gene is represented by two to eight paralogs in the *B. napus* genome [34,35]. Accordingly, six paralogues (Bna.A07.WRKY70, Bna.C06.WRKY70, Bna.A04.WRKY70, Bna.C08.WRKY70, Bna.A09.WRKY70, and Bna.C04.WRKY70) were found in the *B. napus* genome (Figure 1). In the WRKY transcription factor family, the WRKY domain is the major determinant of DNA-binding and specifically binds DNA *cis*-acting element W-box (C/TTGACT/C). Our results showed that all six Bna.WRKY70 had the WRKY protein domain containing the WRKYGQ/KK core motif and a pattern of C2–HC zinc finger-like motif at the C-terminus (Figure 1A). In the present study, among these BnaWRKY70 paralogues, Bna.A07.WRKY70, which was predicted to have the highest identity of protein sequence and the WRKY central conserved domains with AtWRKY70 (Figure 1), was cloned from the *B. napus* cultivar "ZS11" and functionally characterized. Bna.A07.WRKY70 was located in the nucleus of tobacco leaf cells, and we further demonstrated that Bna.A07.WRKY70 could activate the expression of the reporter genes in yeast cells (Figure 2). These results suggested that Bna.A07.WRKY70 functions as a transcription activator. Additionally, the *Bna.A07.WRKY70* transcript was broadly present in different vegetative tissues, with the highest levels observed in leaves (Figure 3), suggesting that *Bna.A07.WRKY70* might regulate a significant function during leaf development. Ectopic expression of *Bna.A07.WRKY70* in the background of *A. thaliana wrky70* mutants significantly delayed the senescence of leaves and restored the chlorophyll content to the wild type level (Figure 5). Moreover, the expression of senescence-associated genes (*AtSEN1*, *AtSAG13*, and *AtCAB1*) was clearly regulated by *Bna.A07.WRKY70* during leaf senescence. Thus, these results may indicate that *Bna.A07.WRKY70* functions as a negative factor in leaf senescence as the *AtWRKY70*.

During leaf senescence, the leaves turned yellow, resulting in photosynthesis deficiency and beginning with chloroplast dismantling, followed by degradation of chlorophyll and chlorophyll-protein complexes. Meanwhile, leaf senescence is accompanied by decreased expression of genes related to photosynthesis and protein synthesis and increased expression of senescence-associated genes (SAGs) [36]. Consistently, our results demonstrated that compared to the wild type, the expression of the photosynthesis-related *AtCAB1* gene was decreased, and the expression of senescence-related gene *AtSAG13* and *AtSEN1* were increased in *A. thaliana wrky70* mutant plants. The expression of these three marker genes was rescued to wild-type in *wrky70 35S:Bna.A07.WRKY70-GFP* transgenic plants, which proved that *Bna.A07.WRKY70* indeed delayed the leaf senescence during plant senescence by affecting the expression of these three senescence genes in *A. thaliana*. Leaf senescence was widely influenced by a variety of external and internal factors, including environmental stresses and phytohormones. Recently, key gene regulatory networks comprising these TFs have been identified, indicating that leaf senescence is controlled by multiple cross-linking pathways, many of which are associated with stress response signaling [37–39]. *Arabidopsis WRKY71* was reported that it is able to directly upregulate the ethylene signaling pathway genes *EIN2* (*ethylene insensitive2*) and *ORE1* (*oresara1*) and promote ethylene synthesis by directly activating the *ACS2* gene to accelerate leaf senescence in *Arabidopsis* [22]. The cotton (*Gossypium hirsutum* L.) *GhWRKY91* directly targets *GhWRKY17*, a gene associated with ABA signals and reactive oxygen species (ROS) production to negatively mediate leaf senescence and provide a foundation for further functional studies on natural and stress-induced leaf senescence [40]. *OsWRKY53* of rice, as a positive regulator, repressed the transcript of ABA catabolic genes (*OsABA8ox1* and *OsABA8ox2*) by directly binding to their promoters to promote ABA accumulation, and modulated ABA-induced leaf senescence [41]. In *Arabidopsis*, *AtWRKY70* transcript levels were more strongly reduced in *npr1* (*non-expressor of PR 1*) and *pad4* (*phytoalexin-deficient 4*) and completely abolished in *NahG* (salicylate hydroxylase gene) plants compared to wild-type at 40 days post germination, among which, the *NahG*, *pad4*, and *npr1* belonged to SA mutants and exhibited a delayed senescence phenotype [3]. These findings support the role of *AtWRKY70* as a senescence-associated gene and indicate a functional requirement of SA for its normal expression. Besides, the preceding research illustrated that the pathway of plant hormones could respond to numerous abiotic stresses; for instance, *GhWRKY17* from upland cotton modulated the increased sensitivity of plants to drought by reducing the level of ABA, and repressed transcript levels of ABA-inducible genes, including *AREB* (*ABA-responsive element binding*), *DREB* (*dehydration-responsive element binding*), *NCED* (*9-cis-epoxycarotenoid dioxygenase*), *ERD* (*early responsive to dehydration*) and *LEA* (*late embryogenesis-abundant protein*) under drought and salt stress conditions, indicating that *GhWRKY17* responds to drought and salt stress through ABA signaling and the regulation of cellular ROS production in plants [42]. With the above findings in mind, whether *Bna.A07.WRKY70* of *B. napus* adjusts the signaling pathways of phytohormone by combining with some key genes during the regulation of leaf senescence and responds to plant stress resistance mediated by the signaling pathways, can be explored further.

Interestingly, it has been reported that the *Arabidopsis wrky70* knockout mutants were slightly reduced in size compared to wild-type plants during the entire period of development in *A. thaliana* [3]. However, from another report, no obvious growth phenotype was observed in a single knockout mutant of *wrky70* compared with the wild-type *A. thaliana* [25]. In this study, our results found that the *A. thaliana wrky70* mutant leaves were smaller than wild-type plants, and the leaf size of the *wrky70* mutant was restored to wild type by the ectopic expression of *Bna.A07.WRKY70*. Whether or not *Bna.A07.WRKY70* plays a role in controlling the size of leaves requires further investigation. Overall, based on the above, the multiple functions and regulation network of *Bna.A07.WRKY70* still has great research potential.

4. Materials and Methods

4.1. Plant Materials and Growth Conditions

The *A. thaliana* wild-type ecotype Columbia (Col-0), the T-DNA mutant *wrky70* (SALK_025198) in the Col-0 background obtained from Arashare (https://www.arashare.cn/index/, accessed on 20 October 2020), and *Brassica napus* L. cultivar "Zhongshuang 11 (ZS11)", were used in this study. The *A. thaliana* plants were grown in a growth chamber at 22 °C under a long day duration (LD, 16 h light/8 h dark) with moderate light intensity (160 μmol m^{-2} s^{-1}). The *B. napus* cultivar "ZS11" was first grown in the greenhouse at 22 °C with a long day duration for six weeks. For vernalization, the plants were transferred to a cold chamber at 4 °C under LD conditions. After vernalization, the plants were returned to the initial greenhouse conditions for 10 weeks until harvest.

4.2. Protein Sequence and Phylogenetic Analysis

The protein sequences of WRKY70 were obtained from the National Center for Biotechnology Information (NCBI) database (https://www.ncbi.nlm.nih.gov/, accessed on 9 September 2022) and the *B. napus* pan-genome information resource (BnPIR) database (http://cbi.hzau.edu.cn/bnapus/index.php, accessed on 9 September 2022). Protein sequence alignment was carried out using MUSCLE (http://www.ebi.ac.uk/Tools/msa/muscle/, accessed on 12 September 2022). The conserved WRKY domain of Bna.A07.WRKY70 was indicated using the conserved domain search program in the National Center for Biotechnology Information (http://www.ncbi.nlm.nih.gov/Structure/cdd/wrpsb.cgi, accessed on 12 September 2022). The phylogenetic tree was conducted using the neighbor-joining tree (Jones–Taylor–Thornton model) by MEGA7. Bootstrap analysis with 1000 replicates was performed to assess the statistical reliability of the tree topology.

4.3. Gene Cloning and Plasmid Construction

The full-length coding domain sequence (CDS) of *Bna.A07.WRKY70* (XP_ 013648025.1) without stop codon was amplified by the specific primer designed in NCBI (https://www.ncbi.nlm.nih.gov/tools/primerblast/, accessed on 20 October 2020). The total RNA was extracted from young leaves of the *B. napus* cultivar "ZS11" by the SteadyPure Plant RNA Extraction Kit (Accurate Biology, Changsha, China), and the RNA concentration was determined by spectrometry (Nano Drop; Thermo Scientific, Wilmington, MA, USA) (Supplementary Table S3) and quality was checked by 1% agarose gel electrophoresis. For cloning, first-stand cDNA was synthesized from total RNA using EasyScript One-Step gDNA Removal and cDNA Synthesis SuperMix (TransGen, Beijing, China). The CDS of *Bna.A07.WRKY70* was isolated from cDNA through PCR (Thermal Cycler Block, Thermo Fisher Scientific) using the high-fidelity thermostable DNA polymerase KOD-Plus-Neo (Toyobo Co., Ltd., Osaka, Japan). The PCR conditions were as follows: pre-denaturation at 94 °C for 2 min, followed by 35 cycles of 98 °C for 10 s, 55 °C for 30 s, and 68 °C for 1 min, and final extension at 68 °C for 7 min. Cloning primers are listed in Supplementary Table S1.

To construct the plasmid *35S:Bna.A07.WRKY70-GFP*, the CDS of *Bna.A07.WRKY70* without stop code was digested with the restriction endonucleases *XbaI* and *BamHI* and cloned into P1300-35S-green fluorescent protein (GFP) vector, which was driven by the CaMV35S (35S) promoter. Similarly, the digested PCR fragment of *Bna.A07.WRKY70* was also cloned into pGreen-35S-eGFP to produce a fusion of *GFP-Bna.A07.WRKY70* under the control of the 35S promoter. To obtain the construct of *pBna.A07.WRKY70:GUS*, the 2600 bp 5′ regulatory region upstream of the ATG start codon, as the *Bna.A07.WRKY70* promoter region was amplified and cloned into pHY107 [43]. The CDS of *Bna.A07.WRKY70* was cloned into the pGBKT7 vector containing the GAL4 DNA binding domain to form a construct of *pBD-Bna.A07.WRKY70*. Eight single colonies of each plasmid were selected randomly and sequenced by Sangon Biotechnology (Shanghai, China). Primers used for plasmid construction are listed in Supplementary Table S1.

4.4. Subcellular Localization of Bna.A07.WRKY70-GFP Protein

The *35S:GFP-Bna.A07.WRKY70* construct was transformed into *Agrobacterium tumefaciens* strain GV3101 and transiently expressed in the leaves of transgenic tobacco (*Nicotiana benthamiana*) carrying a nuclear localization signal as previously described [44]. Images of fluorescent signals were detected through a confocal laser scanning microscope (Leica TCS SP8 SR, Wetzlar, Germany) 72 h after agroinfiltration of the tobacco plants.

4.5. Transcriptional Activation Assays in Yeast

The construct of *pBD-Bna.A07.WRKY70* and the negative control pGBKT7 vector were transformed separately into the yeast strain Y2HGold, including the *HIS3* and *ADE2* reporter genes. The transformed strains were cultured on synthetic dropout nutrient medium without tryptophan (SD/-Trp) plates and then were spotted on SDO (SD/-Trp) and TDO (SD/-Trp/-His/-Ade) plates by diluting to different concentrations. The transcription activation activity of each construct was observed according to the growth conditions of the corresponding yeast cells after incubating for 2–3 days in a 30 °C incubator.

4.6. Generation of A. thaliana Transgenic Plants

The construct of *pBna.A07.WRKY70:GUS* and *35S:Bna.A07.WRKY70-GFP* was transformed into *Agrobacterium tumefaciens* strain GV3101, which was subsequently used to transform the *A. thaliana* wild type and *wrky70* mutant plants using the floral dip method [45]. The transgenic lines of *pBna.A07.WRKY70:GUS* in wild type was selected on soil using Basta® and the transgenic lines of *35S:Bna.A07.WRKY70-GFP* in *wrky70* mutants were screened by hygromycin. All the transgenic plants were genotyped according to DNA and RNA analyses and selfed until T_3 generation homozygous plants, which were generated and used for subsequent experiments.

4.7. RNA Extraction and RT-qPCR Analysis

The total RNA from various tissues of *B. napus* and leaves of *A. thaliana* were extracted using the SteadyPure Plant RNA Extraction Kit (Accurate Biology, Changsha, China). The quality of RNA was assessed using 1% agarose gel electrophoresis, and the concentration was determined by spectrometry (Nano Drop; Thermo Scientific, Wilmington, MA, USA) (Supplementary Table S3). RNA was reverse transcribed by EasyScript One-Step gDNA Removal and cDNA Synthesis SuperMix (TransGen, Beijing, China) according to the manufacturer's instructions, and conditions were 37 °C for 15 min; 85 °C for 5 s, followed by maintaining at 4 °C. Quantitative real-time PCR (RT-qPCR) was utilized to evaluate gene expression with SYBR Green Master Mix (Cofitt, Hongkong, China) using the QuantStudio™ 7 Flex Real-Time PCR System (Thermo Scientific), which were performed by three independent biological replicates with two technical replicates for each biological replicate. Reactions were performed in a total volume of 20 μL containing 100 nM of each primer and 2 μL of diluted cDNA (50 ng/μL) templates and amplified using the following cycling conditions: 95 °C for 2 min, 40 cycles of 95 °C for 15 s, 60 °C for 30 s, and 72 °C for 30 s. *AtACTIN7* (amplified product with 161 bp) and *BnACTIN7* (amplified product with 400 bp) were used as the internal control in *Arabidopsis* and rapeseed, respectively. For each reaction run, the specificity of the amplification was validated, and the threshold cycle (Ct) above the background was calculated using Bio-Rad iCycler (Bio-Rad, Hercules, CA, USA). The relative expression levels of the target genes were calculated using a modified double delta method [46]. Primers used for RT-qPCR analyses are listed in Supplementary Table S2.

4.8. Phenotypic Observation of A. thaliana Leaves

The seeds of the Col-0, the *wrky70* mutant, and three independent lines—*wrky70 35S:Bna.A07.WRKY70-GFP #4, #6, and #12*—were germinated on 1/2 MS agar medium for one week. Subsequently, the seedlings were transplanted into 8 × 8 cm pots. When the *A. thaliana* plants grew 35 days after germination (DAG), the phenotype of leaf senescence was observed and photographed by a camera (D7500, Nikon, Tokyo, Japan).

4.9. Measurement of the Chlorophyll Content

The fifth and sixth rosette leaf samples of Col-0, *wrky70* mutants, and *Bna.A07.WRKY70* transgenic plants from the fourth, fifth, sixth, and seventh weeks were separately collected and weighed and then placed in a 1.5 mL centrifuge tube with 1 mL extraction solution (80% acetone), soaked the leaves in the dark for 24 h until they faded [47]. To calculate the chlorophyll content of leaves, the 0.2 mL supernatant was absorbed into Costar 96 Flat Transparent plate, and the absorbance values at 663 nm and 645 nm were measured using a microplate reader (Infinite M200pro, Tecan, Mannedorf, Switzerland). Each experiment was represented by three biological replicates.

5. Conclusions

As an indispensable portion, leaf senescence spans the latter half of leaf development, which is essential to guarantee better production and survival of the next generation. This study suggested that *Bna.A07.WRKY70* may act as a negative regulator to share a conserved function with *AtWRKY70* in controlling leaf senescence when it is expressed in *A. thaliana*. Thus, *Bna.A07.WRKY70* can be utilized as a potential target to genetically manipulate leaf senescence and to create new stay-green materials to improve the rapeseed yield.

Supplementary Materials: The following supporting information can be downloaded at: https://www.mdpi.com/article/10.3390/plants12020347/s1, Figure S1: Percent identity of full-length protein sequences of the WRKY70 protein from *A. thaliana* and *B. napus*; Table S1: Primers used for gene cloning and various constructs in the present study; Table S2: Primers used for RT-qPCR analysis in the present study; Table S3: The amount of RNA per sample in the *Arabidopsis thaliana* and *Brassica napus*.

Author Contributions: Y.G. and M.C. conceived and designed the experiments. T.L. conducted the experiments and analyzed the data. Y.L., C.W., D.Z., J.L. and M.H. conducted parts of the experiments. T.L. wrote the draft of the manuscript, and M.C. and Y.G. revised the manuscript. All authors have read and agreed to the published version of the manuscript.

Funding: This research was financially supported by a grant from the Yang Ling Seed Industry Innovation Center (Grant no. K3031122024) and the National Natural Science Foundation of China (Grant no. 31801393).

Institutional Review Board Statement: Not applicable.

Informed Consent Statement: Not applicable.

Data Availability Statement: All data included in this study are available upon reasonable request by contact with the corresponding author.

Conflicts of Interest: The authors declare no conflict of interest.

References

1. Chalhoub, B.; Denoeud, F.; Liu, S.; Parkin, I.A.; Tang, H.; Wang, X.; Chiquet, J.; Belcram, H.; Tong, C.; Samans, B.; et al. Early allopolyploid evolution in the post-Neolithic *Brassica napus* oilseed genome. *Science* **2014**, *345*, 950–953. [CrossRef] [PubMed]
2. Sashidhar, N.; Harloff, H.J.; Jung, C. Identification of phytic acid mutants in oilseed rape (*Brassica napus*) by large-scale screening of mutant populations through amplicon sequencing. *New Phytol.* **2020**, *225*, 2022–2034. [CrossRef]
3. Ülker, B.; Shahid Mukhtar, M.; Somssich, I.E. The WRKY70 transcription factor of *Arabidopsis* influences both the plant senescence and defense signaling pathways. *Planta* **2007**, *226*, 125–137. [CrossRef] [PubMed]
4. Chen, M.X.; Maodzeka, A.; Zhou, L.H.; Ali, E.; Wang, Z.; Jiang, L.X. Removal of DELLA repression promotes leaf senescence in *Arabidopsis*. *Plant Sci.* **2014**, *219–220*, 26–34. [CrossRef]
5. Lim, P.O.; Kim, H.J.; Nam, H.G. Leaf Senescence. *Annu. Rev. Plant Biol.* **2007**, *58*, 115–136. [CrossRef]
6. Kim, C.; Kim, S.J.; Jeong, J.; Park, E.; Oh, E.; Park, Y.I.; Lim, P.O.; Choi, G. High ambient temperature accelerates leaf senescence via PHYTOCHROME-INTERACTING FACTOR 4 and 5 in *Arabidopsis*. *Mol. Cells* **2020**, *43*, 645–661. [PubMed]
7. Lee, S.; Seo, P.J.; Lee, H.J.; Park, C.M. A NAC transcription factor NTL_4 promotes reactive oxygen species production during drought-induced leaf senescence in *Arabidopsis*. *Plant J.* **2012**, *70*, 831–844. [CrossRef] [PubMed]

8. Yang, L.; Ye, C.F.; Zhao, Y.T.; Cheng, X.L.; Wang, Y.Q.; Jiang, Y.Q.; Yang, B. An oilseed rape WRKY-type transcription factor regulates ROS accumulation and leaf senescence in *Nicotiana benthamiana* and Arabidopsis through modulating transcription of *RbohD* and *RbohF*. *Planta* **2018**, *247*, 1323–1338. [CrossRef]
9. Schippers, J.H. Transcriptional networks in leaf senescence. *Curr. Opin. Plant Biol.* **2015**, *27*, 77–83. [CrossRef]
10. Kim, J.; Kim, J.H.; Lyu, J.I.; Woo, H.R.; Lim, P.O. New insights into the regulation of leaf senescence in *Arabidopsis*. *J. Exp. Bot.* **2018**, *69*, 787–799. [CrossRef]
11. Zhang, Y.M.; Guo, P.; Xia, X.; Guo, H.; Li, Z. Multiple layers of regulation on leaf senescence: New advances and perspectives. *Front. Plant Sci.* **2021**, *12*, 788996. [CrossRef]
12. Liu, W.; Liang, X.; Cai, W.; Wang, H.; Liu, X.; Cheng, L.; Song, P.; Luo, G.; Han, D. Isolation and functional analysis of *VvWRKY28*, a *Vitis Vinifera* WRKY transcription factor gene, with functions in tolerance to cold and salt stress in transgenic *Arabidopsis thaliana*. *Int. J. Mol. Sci.* **2022**, *23*, 13418. [CrossRef] [PubMed]
13. Eulgem, T.; Rushton, P.J.; Robatzek, S.; Somssich, I.E. The WRKY superfamily of plant transcription factors. *Trends Plant Sci.* **2000**, *5*, 199–206. [CrossRef] [PubMed]
14. Rushton, P.J.; Somssich, I.E.; Ringler, P.; Shen, Q.J. WRKY transcription factors. *Trends Plant Sci.* **2010**, *15*, 247–258. [CrossRef]
15. Fan, Z.Q.; Tan, X.L.; Shan, W.; Kuang, J.F.; Lu, W.J.; Chen, J.Y. BrWRKY65, a WRKY transcription factor, is involved in regulating three leaf senescence-associated genes in Chinese flowering cabbage. *Int. J. Mol. Sci.* **2017**, *18*, 1228. [CrossRef]
16. Devaiah, B.N.; Karthikeyan, A.S.; Raghothama, K.G. WRKY75 transcription factor is a modulator of phosphate acquisition and root development in Arabidopsis. *Plant Physiol.* **2007**, *143*, 1789–1801. [CrossRef]
17. Zhang, C.Q.; Xu, Y.; Lu, Y.; Yu, H.X.; Gu, M.H.; Liu, Q.Q. The WRKY transcription factor OsWRKY78 regulates stem elongation and seed development in rice. *Planta* **2011**, *234*, 541–554. [CrossRef]
18. Pandey, S.P.; Somssich, I.E. The role of WRKY transcription factors in plant immunity. *Plant Physiol.* **2009**, *150*, 1648–1655. [CrossRef]
19. Pan, L.J.; Jiang, L. Identification and expression of the WRKY transcription factors of *Carica papaya* in response to abiotic and biotic stresses. *Mol. Biol. Rep.* **2014**, *41*, 1215–1225. [CrossRef]
20. Miao, Y.; Laun, T.; Zimmermann, P.; Zentgraf, U. Targets of the WRKY53 transcription factor and its role during leaf senescence in *Arabidopsis*. *Plant Mol. Biol.* **2004**, *55*, 853–867. [CrossRef]
21. Chen, L.G.; Xiang, S.Y.; Chen, Y.L.; Li, D.B.; Yu, D.Q. *Arabidopsis* WRKY45 interacts with the DELLA protein RGL1 to positively regulate age-triggered leaf senescence. *Mol. Plant* **2017**, *10*, 1174–1189. [CrossRef] [PubMed]
22. Yu, Y.C.; Qi, Y.A.; Xu, J.P.; Dai, X.H.; Chen, J.C.; Dong, C.H.; Xiang, F.N. Arabidopsis WRKY71 regulates ethylene-mediated leaf senescence by directly activating *EIN2*, *ORE1* and *ACS2* genes. *Plant J.* **2021**, *107*, 1819–1836. [CrossRef] [PubMed]
23. Hu, Y.R.; Dong, Q.Y.; Yu, D.Q. Arabidopsis WRKY46 coordinates with WRKY70 and WRKY53 in basal resistance against pathogen *Pseudomonas syringae*. *Plant Sci.* **2012**, *185–186*, 288–297. [CrossRef]
24. Li, J.; Brader, G.; Kariola, T.; Tapio Palva, E. WRKY70 modulates the selection of signaling pathways in plant defense. *Plant J.* **2006**, *46*, 477–491. [CrossRef]
25. Chen, J.N.; Nolan, T.M.; Ye, H.X.; Zhang, M.C.; Tong, H.N.; Xin, P.Y.; Chu, J.F.; Chu, C.C.; Li, Z.H.; Yin, Y.H. Arabidopsis WRKY46, WRKY54 and WRKY70 transcription factors are involved in brassinosteroid-regulated plant growth and drought response. *Plant Cell* **2017**, *29*, 1425–1439. [CrossRef] [PubMed]
26. Sen, S.; Chakraborty, J.; Ghosh, P.; Basu, D.; Das, S. Chickpea WRKY70 regulates the expression of a homeodomain-leucine zipper (HD-Zip) I transcription factor *CaHDZ12*, which confers abiotic stress tolerance in transgenic tobacco and chickpea. *Plant Cell Physiol.* **2017**, *58*, 1934–1952. [CrossRef]
27. Xiong, X.P.; Sun, S.C.; Zhang, X.Y.; Li, Y.J.; Liu, F.; Zhu, Q.H.; Xue, F.; Sun, J. GhWRKY70D13 regulates resistance to *Verticillium dahliae* in cotton through the ethylene and jasmonic acid signaling pathways. *Front. Plant Sci.* **2020**, *11*, 69. [CrossRef]
28. Jia, Z.Z.; Li, M.Z.; Wang, H.C.; Zhu, B.; Gu, L.; Du, X.Y.; Ren, M.J. *TaWRKY70* positively regulates *TaCAT5* enhanced Cd tolerance in transgenic *Arabidopsis*. *Environ. Exp. Bot.* **2021**, *190*, 104591. [CrossRef]
29. Sun, Q.F.; Lin, L.; Liu, D.X.; Wu, D.W.; Fang, Y.J.; Wu, J.; Wang, Y.P. CRISPR/Cas9-mediated multiplex genome editing of the *BnWRKY11* and *BnWRKY70* genes in *Brassica napus* L. *Int. J. Mol. Sci.* **2018**, *19*, 2716. [CrossRef]
30. Woo, H.R.; Kim, H.J.; Lim, P.O.; Nam, H.G. Leaf senescence: Systems and dynamics aspects. *Annu. Rev. Plant Biol.* **2019**, *70*, 347–376. [CrossRef]
31. Kim, H.J.; Nam, H.G.; Lim, P.O. Regulatory network of NAC transcription factors in leaf senescence. *Curr. Opin. Plant Biol.* **2016**, *33*, 48–56. [CrossRef]
32. Besseau, S.; Li, J.; Palva, E.T. WRKY54 and WRKY70 co-operate as negative regulators of leaf senescence in *Arabidopsis thaliana*. *J. Exp. Bot.* **2012**, *63*, 2667–2679. [CrossRef]
33. Osborn, T.C.; Kole, C.; Parkin, I.A.P.; Sharpe, A.G.; Kuiper, M.; Lydiatet, D.J.; Trick, M. Comparison of flowering time genes in *Brassica rapa*, *B. napus* and *Arabidopsis thaliana*. *Genetics* **1997**, *146*, 1123–1129. [CrossRef] [PubMed]
34. Cavell, A.C.; Lydiate, D.J.; Parkin, I.A.P.; Dean, C.; Trick, M. Collinearity between a 30-centimorgan segment of *Arabidopsis thaliana* chromosome 4 and duplicated regions within the *Brassica napus* genome. *Genome* **1998**, *41*, 62–69. [CrossRef]
35. Scheffler, J.A.; Sharpe, A.G.; Schmidt, H.; Sperling, P.; Parkin, I.A.P.; Lühs, W.; Lydiate, D.J.; Heinz, E. Desaturase multigene families of *Brassica napus* arose through genome duplication. *Theor. Appl. Genet.* **1997**, *94*, 583–591. [CrossRef]

36. Yoshida, S.; Ito, M.; Nishida, I.; Watanabe, A. Isolation and RNA gel blot analysis of genes that could serve as potential molecular markers for leaf senescence in *Arabidopsis thaliana*. *Plant Cell Physiol.* **2001**, *42*, 170–178. [CrossRef] [PubMed]
37. Breeze, E.; Harrison, E.; Mchattie, S.; Hughes, L.; Hickman, R.; Hill, C.; Kiddle, S.; Kim, Y.S.; Penfold, C.A.; Jenkins, D.; et al. High-resolution temporal profiling of transcripts during *Arabidopsis* leaf senescence reveals a distinct chronology of processes and regulation. *Plant Cell* **2011**, *23*, 873–894. [CrossRef]
38. Lee, H.N.; Lee, K.H.; Kim, C.S. Abscisic acid receptor PYRABACTIN RESISTANCE-LIKE 8, *PYL8*, is involved in glucose response and dark-induced leaf senescence in *Arabidopsis*. *Biochem. Biophys. Res. Commun.* **2015**, *463*, 24–28. [CrossRef]
39. Penfold, C.A.; Buchanan-Wollaston, V. Modelling transcriptional networks in leaf senescence. *J. Exp. Bot.* **2014**, *65*, 3859–3873. [CrossRef]
40. Gu, L.J.; Ma, Q.; Zhang, C.; Wang, C.C.; Wei, H.L.; Wang, H.T.; Yu, S.X. The cotton GhWRKY91 transcription factor mediates leaf senescence and responses to drought stress in transgenic *Arabidopsis thaliana*. *Front. Plant Sci.* **2019**, *10*, 1352. [CrossRef]
41. Xie, W.Y.; Li, X.R.; Wang, S.P.; Yuan, M. *OsWRKY53* promotes abscisic acid accumulation to accelerate leaf senescence and inhibit seed germination by downregulating abscisic acid catabolic genes in rice. *Front. Plant Sci.* **2022**, *12*, 816156. [CrossRef] [PubMed]
42. Yan, H.R.; Jia, H.H.; Chen, X.B.; Hao, L.L.; An, H.L.; Guo, X.Q. The cotton WRKY transcription factor GhWRKY17 functions in drought and salt stress in transgenic *Nicotiana benthamiana* through ABA signaling and the modulation of reactive oxygen species production. *Plant Cell Physiol.* **2014**, *55*, 2060–2076. [CrossRef]
43. Liu, C.; Zhou, J.; Bracha-Drori, K.; Yalovsky, S.; Ito, T.; Yu, H. Specification of *Arabidopsis* floral meristem identity by repression of flowering time genes. *Development* **2007**, *134*, 1901–1910. [CrossRef] [PubMed]
44. Yang, Q.S.; Yang, B.; Li, J.Z.; Wang, Y.; Tao, R.Y.; Yang, F.; Wu, X.Y.; Yan, X.H.; Ahmad, M.; Shen, J.Q.; et al. ABA-responsive ABRE-BINDING FACTOR3 activates *DAM3* expression to promote bud dormancy in Asian pear. *Plant Cell Environ.* **2020**, *43*, 1360–1375. [CrossRef] [PubMed]
45. Clough, S.J.; Bent, A.F. Floral dip: A simplified method for *Agrobacterium*-mediated transformation of *Arabidopsis thaliana*. *Plant J.* **1998**, *16*, 735–743. [CrossRef]
46. Pfaffl, M.W.; Horgan, G.W.; Dempfle, L. Relative expression software tool (REST©) for group-wise comparison and statistical analysis of relative expression results in real-time PCR. *Nucleic Acids Res.* **2002**, *30*, e36. [CrossRef] [PubMed]
47. Yang, Z.L.; Ohlrogge, J.B. Turnover of fatty acids during natural senescence of Arabidopsis, *Brachypodium*, and Switchgrass and in Arabidopsis β-Oxidation mutants. *Plant Physiol.* **2009**, *150*, 1981–1989. [CrossRef]

plants MDPI

Article

Genome-Wide Characterization of the Sulfate Transporter Gene Family in Oilseed Crops: *Camelina sativa* and *Brassica napus*

Parviz Heidari [1,*], Soosan Hasanzadeh [1], Sahar Faraji [2], Sezai Ercisli [3] and Freddy Mora-Poblete [4,*]

1 Faculty of Agriculture, Shahrood University of Technology, Shahrood 3619995161, Iran
2 Department of Plant Breeding, Faculty of Crop Sciences, Sari Agricultural Sciences and Natural Resources University (SANRU), Sari 4818168984, Iran
3 Department of Horticulture, Faculty of Agriculture, Ataturk University, 25240 Erzurum, Turkey
4 Institute of Biological Sciences, University of Talca, Talca 3460000, Chile
* Correspondence: heidarip@shahroodut.ac.ir (P.H.); morapoblete@gmail.com (F.M.-P.)

Abstract: Sulfate transporters (SULTRs) are responsible for the uptake of sulfate (SO_4^{2-}) ions in the rhizosphere by roots and their distribution to plant organs. In this study, SULTR family members in the genomes of two oilseed crops (*Camelina sativa* and *Brassica napus*) were identified and characterized based on their sequence structures, duplication events, phylogenetic relationships, phosphorylation sites, and expression levels. In total, 36 and 45 putative *SULTR* genes were recognized in the genomes of *C. sativa* and *B. napus*, respectively. SULTR proteins were predicted to be basophilic proteins with low hydrophilicity in both studied species. According to the observed phylogenetic relationships, we divided the SULTRs into five groups, out of which the SULTR 3 group showed the highest variation. Additionally, several duplication events were observed between the *SULTRs*. The first duplication event occurred approximately five million years ago between three *SULTR 3.1* genes in *C. sativa*. Furthermore, two subunits were identified in the 3D structures of the SULTRs, which demonstrated that the active binding sites differed between *C. sativa* and *B. napus*. According to the available RNA-seq data, the *SULTRs* showed diverse expression levels in tissues and diverse responses to stimuli. *SULTR 3* was expressed in all tissues. *SULTR 3.1* was more upregulated in response to abiotic stresses in *C. sativa*, while *SULTR 3.3* and *SULTR 2.1* were upregulated in *B. napus*. Furthermore, *SULTR 3* and *SULTR 4.1* were upregulated in response to biotic stresses in *B. napus*. Additionally, the qPCR data showed that the *SULTRs* in *C. sativa* were involved in the plant's response to salinity. Based on the distribution of cis-regulatory elements in the promoter region, we speculated that *SULTRs* might be controlled by phytohormones, such as ABA and MeJA. Therefore, it seems likely that *SULTR* genes in *C. sativa* have been more heavily influenced by evolutionary processes and have acquired further diversity. The results reveal new insights of the structures and functions of SULTRs in oilseed crops. However, further analyses, related to functional studies, are needed to uncover the role of SULTRs in the plants' development and growth processes, as well as in their response to stimuli.

Keywords: bioinformatics; biotic stresses; regulatory mechanisms; protein structure; gene expression; evolutionary analysis

Citation: Heidari, P.; Hasanzadeh, S.; Faraji, S.; Ercisli, S.; Mora-Poblete, F. Genome-Wide Characterization of the Sulfate Transporter Gene Family in Oilseed Crops: *Camelina sativa* and *Brassica napus*. *Plants* **2023**, *12*, 628. https://doi.org/10.3390/plants12030628

Academic Editors: Mingxun Chen, Lixi Jiang and Yuan Guo

Received: 3 January 2023
Revised: 23 January 2023
Accepted: 28 January 2023
Published: 31 January 2023

1. Introduction

Sulfur (S) is a macronutrient that is required for the biosynthesis of amino acids (such as cysteine (Cys) and methionine (Met)), vitamins, cofactors, and glutathione (GSH), as well as secondary metabolites; therefore, (S) is a vital element for plant growth, development, and stress response [1–3]. Root cells take up sulfate (SO_4^{2-}) in the form of S through a proton codependent process. The uptake and assimilation of sulfate resources that are available in the environment produce essential sulfur (S) metabolites that are crucial for development and stress responses, which is critical for plants and microbes [4]. The soil sulfate content can be modified by various factors, such as the dissimilation of soil microbes, the weathering of S-containing minerals, human activities that modify the deposition of S into the ecosystem, and climate change [1]. Therefore, the available sulfate content in soil can also be altered because of the ability of plant root systems to absorb nutrient compounds according to their requirements and material accessibility. It has been reported that in comparison to other micronutrients, sulfate only has a gentle and limited effect on root structures [5]. To meet demands of S required for the S-containing metabolite synthesis, plant membrane transport systems and their related metabolic enzymes optimize sulfate uptake, acquisition, storage, and use [6]. The uptake and distribution of sulfate in plants are facilitated by networks of sulfate transporters (SULTRs), which are encoded by a multigene family [7]. The H^+/SO_4^{2-} co-transporter SULTRs have been reported to contain 12 transmembrane domains, along with a carboxyl-terminal region, i.e., the so-called STAS (sulfate transporter/anti-sigma factor), which is suggested to play an important role in transporters' activity and their interactions with other proteins [1,8].

The involvement of SULTRs in the transportation of S within plants was first reported by Smith et al. [9]. SULTRs are characterized by 12 transmembrane domains (TMDs) and an anti-sigma factor antagonist (STAS) domain at the C-terminus, which is critical for sulfate transporter activity [10]. The genomes of higher plants, such as *Arabidopsis thaliana*, rice, wheat, sorghum, and apple, have been reported to have 12, 12, 11, 10, and 9 *SULTR* genes, respectively [11–14]. The SULTR family has been well characterized in *Arabidopsis*, and sulfate transporters can be divided into four main groups based on their sequence resemblance, function, and location. The first group includes AtSULTR 1.1, AtSULTR 1.2, and AtSULTR 1.3, which are all high-affinity S transporters [15]. AtSULTRs 1.1 and 1.2 are co-localized in root hairs and epidermal and cortical cells in roots, and they are both responsible for the uptake of sulfate from soil [16,17]. Nevertheless, despite their common function, AtSULTR 1.1 predominantly operates under the conditions of S deficiency, while AtSULTR 1.2 operates efficiently under the conditions of either sulfur abundance or sulfur deficiency [18]. AtSULTR 1.3 is localized in the phloem, and cooperates in the source-sink distribution of sulfate [19]. The second group consists of two low-affinity transporters, AtSULTR 2.1 and AtSULTR 2.2, which are responsible for the transportation of sulfate from root to shoot [20]. The third group comprises five members (AtSULTR 3.1-5) and is the largest group. However, the precise functions of these members have not been fully established. It has been reported that SULTR 3.1, which transports sulfate to chloroplasts, could have a role in helping plants to withstand abiotic stresses [21]. Additionally, SULTR 3.5 has been reported to co-express with SULTR 2.1 to enhance the uptake of sulfate and facilitate its transportation from root to shoot under conditions of S deficiency [22,23]. The fourth group of transporters, SULTR 4.1 and SULTR 4.2, have been demonstrated to be tonoplast localized transporters that release sulfate from vacuoles into the cytosol [24,25]. As well as the study on *A. thaliana*, many other studies have been conducted to functionally characterize SULTRs in crops. For instance, 14 putative *SULTR* genes have been identified in rapeseed (*Brassica napus*), among which only those from group 1 and group 4 were induced in response to S deficiency [26]. In another study, 28 putative *SULTR* genes were identified in the soybean (*Glycine max*) genome and *GmSULTR 1.2b* was confirmed to have important roles in sulfate uptake and improving plants' tolerance to sulfur deficiency [27]. In the potato (*Solanum tuberosum*) genome, 12 *SULTR* genes have been identified and the members of group 3 (StSULTR3s) were potentially proven to be involved in biotic/abiotic stress responses through MYB TFs, which play crucial roles in the modulation of StSULTR3s under these circumstances [28]. The maize (*Zea mays* L.) genome has been shown to include eight putative *SULTR* genes, all of which were induced by drought and heat stresses, except for *ZmSULTR 3.3* [29]. In addition, various studies have confirmed that *SULTRs* can be responsive to heavy metal exposure [30,31]. Despite the progress that has been made in the functional characterization of plant *SULTRs*, there are still more important crops that need to be investigated. *Camelina sativa* is an oilseed crop from the Brassicaceae family that has many qualities, including low inputs, great adaptation and resistance abilities, short life cycles, and easy genetic transformation, which have turned *C. sativa* into an ideal model plant [32,33]. Moreover, *C. sativa* is becoming more important as a biofuel [34,35]. Although oilseed plants typically have very high S demands [36], a study on the response of *C. sativa* to various fertilizers showed that the seed yields and oil contents of *camelina* seeds were not affected by sulfur fertilization [37]. In order to develop S-efficient crops and crop varieties that are tolerant to S deficiency, it is necessary to identify and characterize SULTRs, especially in low-input crops, such as *C. sativa*. To the best of our knowledge, there are no available reports on the genome-wide analysis of *SULTR* genes in *C. sativa*, except for one study that reported the upregulation of *SULTR 3.4* in *C. sativa* under salinity stress [38]. In this study, resources were employed to distinguish the regulation roles of *SULTR* genes in various cellular processes, especially in response to stimuli. *B. napus* is another well-known oilseed plant containing appreciable amounts of erucic acid. In the present study, we focused on SULTR sequences in the *C. sativa* and *B. napus* genomes, and compared and discussed their adjustments and their possible engagement in protection mechanisms against unfavorable environmental stimuli. We also highlighted the potential properties of these genes that could help to facilitate sulfate uptake.

2. Results

2.1. SULTR Properties in Camelina sativa and Brassica napus

In the current study, 36 and 45 putative SULTR genes were recognized in the genomes of *C. sativa* and *B. napus*, respectively (Table S1). The SULTRs of the two oilseed crops were characterized and compared according to their coding DNA sequences (CDS) and protein lengths, exon numbers, isoelectric points (pIs), molecular weights (MWs), grand average of hydropathy (GRAVY) values, and instability indices (Table S1 and Table 1). Our results showed that the physicochemical properties of the SULTR proteins in the two studied plants were almost identical to each other. For instance, the MWs ranged from 29.07 to 91.99 kDa in *C. sativa*, and from 28.94 to 83.86 kDa in *B. napus*. Additionally, the pI values ranged from 7.41 to 9.93 in C. sativa, and from 7.11 to 10.71 in B. napus. Moreover, the GRAVY values varied from 0.271 to 0.624 in C. sativa, and from 0.108 to 0.621 in B. napus. Based on the instability indices, 83% and 73% of SULTR proteins were predicted to be stable proteins in *C. sativa* and *B. napus*, respectively. In addition, the exon numbers varied from 4 to 20 in *C. sativa* and from 4 to 19 in B. napus (Figure 1 and Table 1). Overall, the SULTR proteins were predicted to be basophilic proteins with low hydrophilicity.

Table 1. Summary of SULTRs properties in *Camelina sativa* and *Brassica napus*. Full details of SULTRs properties are provided in Table S1.

Attributes	*C. sativa*	*B. napus*
CDS length (bp)	801–3428	878–3428
Protein length (aa)	266–829	264–758
Exon number	4–20	4–19
pI	7.41–9.93	7.11–10.71
MW (KDa)	29.07–91.99	28.94–83.86
GRAVY	0.271–0.624	0.108–0.621
Instability index	83% stable	73% stable

2.2. Phylogenetic Analysis and Classification of the SULTR Gene Family

In the present study, a phylogenetic tree of the SULTR proteins was created, including 45 SULTRs from *B. napus*, 36 SULTRs from *C. sativa*, 28 SULTRs from *Glycine max*, 12 SULTRs from *Oryza sativa*, and 12 SULTRs from *Arabidopsis thaliana* (Figure 1). The studied SULTRs were classified into five main groups: 16 SULTRs from SULTRs 4.1 and 4.2 were categorized into group 1; SULTRs 2.1 and 2.2 were clustered into group 2; 30 SULTRs from SULTRs 1.1, 1.2, and 1.3 were assigned to group 3; 28 proteins from SULTRs 3.3 and 3.4 were included in group 4; 34 SULTRs from SULTRs 3.1, 3.2, and 3.5 were located in group 5 (Figure 1). The SULTRs from monocot model plant (rice) were very different from the dicot samples. Moreover, the SULTRs from *C. sativa* and *B. napus* were evaluated and compared according to the conserved motifs. Overall, 10 conserved motifs were recognized in the protein sequences of the SULTRs, among which motif 6 was not observed in the SULTRs in group 1 (Figure 2). Additionally, 10 conserved motifs were identified in SULTRs 2.1 and 2.2, except the SULTR 2.1 from *C. sativa* only showed eight conserved motifs. Furthermore, SULTRs 1.1, 1.2, and 1.3 and 3.1, 3.2, and 3.5 were very diverse, according to the patterns of their motif distributions (Figure 2). Motifs 7 and 2 were frequently observed in the SULTR proteins and showed potential as screening markers for members of this family.

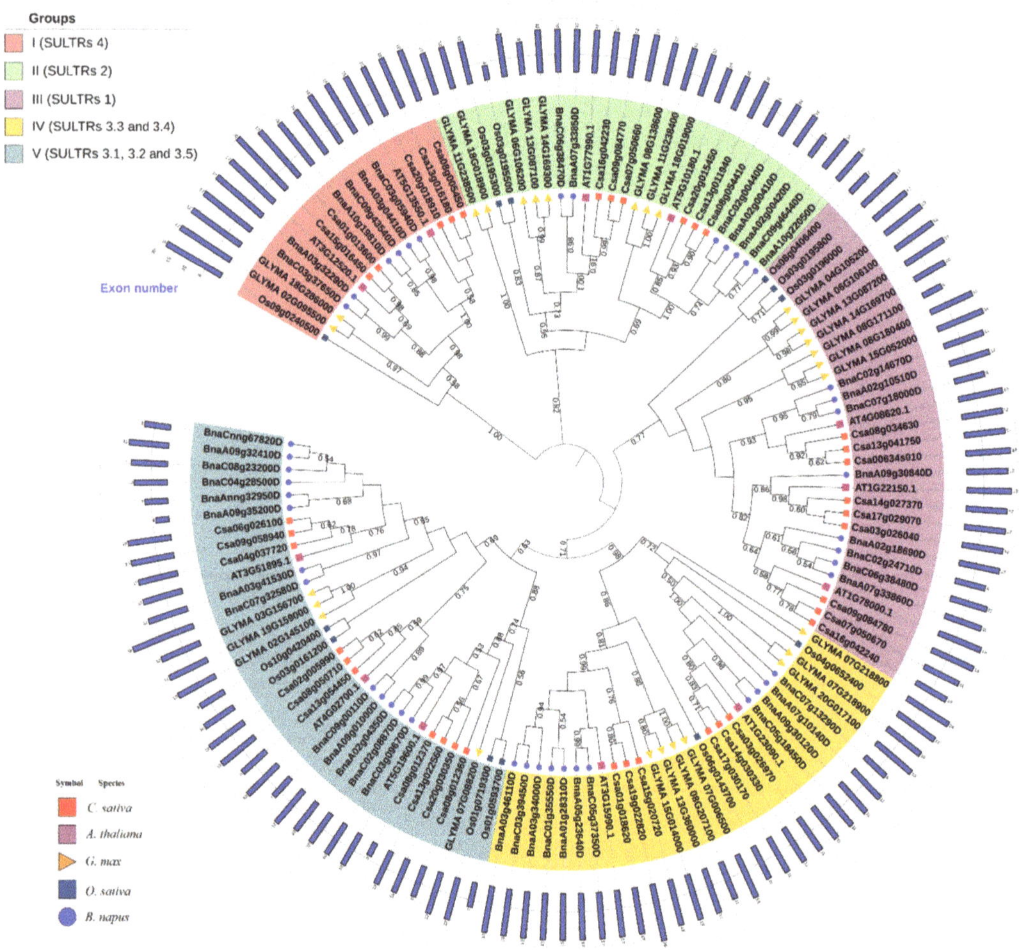

Figure 1. The phylogenetic tree of the SULTRs from *Camelina sativa, Brassica napus, Arabidopsis thaliana, Glycine max*, and *Oryza sativa*. The exon numbers for the SULTR coding genes are shown in the blue bar (more details related to the gene structures are provided in Table S1).

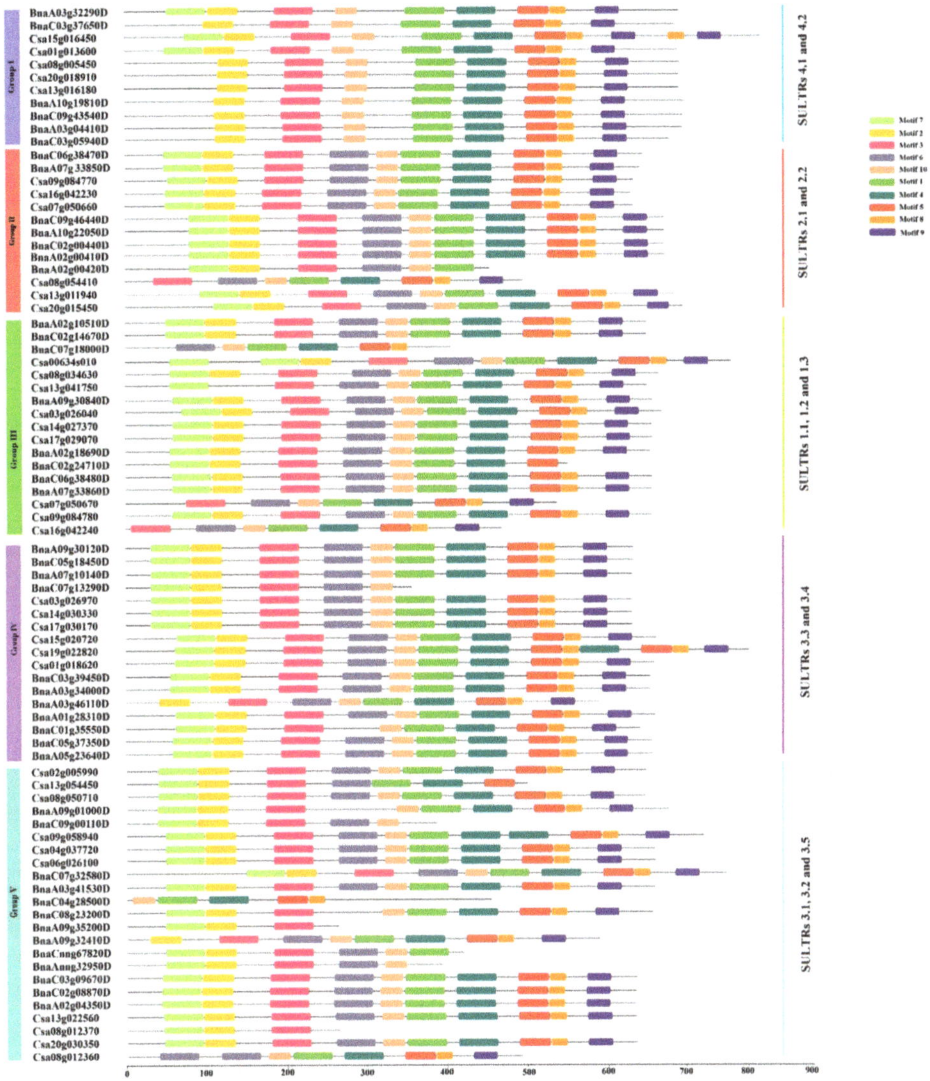

Figure 2. The distributions of the conserved motifs in the SULTRs from *Camelina sativa* and *Brassica napus*. The grouping was based on the phylogenetic tree. The sequences of the conserved motifs are presented in Figure S1.

2.3. Evolutionary Processes in the MGT Genes of Citrullus lanatus and Cucumis sativus

In this study, to investigate the duplication events that have occurred in the SULTR gene family in *C. sativa* and *B. napus*, the synonymous (Ks), non-synonymous (Ka), and Ka/Ks values of each duplicated gene pair were calculated (Figure 3 and Table S2). The Ks values of the *SULTRs* in *C. sativa* were frequently between 0.6 and 1.0 (Figure 3a), while the Ka/Ks values were frequently between 0.7 and 0.9 (Figure 3b). In contrast, the Ks and Ka/Ks values of the *SULTRs* in the *B. napus* genome differed from those in *C. sativa*, with the Ks values frequently being between 1.2 and 1.6 (Figure 3c) and the Ka/Ks values frequently ranging from 0.3 to 0.5 (Figure 3d). In *C. sativa*, the first duplication event was predicted to have occurred around five million years ago (MYA) between three *SULTR*

3.1 genes, including *Csa06g026100-Csa04g037720* and *Csa09g058940-Csa04g037720*, while the first duplication event in *B. napus* occurred approximately three MYA between two *SULTR 3.1 genes*, *BnaA03g41530* and *BnaA09g35200* (Table S2). Several synteny blocks were observed between the *SULTRs* from *C. sativa* and *B. napus* (Figure S2). Additionally, three *SULTR* 1.3 genes (*Csa17g029070, Csa14g027370*, and *Csa03g026040*), four *SULTR* 3 genes (*Csa13g054450, Csa08g050710, Csa02g005990*, and *Csa08g012360*), and a *SULTR 1.1* gene (*Csa08g034630*) from *C. sativa* showed fewer synteny relationships with *SULTRs* from *B. napus* (Figure 4).

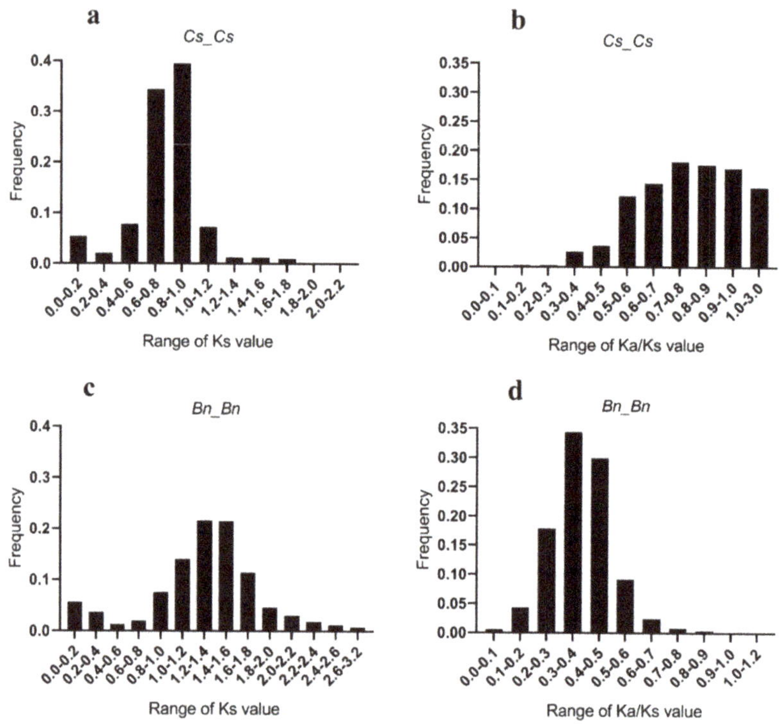

Figure 3. The frequency of Ks and Ka/Ks values in the *SULTRs*: (**a**) the frequency of Ks values in the *SULTRs* of *C. sativa* (Cs); (**b**) the frequency of the Ka/Ks values in the *SULTRs* of *C. sativa* (Cs); (**c**) the frequency of Ks values in the *SULTRs* of *Brassica napus* (Bn); (**d**) the frequency of the Ka/Ks values in the *SULTRs* of *Brassica napus* (Bn). The full details of the duplicated *SULTRs* are provided in Table S2.

2.4. Transmembrane Structures of SULTRs

The SULTR proteins from different groups were compared based on their transmembrane structures *in C. sativa* and *B. napus* (Figure 5). In group 1, 12 transmembrane helices and 11 pores were identified in all SULTRs. However, the SULTRs in *B. napus* showed similar structures based on the positions of the transmembrane helices while the structures in *C. sativa* were diverse. Additionally, the number of transmembrane helices in the group 2 SULTRs ranged from 10 to 12 in *B. napus* and from 8 to 10 in *C. sativa*. Most of the SULTRs in *B. napus* showed 10 transmembrane helices with nine pores (except for BnaC07g18000D with seven transmembrane helices), while the number of transmembrane helices in *C. sativa* varied between 8 and 11. In group 4, the number of transmembrane helices in the SULTRs of *B. napus* ranged from 6 to 11, while the number of transmembrane helices in *C. sativa* ranged from 9 and 13. The SULTRs in group 5 were very diverse in terms of their transmembrane structures, in which between 4 and 14 transmembrane helices were observed.

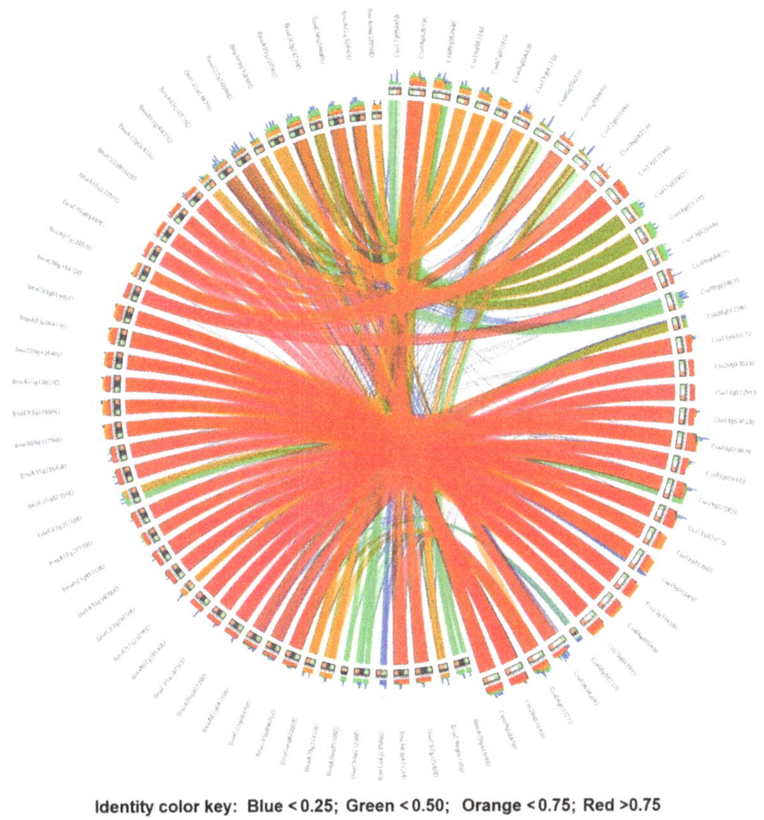

Figure 4. The synteny relationships between the *SULTRs* from *Camelina sativa* and *Brassica napus*.

2.5. 3D Structure Analysis of SULTRs

Our analysis of the 3D structures revealed that the SULTRs in *C. sativa* and *B. napus* had two domains and that the active binding sites could be located in small or large subunits (Figure 6). These results showed that the SULTRs in *C. sativa* were different from those in *B. napus* (Figure 6). In the group 1 SULTRs, the valine (VAL), proline (PRO), phenylalanine (PHE), asparagine (ASN), lysine (LYS), glycine (GLY), and serine (SER) amino acids were frequently observed in the binding sites of SULTRs from *C. sativa*, while PHE, GLY, and leucine (LEU) were frequently observed in the binding sites of SULTRs from *B. napus* (Figure 6). In the group 2 SULTRs, PHE, GLY, and alanine (ALA) were more frequently observed in the binding sites of *C. sativa*, while PHE, SER, and isoleucine (ILE) were frequently observed in the binding sites of *B. napus*. Additionally, six amino acids, including SER, aspartic acid (ASP), LYS, ILE, ALA, and tyrosine (TYR), were more frequently observed in the binding sites of group 3 SULTRs in *C. sativa*, while PHE and threonine (THR) were frequently observed in the binding sites of *B. napus*. In the group 4 SULTRs, SER, GLY, histidine (HIS), and TYR were more commonly identified in the binding sites in *C. sativa*, while LEU, ILE, glutamate (GLU), and arginine (ARG) were frequently observed in the binding sites of *B. napus*. In the group 5 SULTRs, SER, PHE, ILE, ALA, VAL, LEU, and TYR were more frequently observed in the binding sites in *C. sativa*, while ALA, ILE, methionine (MET), VAL, and THR were frequently observed in the binding sites of *B. napus*.

Figure 5. The transmembrane structures of the SULTRs in *C. sativa* and *B. napus*. The grouping was based on the phylogenetic tree.

2.6. SULTR Expression Analysis

In this study, the expression patterns of *SULTRs* in *C. sativa* and *B. napus* were evaluated in different tissues and in response to stress (Figures 7 and 8). We found that two *SULTR* 3.5 genes (*Csa20g030350* and *Csa13g022560*) and two *SULTR* 1.2 genes (*Csa09g084780* and *Csa07g050670*) were expressed more in the roots of *C. sativa*, while three *SULTR* 3.1 genes (*Csa06g026100*, *Csa09g58940*, and *Csa04g0377720*) and three *SULTR 2.1* genes (*Csa13g011940*, *Csa08g054410*, and *Csa20g015450*) were highly expressed in stem tissues (Figure 7a). In the leaf tissues of *C. sativa*, three *SULTR* 3.3 genes (*Csa17g030170*, *Csa14g030330*, and *Csa03g026970*), two *SULTR* 2.2 genes (*Csa16g042230* and *Csa09g084770*), and a *SULTR 4.1* gene (*Csa20g018910*) were highly expressed (Figure 7a). In response to abiotic stresses, *SULTR 3.1* was induced in *C. sativa* (Figure 7b). For example, *Csa06g026100* and *Csa04g037720* were expressed more in response to cold and salt stresses, while *Csa09g058940* was expressed more in response to drought, cold, and cadmium stresses. In addition, *Csa20g018910* (which is a chloroplast *SULTR 4.1*) was expressed more under cold stress (Figure 7b). Additionally, the *SULTRs* of *B. napus* showed diverse expression levels in tissues and in response to abiotic and biotic stresses (Figure 8). We found that two *SULTR 2.1* genes (*BnaA02g00410D* and *BnaC02g00440D*), a *SULTR 3.4* gene (*BnaC01g35550D*), and a *SULTR 3.5* gene (*BnaC02g08870D*) were highly expressed in the root tissues of *B. napus*, while two *SULTR 3.2* genes (*BnaC09g00110D* and *BnaA09g01000D*), two *SULTR 3.1* genes (*BnaA03g41530D* and *BnaC07g32580D*), a *SULTR 3.3* gene (*BnaC05g18450D*), and a *SULTR 2.2* gene (*BnaC06g38470D*) were expressed in seeds (Figure 8a). In the stem tissues of *B. napus*, two *SULTR* 3 genes (*BnaA03g41530D* and *BnaC04g28500D*) were highly expressed, while three *SULTR* 3 genes (*BnaA09g32410D*, *BnaA07g10140D*, and *BnaC07g13290D*), a *SULTR 2.1* gene (*BnaC09g46440D*), and a *SULTR 4.1* gene (*BnaA03g04410D*) were expressed in leaf tissues (Figure 8a). Furthermore, two *SULTR 3.3* genes (*BnaC05g18450D* and *BnaA09g30120*) and two *SULTR 2.1* genes (*BnaA10g22050D* and *BnaC09g46440D*) were more upregulated in response to PEG, NaCl, and ABA treatment (Figure 8b). Interestingly, two *SULTR 2.1* genes (*BnaC06g38470D* and *BnaA07g33850D*) were differentially expressed in response to cold stress in *B. napus*. However, *BnaA07g10140D* (which is a *SULTR 3.3*) and *BnaC09g46440D* (which is a *SULTR 2.1*) were also upregulated under cold stress. In response to biotic stresses, two *SULTR 4.1* genes (*BnaC03g05940D* and *BnaA03g04410D*) were upregulated in response to the fungal pathogen *Leptosphaeria maculans*. In addition, a *SULTR 3.4* gene (*BnaC01g3550D*) and a *SULTR 3.3* gene (*BnaA07g10140D*) were expressed more in response to *Sclerotinia sclerotiorum* and *Bacillus thuringiensis* strain 4f5, respectively (Figure 8b).

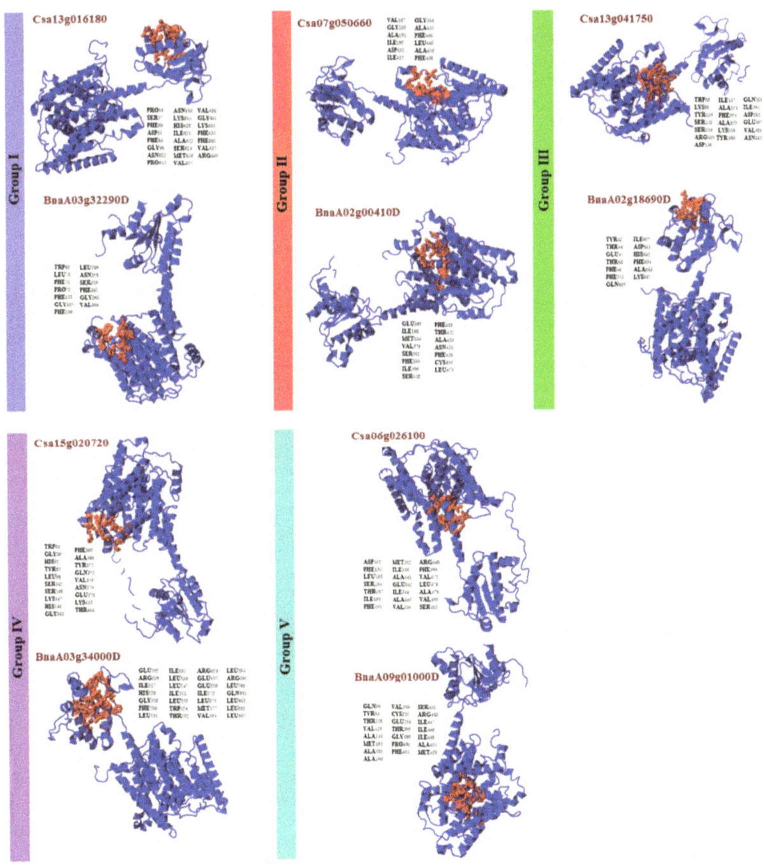

Figure 6. The three-dimensional docking analysis of the SULTRs in *C. sativa* and *B. napus*. The ligand binding sites are highlighted in red and lists of the binding sites are provided next to the protein structures.

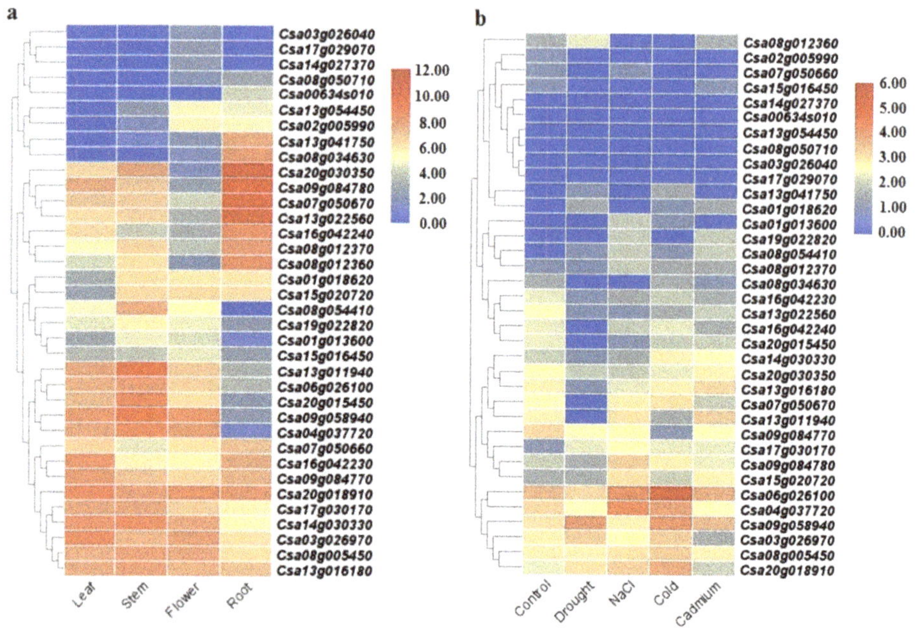

Figure 7. The expression levels of the *SULTRs* in *C. sativa*, based on the available RNA-seq data: (**a**) in different tissues; (**b**) in response to abiotic stresses.

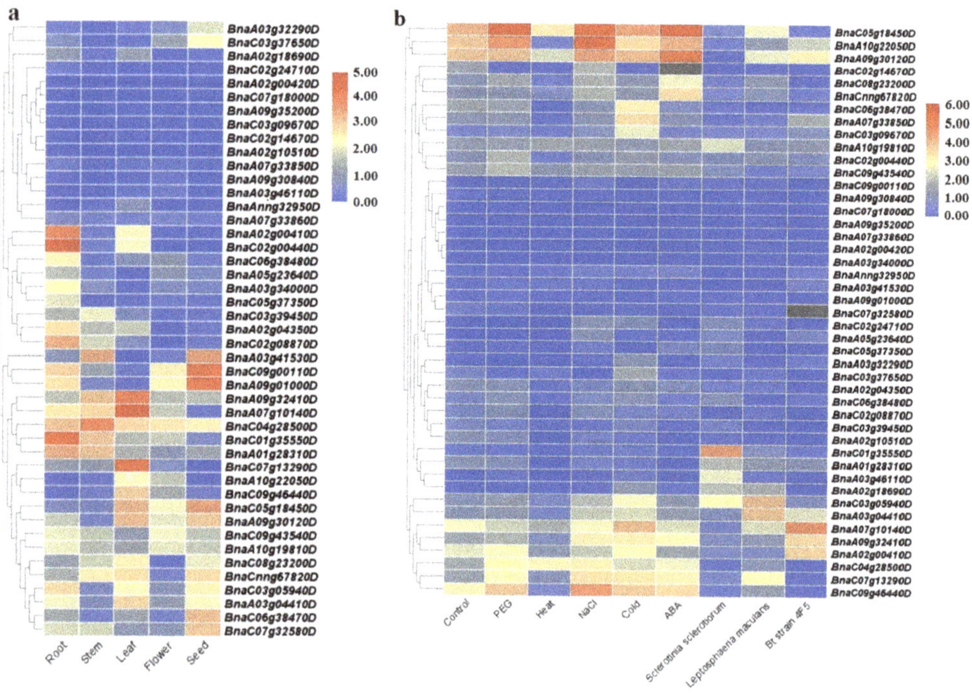

Figure 8. The expression levels of the *SULTRs* in *B. napus*, based on the available RNA-seq data: (**a**) in different tissues; (**b**) in response to abiotic and biotic stresses.

2.7. SULTR Phosphorylation Prediction

The potential phosphorylation sites of the SULTRs in *C. sativa* and *B. napus* were predicted based on serine, threonine, and tyrosine amino acids (Figure 9). The potential phosphorylation sites in the SULTRs ranged from 3 (in Csa13g054450, which is a SULTR 3.2) to 21 (in Csa08g005450, which is a SULTR 4.1 from group 1), with an average of 10.28 sites per protein in *C. sativa* (Figure 9a). Interestingly, SULTR 4.1 showed a high potential for phosphorylation events in *C. sativa*. Additionally, the potential phosphorylation sites in the SULTRs in *B. napus* ranged from a site in BnaC07g18000D (which is a SULTR 1.1) to 23 sites in BnaA10g19810D (which is a SULTR 4.1), with an average of 9.71 sites per protein (Figure 9b). In addition, more phosphorylation sites were predicted in SULTR 4.1 in *B. napus*.

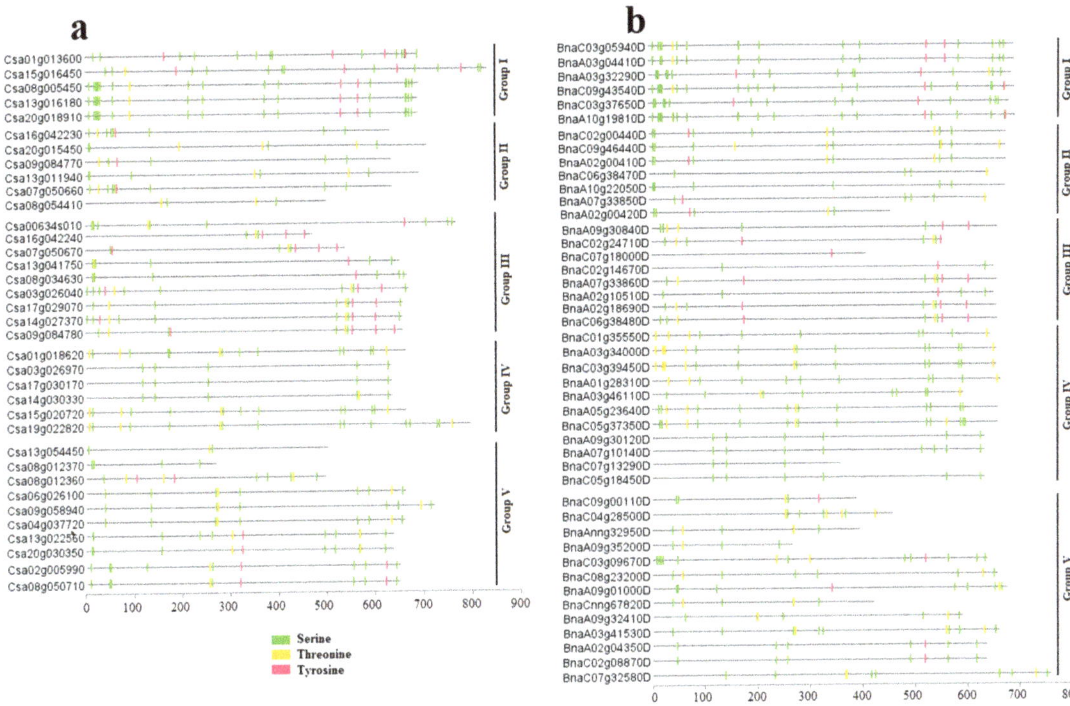

Figure 9. The prediction of phosphorylation sites in the SULTRs with scores ≥ 0.90 using the NetPhos 3.1 server: (**a**) *C. sativa*; (**b**) *B. napus*. The grouping was based on the phylogenetic tree.

2.8. Distribution of Cis-Regulatory Elements in Promoter Sites

In this study, the distribution of cis-regulatory elements in the promoter sites of the *SULTRs* in *C. sativa* and *B. napus* was investigated (Figure 10, Figure S3, and S4). The *SULTRs* in *C. sativa* and *B. napus* were compared based on the cis-regulatory elements that were related to their responses to stress and hormones (Figure 10). The cis-regulatory elements associated with auxin, ABA, MeJA, GA, and SA responses were observed in the promoter regions of the *SULTRs*. The results revealed that the cis-regulatory elements of the ABA response were frequently distributed in the *SULTRs* from *C. sativa*, while the MeJA response elements were more commonly observed in *B. napus* (Figure 10). Additionally, the cis-regulatory elements related to biotic and cold stresses were more frequently observed in the *SULTRs* from *B. napus*, while those related to drought stress were more commonly observed in the promoter sites of the *SULTRs* from *C. sativa*.

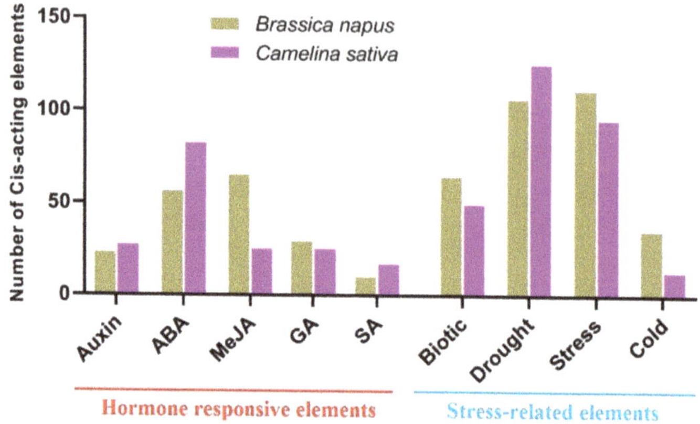

Figure 10. A comparison between the *SULTRs* from *C. sativa* and *B. napus* based on the number of cis-regulatory elements related to hormone and stress responses in promoter sites. More details are provided in Figures S3 and S4.

2.9. Expression Patterns of SULTRs in Camelina in Response to Salinity Stress

To understand the potential roles of the *SULTR* genes in camelina plants, the expression levels of five selected genes were analyzed in response to salt stress (i.e., 200 mM of NaCl). The camelina *SULTR* genes illustrated different expression patterns under salinity (Figure 11). For instance, *Csa01g013600* (which is a SULTR 4.2) was downregulated after 6 h of salinity stress, while its expression was upregulated after 24 h. Moreover, *Csa16g042230* (which is a SULTR 2.2) and *Csa06g026100* (which is a SULTR 3.1) had similar expression patterns. Both genes were upregulated in response to salt stress and the maximum expression was observed after 72 h. In contrast, *Csa07g050670* (which is a SULTR 1.2) was not induced by salinity stress. The expression levels of one SULTR 3.4 gene (*Csa15g020720*) were significantly reduced after 24 h and 72 h of salt stress. Overall, these data showed that some SULTR family members were involved in the response to salt stress.

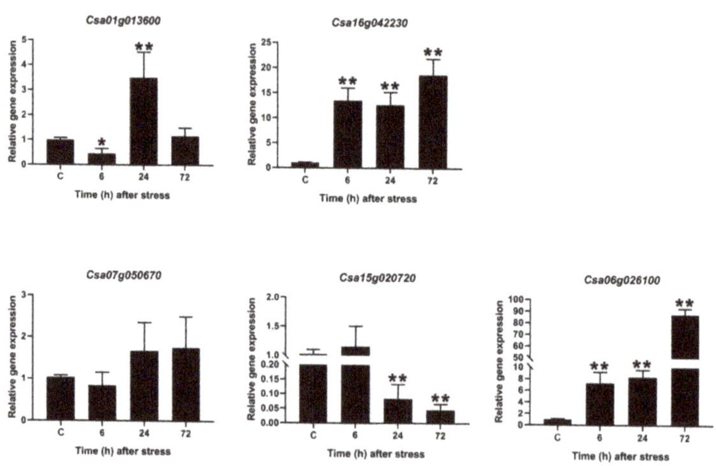

Figure 11. The expression levels of the *SULTRs* in *C. sativa* in response to salinity stress (i.e., 200 mM of NaCl) at three timepoints (6, 24, and 72 h after salt stress) and under control conditions (C, i.e., irrigation without NaCl), based on the qPCR data. Note: * and ** indicate significant differences between the expression levels following the salt treatment and those under normal conditions (based on a Student's t-test) at $p < 0.05$ and $p < 0.01$, respectively.

3. Discussion

The uptake and distribution of sulfate in plants are facilitated by networks of sulfate transporters, which are encoded by a multigene family (SULTRs) [7]. Due to the important role of sulfate in plants, the SULTRs in several plant species have been studied. For instance, the genomes of higher plants, such as *Arabidopsis thaliana*, rice (12 *SULTRs*), wheat (11 *SULTRs*), sorghum (10 *SULTRs*), and apple (9 *SULTRs*), have been identified [11–14]. In this study, we identified and characterized 36 and 45 putative *SULTR* genes in the genomes of *C. sativa* and *B. napus*, respectively (Table S1). More members of this gene family could be associated with changes in ploidy levels and genome sizes in *C. sativa* and *B. napus*, as well as duplication events in evolutionary processes [35,39]. Our investigations revealed that the SULTR proteins in the two studied plants had the same ranges for their physicochemical properties, i.e., MWs, pIs, GRAVY values, and instability indices. In addition, the exon numbers ranged from 4 to 20 in *C. sativa* and from 4 to 19 in *B. napus*. The similarities in their gene structures could indicate that significant evolutionary events have occurred in the plant genomes [40,41]. Our findings also suggested that the exon/intron patterns could provide new insights into the evolutionary relationships among the members of the gene family and that they could have originated from a common ancestor. Moreover, it has been reported that the exon number can affect expression levels, and that genes with lower exon numbers can be expressed quickly in response to environmental stresses [42,43]. SULTRs have been divided into four main classes based on their locations and functions [4]. In this study, the different SULTR classes were further separated based on our phylogenetic analysis. The SULTR 4 genes were very distinct from the other classes, while the SULTR 3 members varied significantly (Figure 1). Differences have also been observed between the SULTRs in the model monocot plant, rice, and dicot plants, indicating that the diversity in the SULTR gene family has occurred after the divergence of monocots and dicots [44,45]. According to our results for the conserved motifs in the SULTRs, some conserved sites were common between SULTR groups, which could be used to distinguish between specific groups.

According to our phylogenetic results, the camellia SULTRs were similar to the SULTRs of *B. napus*, although their evolutionary trends were different. Based on the Ka/Ks indices, the first duplication events in the SULTR genes in *C. sativa* occurred about five million years ago, while those in *B. napus* occurred three million years ago. Furthermore, it seemed that other members of the SULTR gene family originated from SULTR 3. Additionally, the Ka/Ks values revealed that the duplicated SULTRs in *B. napus* occurred under purifying (negative) selection, while both adaptive (positive) selection and purifying selection were observed in the *SULTRs* of *C. sativa* [46]. This suggested that the duplicated genes with conserved functions, pseudogenization, or both were possibly produced via purifying selection [47]. Interestingly, the results of our comparative synteny analysis revealed that several *SULTRs* from *C. sativa*, including three *SULTR 1.3* genes (*Csa17g029070*, *Csa14g027370*, and *Csa03g026040*), four *SULTR 3* genes (*Csa13g054450*, *Csa08g050710*, *Csa02g005990*, and *Csa08g012360*), and a *SULTR 1.1* gene (*Csa08g034630*), had fewer synteny relationships with the *SULTRs* from *B. napus* (Figure 4). It seemed that these genes could have been specifically developed during the evolution of the camellia, although more research is needed to determine their functions.

SULTRs can be classified into four groups based on their sequence structures, locations, and functions [48]. For instance, the genes in group 1 and group 2 are expressed more in root cells and vacuolar tissues, respectively [48,49]. In this study, the *SULTRs* in *C. sativa* and *B. napus* showed diverse expression levels in different tissues and in response to stresses. In the roots of *C. sativa*, two *SULTR 1.2* genes and two *SULTR 3.5* genes were expressed more, while two *SULTR 2.1* genes (*SULTR 3.4*, and *SULTR 3.5)* were highly expressed in the root tissues of *B. napus*. In the shoot tissues, *SULTRs 2, 3*, and 4 were expressed more. Interestingly, *SULTR* 3 showed a diverse range of functions and was expressed in all tissues, indicating that the members of this class were not specific to a tissue or organ. In addition, the members of *SULTR 3* varied greatly in terms of their transmembrane structure. Moreover, different expression patterns were observed between the members of the SULTR gene family in *B. napus* and camellia in response to stimuli. The *SULTR 3.1* genes were expressed more in response to abiotic stresses in *C. sativa*, while the *SULTR 3.3* and *SULTR 2.1* genes were more upregulated in *B. napus*. Several members of SULTR 3 play multiple roles and interact with abscisic acid (ABA) metabolism [21–23]. In the present study, SULTR 3 and SULTR 4.1 were upregulated in response to biotic stresses in *B. napus*, including bacterial and fungal pathogens. Additionally, the cis-regulatory elements related to ABA and MeJA responses were frequently observed in the promoter sites of the SULTRs. We concluded that the SULTRs could be controlled by phytohormones, especially the hormones related to stress, such as ABA and MeJA. These interactions could effectively induce the expression of the members of this gene family in response to stress. It can also be stated that the expression levels of different *SULTRs* could be correlated with hormone and stress response elements observed in the promoter regions. Additionally, the real-time PCR data revealed that the *SULTRs* in *C. sativa* had diverse expression patterns and were involved in the response to salt stress. This indicates that SULTRs could possibly interact with some transcription factors, such as MYB, and be indirectly involved in responses to abiotic stresses [28]. The prediction of the 3D structures revealed two subunits in the SULTRs and that the active binding sites differed between the subgroups (Figure 6). PHE, ALA, ILE, and VAL were identified as key amino acids in the binding sites, playing critical roles in the function and regulation of the SULTRs. Post-translational phosphorylation modifications can affect the function and possible interaction of proteins [50,51]. The prediction of the phosphorylation sites in the SULTRs revealed that the SULTR 4.1 genes had a high potential for influencing post-translation modifications, such as phosphorylation. The SULTR 4.1 and SULTR 4.2 genes have been reported to be tonoplast transporters, which allow sulfate to leave vacuoles to reach cytosol [24,25]. It seems that phosphorylation modifications play key roles in the activity of these transporters.

4. Materials and Methods

4.1. Identification of SULTR Genes in C. sativa and B. napus

To identify all sequences related to the SULTR family, the amino acid sequences of two conserved domains, including Sulfate_transp (PF00916) and STAS (PF01740), were used as queries in a BLASTP search of Ensembl Plants (https://plants.ensembl.org/index.html, accessed: 20 September 2022) in the protein databases of *C. sativa* and *B. napus*. Furthermore, orthologue genes were identified by following the same procedure for *Arabidopsis thaliana*, *Oryza sativa*, and *Glycine max*. All collected sequences were checked using the NCBI Conserved Domain Database (CDD) [52] and the Pfam database [53] to confirm the presence of domains related to the SULTRs [54]. The physiochemical properties, including molecular weight (MW), instability index, isoelectric point (pI), and GRAVY value, of the SULTRs were predicted using the ProtParam tool [55]. The TMHMM version 2.0 server was used to predict the transmembrane structures of the SULTRs in *C. sativa* and *B. napus* [56].

4.2. Phylogenetic and Conserved Motif Analyses

The amino acid sequences of all the identified *SULTRs* from five plant species, i.e., *C. sativa*, *B. napus*, *A. thaliana*, *O. sativa*, and *G. max*, were aligned using the online tool Clustal-Omega [57]. The entire phylogenetic relationships were constructed using the maximum likelihood (ML) method with 1000 bootstrap replicates using the IQ-TREE server [58]. Finally, a phylogenetic tree was created using the interactive tree of life tool (iTOL version 5) [59]. The conserved protein motifs in the SULTRs in *C. sativa* and *B. napus* were identified using the Multiple Expectation Maximization for Motif Elicitation program (MEME version 5.0.5) [60].

4.3. Promoter Analysis

In this study, 1500 bp upstream of the start codon in the *SULTRs* was selected as the promoter site, and these regions in *C. sativa* and *B. napus* were downloaded from Ensembl Plants. The sequence of each promoter site was screened to identify the conserved cis-regulatory elements using the PlantCARE tool [61]. Then, the cis-regulatory elements were classified based on their functions.

4.4. Ka/Ks Ratio and Duplication Analysis

In the present study, pairs of *SULTR* genes from each species (*C. sativa* and *B. napus*) with similarities of more than 85% were considered to be duplicated genes [62]. Additionally, the synonymous (Ks) and non-synonymous (Ka) indices were calculated for all gene pairs using the MEGAX software [63]. The time of divergence of the duplicated *SULTR* genes was estimated using the following equation: $T = (Ks/2\lambda) \times 10^{-6}$. ($\lambda = 6.5 \times 10^{-9}$) [64]. In addition, the synteny relationships between the SULTRs in each species, and between the orthologous genes of *C. sativa* and *B. napus*, were drawn using the Circos tool [65].

4.5. Gene Expression Analysis

In this study, the available RNA-seq data for *C. sativa* and *B. napus* were screened to extract the expression levels of the *SULTR* genes. In total, four RNA-seq datasets for *C. sativa*, including SRR935368 (root tissue), SRR935362 (leaf tissue), SRR935365 (stem tissue), and SRR935369 (flower tissue) were retrieved from the NCBI gene bank and analyzed. To extract the expression patterns of the *SULTRs* in response to stresses, the RNA-seq datasets related to salt stress (SRR935382), drought stress (SRR935380), cadmium stress (SRR935383), cold stress (SRR935372), and control conditions (SRR935385) were used. For the raw data analysis, we used FastQC software (version 0.11.6) (http://www.bioinformatics.babraham.ac.uk/projects/fastqc/, accessed: 20 September 2022) to check the quality of the data and HISAT [66] to map the sequences. The FPKM (fragments per kilobase of exon model per million mapped reads) metric was used to evaluate the transcription levels of each *SULTR* gene in *C. sativa*. To illustrate the expression levels of the *SULTRs* in *B. napus*, we utilized RNA-seq data for the rapeseed cultivar ZhongShuang11 (ZS11), which were related to 18 tissues and responses to biotic and abiotic stresses, from the Brassica Expression Database [67]. The expression patterns of the target genes were extracted based on their FPKM values. Finally, heatmaps were constructed using the log2 transformed method in TBtools software (version 0.665) [68].

4.6. Prediction of 3D Structures, Modeling, Binding Sites, and Phosphorylation

In this study, five proteins from each species (*C. sativa* and *B. napus*) were selected, based on the phylogenetic tree. Additionally, the three-dimensional structures of 10 SULTRs were predicted using the Phyre2 server [69]. In the next step, the predicted structures were checked using a Ramachandran plot analysis [70]. The binding sites of each model were highlighted on the predicted structures. The NetPhos 3.1 server [71], with a potential value of more than 0.90, was used to predict the phosphorylation sites of the SULTRs in *C. sativa* and *B. napus*.

4.7. Expression Patterns of SULTR Genes in C. sativa under Salinity Stress

Sterilized camelina seeds were planted at a depth of 2 cm in pots containing peat moss and were kept under the conditions of 16 h of light and a temperature of 25 °C with irrigation every three days. Then, the five-week-old seedlings were treated with salt (200 mM of NaCl) via irrigation, which was repeated after 24 h. After the salt treatment, leaves were collected at different time points (after 6, 24, and 72 h). The total RNA samples were extracted using an RNX kit (Sinaclon, Iran) and the cDNA was synthesized using a reverse transcriptase kit (Roche, Germany), according to manufacturer protocols. In the present study, five members of the SULTR family were selected for real-time PCR analysis. The genes were selected based on the phylogeny analysis. In addition, *actin-2* (Csa15g026420) was used as a reference gene to normalize the expression data. Specific primers were designed using the Primer3 online software (version 4.1.0) [72], based on the coding sequences of the selected SULTR genes (Table S3). The expression patterns of the SULTR genes were analyzed using a Maxima SYBR Green/ROX qPCR Master Mix kit (Thermo Fisher, France) and the ABI Step One, according to manufacturer protocols. The expression levels of each SULTR gene were calculated using the delta Ct method [73], using three biological replicates.

5. Conclusions

In this study, we identified and characterized 36 and 45 putative *SULTR* genes in two important oilseed crops, *Camelina sativa* and *Brassica napus*, respectively. We found that the first duplication event occurred in the *SULTR* genes of *C. sativa* and that members of this family showed diverse structures and functions. Additionally, several *SULTR* genes in *C. sativa* were uniquely developed under evolutionary processes. SULTR 3 was identified as the class of sulfate transporter family genes with the highest diversity. Overall, our results revealed new insights into the structures and functions of SULTRs in oilseed crops. However, further functional studies are needed to evaluate the roles of SULTRs in development and growth processes, as well as in responses to stimuli. Also, investigation of upstream key proteins/enzymes that affect the activity of SULTRs, can reveal the pathways linked to SULTR.

Supplementary Materials: The following are available online at https://www.mdpi.com/article/10.3390/plants12030628/s1, Table S1: A list of the identified SULTRs and their characteristics in Camelina sativa and Brassica napus, Table S2: The predicted Ka/Ks values in the duplicated gene pairs from the sulfate transporter family in the Camelina sativa and Brassica napus genomes, Table S3: A list of the primers for the camelina SULTR genes that were used in our real-time PCR, Figure S1: The logos of 10 conserved motifs in the sulfate transporter family proteins in Camelina sativa and Brassica napus, Figure S2: A synteny analysis of the SULTR genes in the (a) Camelina sativa and (b) Brassica napus genomes, Figure S3: The distribution of cis-regulatory elements in the SULTR promoter site of Camelina sativa, Figure S4: The distribution of cis-regulatory elements in the SULTR promoter site of Brassica napus.

Author Contributions: Conceptualization, P.H., S.H. and S.F.; methodology, P.H. and S.F.; software, P.H., S.E. and F.M.-P.; writing—original draft preparation, P.H. and F.M.-P.; writing—review and editing, P.H. and F.M.-P. All authors have read and agreed to the published version of the manuscript.

Funding: This research received no external funding.

Institutional Review Board Statement: Not applicable.

Informed Consent Statement: Not applicable.

Data Availability Statement: Not applicable.

Acknowledgments: F.M.-P. acknowledges the support from ANID FONDECYT grant No. 1201973.

Conflicts of Interest: The authors declare no conflict of interest.

References

1. Takahashi, H. Sulfate transport systems in plants: Functional diversity and molecular mechanisms underlying regulatory coordination. *J. Exp. Bot.* **2019**, *70*, 4075–4087. [CrossRef]
2. Li, Q.; Gao, Y.; Yang, A. Sulfur Homeostasis in Plants. *Int. J. Mol. Sci.* **2020**, *21*, 8926. [CrossRef] [PubMed]
3. Faraji, S.; Heidari, P.; Amouei, H.; Filiz, E.; Poczai, P. Investigation and Computational Analysis of the Sulfotransferase (SOT) Gene Family in Potato (*Solanum tuberosum*): Insights into Sulfur Adjustment for Proper Development and Stimuli Responses. *Plants* **2021**, *10*, 2597. [CrossRef] [PubMed]
4. Takahashi, H.; Buchner, P.; Yoshimoto, N.; Hawkesford, M.J.; Shiu, S.-H. Evolutionary relationships and functional diversity of plant sulfate transporters. *Front. Plant Sci.* **2012**, *2*, 119. [CrossRef]
5. Gruber, B.D.; Giehl, R.F.H.; Friedel, S.; von Wirén, N. Plasticity of the Arabidopsis root system under nutrient deficiencies. *Plant Physiol.* **2013**, *163*, 161–179. [CrossRef] [PubMed]
6. Koprivova, A.; Kopriva, S. Sulfation pathways in plants. *Chem. Biol. Interact.* **2016**, *259*, 23–30. [CrossRef] [PubMed]
7. Leustek, T.; Saito, K. Sulfate transport and assimilation in plants. *Plant Physiol.* **1999**, *120*, 637–644. [CrossRef]
8. Shibagaki, N.; Grossman, A.R. Binding of cysteine synthase to the STAS domain of sulfate transporter and its regulatory consequences. *J. Biol. Chem.* **2010**, *285*, 25094–25102. [CrossRef]

9. Smith, F.W.; Ealing, P.M.; Hawkesford, M.J.; Clarkson, D.T. Plant members of a family of sulfate transporters reveal functional subtypes. *Proc. Natl. Acad. Sci. USA* **1995**, *92*, 9373–9377. [CrossRef]
10. Shibagaki, N.; Rose, A.; McDermott, J.P.; Fujiwara, T.; Hayashi, H.; Yoneyama, T.; Davies, J.P. Selenate-resistant mutants of Arabidopsis thaliana identify Sultr1; 2, a sulfate transporter required for efficient transport of sulfate into roots. *Plant J.* **2002**, *29*, 475–486. [CrossRef]
11. Kumar, S.; Asif, M.H.; Chakrabarty, D.; Tripathi, R.D.; Trivedi, P.K. Differential expression and alternative splicing of rice sulphate transporter family members regulate sulphur status during plant growth, development and stress conditions. *Funct. Integr. Genom.* **2011**, *11*, 259–273. [CrossRef] [PubMed]
12. Buchner, P.; Parmar, S.; Kriegel, A.; Carpentier, M.; Hawkesford, M.J. The sulfate transporter family in wheat: Tissue-specific gene expression in relation to nutrition. *Mol. Plant* **2010**, *3*, 374–389. [CrossRef] [PubMed]
13. Akbudak, M.A.; Filiz, E.; Kontbay, K. Genome-wide identification and cadmium induced expression profiling of sulfate transporter (SULTR) genes in sorghum (*Sorghum bicolor* L.). *Biometals* **2018**, *31*, 91–105. [CrossRef] [PubMed]
14. Xun, M.; Song, J.; Shi, J.; Li, J.; Shi, Y.; Yan, J.; Zhang, W.; Yang, H. Genome-Wide Identification of Sultr Genes in Malus domestica and Low Sulfur-Induced MhSultr3; 1a to Increase Cysteine-Improving Growth. *Front. Plant Sci.* **2021**, *12*, 2114. [CrossRef] [PubMed]
15. Rouached, H.; Secco, D.; Arpat, A.B. Getting the most sulfate from soil: Regulation of sulfate uptake transporters in Arabidopsis. *J. Plant Physiol.* **2009**, *166*, 893–902. [CrossRef] [PubMed]
16. Zheng, Z.-L.; Zhang, B.; Leustek, T. Transceptors at the boundary of nutrient transporters and receptors: A new role for Arabidopsis SULTR1; 2 in sulfur sensing. *Front. Plant Sci.* **2014**, *5*, 710. [CrossRef]
17. Aarabi, F.; Naake, T.; Fernie, A.R.; Hoefgen, R. Coordinating sulfur pools under sulfate deprivation. *Trends Plant Sci.* **2020**, *25*, 1227–1239. [CrossRef]
18. Maruyama-Nakashita, A.; Nakamura, Y.; Yamaya, T.; Takahashi, H. Regulation of high-affinity sulphate transporters in plants: Towards systematic analysis of sulphur signalling and regulation. *J. Exp. Bot.* **2004**, *55*, 1843–1849. [CrossRef]
19. Takahashi, H.; Kopriva, S.; Giordano, M.; Saito, K.; Hell, R. Sulfur assimilation in photosynthetic organisms: Molecular functions and regulations of transporters and assimilatory enzymes. *Annu. Rev. Plant Biol.* **2011**, *62*, 157–184. [CrossRef]
20. Takahashi, H.; Watanabe-Takahashi, A.; Smith, F.W.; Blake-Kalff, M.; Hawkesford, M.J.; Saito, K. The roles of three functional sulphate transporters involved in uptake and translocation of sulphate in *Arabidopsis thaliana*. *Plant J.* **2000**, *23*, 171–182. [CrossRef]
21. Cao, M.; Wang, Z.; Zhao, Q.; Mao, J.; Speiser, A.; Wirtz, M.; Hell, R.; Zhu, J.; Xiang, C. Sulfate availability affects ABA levels and germination response to ABA and salt stress in *Arabidopsis thaliana*. *Plant J.* **2014**, *77*, 604–615. [CrossRef] [PubMed]
22. Kataoka, T.; Hayashi, N.; Yamaya, T.; Takahashi, H. Root-to-shoot transport of sulfate in Arabidopsis. Evidence for the role of SULTR3; 5 as a component of low-affinity sulfate transport system in the root vasculature. *Plant Physiol.* **2004**, *136*, 4198–4204. [PubMed]
23. Zuber, H.; Davidian, J.-C.; Aubert, G.; Aimé, D.; Belghazi, M.; Lugan, R.; Heintz, D.; Wirtz, M.; Hell, R.; Thompson, R. The seed composition of *Arabidopsis mutants* for the group 3 sulfate transporters indicates a role in sulfate translocation within developing seeds. *Plant Physiol.* **2010**, *154*, 913–926. [CrossRef] [PubMed]
24. Kataoka, T.; Watanabe-Takahashi, A.; Hayashi, N.; Ohnishi, M.; Mimura, T.; Buchner, P.; Hawkesford, M.J.; Yamaya, T.; Takahashi, H. Vacuolar sulfate transporters are essential determinants controlling internal distribution of sulfate in Arabidopsis. *Plant Cell* **2004**, *16*, 2693–2704. [CrossRef]
25. Wang, L.; Chen, K.; Zhou, M. Structure and function of an *Arabidopsis thaliana* sulfate transporter. *Nat. Commun.* **2021**, *12*, 4455. [CrossRef]
26. Parmar, S.; Buchner, P.; Hawkesford, M.J. Leaf developmental stage affects sulfate depletion and specific sulfate transporter expression during sulfur deprivation in *Brassica napus* L. *Plant Biol.* **2007**, *9*, 647–653. [CrossRef]
27. Ding, Y.; Zhou, X.; Zuo, L.; Wang, H.; Yu, D. Identification and functional characterization of the sulfate transporter gene GmSULTR1; 2b in soybean. *BMC Genom.* **2016**, *17*, 373. [CrossRef] [PubMed]
28. Vatansever, R.; Koc, I.; Ozyigit, I.I.; Sen, U.; Uras, M.E.; Anjum, N.A.; Pereira, E.; Filiz, E. Genome-wide identification and expression analysis of sulfate transporter (SULTR) genes in potato (*Solanum tuberosum* L.). *Planta* **2016**, *244*, 1167–1183. [CrossRef]
29. Huang, Q.; Wang, M.; Xia, Z. The SULTR gene family in maize (*Zea mays* L.): Gene cloning and expression analyses under sulfate starvation and abiotic stress. *J. Plant Physiol.* **2018**, *220*, 24–33. [CrossRef]
30. Huang, S.Q.; Xiang, A.L.; Che, L.L.; Chen, S.; Li, H.; Song, J.B.; Yang, Z.M. A set of miRNAs from Brassica napus in response to sulphate deficiency and cadmium stress. *Plant Biotechnol. J.* **2010**, *8*, 887–899. [CrossRef]
31. Kumar, S.; Asif, M.H.; Chakrabarty, D.; Tripathi, R.D.; Dubey, R.S.; Trivedi, P.K. Comprehensive analysis of regulatory elements of the promoters of rice sulfate transporter gene family and functional characterization of OsSul1; 1 promoter under different metal stress. *Plant Signal. Behav.* **2015**, *10*, e990843. [CrossRef] [PubMed]
32. Brock, J.R.; Dönmez, A.A.; Beilstein, M.A.; Olsen, K.M. Phylogenetics of *Camelina Crantz.* (Brassicaceae) and insights on the origin of gold-of-pleasure (*Camelina sativa*). *Mol. Phylogenet. Evol.* **2018**, *127*, 834–842. [CrossRef] [PubMed]
33. Yuan, L.; Li, R. Metabolic engineering a model oilseed Camelina sativa for the sustainable production of high-value designed oils. *Front. Plant Sci.* **2020**, *11*, 11. [CrossRef] [PubMed]
34. Ahmadizadeh, M.; Rezaee, S.; Heidari, P. Genome-wide characterization and expression analysis of fatty acid desaturase gene family in *Camelina sativa*. *Gene Rep.* **2020**, *21*, 100894. [CrossRef]

35. Faraji, S.; Ahmadizadeh, M.; Heidari, P. Genome-wide comparative analysis of Mg transporter gene family between *Triticum turgidum* and *Camelina sativa*. *Biometals* **2021**, *34*, 639–660. [CrossRef]
36. Lošák, T.; Hlusek, J.; Martinec, J.; Vollmann, J.; Peterka, J.; Filipcik, R.; Varga, L.; Ducsay, L.; Martensson, A. Effect of combined nitrogen and sulphur fertilization on yield and qualitative parameters of *Camelina sativa* [L.] Crtz.(false flax). *Acta Agric. Scand. Sect. B-Soil Plant Sci.* **2011**, *61*, 313–321.
37. Solis, A.; Vidal, I.; Paulino, L.; Johnson, B.L.; Berti, M.T. Camelina seed yield response to nitrogen, sulfur, and phosphorus fertilizer in South Central Chile. *Ind. Crops Prod.* **2013**, *44*, 132–138. [CrossRef]
38. Heydarian, Z.; Yu, M.; Gruber, M.; Coutu, C.; Robinson, S.J.; Hegedus, D.D. Changes in gene expression in *Camelina sativa* roots and vegetative tissues in response to salinity stress. *Sci. Rep.* **2018**, *8*, 9804. [CrossRef]
39. Abdullah; Faraji, S.; Mehmood, F.; Malik, H.M.T.; Ahmed, I.; Heidari, P.; Poczai, P. The GASA Gene Family in Cacao (*Theobroma cacao*, Malvaceae): Genome Wide Identification and Expression Analysis. *Agronomy* **2021**, *11*, 1425. [CrossRef]
40. Faraji, S.; Filiz, E.; Kazemitabar, S.K.; Vannozzi, A.; Palumbo, F.; Barcaccia, G.; Heidari, P. The AP2/ERF Gene Family in Triticum durum: Genome-Wide Identification and Expression Analysis under Drought and Salinity Stresses. *Genes* **2020**, *11*, 1464. [CrossRef]
41. Musavizadeh, Z.; Najafi-Zarrini, H.; Kazemitabar, S.K.; Hashemi, S.H.; Faraji, S.; Barcaccia, G.; Heidari, P. Genome-Wide Analysis of Potassium Channel Genes in Rice: Expression of the OsAKT and OsKAT Genes under Salt Stress. *Genes* **2021**, *12*, 784. [CrossRef]
42. Koralewski, T.E.; Krutovsky, K. V Evolution of exon-intron structure and alternative splicing. *PLoS ONE* **2011**, *6*, e18055. [CrossRef] [PubMed]
43. Heidari, P.; Puresmaeli, F.; Mora-Poblete, F. Genome-Wide Identification and Molecular Evolution of the Magnesium Transporter (MGT) Gene Family in *Citrullus lanatus* and *Cucumis sativus*. *Agronomy* **2022**, *12*, 2253. [CrossRef]
44. Rezaee, S.; Ahmadizadeh, M.; Heidari, P. Genome-wide characterization, expression profiling, and post- transcriptional study of GASA gene family. *Gene Rep.* **2020**, *20*, 100795. [CrossRef]
45. Heidari, P.; Faraji, S.; Poczai, P. Magnesium transporter Gene Family: Genome-Wide Identification and Characterization in *Theobroma cacao*, *Corchorus capsularis* and *Gossypium hirsutum* of Family Malvaceae. *Agronomy* **2021**, *11*, 1651. [CrossRef]
46. Zhang, Z.; Li, J.; Zhao, X.-Q.; Wang, J.; Wong, G.K.-S.; Yu, J. KaKs_Calculator: Calculating Ka and Ks through model selection and model averaging. *Genom. Proteom. Bioinform.* **2006**, *4*, 259–263. [CrossRef]
47. Visser, R.G.F.; Bachem, C.W.B.; de Boer, J.M.; Bryan, G.J.; Chakrabati, S.K.; Feingold, S.; Gromadka, R.; van Ham, R.C.H.J.; Huang, S.; Jacobs, J.M.E. Sequencing the potato genome: Outline and first results to come from the elucidation of the sequence of the world's third most important food crop. *Am. J. Potato Res.* **2009**, *86*, 417–429. [CrossRef]
48. Yoshimoto, N.; Takahashi, H.; Smith, F.W.; Yamaya, T.; Saito, K. Two distinct high-affinity sulfate transporters with different inducibilities mediate uptake of sulfate in Arabidopsis roots. *Plant J.* **2002**, *29*, 465–473. [CrossRef] [PubMed]
49. Gigolashvili, T.; Kopriva, S. Transporters in plant sulfur metabolism. *Front. Plant Sci.* **2014**, *5*, 442. [CrossRef]
50. Heidari, P.; Mazloomi, F.; Nussbaumer, T.; Barcaccia, G. Insights into the SAM Synthetase Gene Family and Its Roles in Tomato Seedlings under Abiotic Stresses and Hormone Treatments. *Plants* **2020**, *9*, 586. [CrossRef]
51. Heidari, P.; Ahmadizadeh, M.; Izanlo, F.; Nussbaumer, T. In silico study of the CESA and CSL gene family in *Arabidopsis thaliana* and *Oryza sativa*: Focus on post-translation modifications. *Plant Gene* **2019**, *19*, 100189. [CrossRef]
52. Marchler-Bauer, A.; Derbyshire, M.K.; Gonzales, N.R.; Lu, S.; Chitsaz, F.; Geer, L.Y.; Geer, R.C.; He, J.; Gwadz, M.; Hurwitz, D.I. CDD: NCBI's conserved domain database. *Nucleic Acids Res.* **2015**, *43*, D222–D226. [CrossRef] [PubMed]
53. Finn, R.D.; Bateman, A.; Clements, J.; Coggill, P.; Eberhardt, R.Y.; Eddy, S.R.; Heger, A.; Hetherington, K.; Holm, L.; Mistry, J. Pfam: The protein families database. *Nucleic Acids Res.* **2014**, *42*, D222–D230. [CrossRef] [PubMed]
54. Li, D.; Zaman, W.; Lu, J.; Niu, Q.; Zhang, X.; Ayaz, A.; Saqib, S.; Yang, B.; Zhang, J.; Zhao, H.; et al. Natural lupeol level variation among castor accessions and the upregulation of lupeol synthesis in response to light. *Ind. Crops Prod.* **2023**, *192*, 116090. [CrossRef]
55. Gasteiger, E.; Hoogland, C.; Gattiker, A.; Duvaud, S.; Wilkins, M.R.; Appel, R.D.; Bairoch, A. Protein identification and analysis tools on the ExPASy server. In *The Proteomics Protocols Handbook*; Humana Press: Totowa, NJ, USA, 2005; pp. 571–607.
56. Möller, S.; Croning, M.D.R.; Apweiler, R. Evaluation of methods for the prediction of membrane spanning regions. *Bioinformatics* **2001**, *17*, 646–653. [CrossRef]
57. Sievers, F.; Wilm, A.; Dineen, D.; Gibson, T.J.; Karplus, K.; Li, W.; Lopez, R.; McWilliam, H.; Remmert, M.; Söding, J.; et al. Fast, scalable generation of high-quality protein multiple sequence alignments using Clustal Omega. *Mol. Syst. Biol.* **2011**, *7*, 539. [CrossRef]
58. Nguyen, L.-T.; Schmidt, H.A.; von Haeseler, A.; Minh, B.Q. IQ-TREE: A fast and effective stochastic algorithm for estimating Maximum-likelihood phylogenies. *Mol. Biol. Evol.* **2015**, *32*, 268–274. [CrossRef]
59. Letunic, I.; Bork, P. Interactive Tree of Life (iTOL) v4: Recent updates and new developments. *Nucleic Acids Res.* **2019**, *47*, W256–W259. [CrossRef]
60. Bailey, T.L.; Boden, M.; Buske, F.A.; Frith, M.; Grant, C.E.; Clementi, L.; Ren, J.; Li, W.W.; Noble, W.S. MEME SUITE: Tools for motif discovery and searching. *Nucleic Acids Res.* **2009**, *37*, W202–W208. [CrossRef]
61. Lescot, M.; Déhais, P.; Thijs, G.; Marchal, K.; Moreau, Y.; Van de Peer, Y.; Rouzé, P.; Rombauts, S. PlantCARE, a database of plant cis-acting regulatory elements and a portal to tools for in silico analysis of promoter sequences. *Nucleic Acids Res.* **2002**, *30*, 325–327. [CrossRef]

62. Heidari, P.; Faraji, S.; Ahmadizadeh, M.; Ahmar, S.; Mora-Poblete, F. New insights into structure and function of TIFY genes in *Zea mays* and *Solanum lycopersicum*: A genome-wide comprehensive analysis. *Front. Genet.* **2021**, *12*, 534. [CrossRef]
63. Kumar, S.; Stecher, G.; Li, M.; Knyaz, C.; Tamura, K. MEGA X: Molecular evolutionary genetics analysis across computing platforms. *Mol. Biol. Evol.* **2018**, *35*, 1547–1549. [CrossRef] [PubMed]
64. Yang, S.; Zhang, X.; Yue, J.-X.; Tian, D.; Chen, J.-Q. Recent duplications dominate NBS-encoding gene expansion in two woody species. *Mol. Genet. Genom.* **2008**, *280*, 187–198. [CrossRef] [PubMed]
65. Krzywinski, M.; Schein, J.; Birol, I.; Connors, J.; Gascoyne, R.; Horsman, D.; Jones, S.J.; Marra, M.A. Circos: An information aesthetic for comparative genomics. *Genome Res.* **2009**, *19*, 1639–1645. [CrossRef] [PubMed]
66. Kim, D.; Langmead, B.; Salzberg, S.L. HISAT: A fast spliced aligner with low memory requirements. *Nat. Methods* **2015**, *12*, 357–360. [CrossRef] [PubMed]
67. Chao, H.; Li, T.; Luo, C.; Huang, H.; Ruan, Y.; Li, X.; Niu, Y.; Fan, Y.; Sun, W.; Zhang, K. BrassicaEDB: A gene expression database for Brassica crops. *Int. J. Mol. Sci.* **2020**, *21*, 5831. [CrossRef]
68. Chen, C.; Chen, H.; Zhang, Y.; Thomas, H.R.; Frank, M.H.; He, Y.; Xia, R. TBtools: An Integrative Toolkit Developed for Interactive Analyses of Big Biological Data. *Mol. Plant* **2020**, *13*, 1194–1202. [CrossRef]
69. Kelley, L.A.; Mezulis, S.; Yates, C.M.; Wass, M.N.; Sternberg, M.J.E. The Phyre2 web portal for protein modeling, prediction and analysis. *Nat. Protoc.* **2015**, *10*, 845–858. [CrossRef]
70. Lovell, S.C.; Davis, I.W.; Arendall, W.B., III; De Bakker, P.I.W.; Word, J.M.; Prisant, M.G.; Richardson, J.S.; Richardson, D.C. Structure validation by Cα geometry: Φ, Ψ and Cβ deviation. *Proteins Struct. Funct. Bioinform.* **2003**, *50*, 437–450. [CrossRef]
71. Blom, N.; Sicheritz-Pontén, T.; Gupta, R.; Gammeltoft, S.; Brunak, S. Prediction of post-translational glycosylation and phosphorylation of proteins from the amino acid sequence. *Proteomics* **2004**, *4*, 1633–1649. [CrossRef]
72. Untergasser, A.; Cutcutache, I.; Koressaar, T.; Ye, J.; Faircloth, B.C.; Remm, M.; Rozen, S.G. Primer3—New capabilities and interfaces. *Nucleic Acids Res.* **2012**, *40*, e115. [CrossRef] [PubMed]
73. Livak, K.J.; Schmittgen, T.D. Analysis of relative gene expression data using real-time quantitative PCR and the $2^{-\Delta\Delta CT}$ method. *Methods* **2001**, *25*, 402–408. [CrossRef] [PubMed]

plants

MDPI

Article

Foliar Application of Oil Palm Wood Vinegar Enhances *Pandanus amaryllifolius* Tolerance under Drought Stress

Muhammad Asyraf Mohd Amnan [1], Wee Fei Aaron Teo [1], Wan Mohd Aizat [2], Fiqri Dizar Khaidizar [1] and Boon Chin Tan [1,*]

1 Centre for Research in Biotechnology for Agriculture (CEBAR), Universiti Malaya, Kuala Lumpur 50603, Malaysia
2 Institute of Systems Biology (INBIOSIS), Universiti Kebangsaan Malaysia, Bangi 43600, Malaysia
* Correspondence: boonchin@um.edu.my; Tel.: +60-37967-7982

Abstract: Drought stress severely threatens plant growth, yield and survivability. Wood vinegar, formed by the condensation of smoke produced during biochar production, has been shown to promote plant growth and enhance stress tolerance. They have now been recognized as a sustainable alternative and are frequently used exogenously to support plants coping with environmental stress. This study aimed to evaluate the efficacy of oil palm wood vinegar (OPWV) in mitigating the adverse effects of drought stress on *Pandanus amaryllifolius*. The optimal concentrations and frequencies of OPWV application were determined before the drought treatment. The results showed that the imposed drought stress negatively affected the plant growth parameters but applying OPWV at 1:500 dilution at 3-day intervals for 12 days increased its tolerance. These include increased leaf relative water content, root-to-shoot ratio, relative stem circumference, chlorophyll pigments and antioxidant enzyme activities. In contrast, the drought-stressed plants treated with OPWV showed decreased relative electrolyte leakage, hydrogen peroxide, proline, malondialdehyde, and enhanced drought-responsive gene expressions, such as *HSP70*, *GAPDH*, and *Thau*, while *ENO* and *β-Fruc* were reduced. These biostimulatory effects of OPWV might be due to several antioxidant compounds, such as anthranilic acid, tetrasiloxane, syringol, guaiacol, and catechol. Altogether, our results showed the effectiveness of OPWV in alleviating the adverse effects of drought stress, and as such, OPWV could be potentially applied in agriculture.

Keywords: agriculture; antioxidant enzymes; biostimulants; drought stress; wood vinegar; drought-responsive genes

Citation: Mohd Amnan, M.A.; Teo, W.F.A.; Aizat, W.M.; Khaidizar, F.D.; Tan, B.C. Foliar Application of Oil Palm Wood Vinegar Enhances *Pandanus amaryllifolius* Tolerance under Drought Stress. *Plants* **2023**, *12*, 785. https://doi.org/10.3390/plants12040785

Academic Editors: Lixi Jiang, Mingxun Chen and Yuan Guo

Received: 31 December 2022
Revised: 7 February 2023
Accepted: 8 February 2023
Published: 9 February 2023

1. Introduction

Climate change has increased the occurrence, intensity, and complexity of the multifactorial combination of stresses, affecting agricultural production. For example, the flooding event in Malaysia in 2006 resulted in a loss of USD 18.9 million in agriculture damages [1], while drought has caused USD 30 billion in global crop yield losses over the past decade [2]. Among the environmental stresses, drought stress is a significant threat to crop production. It affects about 55 million people's socioeconomic activity globally [3].

Drought stress negatively affects plant growth and development. When plants are exposed to drought stress, their biochemical and metabolic activities are disrupted due to the limited water supply. This hinders some critical processes, such as photosynthesis, nutrient acquisition, transportation of minerals and translocation, leading to excessive reactive oxygen species (ROS) accumulation [4]. Although plants have defense machinery to protect themselves against oxidative stress damages, exploring innovative and sustainable interventions to mitigate the drought stress effects on plant growth is indispensable.

Biostimulants are any substance or microorganism applied to plants, aiming to increase nutrition efficiency, environmental stress tolerance and/or crop quality traits [5]. They have gained much interest because of their ability to stimulate growth and mitigate stress-induced limitations. Moreover, biostimulants have been recognized as a sustainable alternative to support plants coping with environmental stress. For instance, seaweed extract [6] and chitosan [7] have enhanced the plant's yield and tolerance against drought stress. In addition, the exogenous application of wood vinegar [8] and smoked water [9] elevated the drought tolerance of rice and cowpea, respectively.

Wood vinegar or pyroligneous acid is a reddish-brown translucent liquid produced as a by-product of pyrolysed wood heated from 200 °C to 450 °C under low oxygen conditions. The accumulated smoke from the kiln is then channeled into a condensation tube to form a condensed smoked liquid [10]. Wood vinegar consists of many beneficial compounds. For example, wood vinegar derived from lychee wood consists of three major components, which are 2,6-dimethoxyphenol (syringol) and 2-methoxyphenol (guaiacol), and 3,5-dimethoxy-4-hydroxytoluene (pyrogallol) [11]. These compounds were reported for their ability as an antioxidant to scavenge ROS [12–14]. Previous studies reported that wood vinegar could be used as a biostimulant to promote plant performance and growth [15,16]. Additionally, Wang et al. reported that wood vinegar enhanced wheat tolerance against drought by increasing abscisic acid accumulation, antioxidant activities and reducing ROS [17]. Despite its potential, applying wood vinegar to mitigate the adverse effect of environmental stresses is scarce. Therefore, this study explored the benefits of wood vinegar produced from oil palm kernel shells on plant growth and development.

Pandanus amaryllifolius, a member of the botanical family Pandanaceae, is a tropical plant species commonly referred to as fragrant screw pine or pandan [18]. This species is widely cultivated in Southeast Asia. The leaves of *P. amaryllifolius* are commonly used as flavoring agents, natural colorants, and in traditional medicine [18]. Phytochemical analysis of *P. amaryllifolius* leaves has revealed the presence of phenolic compounds, such as gallic acid and cinnamic acid, which have been shown to inhibit the growth of breast cancer cells in laboratory studies [19]. Additionally, the leaf extract of *P. amaryllifolius* has been found to have potentially beneficial effects on metabolic syndromes, such as weight gain and blood pressure.

We previously determined the effects of mild and severe drought stress on morphological, biochemical, and proteomics changes of *P. amaryllifolius* [18]. Our results showed that *Pandanus* plants could endure up to 10 days of drought in clay silt loam soil. Since wood vinegar may act by priming the treated crops against environmental stresses, we hypothesized that the exogenous application of oil palm wood vinegar (OPWV) would improve *P. amaryllifolius* growth and enhance its tolerance to drought stress. To address our hypotheses, we investigated the responses of *P. amaryllifolius* plants under well-watered and drought-stress conditions with or without OPWV treatment. In addition, we also determined the optimal dilution factors and application frequencies of OPWV and its compound profile. The novelty of our study is that OPWV has not been explored to enhance plant growth and stress tolerance. Since OPWV is made from the waste products of oil palm trees, this makes it a more sustainable and eco-friendly option compared to other wood vinegar. Additionally, OPWV may have unique chemical properties and composition compared to wood vinegar made from other types of wood, which could make it useful for different applications.

2. Results

2.1. Different Oil Palm Wood Vinegar (OPWV) Dilution Factors and Application Frequencies on *Pandanus amaryllifolius* Growth

P. amaryllifolius plants were sprayed with OPWV at different dilution factors. All treated plants did not show visible differences (Figure S1A). However, the relative stem circumference and root dry weight (DW) of plants treated with OPWV at 1:500 dilution were significantly higher than other treatments (Figure 1A,B). Although this treatment also produced a higher shoot DW than others, the shoot-to-root ratio of all treatments showed no significant changes (Figure 1C,D). Regarding chlorophyll contents, although the 1:250 OPWV treatment increased the chlorophyll contents (Figure 1E–H), OPWV at 1:500 dilution was chosen for the subsequent experiment, as it showed enhanced stem circumference and biomass.

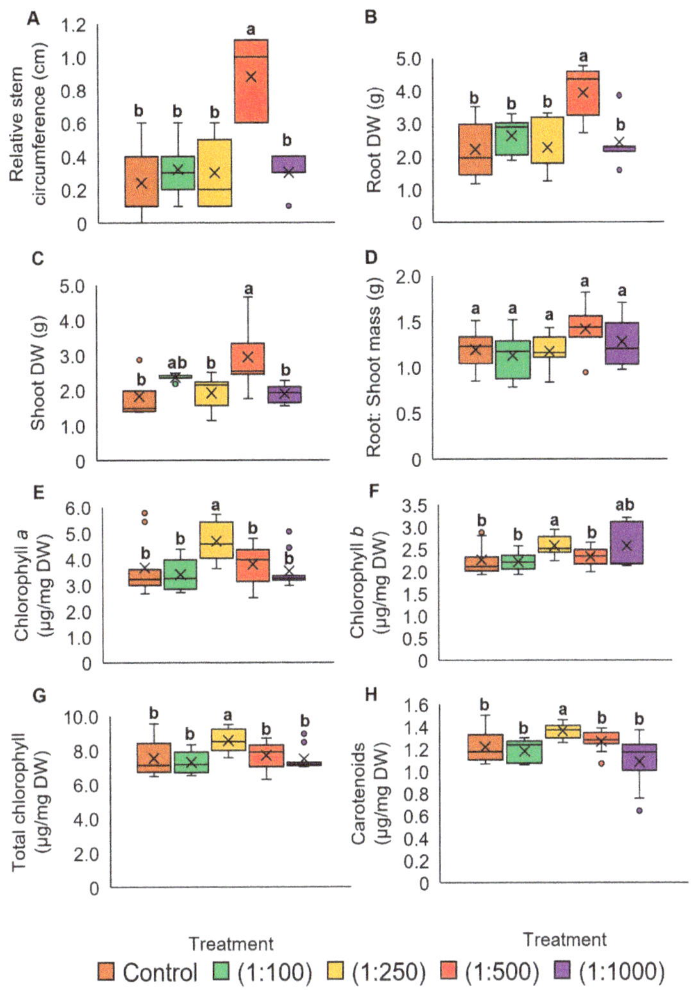

Figure 1. The plant morphological and pigment analyses of the *Pandanus amaryllifolius* treated with different dilution factors of oil palm wood vinegar (OPWV). (**A**) Relative stem circumference. (**B**) Root dry weight (DW). (**C**) Shoot DW. (**D**) The root-to-shoot ratio. (**E**) Chlorophyll *a* (μg mg^{-1} DW). (**F**) Chlorophyll *b* (μg mg^{-1} DW). (**G**) Total chlorophyll (μg mg^{-1} DW). (**H**) Carotenoids (μg mg^{-1} DW). The letter labelled on the mean value indicates a significant level between treatments based on the one-way ANOVA, followed by the post hoc Tukey test when its p-value < 0.05. The letter 'a' above the bars indicates the highest value, followed by 'ab' and 'b'. The (×) labelled in the box plot indicates the mean value of the treatment, while the (•) refers to the outlier value of the replicates.

Different application frequencies of the diluted OPWV at 1:500 were evaluated. In general, plants treated with OPWV at different frequencies showed no visible differences (Figure S1B). However, F2-treated plants showed the highest mean relative stem circumference (0.9 cm) and shoot (2.9 g) and root DW (3.9 g) (Figure 2A–C). In contrast, the root-to-shoot ratio for all treatments was not significantly different (Figure 2D). The plants treated with F1 and F2 frequencies showed a higher accumulation of pigment constituents than those treated with F3 (Figure 2E–H). Therefore, a combination of the 1:500 dilution and F2 application frequency was selected for the subsequent experiments.

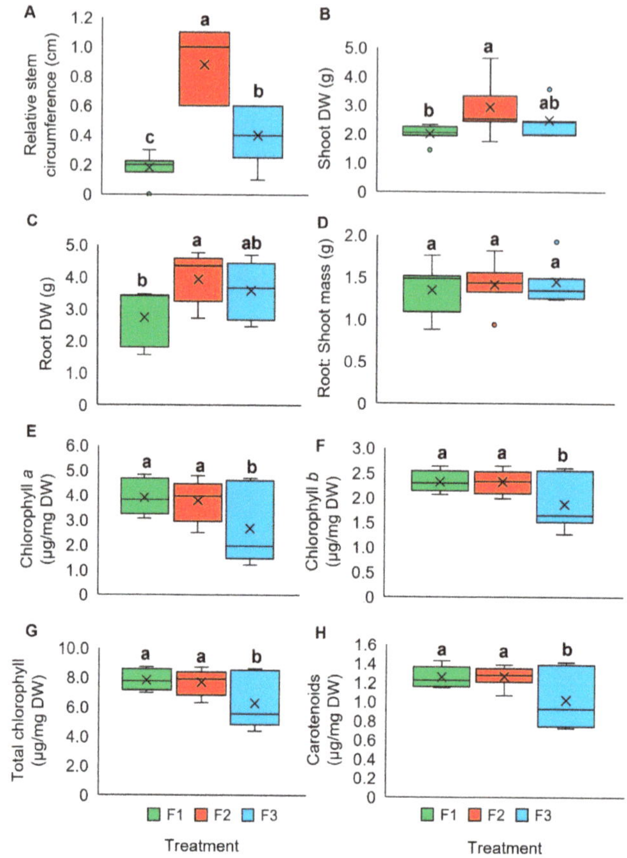

Figure 2. The plant morphological and pigment analyses of the *Pandanus amaryllifolius* were subjected to different application frequencies. F1: applied once at 6-day intervals, F2: applied once at 3-day intervals, and F3: applied once at 1-day intervals. (**A**) Relative stem circumference. (**B**) Shoot dry weight (DW). (**C**) Root DW. (**D**) The root-to-shoot ratio (**E**) Chlorophyll *a* (µg mg^{-1} DW). (**F**) Chlorophyll *b* (µg mg^{-1} DW). (**G**) Total chlorophyll (µg mg^{-1} DW). (**H**) Carotenoids (µg mg^{-1} DW). The letter labelled on the mean value indicates a significant level between treatments based on the one-way ANOVA, followed by the post hoc Tukey test when its p-value < 0.05. The letter 'a' above the bars indicates the highest value, followed by 'ab', 'b', and 'c'. The (×) labelled in the box plot indicates the mean value of the treatment, while the (•) refers to the outlier value of the replicates.

2.2. Oil Palm Wood Vinegar (OPWV) Improved Drought-Stressed Pandanus amaryllifolius

To determine the biostimulant effects of OPWV against a drought environment, the morphological changes of plants subjected to water withholding for 7 and 10 days were recorded (Figure 3A). The 7-day drought-stressed plants without OPWV treatment showed notable stress symptoms, such as wilting and leaf folding. However, the plants showed more severe effects when prolonged drought stress conditions to 10 days. These include losing their structural turgor, and some completely collapsed (Figure 3A). In contrast, plants with OPWV treatment could sustain their leaf structure, showing minimal wilting and leaf folding despite water withholding for 7 days, and retain their turgor pressure, sustaining their upright structure (Figure 3A).

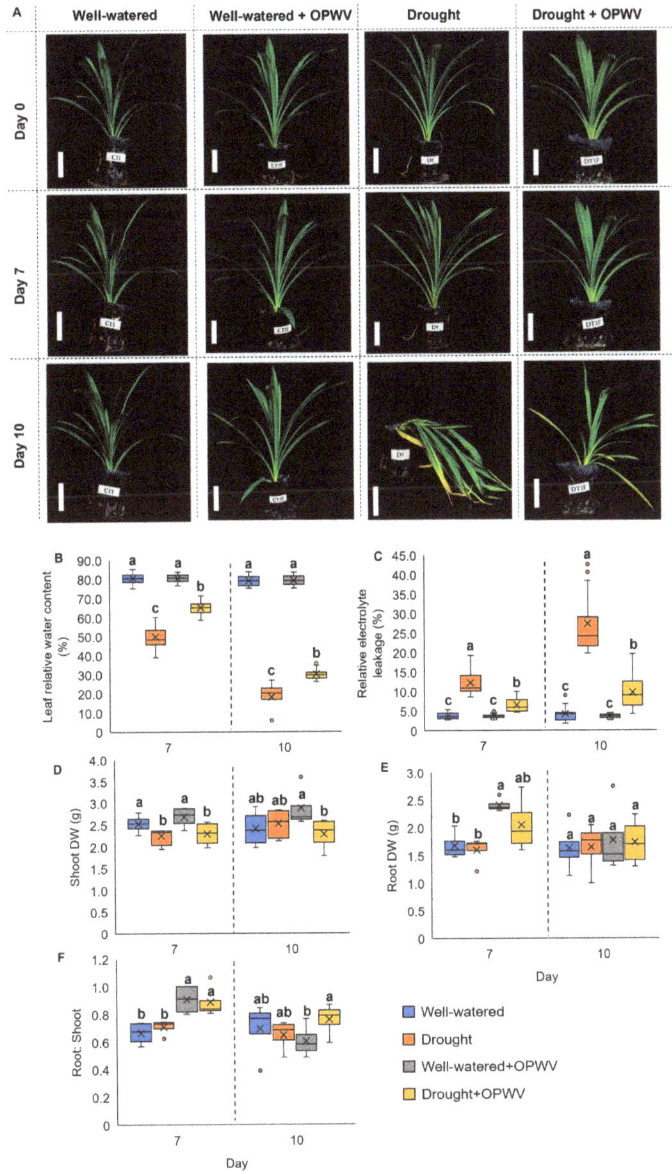

Figure 3. The morphological analysis of the well-watered and drought-stressed *Pandanus amaryllifolius* plants treated with or without OPWV for 7 and 10 days. (**A**) Representative photographs of *P. amaryllifolius* under different conditions. Bar = 10 cm. (**B**) Leaf relative water content (LRWC). (**C**) Leaf relative electrolyte leakage (REL). (**D**) The shoot dry weight (DW). (**E**) Root DW. (**F**) Root-to-shoot mass ratio. The letter labelled on the mean value indicates a significant level between treatments based on the one-way ANOVA, followed by the post hoc Tukey test when its p-value < 0.05. The letter 'a' above the bars indicates the highest value, followed by 'ab', 'b', and 'c'. The (×) labelled in the box plot indicates the mean value of the treatment, while the (•) refers to the outlier value of the replicates.

The plant leaf relative water content (LRWC) and relative electrolyte leakage (REL) were analyzed to investigate the degree of water loss and cell damage within the plant cells. The well-watered plants, with or without OPWV treatment, showed more than 80% LRWC for both days 7 and 10 (Figure 3B). However, a significant 50% and 18% decrease was observed in the drought-stressed plants without OPWV treatment on days 7 and 10, respectively. On the contrary, despite drought stress, OPWV-treated *P. amaryllifolius* retained 65% and 30% LRWC on days 7 and 10, respectively. Regarding REL, the well-watered plants, with or without OPWV treatment, showed the lowest REL (4%) on days 7 and 10 (Figure 3C). In contrast, the drought-stressed plants without OPWV treatment recorded the highest REL of 12% and 27% on days 7 and 10, respectively, followed by the OPWV-treated drought-stressed plants with 7% and 10% REL on days 7 and 10, respectively.

The DW of shoots and roots and the root-to-shoot ratio of the plant samples were recorded. The well-watered and OPWV-treated plants showed higher shoot DW than the drought-stressed plants on day 7, although it was not significantly different on day 10 (Figure 3D). On the other hand, the OPWV-treated plants showed increased root DW on day 7 (Figure 3E). Similar to shoot DW, all treatments showed no significant difference on day 10 (Figure 3E). However, having a higher root and shoot DW, it is no surprise that the OPWV-treated plants recorded the highest root-to-shoot ratio on day 7 (Figure 3F).

The highest relative stem circumference was recorded in the well-watered *P. amaryllifolius* treated with OPWV for days 7 (0.3 cm) and 10 (0.4 cm), followed by well-watered (0.1 cm for both 7 and 10 days), OPWV-treated drought-stressed (−0.2 and −0.7 cm for 7 and 10 days, respectively), and drought-stressed samples (−0.5 and −1.4 cm for 7 and 10 days, respectively) (Figure 4A).

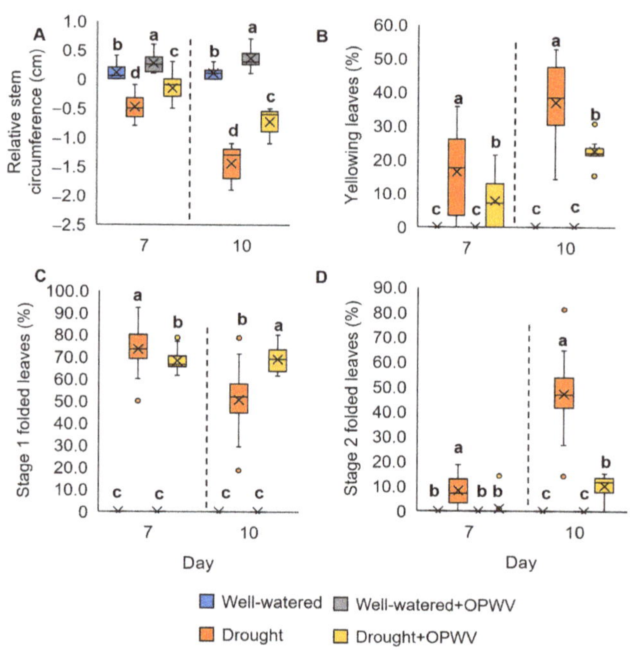

Figure 4. The stem and leaf morphological analyses of the *Pandanus amaryllifolius* treated with 1:500 diluted OPWV at 3-day intervals under well-watered and drought stress conditions for 7 and 10 days. (**A**) The relative stem circumference. (**B**) The percentage of yellowing leaves. (**C**) The percentage of Stage 1 leaf folding. (**D**) The percentage of Stage 2 leaf folding. The letter labelled on the mean value indicates a significant level between treatments based on the one-way ANOVA, followed by the post hoc Tukey test when its p-value < 0.05. The letter 'a' above the bars indicates the highest value, followed by 'b', 'c', and 'd'. The (×) labelled in the box plot indicates the mean value of the treatment, while the (•) refers to the outlier value of the replicates.

The percentage of leaf yellowing was analyzed based on the leaf color reference (Figure S2E). Both OPWV-treated and non-treated well-watered samples showed no yellowing symptoms (Figure 4B). On the contrary, the drought-stressed *P. amaryllifolius* showed 16% and 37% yellowing leaves on days 7 and 10, respectively (Figure 4B). In comparison, the percentage of leaf yellowing for the OPWV-treated samples was lower, 8% and 23% on days 7 and 10, respectively.

The percentage of leaves folding was determined according to the references (Figure S2D). After 7 days of drought stress, the plants showed a higher percentage of stage 1 folding (73%) compared to the OPWV-treated drought-stressed samples (68%) (Figure 4C). However, the 10-day drought-stressed plants showed a lower percentage of stage 1 folding (51%) than the OPWV-treated samples (69%) (Figure 4C). This result indicates the shift of leaf folding severity from Stage 1 to Stage 2. On the other hand, the OPWV-treated drought-stressed samples maintained the Stage 1 leaf folding and had lower Stage 2 leaf folding than those without OPWV treatment (Figure 4D).

2.3. *Leaf Pigment Constituents in Pandanus amaryllifolius in Response to OPWV*

In this study, the drought-stressed *P. amaryllifolius* exhibited a significant reduction in chlorophyll *a*, chlorophyll *b*, total chlorophyll, and carotenoids on days 7 and 10 compared to well-watered samples (Figure 5A–D). In addition, when treated with OPWV, the pigmentation constituents of the drought-stressed samples were higher than those without OPWV (Figure 5A–D).

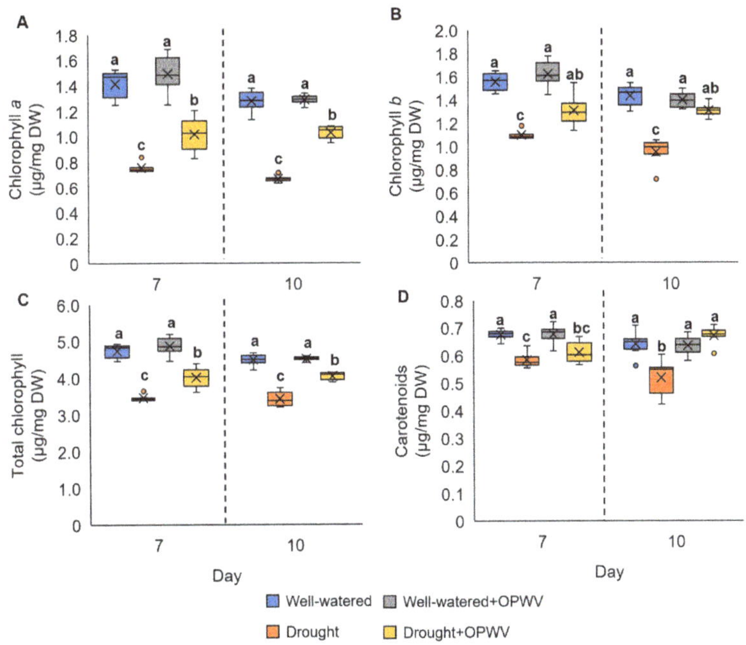

Figure 5. The pigmentation constituents of the well-watered and drought-stressed plants with or without OPWV for 7 and 10 days. (**A**) Chlorophyll *a* (µg mg^{-1} DW). (**B**) Chlorophyll *b* (µg mg^{-1} DW). (**C**) Total chlorophyll (µg mg^{-1} DW). (**D**) Carotenoids (µg mg^{-1} DW). The letter labelled on the mean value indicates a significant level between treatments based on the one-way ANOVA, followed by the post hoc Tukey test when its p-value < 0.05. The letter 'a' indicates the highest value, followed by 'ab', 'b', and 'c'. The (×) labelled in the box plot indicates the mean value of the treatment, while the (•) refers to the outlier value of the replicates.

2.4. Hydrogen Peroxide, Osmolyte and Lipid Peroxidation of OPWV-Treated Pandanus amaryllifolius

In this experiment, the drought-stressed *P. amaryllifolius* on both days 7 and 10 showed a significant increment in hydrogen peroxide (H_2O_2), proline and MDA compared to the well-watered samples (Figure 6A–C). In contrast, applying OPWV to the drought-stressed *P. amaryllifolius* reduced the content of H_2O_2, proline and MDA (Figure 6A–C).

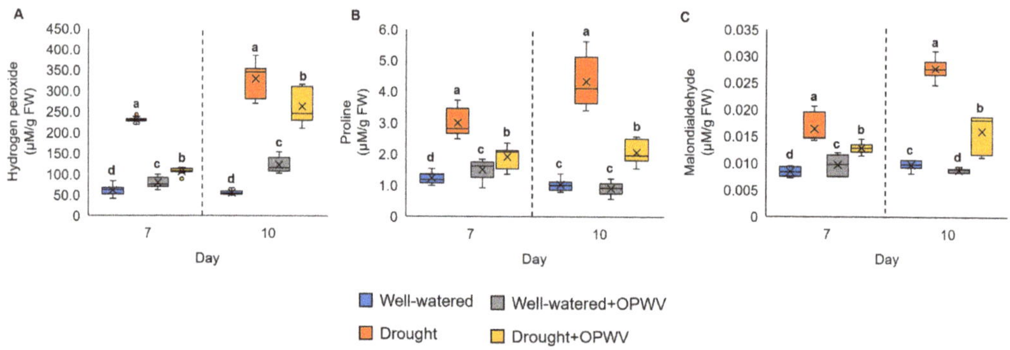

Figure 6. The hydrogen peroxide (H_2O_2), proline and malondialdehyde (MDA) contents in the well-watered and drought-stressed *Pandanus amaryllifolius* plants with or without OPWV for 7 and 10 days. (**A**) H_2O_2. (**B**) Proline. (**C**) MDA. The letter labelled on the mean value indicates a significant level between treatments based on the one-way ANOVA, followed by the post hoc test when its p-value < 0.05. The letter 'a' above the bars indicates the highest value, followed by 'b', 'c', and 'd'. The (×) labelled in the box plot indicates the mean value of the treatment, while the (•) refers to the outlier value of the replicates.

2.5. Antioxidant Enzymes Altered in Pandanus amaryllifolius in Response to OPWV and Drought Stress

Drought stress induced the antioxidant enzyme activities of *P. amaryllifolius* on both days 7 and 10 (Figure 7). However, after being treated with OPWV, the superoxide dismutase (SOD), catalase (CAT), peroxidase (POD), ascorbate peroxidase (APX) and glutathione reductase (GR) activities of the drought-stressed samples were significantly reduced (Figure 7A–E), except for SOD and APX activities on day 10. Interestingly, applying OPWV to well-watered Pandanus plants increased their SOD, APX and GR activities (Figure 7A,D,E).

2.6. Pandanus amaryllifolius Drought-Stressed Responsive Genes Triggered with OPWV Application

The expression of several drought-responsive genes was determined to understand the OPWV effects on *P. amaryllifolius* under different conditions. These genes were selected based on our previous findings [18]. Overall, all gene expressions were significantly up-regulated on 7-day drought stress (Figure 8A–E). However, the OPWV-treated drought-stressed plants showed increased *PaHSP70*, *PaGAPDH* and *PaThau* expression (Figure 8A,B,D) but decreased expression for PaENO and Paβ-Fruc (Figure 8C,E). In contrast, applying OPWV to the well-watered plants increased the expression of *PaHSP70*, *PaGAPDH*, and *Paβ-Fruc* (Figure 8A,B,E) but decreased *PaENO* and *PaThau* expressions (Figure 8C,D).

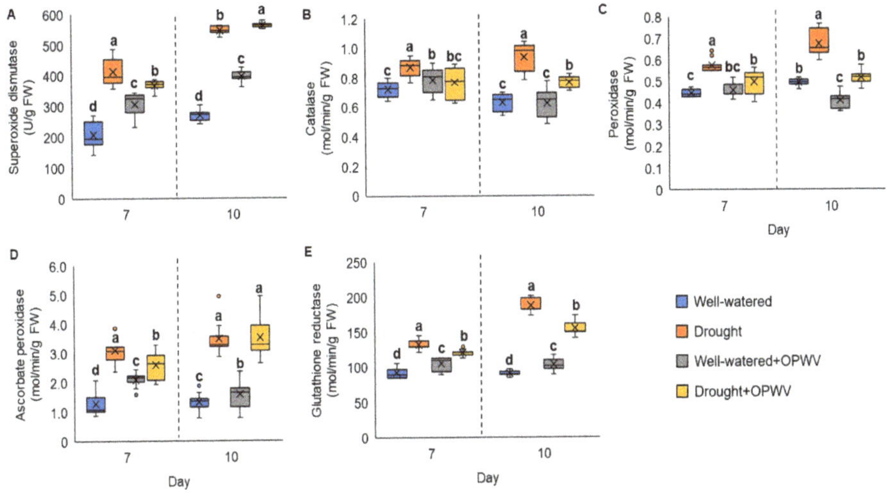

Figure 7. The antioxidant enzyme activities of the well-watered and drought-stressed *Pandanus amaryllifolius* plants with or without OPWV for 7 and 10 days. (**A**) Superoxide dismutase (SOD). (**B**) Catalase (CAT). (**C**) Peroxidase (POD). (**D**) Ascorbate peroxidase (APX). (**E**) Glutathione reductase (GR). The letter labelled on the mean value indicates a significant level between treatments based on the one-way ANOVA, followed by the post hoc Tukey test when its p-value < 0.05. The letter 'a' indicates the highest value, followed by 'b', 'bc', 'c' and 'd'. The (×) labelled in the box plot indicates the mean value of the treatment, while the (•) refers to the outlier value of the replicates.

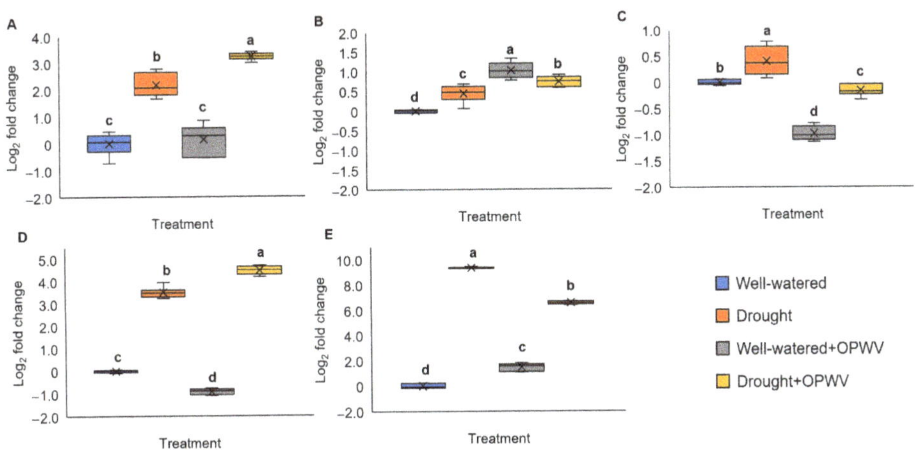

Figure 8. The drought-responsive gene expression of the well-watered or drought-stressed *Pandanus amaryllifolius* treated with or without OPWV for 7 days. (**A**) *Heat shock protein 70* (*PaHSP70*). (**B**) *Glyceraldehyde-3-phosphate dehydrogenase* (*PaGAPDH*). (**C**) *Enolase* (*PaENO*). (**D**) *Thaumatin* (*PaThau*). (**E**) β-fructofuranosidase (Paβ-Fruc). Actin and elongation factor-1 of *P. amaryllifolius* were reference genes for gene expression normalization. The letter labelled on the mean value indicates a significant level between treatments based on the one-way ANOVA, followed by the post hoc Tukey test when its p-value < 0.05. The letter 'a' indicates the highest value, followed by 'b', 'c', and 'd'. The (×) labelled in the box plot indicates the mean value of the treatment, while the (•) refers to the outlier value of the replicates.

2.7. OPWV GC-MS Profiling

GC-MS analysis was conducted to profile the compounds present in the OPWV diluent of 1:500. In total, there are 25 compound peaks identified in the 25 min GC-MS run (Figure S3). As shown in Table 1, the OPWV major component was phenyl carbamate (64%), with other minor components of phenol (16%), guaiacol (4%) and syringol (3%).

Table 1. GC-MS analysis of oil palm wood vinegar.

No.	Identified Compound	Molecular Formula	Retention Time (RT)	m/z	Area under Peak	Concentration (ppm/μL)	Concentration (ppm/mL)
1	Phenyl carbamate	$C_7H_7NO_2$	7.111	94.15	48,765,158	340.55	340,550
2	Phenol	C_6H_6O	6.868	94.05	12,407,245	86.65	86,650
3 *	Pyridine, 2,4,6-trimethyl-	$C_8H_{11}N$	7.191	121.05	7,159,676	50.00	50,000
4	Guaiacol	$C_7H_8O_2$	9.079	109.05	3,369,657	23.53	23,530
5	Syringol	$C_8H_{10}O_3$	13.856	154.05	2,215,362	15.47	15,470
6	2-(2′,4′,4′,6′,6′,8′,8′-Heptamethyltetrasiloxan-2′-yloxy)-2,4,4,6,6,8,8,10,10-nonamethylcyclopentasiloxane	$C_{16}H_{48}O_{10}Si_9$	19.103	73.1	1,836,709	12.83	12,830
7	3-Isopropoxy-1,1,1,7,7,7-hexamethyl-3,5,5-tris(trimethylsiloxy) tetrasiloxane	$C_{18}H_{52}O_7Si_7$	23.366	73.1	1,741,464	12.16	12,160
8	Catechol	$C_6H_6O_2$	11.458	110.05	1,096,505	7.66	7660
9	3-Oxabicyclo[3.3.0]oct-7-en-2-one,4-methoxy-	$C_9H_{12}O_3$	8.946	109.05	961,798	6.72	6720
10	o-Creosol	C_7H_8O	8.271	108.05	803,567	5.61	5610
11	1,2-Cyclopentanedione, 3-methyl-	$C_6H_8O_2$	8.011	112.05	753,407	5.26	5260
12	1,2,4-Trimethoxybenzene	$C_9H_{12}O_3$	15.875	168.1	546,464	3.82	3820
13	4-Ethylguaiacol	$C_9H_{12}O_2$	12.918	137.05	513,484	3.59	3590
14	3-Methoxyycatechol	$C_7H_8O_3$	12.644	140.05	508,764	3.55	3550
15	Methoxyacetylene	$C_{10}H_{14}O_3$	17.18	167.1	242,644	1.69	1690
16	Creosol	$C_8H_{10}O_2$	10.938	123	190,595	1.33	1330
17	Methyl palmitate	$C_{17}H_{34}O_2$	22.761	74.1	116,444	0.81	810

The asterisk indicates the internal standard spiked prior to injection.

3. Discussion

3.1. OPWV Induces Morphological Alteration and Improves Drought Tolerance

Drought affects plant growth and development. Plants adapt to such detrimental conditions by exploiting their complex morpho-physiological, biochemical and molecular mechanisms [20]. In this study, the imposed drought stress negatively affected the plant growth parameters, such as LRWC, leaf yellowing and folding, stem circumference, and root-to-shoot ratio. Conversely, spraying OPWV on plants prior to drought stress improved their tolerance by increasing LRWC and reducing REL, leaf wilting and leaf yellowing. This finding suggests that the OPWV-treated plants have better cell water holding capacity and structural integrity, thus enhancing their survivability under drought conditions. Plants with a lesser or slower leaf wilting are crucial for maintaining their yield and drought tolerance [21]. Similar findings have been shown in wheat, banana, and Arabidopsis when primed with wood vinegar, sodium nitroprusside and seaweed extract, respectively [17,22,23]. Although the exact mechanism is unclear, we speculate that the improved growth might be due to the reduced leaf folding and wilting by OPWV.

3.2. OPWV Ameliorated Photosynthetic Pigments

Photosynthesis is crucial in providing material and energy for plant growth, yet it is very susceptible to environmental stress. Generally, drought stress reduces the plant photosynthesis rate due to damage that occurs to its machinery, such as chlorophyll pigments [24,25]. In this study, OPWV treatment induced a higher accumulation of total chlorophyll content, including chlorophyll *a* and *b*, in either well-watered or drought-stressed plants. Our results correspond to the study by Vannini et al., where sweet chestnut wood distillate-primed lettuce plants recorded higher chlorophyll content than non-treated samples [16]. In addition, another study showed that rapeseed plants primed with peach wood vinegar before being exposed to salt stress produced higher chlorophyll content than salt-stressed non-primed rapeseed [26].

These findings evince the potential of OPWV as a biostimulant to boost and maintain plant photosynthesis machinery, even under stress conditions.

3.3. OPWV Mitigated Drought Stress Indicators

Drought stress induces oxidative damage to plants. Over-accumulation of ROS species, such as H_2O_2, causes damage to cellular molecules, such as lipids, and alters intrinsic membrane properties. MDA, a product formed via lipid peroxidation of arachidonic acid and large polyunsaturated fatty acid (PUFA) [27], is often used as a marker to indicate cell damage due to ROS. In this study, the drought-stressed plants exhibited increased H_2O_2 and MDA levels but were reduced with OPWV application, suggesting OPWV could mitigate ROS accumulation and cell damage. Other studies showed similar findings. For instance, the wheat and carrot plants treated with H_2O_2 as well as sodium nitroprusside [28] and α-tocopherol [29], respectively, showed lower ROS and MDA accumulation than the control. Interestingly, the well-watered *P. amaryllifolius* treated with OPWV showed increased H_2O_2 and MDA. This might be due to a sudden boost of H_2O_2, causing damage to the cell's lipids and increased MDA. A similar finding has been reported by Gohari et al., where the authors found that biostimulant-treated well-watered basil plants had higher H_2O_2 and MDA levels than the control [30].

Proline, a free amino acid, is often used as an indicator of stress. Under normal conditions, proline biosynthesis predominantly occurs in the cytosol, while an imbalance of osmotic pressure in the cell during drought stress causes proline production in both cytosol and chloroplast, which preferred the glutamate pathway proline metabolism instead of the ornithine pathway [31]. In this study, a significant build-up of proline in drought-stressed *P. amaryllifolius* was observed. However, plants treated with OPWV prior to the drought stress experienced lower proline accumulation. This finding was in agreement with previous studies [32,33]. For instance, foliar silicon treatment on salt-stressed-sweet pepper showed reduced proline accumulation [34]. Furthermore, drought-stressed sugar beet recorded lower proline content when treated with silicone and proline [35]. Although the biostimulatory mechanisms of OPWV are unclear, this finding implies that OPWV could alleviate the adverse effects of drought stress.

3.4. Enhanced Antioxidant Responses with OPWV Application

To counteract the harmful effects of ROS, plants have developed antioxidant defense mechanisms that include enzymatic components. When stressed, plants may produce excessive levels of ROS, which in turn increases the activity of antioxidant enzymes. This investigation observed increased antioxidant activities in drought-stressed *P. amaryllifolius* plants, indicating the plants were coping with the stress by scavenging harmful ROS. The initial line of defense in the antioxidant defense system appears to be provided by an increase in the activity of SOD, an enzyme that converts superoxide anion into H_2O_2. The increase in SOD activity under drought stress conditions is consistent with several previous observations [28,35]. An increase in the activity of CAT, APX, POD and GR by drought treatment indicates their role in H_2O_2 detoxification. CAT, POD and APX convert H_2O_2 into water and oxygen with the help of GR.

On the other hand, OPWV treatment reduced the antioxidant activities, such as SOD, CAT, POD, APX, and GR, in drought-stressed *P. amaryllifolius*. This finding was in contrast with other previous studies. For instance, applying diluted honey to fava beans enhanced their antioxidant activities in both well-watered and drought conditions [36]. It is noteworthy that OPWV treatment increased SOD, CAT, APX, and GR on well-watered *P. amaryllifolius*. Similar results have been reported by Zhu et al., where rapeseed plants primed with poplar-derived wood vinegar showed increased antioxidant activities [15]. Taken together, our findings suggest that priming plants with OPWV might activate their antioxidant defense mechanism, and OPWV-primed plants could overcome the drought stress effect since they recorded decreased antioxidant activities.

3.5. Carbohydrate-Related and Drought-Responsive Genes Altered after OPWV Treatment

Several carbohydrate-related, such as *PaGAPDH*, *PaENO* and *Paβ-Fruc*, and drought-responsive gene expressions, such as *PaHSP70* and *PaThau*, were determined. As expected, the drought-stressed *P. amaryllifolius* showed increased expression of *PaHSP70*, *PaENO*, *PaThau*, *PaGAPDH*, and *Paβ-Fruc* compared to well-watered samples. It is unsurprised that plants respond to stressors by activating their stress defense and carbohydrate-related genes to improve their survival. Increased carbohydrate metabolism is needed to meet the energy demands in drought tolerance. These findings were aligned with previous studies [37–39].

HSP70 and *Thau* genes have been shown to play an important role in plants responding to environmental stresses, such as heat and drought stress. Heat shock proteins (HSPs) are molecular chaperones that play a crucial role in maintaining proper protein folding and preventing aggregation in plants. These processes are vital for cell survival under stress conditions, such as drought. Several members of the *HSP70* gene family have been identified in various plant species. Some of these *HSP70* members have been reported to be involved in the plant's response to drought stress. For example, research has shown that the expression of *HSP70* in soybean plants increases in response to drought stress. On the other hand, thaumatin is a group of small and highly hydrophilic proteins found in various plant species. These proteins are involved in various plant stress responses, including drought stress. For instance, the drought-stressed durum wheat plants revealed an increased *Thau*, suggesting its vital role in the plant's adaptive response to drought stress [40].

Applying OPWV, however, reduced the expression of *PaENO* and *Paβ-Fruc*. Although the exact mechanism is unclear, the downregulation of these genes did not affect the overall plant tolerance against drought stress. On the contrary, *PaGAPDH*, *PaHSP70* and *PaThau* showed elevated expression in the OPWV-treated drought-stressed samples. In addition, several studies showed that *GAPDH*, *HSP70* and *Thau* play an essential role in plant biotic and abiotic tolerances [41–44]. These findings suggest that OPWV could modulate these drought-responsive genes to improve plant drought tolerance.

3.6. OPWV Compound Profile Elucidated

Wood vinegar is a complex mixture of water and various organic compounds, including acids, alcohols, phenols, aldehydes, and esters [45]. However, the composition of wood vinegar can vary depending on factors such as the pyrolysis process employed, the moisture content of the feedstock, and the type of biomass used, although differences in this last case are usually scarce. In this study, we identified 16 different compounds in OPWV. Of these, phenyl carbamate or anthranilic acid ($C_7H_7NO_2$), phenol, syringol, guaiacol, catechol, and tetrasiloxane were the major constituents in OPWV. These compounds have been reported in other wood vinegar [11,46,47] and have shown health-beneficial properties. For instance, phenyl carbamate, syringol, guaiacol, catechol, and tetrasiloxane exhibited antioxidant properties [12,13,48,49], which might be necessary for drought tolerance in plants. High phenols and their derivatives content in wood vinegar originated from the pyrolytic degradation of lignin. During the thermal reaction, lignin degradation releases aromatic compounds, such as guaiacyl and syringyl, resulting from the cleavage of alpha-ether and beta-ether weak bonds. Lignin is typically converted into derivatives, such as phenol, syringol, and catechol. Other chemical compounds present in wood vinegar, including alcohols, aldehydes, ketones, carboxylic acids, and ethers, are primarily derived from the decomposition of cellulose and hemicelluloses. Studies have also shown that wood vinegar contains high phenolic compounds, indicating its potential use as an antioxidant agent [11,47]. Since phenolic compounds are the largest group in OPWV, we speculate that adding them to plants might help them cope with drought stress, as shown in reduced antioxidant enzyme activities in OPWV-treated plants. However, how plants interact with these compounds or does this involve their intrinsic antioxidant systems remains unknown.

The improved plant morphological structure, such as root biomass, by OPWV might also be due to the compounds in OPWV. For example, anthranilic acid is an early precursor of auxin indole-3-acetic acid (IAA) in root cells [50]. Therefore, applying OPWV-containing anthranilic acid on Pandanus leaves could probably induce IAA synthesis, which is critical for root formation. On the other hand, syringol and guaiacol, the main components of lignin in OPWV, might improve the plant cell structural integrity and induce lignin biosynthesis [51].

4. Materials and Methods

4.1. Plant Material

Three-month-old uniform-size and disease-free *P. amaryllifolius* plants obtained from Green Nursery Sdn. Bhd., Muar, Johor, Malaysia, were transferred to polybags (20 × 11 cm^2) consisting of 600 g clay slit loam soil and acclimatized for 2 weeks in a growth room at Universiti Malaya, Malaysia. Whilst acclimatizing, the plants were treated with foliar fertilizer, Mr Garnick 20X (Baba, Pulau Pinang, Malaysia) once a week. All plants were maintained at 28 ± 2 °C, under a relative humidity of 80 ± 5% and 1500 lux light intensity for a 12:12 h photoperiod cycle. About 20 mL of distilled water was supplemented to all plants once a day.

4.2. Experimental Design and Foliar Application

OPWV obtained from Palm Leaf Trader Sdn. Bhd., Malaysia, was diluted to 1:100, 1:250, 1:500, and 1:1000 with distilled water containing 0.01% (*v*/*v*) Tween-20. These dilutions were selected based on the manufacturer's recommendations. The diluted OPWV (20 mL) was applied to each plant using a small handheld sprayer (Figure S2A) on the leaves adaxially and abaxially until runoff. After being air-dried, the plants were randomly placed in the growing plot. Foliar application of OPWV was repeated at 3-day intervals until 12 days (Figure 9A). The plants were allowed to grow for an additional 7 days before harvesting (Figure 9A,B). Distilled water containing 0.01% Tween-20 served as control. All plants received 20 mL of distilled water daily. Each treatment consisted of 5 plants, and the experiment was repeated thrice.

To determine the frequency of OPWV application, the determined diluted OPWV was applied to *P. amaryllifolius* plants once at 6-day intervals (F1), 3-day intervals (F2), or 1-day intervals (F3) using a similar method, as previously described (Figure 9B). Distilled water containing 0.01% Tween-20 served as control. All plants received 20 mL of distilled water daily. Each treatment consisted of 5 plants, and the experiment was repeated thrice.

The plant morphological changes for both experiments, such as DW, root-to-shoot ratio, and relative stem circumference, were recorded, and chlorophyll content was measured as described below.

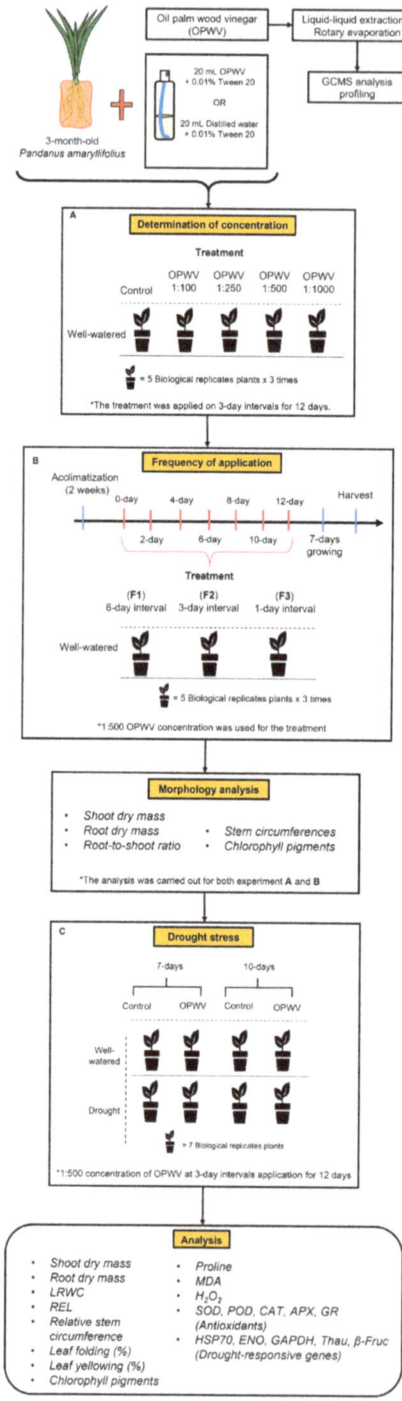

Figure 9. The schematic diagram of the experimental setup. (**A**) *Pandanus amaryllifolius* plants were sprayed with oil palm wood vinegar (OPWV) at 1:100, 1:250, 1:500, and 1:1000 dilutions in 3-day intervals until 12 days. The plants were allowed to grow for an additional 7 days before harvesting.

(**B**) *P. amaryllifolius* plants were sprayed with OPWV at 1:500 dilution at 6-day intervals (F1), 3-day intervals (F2), and 1-day intervals (F3). Well-watered plants served as control. (**C**) The optimized concentration and application frequency of OPWV was used to determine its potential to mitigate the drought stress effects. *P. amaryllifolius* plants were well-watered or treated with 7- or 10-day drought stress with or without OPWV.

4.3. Determination of Photosynthetic Pigment Contents

Chlorophyll pigments were measured according to Lichtenthaler [52]. The fresh leaves were ground and freeze-dried overnight to remove water content in the cells prior to the pigment's extraction. About 100 mg of the lyophilized leaf powder was added with 2 mL 90% (*v*/*v*) acetone in a 2 mL centrifuge tube. The mixture was kept in the dark for about 20 min before centrifugation at 10,000× *g* for 15 min to collect the supernatant. Then, 100 µL of the supernatant was diluted with 900 µL 80% (*v*/*v*) acetone before the absorbance of the mixture was measured using a spectrophotometer, NanoPhotometer P300 (Implen GmbH, Munich, Germany) at 470, 647, and 663 nm wavelengths. Each sample was analyzed in triplicate. The chlorophyll and carotenoid contents were calculated as follows:

$$\text{Chlorophyll } a\ (\mu\text{g mg}^{-1}) = 12.25\text{A}_{663} - 2.79\text{A}_{647} \tag{1}$$

$$\text{Chlorophyll } b\ (\mu\text{g mg}^{-1}) = 21.50\text{A}_{647} - 5.10\text{A}_{663} \tag{2}$$

$$\text{Total chlorophyll}\ (\mu\text{g mg}^{-1}) = 7.15\text{A}_{663} + 18.71\text{A}_{647} \tag{3}$$

$$\text{Carotenoids}\ (\mu\text{g mg}^{-1}) = (1000\text{A}_{470} - 1.82\text{C}_a - 85.02\text{C}_b)/198 \tag{4}$$

C_a = Chlorophyll *a*
C_b = Chlorophyll *b*

4.4. Drought Treatment

P. amaryllifolius plants were well-watered or treated drought stress with or without OPWV at 1:500 dilution. The OPWV was foliar applied to *P. amaryllifolius* plants at 3-day intervals (F2). Drought treatment was conducted by withholding water for 7 or 10 days, according to Amnan et al. [18]. After drought treatment, each plant's root and shoot DW, root-to-shoot ratio, and relative stem circumference were recorded (Figure 9C). Chlorophyll content was measured as previously described. The leaf numbers 3, 4, and 5 were harvested and wiped with distilled water and tissue paper before being used for LRWC, REL, gene expression and biochemical analyses to ensure consistency and standardization of data recording. Each treatment consisted of 7 independent biological replicates (three technical replicates within each) (Figure S2B). The overall experimental design is shown in Figure 9.

4.5. Leaf Relative Water Content

The harvested leaf samples were immediately processed to measure the LRWC, according to Turner [53]. In brief, the leaf samples were wiped with 70% ethanol and cut into several pieces (2 cm^2) prior to fresh weight (FW) weighing. The cut leaf pieces were then transferred into a Petri dish and submerged in 20 mL of distilled water for 6 h at room temperature. After being submerged, the saturated weight (SW) of the leaf pieces was measured before being subjected to oven drying for 2 days. The DW of the leaf samples was then measured, and LRWC was calculated using the following formula:

$$\text{LRWC (\%)} = [(\text{FW} - \text{DW})/(\text{SW} - \text{DW})] \times 100 \tag{5}$$

4.6. Relative Electrolyte Leakage

The REL was analyzed according to Quan et al. by immersing 100 mg leaves pieces in 10 mL deionized water before incubating at room temperature for 6 h under shaking conditions of 150 rpm [54]. Initial electrical conductivity (C_i) was measured using a conductivity meter, Cyberscan CON 11 (Thermo Fisher Scientific Inc., Waltham, MA, USA) before boiling the leaf samples for 20 min. The conductivity of lysed cells (C_{max}) was then measured. The REL was calculated using the following formula:

$$\text{REL (\%)} = C_i / C_{max} \times 100 \quad (6)$$

4.7. Plant Weight and Relative Stem Circumference

The fresh weight of the plants was analyzed immediately after harvesting, whereby the shoots and roots were separately weighed. The plants were collected and stored in a paper bag before drying in an oven for at least 7 days or until a constant weight was achieved to measure the DW. The root-to-shoot ratio was then calculated by dividing the DW of roots and shoots.

The stem circumference of each plant was recorded on days 0, 12 (end of foliar application treatment), 19 (end of the 7-day drought), and 22 (end of the 10-day drought) to analyze and compare the relative stem circumference between each treatment. The stem circumference was measured at the meeting point between the shoot and stem using a flexible one-meter measuring tape (Figure S2C).

4.8. Examination of Folded and Yellowing Leaves

Throughout the experiment, the leaf shape and color of *P. amaryllifolius* were observed and recorded. The leaf shape was categorized into three groups: No folding (Normal), less than 50% folding (Stage 1), and more than 50% folding (Stage 2) (Figure S2D). The number of folded leaves per plant was calculated accordingly to its stage as follows:

$$\text{Percentage of folding leaves (\%)} = (\text{Number of folded leaves per plant/Total number of leaves per plant}) \times 100 \quad (7)$$

The leaf color was recorded based on the number of yellowing leaves per plant according to the yellowing percentage scale (Figure S1E). The leaves with a 25% yellowing area and above were considered yellowing. The number of yellowing leaves per plant was calculated as follows:

$$\text{Percentage of yellowing leaves (\%)} = (\text{Number of yellowing leaves per plant/Total number of leaves per plant}) \times 100 \quad (8)$$

4.9. Malondialdehyde Content

MDA level in the leaf samples was analyzed according to Heath and Packer [55]. About 100 mg of fresh leaves were ground into a fine powder in a mortar with 1.5 mL 0.1% trichloroacetic acid (TCA) and transferred into a 2 mL centrifuge tube. The ground sample was centrifuged at 13,000× *g* for 10 min at 4 °C. The supernatant (300 μL) was mixed with 1 mL mixture containing 0.5% (*v*/*v*) thiobarbituric acid (TBA) in 20% (*w*/*v*) TCA. The mixture was heated at 95 °C for 30 min, ice-cooled, and centrifuged at 10,000× *g* for 10 min. The MDA content was calculated with the formula below from the absorbance measurement of the supernatant at 532 and 600 nm wavelengths using a spectrophotometer, NanoPhotometer P300 (Implen GmbH, Munich, Germany).

$$\text{MDA} = [(A_{532}-A_{600}) \times \text{VTr} \times 1000]/(\text{Extinction coefficient MDA} \times 1\text{ cm} \times \text{Ve})/\text{g FW} \quad (9)$$

$A_{532}-A_{600}$ = Absorbance of MDA–TBA
VTr = Volume of reaction (mL)
Ve = Volume of enzyme extract (mL)

The extinction coefficient of this MDA-TBA abduct at 532 nm is 155 mM^{-1} cm^{-1}.

4.10. Proline Content

Proline content was determined, as described by Bates et al. [56]. About 2 mL 70% (*v*/*v*) ethanol was added to 200 mg lyophilized leaf powder. After grinding, the mixture was centrifuged at 13,000× *g* for 20 min. The collected supernatant was mixed with 500 μL sample extract or proline standards (500 μL glacial acetic acid: 500 μL freshly prepared acid-ninhydrin reagent). The mixture was vortexed and boiled at 100 °C in a heat block for 1 h and cooled in ice for 30 min. The pigmentation developed during the reaction was extracted by adding 1 mL toluene, vortexed and centrifuged at

13,000× *g* for 20 min. The toluene phase was carefully collected into a new centrifuge tube before being measured at a 520 nm wavelength with toluene as a blank. A series of proline standards with the concentration of 3, 5, 10, 25, 50, 75, and 100 µM were measured to construct a proline standard curve which was then used to calculate the proline content in the leaf samples based on the following formula:

$$\text{Proline } (\mu\text{M g}^{-1}\text{ FW}) = [(\mu\text{g proline/mL} \times \text{mL toluene})/\mu\text{g } 115.5/\mu\text{mole}]/(\text{g FW}/5) \quad (10)$$

4.11. Antioxidant Enzyme Assays

The leaf samples were extracted in an extraction buffer containing (100 mM phosphate buffer, pH 7.0, 0.1 mM disodium ethylenediaminetetraacetic acid, and 0.1 g polyvinylpyrrolidone). About 200 mg of lyophilized leaf powder was mixed with 2 mL of the cold extraction buffer before centrifuging at 13,000× *g* for 10 min at 4 °C. The supernatant was collected and used for the subsequent antioxidant enzymatic assays. The enzyme activity for each assay was calculated using the formula below:

$$\text{Enzyme activity } (\text{M min}^{-1}\text{ g}^{-1}\text{ FW}) = (\Delta\text{A} \times \text{VTr})/(\varepsilon \times \Delta\text{t} \times 1\text{ cm} \times \text{Ve} \times \text{g FW}) \times 1000 \quad (11)$$

ΔA = Difference in absorbance
VTr = Volume of reaction (mL)
Ve = Volume of enzyme extract (mL)
Δt = Difference in time of absorbance (min)
For CAT, ε(Hydrogen peroxide) = 36.0 $\text{mol}^{-1}\text{ cm}^{-1}$
For APX, ε(Ascorbic acid) = 2.8 $\text{mmol}^{-1}\text{ cm}^{-1}$
For POD, ε(Tetraguaiacol) = 26.6 $\text{mol}^{-1}\text{ cm}^{-1}$
For GR, ε(NADPH) = 6220 $\text{mol}^{-1}\text{ cm}^{-1}$

4.11.1. Catalase

CAT activity in the leaf sample was analyzed, as described by Aebi with minor modifications [57]. The 3 mL reaction mixture comprised 50 mM phosphate buffer (pH 7.0), freshly prepared 8.33 mM H_2O_2, and 100 µL enzyme extract, which was added last to initiate the reaction. The enzyme activity was monitored and measured at 240 nm for 2 min with a 15 s reading interval.

4.11.2. Ascorbate Peroxidase

The APX assay was determined by Chen and Asada [58]. The 1 mL reaction mixture consisted of 50 mM phosphate buffer (pH 7.0), 200 µL enzyme extract, 0.5 mM ascorbic acid, and 0.42 mM H_2O_2. H_2O_2 was added last to initiate the reaction. The enzyme activity was monitored and measured at 290 nm for 2 min with a 15 s reading interval.

4.11.3. Peroxidase

POD activity was analyzed as described by Chance and Maehly with minor modifications [59]. The 1 mL reaction mixture contained 100 mM phosphate buffer (pH 7.0), 0.5 mM guaiacol, 0.0833 mM H_2O_2, and 100 µL enzyme extract. H_2O_2 was added last to initiate the reaction. The POD activity was measured at 470 nm wavelength for 2 min with a 15 s reading interval.

4.11.4. Glutathione Reductase

GR assay was conducted according to Mannervik [60]. The reaction mixture comprised of 500 µL assay buffer (0.2 M potassium phosphate buffer, pH 7.0, 0.2 mM EDTA), 50 µL 20 mM freshly prepared oxidized glutathione, 50 µL 2 mM NADPH solution, distilled water, and 300 µL enzyme extract. The decrease in absorbance at 340 nm after adding enzyme extract was monitored for 1 min with a 15 s reading interval.

4.11.5. Superoxide Dismutase

According to Dhindsa et al., the SOD activity in the sample was determined by a 3 mL reaction mixture containing 50 mM phosphate buffer (pH 7.0), 9.9 mM L-methionine, 55 µM nitro blue tetrazolium, 0.025% (*v*/*v*) Triton X-100 (Thermo Fisher Scientific, Inc., Waltham, MA, USA), 100 µL enzyme extract, and 4.8 µM riboflavin [61]. This reaction was prepared in an aluminium-covered test tube. The reaction was initiated by adding the riboflavin, mixing by shaking, and incubating at 30 °C for 10 min under a white light source (35 W) placed at 20 cm height above the test tubes. After the reaction, the mixture was measured at 560 nm with a blank prepared using the extraction buffer to replace the sample. The SOD activity was calculated based on the formula below:

$$\text{SOD (Unit g}^{-1}\text{ FW)} = [(\text{Blank} - \text{Sample})\ A_{560nm}/(\text{Blank}\ A_{560nm})] \times (\text{Volume reaction})/(\text{Volume enzyme}) \times 100 \times 1/50/0.1\ \text{g FW} \quad (12)$$

4.11.6. Hydrogen Peroxide

H_2O_2 content was analyzed according to Velikova et al. with some modifications [62]. In an ice-cold mortar, 100 mg of leaf powder was homogenized with 1.5 mL 0.1% (*w*/*v*) TCA and transferred into a tube. The mixture was centrifuged at 10,000× *g* for 15 min at 4 °C. About 250 μL of the collected supernatant was mixed with a 1 mL reaction mixture containing 2.5 mM potassium phosphate buffer, pH 7.0, and 0.5 M potassium iodide. The H_2O_2 level was calculated based on the reaction absorbance of the leaf samples and H_2O_2 standards of 2.5 to 100 μM measured at 390 nm wavelength.

4.12. RNA Extraction and cDNA Synthesis

The RNA was extracted using the cetyltrimethylammonium bromide conventional method, according to Asif et al. [63]. Briefly, 100 mg of finely ground leaf powder was used to extract the RNA. The final pellet was eluted in 30 μL of nuclease-free water to determine its concentration (ng/μL) and purity at A260/280 and A260/230 wavelengths using a spectrophotometer, NanoPhotometer P300 (Implen GmbH, Munich, Germany). Next, the extracted RNA was treated with DNase by RapidOut DNA removal kit (Thermo Scientific) to remove genomic DNA, followed by an RNA precipitation step. The precipitated DNA-free and DNase-free RNA was air dried before eluting with 30 μL of RNase-free water before measuring its concentration with a spectrophotometer. The RNA was then converted to cDNA using an NxGen M-MuLV Reverse Transcriptase (Lucigen) following the manufacturer's protocol. The RNA samples were stored at −80 °C until further use.

4.13. Quantitative Real-Time PCR

Quantitative real-time PCR (qPCR) was performed to analyze the expression of drought-stressed responsive genes (Table 2). The qPCR consisted of a reaction volume of 10 μL containing 40 ng of cDNA, 0.2 μM primers, and 1× SG Fast qPCR Master Mix (Sangon Biotech Co., Ltd., Shanghai, China), with the elongation factor-1 and actin as reference genes (Table 2). The qPCR assay was conducted according to the manufacturer's protocol. The relative expression levels were calculated according to Pfaffl [64]. The qPCR analysis was conducted with three biological replicates and three technical replicates for each gene.

Table 2. List of genes and their primer sequences used for relative expression analysis.

Gene	Sequences, 5′–3′
HSP70_Pandan (*PanHSP70*)	F-ACCTACAAGGGTGAGGAGAAG R-GAAATAGGCAGGGACAGTGATG
GAPDH_Pandan (*PanGAPDH*) [65]	F-AGGGTGGTGCCAAGAAGGT R-CCACCTCTCCAGTCCTT
Enolase_Pandan (*PanENO*)	F-TGAGTGATGGCACTTACGCC R-ACGTTCTCCACAGCCTTGAG
Thaumatin_Pandan (*PanThau*)	F-TCGCTGTCCTTCTCCTTTGG R-CACCTTGTGAGGAATGCAGC
β-fructofuranosidase_Pandan (*Panβ-Fruc*)	F-GAACCCTGGATGGTATCGGG R-CCGGCAAATGCTCCTAAGTG
* Actin_Pandan	F-GAGGCTATTCCTTCACCACTAC R-GTCTCAAGCTCCTCCTCATAATC
* Elongation factor-1_Pandan	F-TCTTCACAAAGCCAGCATCTC R-GACTGCCACACCTCTCATATTG

The asterisk (*) indicates the constitutive genes used as reference genes for gene expression normalization.

4.14. Oil Palm Wood Vinegar (OPWV) Gas Chromatography-Mass Spectrometry (GC-MS) Profiling

The liquid–liquid extraction method using dichloromethane (DCM) was used to extract OPWV compounds [66,67]. Briefly, 50 mL of the crude OPWV was extracted with 50 mL DCM in a separating funnel. The organic layer was collected, whereas the aqueous layer was extracted twice with DCM. All organic layers were combined to a final volume of 150 mL while the aqueous layer was discarded. The solvent was removed using a rotary evaporator at 40 °C for 1 h. The dried residue was weighed and reconstituted in 10 mL methanol prior to GC-MS profiling. About 100 ppm of collidine was added to the methanol solution with a final volume of 1 mL. The mixture was then diluted at 1:30 with methanol before injecting into a GC-MS (Shimadzu Manufacturing Co., Ltd., Kyoto, Japan, QP-2010). The GC-MS analysis parameters were set as follows: capillary columns (SH-5MS); 30 m $\times$ 0.25 mm diameter $\times$ 0.25 μm thickness; a temperature of injection: 250 °C; column temperature program: 50–220 °C and helium flow rate: 4.7 mL/min. The GC-MS was arranged in the electron ionization mode at 70 eV with an ion source and interface temperature of 250 °C and 300 °C, respectively. About 1 μL of the sample was injected into a column and held at 50 °C for 2 min with an increasing rate of 8 °C/min until 250 °C at which the temperature was held for another 2 min. The compounds were identified by comparison with the standard library data calculated based on the integrated peak areas relative to the internal standard peak area.

4.15. Statistical Analyses

The morphology and biochemical assays data were analyzed by one-way analysis of variance (ANOVA) followed by a post hoc Tukey range using SPSS Statistics software (version 23.0; IBM). The analyzed data were considered statistically significant when its p-value < 0.05.

5. Conclusions

In summary, our investigation unveils the potential of OPWV as a biostimulant to mitigate drought stress in *P. amaryllifolius*. Applying OPWV at 1:500 dilution at 3-day intervals for 12 days improved growth parameters in *P. amaryllifolius*. Although the imposed drought decreased stem circumference, leaf structure and pigmentation, applying OPWV alleviated these adverse effects. Furthermore, increased ROS, proline and MDA contents, antioxidant activities and drought-responsive gene expression in drought-stressed plants were reduced by OPWV. In addition, several compounds in OPWV, such as phenyl carbamate or anthranilic acid, tetrasiloxane, syringol, guaiacol, and catechol, might be responsible for their biostimulant properties that have been identified. However, although OPWV showed beneficial effects to plant growth, further studies on its effects on open field conditions and/or other crops are desirable. Identifying each compound that gives such biostimulatory effects to plants might be helpful for product development.

Supplementary Materials: The following supporting information can be downloaded at: https://www.mdpi.com/article/10.3390/plants12040785/s1, Figure S1. (**A**) Photographs of representative *Pandanus amaryllifolius* treated with 1:100, 1:250, 1:500, 1:1000 oil palm wood vinegar (OPWV) and distilled water (control) on days 0 and 19. (**B**) Photographs of representative *P. amaryllifolius* sprayed with OPWV at different frequencies on days 0 and 19. F1: applied at 6-day intervals, F2: applied at 3-day intervals, and F3: applied at 1-day interval. Bar = 10 cm; Figure S2. (**A**) Commercial handheld sprayer with 120 mL capacity. (**B**) The position of the harvested leaves. Leaf numbers 3, 4 and 5 were harvested for all assays. (**C**) The region for measuring the stem circumference. (**D**) Normal leaf and different stages of leaf folding. (**E**) The percentage of yellowing leaves from 0% to 100% yellowing. Figure S3. The GC-MS profile of oil palm wood vinegar (OPWV) at 1:500 dilution.

Author Contributions: Conceptualization, B.C.T., W.M.A. and F.D.K.; methodology and data analysis, M.A.M.A. and W.F.A.T.; writing—original draft preparation, M.A.M.A., B.C.T., W.M.A. and F.D.K.; supervision, B.C.T., W.M.A. and F.D.K.; funding acquisition, B.C.T. All authors have read and agreed to the published version of the manuscript.

Funding: This research was funded by Ministry of Higher Education, Malaysia, under the Fundamental Research Grant Scheme (FRGS/1/2018/STG03/UM/02/2) and Universiti Malaya under RU Fund (ST003-2021).

Data Availability Statement: The data presented in this study are available in Supplementary Materials.

Conflicts of Interest: The authors declare no competing interest.

References

1. Teoh, E.Y.; Teo, C.H.; Baharum, N.A.; Pua, T.-L.; Tan, B.C. Waterlogging Stress Induces Antioxidant Defense Responses, Aerenchyma Formation and Alters Metabolisms of Banana Plants. *Plants* **2022**, *11*, 2052. [CrossRef] [PubMed]
2. Gupta, A.; Rico-Medina, A.; Caño-Delgado, A.I. The Physiology of Plant Responses to Drought. *Science* **2020**, *368*, 266–269. [CrossRef] [PubMed]

3. Fadiji, A.E.; Santoyo, G.; Yadav, A.N.; Babalola, O.O. Efforts towards Overcoming Drought Stress in Crops: Revisiting the Mechanisms Employed by Plant Growth-Promoting Bacteria. *Front. Microbiol.* **2022**, *13*, 962427. [CrossRef]
4. Yang, X.; Lu, M.; Wang, Y.; Wang, Y.; Liu, Z.; Chen, S. Response Mechanism of Plants to Drought Stress. *Horticulturae* **2021**, *7*, 50. [CrossRef]
5. Lau, S.-E.; Teo, W.F.A.; Teoh, E.Y.; Tan, B.C. Microbiome Engineering and Plant Biostimulants for Sustainable Crop Improvement and Mitigation of Biotic and Abiotic Stresses. *Discov. Food* **2022**, *2*, 9. [CrossRef]
6. Frioni, T.; VanderWeide, J.; Palliotti, A.; Tombesi, S.; Poni, S.; Sabbatini, P. Foliar vs. Soil Application of *Ascophyllum nodosum* Extracts to Improve Grapevine Water Stress Tolerance. *Sci. Hortic.* **2021**, *277*, 109807. [CrossRef]
7. Ali, E.F.; El-Shehawi, A.M.; Ibrahim, O.H.M.; Abdul-Hafeez, E.Y.; Moussa, M.M.; Hassan, F.A.S. A Vital Role of Chitosan Nanoparticles in Improvisation the Drought Stress Tolerance in *Catharanthus roseus* (L.) through Biochemical and Gene Expression Modulation. *Plant Physiol. Biochem.* **2021**, *161*, 166–175. [CrossRef]
8. Zhang, K.; Khan, Z.; Liu, J.; Luo, T.; Zhu, K.; Hu, L.; Bi, J.; Luo, L. Germination and Growth Performance of Water-Saving and Drought-Resistant Rice Enhanced by Seed Treatment with Wood Vinegar and Biochar under Dry Direct-Seeded System. *Agronomy* **2022**, *12*, 1223. [CrossRef]
9. Voko, M.P.; Kulkarni, M.G.; Ngoroyemoto, N.; Gupta, S.; Finnie, J.F.; Van Staden, J. Vermicompost Leachate, Seaweed Extract and Smoke-Water Alleviate Drought Stress in Cowpea by Influencing Phytochemicals, Compatible Solutes and Photosynthetic Pigments. *Plant Growth Regul.* **2022**, *97*, 327–342. [CrossRef]
10. Wei, Q.; Ma, X.; Dong, J. Preparation, Chemical Constituents and Antimicrobial Activity of Pyroligneous Acids from Walnut Tree Branches. *J. Anal. Appl. Pyrolysis* **2010**, *87*, 24–28. [CrossRef]
11. Yang, J.-F.; Yang, C.-H.; Liang, M.-T.; Gao, Z.-J.; Wu, Y.-W.; Chuang, L.-Y. Chemical Composition, Antioxidant, and Antibacterial Activity of Wood Vinegar from *Litchi chinensis*. *Molecules* **2016**, *21*, 1150. [CrossRef]
12. Loo, A.Y.; Jain, K.; Darah, I. Antioxidant Activity of Compounds Isolated from the Pyroligneous Acid, *Rhizophora apiculata*. *Food Chem.* **2008**, *107*, 1151–1160. [CrossRef]
13. Azadfar, M.; Gao, A.H.; Bule, M.V.; Chen, S. Structural Characterization of Lignin: A Potential Source of Antioxidants Guaiacol and 4-Vinylguaiacol. *Int. J. Biol. Macromol.* **2015**, *75*, 58–66. [CrossRef]
14. Ozturk Sarikaya, S.B. Acethylcholinesterase Inhibitory Potential and Antioxidant Properties of Pyrogallol. *J. Enzyme Inhib. Med. Chem.* **2015**, *30*, 761–766. [CrossRef]
15. Zhu, K.; Gu, S.; Liu, J.; Luo, T.; Khan, Z.; Zhang, K.; Hu, L. Wood Vinegar as a Complex Growth Regulator Promotes the Growth, Yield, and Quality of Rapeseed. *Agronomy* **2021**, *11*, 510. [CrossRef]
16. Vannini, A.; Moratelli, F.; Monaci, F.; Loppi, S. Effects of Wood Distillate and Soy Lecithin on the Photosynthetic Performance and Growth of Lettuce (*Lactuca sativa* L.). *SN Appl. Sci.* **2021**, *3*, 113. [CrossRef]
17. Wang, Y.; Qiu, L.; Song, Q.; Wang, S.; Wang, Y.; Ge, Y. Root Proteomics Reveals the Effects of Wood Vinegar on Wheat Growth and Subsequent Tolerance to Drought Stress. *Int. J. Mol. Sci.* **2019**, *20*, 943. [CrossRef]
18. Amnan, M.A.M.; Aizat, W.M.; Khaidizar, F.D.; Tan, B.C. Drought Stress Induces Morpho-Physiological and Proteome Changes of *Pandanus amaryllifolius*. *Plants* **2022**, *11*, 221. [CrossRef]
19. Ghasemzadeh, A.; Jaafar, H.Z. Profiling of phenolic compounds and their antioxidant and anticancer activities in pandan (*Pandanus amaryllifolius* Roxb.) extracts from different locations of Malaysia. *BMC Complement. Altern. Med.* **2013**, *13*, 341. [CrossRef]
20. Lau, S.-E.; Hamdan, M.F.; Pua, T.-L.; Saidi, N.B.; Tan, B.C. Plant Nitric Oxide Signaling under Drought Stress. *Plants* **2021**, *10*, 360. [CrossRef]
21. Ye, H.; Song, L.; Schapaugh, W.T.; Ali, M.L.; Sinclair, T.R.; Riar, M.K.; Mutava, R.N.; Li, Y.; Vuong, T.; Valliyodan, B.; et al. The Importance of Slow Canopy Wilting in Drought Tolerance in Soybean. *J. Exp. Bot.* **2020**, *71*, 642–652. [CrossRef] [PubMed]
22. Amnan, M.A.M.; Pua, T.-L.; Lau, S.-E.; Tan, B.C.; Yamaguchi, H.; Hitachi, K.; Tsuchida, K.; Komatsu, S. Osmotic Stress in Banana Is Relieved by Exogenous Nitric Oxide. *PeerJ* **2021**, *9*, e10879. [CrossRef] [PubMed]
23. Rasul, F.; Gupta, S.; Olas, J.J.; Gechev, T.; Sujeeth, N.; Mueller-Roeber, B. Priming with a Seaweed Extract Strongly Improves Drought Tolerance in Arabidopsis. *Int. J. Mol. Sci.* **2021**, *22*, 1469. [CrossRef]
24. Adiba, A.; Hssaini, L.; Haddioui, A.; Hamdani, A.; Charafi, J.; El Iraqui, S.; Razouk, R. Pomegranate Plasticity to Water Stress: Attempt to Understand Interactions between Cultivar, Year and Stress Level. *Heliyon* **2021**, *7*, e07403. [CrossRef] [PubMed]
25. Zahedi, S.M.; Hosseini, M.S.; Daneshvar Hakimi Meybodi, N.; Abadía, J.; Germ, M.; Gholami, R.; Abdelrahman, M. Evaluation of Drought Tolerance in Three Commercial Pomegranate Cultivars Using Photosynthetic Pigments, Yield Parameters and Biochemical Traits as Biomarkers. *Agric. Water Manag.* **2022**, *261*, 107357. [CrossRef]
26. Ma, J.; Islam, F.; Ayyaz, A.; Fang, R.; Hannan, F.; Farooq, M.A.; Ali, B.; Huang, Q.; Sun, R.; Zhou, W. Wood Vinegar Induces Salinity Tolerance by Alleviating Oxidative Damages and Protecting Photosystem II in Rapeseed Cultivars. *Ind. Crops Prod.* **2022**, *189*, 115763. [CrossRef]
27. Esterbauer, H.; Schaur, R.J.; Zollner, H. Chemistry and Biochemistry of 4-Hydroxynonenal, Malonaldehyde and Related Aldehydes. *Free Radic. Biol. Med.* **1991**, *11*, 81–128. [CrossRef]
28. Habib, N.; Ali, Q.; Ali, S.; Javed, M.T.; Zulqurnain Haider, M.; Perveen, R.; Shahid, M.R.; Rizwan, M.; Abdel-Daim, M.M.; Elkelish, A.; et al. Use of Nitric Oxide and Hydrogen Peroxide for Better Yield of Wheat (*Triticum aestivum* L.) under Water Deficit Conditions: Growth, Osmoregulation, and Antioxidative Defense Mechanism. *Plants* **2020**, *9*, 285. [CrossRef]

29. Hameed, A.; Akram, N.A.; Saleem, M.H.; Ashraf, M.; Ahmed, S.; Ali, S.; Abdullah Alsahli, A.; Alyemeni, M.N. Seed Treatment with α-Tocopherol Regulates Growth and Key Physio-Biochemical Attributes in Carrot (*Daucus carota* L.) Plants under Water Limited Regimes. *Agronomy* **2021**, *11*, 469. [CrossRef]
30. Gohari, G.; Alavi, Z.; Esfandiari, E.; Panahirad, S.; Hajihoseinlou, S.; Fotopoulos, V. Interaction between Hydrogen Peroxide and Sodium Nitroprusside Following Chemical Priming of *Ocimum basilicum* L. against Salt Stress. *Physiol. Plant.* **2020**, *168*, 361–373. [CrossRef]
31. Per, T.S.; Khan, N.A.; Reddy, P.S.; Masood, A.; Hasanuzzaman, M.; Khan, M.I.R.; Anjum, N.A. Approaches in Modulating Proline Metabolism in Plants for Salt and Drought Stress Tolerance: Phytohormones, Mineral Nutrients and Transgenics. *Plant Physiol. Biochem.* **2017**, *115*, 126–140. [CrossRef]
32. Dien, D.C.; Mochizuki, T.; Yamakawa, T. Effect of Various Drought Stresses and Subsequent Recovery on Proline, Total Soluble Sugar and Starch Metabolisms in Rice (*Oryza sativa* L.) Varieties. *Plant Prod. Sci.* **2019**, *22*, 530–545. [CrossRef]
33. Furlan, A.L.; Bianucci, E.; Giordano, W.; Castro, S.; Becker, D.F. Proline Metabolic Dynamics and Implications in Drought Tolerance of Peanut Plants. *Plant Physiol. Biochem.* **2020**, *151*, 566–578. [CrossRef]
34. Abdelaal, K.A.A.; Attia, K.A.; Alamery, S.F.; El-Afry, M.M.; Ghazy, A.I.; Tantawy, D.S.; Al-Doss, A.A.; El-Shawy, E.-S.E.; M. Abu-Elsaoud, A.; Hafez, Y.M. Exogenous Application of Proline and Salicylic Acid Can Mitigate the Injurious Impacts of Drought Stress on Barley Plants Associated with Physiological and Histological Characters. *Sustainability* **2020**, *12*, 1736. [CrossRef]
35. AlKahtani, M.D.F.; Hafez, Y.M.; Attia, K.; Rashwan, E.; Husnain, L.A.; AlGwaiz, H.I.M.; Abdelaal, K.A.A. Evaluation of Silicon and Proline Application on the Oxidative Machinery in Drought-Stressed Sugar Beet. *Antioxidants* **2021**, *10*, 398. [CrossRef]
36. Rady, M.M.; Boriek, S.H.K.; Abd El-Mageed, T.A.; Seif El-Yazal, M.A.; Ali, E.F.; Hassan, F.A.S.; Abdelkhalik, A. Exogenous Gibberellic Acid or Dilute Bee Honey Boosts Drought Stress Tolerance in Vicia Faba by Rebalancing Osmoprotectants, Antioxidants, Nutrients, and Phytohormones. *Plants* **2021**, *10*, 748. [CrossRef]
37. Xu, J.; Zhou, Y.; Xu, Z.; Chen, Z.; Duan, L. Physiological and Transcriptome Profiling Analyses Reveal Important Roles of Coronatine in Improving Drought Tolerance of Tobacco. *J. Plant Growth Regul.* **2020**, *39*, 1346–1358. [CrossRef]
38. Chen, L.; Zheng, F.; Feng, Z.; Li, Y.; Ma, M.; Wang, G.; Zhao, H. A Vacuolar Invertase CsVI2 Regulates Sucrose Metabolism and Increases Drought Tolerance in *Cucumis sativus* L. *Int. J. Mol. Sci.* **2022**, *23*, 176. [CrossRef]
39. Augustine, S.M.; Cherian, A.V.; Syamaladevi, D.P.; Subramonian, N. *Erianthus arundinaceus* HSP70 (EaHSP70) Acts as a Key Regulator in the Formation of Anisotropic Interdigitation in Sugarcane (*Saccharum* spp. Hybrid) in Response to Drought Stress. *Plant Cell Physiol.* **2015**, *56*, 2368–2380. [CrossRef]
40. Djemal, R.; Bahloul, O.; Khoudi, H. A novel Thaumatin-Like Protein From Durum Wheat, TdPR-5, is Homologous To Known Plant Allergens, Responsive to Stress Exposure, and Confers Multiple-Abiotic Stress Tolerances to Transgenic Yeast. *Plant Gene* **2022**, *31*, 100360. [CrossRef]
41. Kappachery, S.; Baniekal-Hiremath, G.; Yu, J.W.; Park, S.W. Effect of Over-and under-Expression of Glyceraldehyde 3-Phosphate Dehydrogenase on Tolerance of Plants to Water-Deficit Stress. *Plant Cell Tissue Organ Cult.* **2015**, *121*, 97–107. [CrossRef]
42. Misra, R.C.; Sandeep; Kamthan, M.; Kumar, S.; Ghosh, S. A Thaumatin-like Protein of *Ocimum basilicum* Confers Tolerance to Fungal Pathogen and Abiotic Stress in Transgenic Arabidopsis. *Sci. Rep.* **2016**, *6*, 25340. [CrossRef] [PubMed]
43. Odeny Ojola, P.; Nyaboga, E.N.; Njiru, P.N.; Orinda, G. Overexpression of Rice Thaumatin-like Protein (*Ostlp*) Gene in Transgenic Cassava Results in Enhanced Tolerance to *Colletotrichum gloeosporioides* f. sp. Manihotis. *J. Genet. Eng. Biotechnol.* **2018**, *16*, 125–131. [CrossRef] [PubMed]
44. Sharma, A.; Sharma, H.; Rajput, R.; Pandey, A.; Upadhyay, S.K. Molecular Characterization Revealed the Role of Thaumatin-Like Proteins of Bread Wheat in Stress Response. *Front. Plant Sci.* **2022**, *12*, 807448. [CrossRef]
45. Liu, X.; Wang, J.; Feng, X.; Yu, J. Wood Vinegar Resulting From the Pyrolysis of Apple Tree Branches for Annual Bluegrass Control. *Ind. Crops Prod.* **2021**, *147*, 114193. [CrossRef]
46. Oramahi, H.A.; Yoshimura, T.; Diba, F.; Setyawati, D. Nurhaida Antifungal and Antitermitic Activities of Wood Vinegar from Oil Palm Trunk. *J. Wood Sci.* **2018**, *64*, 311–317. [CrossRef]
47. Xue, R.; Cui, E.-L.; Hu, G.-Q.; Zhu, M.-Q. The Composition, Physicochemical Properties, Antimicrobial and Antioxidant Activity of Wood Vinegar Prepared by Pyrolysis of *Eucommia ulmoides* Oliver Branches under Different Refining Methods and Storage Conditions. *Ind. Crops Prod.* **2022**, *178*, 114586. [CrossRef]
48. Li, Y.; Liang, L.; Huang, S.; Li, W.; Ashraf, U.; Ma, L.; Mo, Z. Exogenous Melatonin and Catechol Application Modulate Physio-Biochemical Attributes and Early Growth of Fragrant Rice Under Cd Toxicity. *J. Soil Sci. Plant Nutr.* **2021**, *21*, 2285–2296. [CrossRef]
49. Singh, N.; Mansoori, A.; Jiwani, G.; Solanke, A.U.; Thakur, T.K.; Kumar, R.; Chaurasiya, M.; Kumar, A. Antioxidant and Antimicrobial Study of *Schefflera vinosa* Leaves Crude Extracts against Rice Pathogens. *Arab. J. Chem.* **2021**, *14*, 103243. [CrossRef]
50. Doyle, S.M.; Rigal, A.; Grones, P.; Karady, M.; Barange, D.K.; Majda, M.; Pařízková, B.; Karampelias, M.; Zwiewka, M.; Pěnčík, A.; et al. A Role for the Auxin Precursor Anthranilic Acid in Root Gravitropism via Regulation of PIN-FORMED Protein Polarity and Relocalisation in Arabidopsis. *New Phytol.* **2019**, *223*, 1420–1432. [CrossRef]
51. Liu, Q.; Luo, L.; Zheng, L. Lignins: Biosynthesis and Biological Functions in Plants. *Int. J. Mol. Sci.* **2018**, *19*, 335. [CrossRef] [PubMed]
52. Lichtenthaler, H.K. Chlorophylls and Carotenoids: Pigments of Photosynthetic Biomembranes. In *Methods in Enzymology*; Plant Cell Membranes; Academic Press: Cambridge, MA, USA, 1987; Volume 148, pp. 350–382.

53. Turner, N.C. Techniques and Experimental Approaches for the Measurement of Plant Water Status. *Plant Soil* **1981**, *58*, 339–366. [CrossRef]
54. Quan, W.; Hu, Y.; Mu, Z.; Shi, H.; Chan, Z. Overexpression of *AtPYL5* under the Control of Guard Cell Specific Promoter Improves Drought Stress Tolerance in Arabidopsis. *Plant Physiol. Biochem.* **2018**, *129*, 150–157. [CrossRef]
55. Heath, R.L.; Packer, L. Photoperoxidation in Isolated Chloroplasts: I. Kinetics and Stoichiometry of Fatty Acid Peroxidation. *Arch. Biochem. Biophys.* **1968**, *125*, 189–198. [CrossRef] [PubMed]
56. Bates, L.S.; Waldren, R.P.; Teare, I.D. Rapid Determination of Free Proline for Water-Stress Studies. *Plant Soil* **1973**, *39*, 205–207. [CrossRef]
57. Aebi, H. Catalase. In *Methods of Enzymatic Analysis*, 2nd ed.; Bergmeyer, H.U., Ed.; Academic Press: Cambridge, MA, USA, 1974; pp. 673–684. ISBN 978-0-12-091302-2.
58. Chen, G.-X.; Asada, K. Inactivation of Ascorbate Peroxidase by Thiols Requires Hydrogen Peroxide. *Plant Cell Physiol.* **1992**, *33*, 117–123. [CrossRef]
59. Chance, B.; Maehly, A.C. *Assay of Catalases and Peroxidases: In Methods in Enzymology*; Academic Press: Cambridge, MA, USA, 1955; Volume 2, pp. 764–775.
60. Mannervik, B. Measurement of Glutathione Reductase Activity. *Curr. Protoc. Toxicol.* **1999**, 7.2.1–7.2.4. [CrossRef]
61. Dhindsa, R.S.; Plumb-Dhindsa, P.; Thorpe, T.A. Leaf Senescence: Correlated with Increased Levels of Membrane Permeability and Lipid Peroxidation, and Decreased Levels of Superoxide Dismutase and Catalase. *J. Exp. Bot.* **1981**, *32*, 93–101. [CrossRef]
62. Velikova, V.; Yordanov, I.; Edreva, A. Oxidative Stress and Some Antioxidant Systems in Acid Rain-Treated Bean Plants: Protective Role of Exogenous Polyamines. *Plant Sci.* **2000**, *151*, 59–66. [CrossRef]
63. Asif, M.H.; Dhawan, P.; Nath, P. A Simple Procedure for the Isolation of High Quality RNA from Ripening Banana Fruit. *Plant Mol. Biol. Report.* **2000**, *18*, 109–115. [CrossRef]
64. Pfaffl, M.W. A New Mathematical Model for Relative Quantification in Real-Time RT–PCR. *Nucleic Acids Res.* **2001**, *29*, e45. [CrossRef] [PubMed]
65. Arora, V.; Sultana, M.; Kumar, V.; Gangopadhyay, G. Isolation and Characterization of *BADH2* Gene from in Vitro Propagated *Pandanus amaryllifolius* Roxb. *Plant Cell Tissue Organ Cult.* **2017**, *130*, 131–140. [CrossRef]
66. Mahmud, K.N.; Yahayu, M.; Siti Hajar, M.S.; Rizan, N.H.; ChaiBing, M.; Mustafa, N.F.; Ngadiran, S.; Ujang, S.; Zakaria, Z.A. Evaluation on Efficiency of Pyroligneous Acid from Palm Kernel Shell as Antifungal and Solid Pineapple Biomass as Antibacterial and Plant Growth Promoter. *Sains Malays.* **2016**, *45*, 1423–1434.
67. Mohd Hamzah, M.A.A.; Hasham, R.; Nik Malek, N.A.N.; Raja Sulong, R.S.; Yahayu, M.; Abdul Razak, F.I.; Zakaria, Z.A. Structural-Based Analysis of Antibacterial Activities of Acid Condensate from Palm Kernel Shell. *Biomass Convers. Biorefin.* **2022**. [CrossRef]

plants MDPI

Article

Unexpected Effects of Sulfate and Sodium Chloride Application on Yield Qualitative Characteristics and Symmetry Indicators of Hard and Soft Wheat Kernels

Tatiana S. Aniskina [1,*], Ekaterina N. Baranova [1,2,*], Svyatoslav V. Lebedev [3], Nelli S. Reger [3], Ishen N. Besaliev [3], Alexander A. Panfilov [3], Viktoriya A. Kryuchkova [1] and Alexander A. Gulevich [2,*]

1 N.V. Tsitsin Main Botanical Garden of Russian Academy of Sciences, Botanicheskaya 4, 127276 Moscow, Russia
2 All-Russia Research Institute of Agricultural Biotechnology, Timiryazevskaya 42, 127550 Moscow, Russia
3 Federal Scientific Center of Biological Systems and Agrotechnology of the Russian Academy of Sciences, 9 Yanvarya 29, 460000 Orenburg, Russia
* Correspondence: tatianiskina@gmail.com (T.S.A.); greenpro2007@rambler.ru (E.N.B.); a_gulevich@mail.ru (A.A.G.)

Abstract: The heterogeneity of grain quality can lead to limited predictability of qualitative and quantitative characteristics of the wheat yield, especially with an increase in the importance of drought and salinity caused by climate change. This study was undertaken with the aim of creating basic tools for phenotyping and assessing the sensitivity of genotypes to salt effects at the level of some wheat kernel attributes. The study considers 36 variants of the experiment, including four wheat cultivars—Zolotaya, Ulyanovskaya 105, Orenburgskaya 10, Orenburgskaya 23; three treatment variants—control (without salt) and two salts exposure (NaCl at a concentration of 1.1 g L^{-1} and Na_2SO_4 at a concentration of 0.4 g L^{-1}); as well as three options for the arrangement of kernels in a simple spikelet—left, middle, and right. It has been established that the salt exposure had a positive effect on the percentage of kernel fulfilling in the cultivars Zolotaya, Ulyanovskaya 105, and Orenburgskaya 23 compared to control. The kernels of the Orenburgskaya 10 variety matured better in the experiment with Na_2SO_4 exposure, while the control variant and NaCl gave the same effect. When exposed to NaCl, significantly greater values of weight, transverse section area, and transverse section perimeter of the kernel were noted in the cv Zolotaya and Ulyanovskaya 105. Cv Orenburgskaya 10 responded positively to the use of Na_2SO_4. This salt caused an increase in the area, length, and width of the kernel. The fluctuating asymmetry of the left, middle, and right kernels in the spikelet was calculated. In the cv Orenburgskaya 23 the salts affected only the kernel perimeter among parameters examined. The indicators of the general (fluctuating) asymmetry were lower in the experiments with the use of salts, i.e., kernels were more symmetrical than in the control variant, both for the cultivar as a whole and when compared taking into account the kernel location in spikelet. However, this result was unexpected, since salt stress inhibited a number of morphological parameters: the number and average length of embryonic, adventitious, and nodal roots, flag leaf area, plant height, dry biomass accumulation, and plant productivity indicators. The study showed that low concentrations of salts can positively affect the fulfilling of kernels (the absence of a cavity inside the kernel) and the symmetry of the left and right sides of the kernel.

Keywords: wheat kernel; salt effect; NaCl; Na_2SO_4; kernel phenotype; fluctuating asymmetry

Citation: Aniskina, T.S.; Baranova, E.N.; Lebedev, S.V.; Reger, N.S.; Besaliev, I.N.; Panfilov, A.A.; Kryuchkova, V.A.; Gulevich, A.A. Unexpected Effects of Sulfate and Sodium Chloride Application on Yield Qualitative Characteristics and Symmetry Indicators of Hard and Soft Wheat Kernels. *Plants* **2023**, *12*, 980. https://doi.org/10.3390/plants12050980

Academic Editors: Mingxun Chen, Lixi Jiang and Yuan Guo

Received: 31 December 2022
Revised: 16 February 2023
Accepted: 17 February 2023
Published: 21 February 2023

1. Introduction

Wheat is the most important crop in the world; it provides the main food in most regions of the planet and is sown on more than 220 million hectares [1]. A variety of genotypes and growing conditions allows breeders to create varieties with different types of resistance, different features of ontogenesis, and, hence, different qualitative characteristics of the most valuable wheat product—single-seeded fruits or caryopses (kernels) [2]. The formation of a kernel is a complex process that includes many transformations of various tissues in the developing wheat fruit. These processes occur over several weeks and include both the death of a number of tissues and the formation of new tissues that form the kernel (caryopsis) and its integuments [3,4]. The impact of abiotic factors on such a system should hypothetically cause significant changes both in the formation processes themselves and in the storage processes in specialized tissues such as the endosperm and the aleurone layer.

The structure of the wheat spike creates a deceptive feeling of the presence of bilateral symmetry [5]. The features of the spiral geometry characteristic of the reproductive organs, as well as for the entire wheat plant as a whole, become apparent only with a detailed study of ontogenesis and the sequential transformation of the phenotypic pattern of the flowering organ from the primary tubercle to a full-fledged multi-spikelet inflorescence, traditionally described as a spike (ear) [6]. The very sequence of laying flowers in spikelets makes it obvious that there is a certain underlying asymmetry in this seemingly ideal model—the kernel. The performance of asymmetry in the spikelets and spikes of several wheat genotypes with different characteristic arrangement of kernels in the ear was described earlier [7]. The physical reasons for the asymmetry of kernels were predicted at the beginning of the 20th century [8] and have not yet been experimentally confirmed due to the complexity of the qualitative setting of the experiment and the lack of an explicit experimental approach [9]. Meanwhile, a higher degree of symmetry is considered by breeders as an indication of a genetic advantage [10].

Climate change in the near future will lead to a significant reduction in yields due to increased adverse impacts such as drought, excessive humidity, and primary and secondary soil salinization [11]. Currently, salinity is of concern due to the fact that sodium salts are not able to form stable insoluble compounds and often cause an increase in osmotic potential and accumulation of ions to critical concentrations that damage plant roots when accumulated by groundwater during reclamation and irrigation [12]. Additionally, in some cases, salinization can occur due to the use of road de-icing agents, which are then washed out into the environment [13,14]. In regions close to large saline lakes, seas, or oceans, this effect can be caused by climate change causing salt transport by wind [15], droughts [16], and storms from the sea leading to salinization of coastal wetlands [17,18].

Wheats with a different genotype and origin may show different tolerance and sensitivity to salinity [19–21]. The effects of the action of salts on seedlings, on plant biomass, on the ability for effective photosynthesis are exceedingly well studied; changes are shown at the ultrastructure level of the nucleus, plastids, and cytoskeleton [22,23]. Salinity causes intracellular processes characteristic of oxidative, osmotic, and toxic stress. Cells cannot cope with the excess supply of toxic sodium ions; as a result, intra- and intercellular transport in various parts of the plant is disrupted. A specific system of various transporters causes a redistribution of other ions, disrupting the processes of synthesis, transport, and accumulation of primary metabolites and stored compounds. Additionally, significant damage is observed in the activity of a number of important enzymes and gene expression [19,24]. However, the final results for evaluating the productivity, namely, the quality characteristics of the grains and the yield, are important. Selection and forecasting require both qualitative and quantitative evaluation methods capable of predicting the consequences of using a variety or breeding line. It is possible that the assessment of the phenotype of kernels in the spike can become reliable evaluation criterion in the future agronomic studies of wheat cultivated in saline areas.

The present study was carried out to apply the previously estimated indicators of quantitative determination of the degree of kernel asymmetry in two cultivars of soft wheat and two cultivars of durum wheat when cultivated in the field under salinity conditions due to sulfate and sodium chloride. In addition, the goal was to evaluate the quality indicators and the fluctuating asymmetry in the kernels.

2. Results

2.1. Salt Exposure Effect on Morphological Parameters of Wheat Plants

Due to the presence of moisture in the root layer and precipitation that fell in the initial period of development (shooting–tillering) of spring wheat, the root systems of the studied cultivars were well developed. They consisted of both primary (embryonic and adventitious) and secondary (nodal) roots. The embryonic roots were subject to a more significant decrease under the influence of salinity, with the greatest manifestation in the cv Orenburgskaya 23 and Zolotaya (Table 1). The cv Orenburgskaya 10 was more stable in this indicator. In the first two cultivars, not only their number decreased, but also the average length. In the cv Ulyanovskaya 105, only the number of embryonic roots decreased.

Table 1. Characteristics of the root system of spring wheat cultivars under salt stress caused by different salts.

Cultivar	Experiment Variant	Embryonic Roots Number, pcs.	Embryonic Roots Length, cm	Adventitious Roots, pcs.	Nodal Roots, pcs.
Orenburgskaya 23	Control	4.4 ± 1.4	5.2 ± 3.0	1.5 ± 0.7	0.7 ± 0.3
	NaCl—1.1 g L^{-1}	3.3 ± 1.4	4.5 ± 3.1	2.2 ± 0.6	1.5 ± 0.6
	Na_2SO_4—0.40 g L^{-1}	3.9 ± 1.0	4.0 ± 2.1	1.9 ± 0.3	1.0 ± 0.4
Ulyanovskaya 105	control	4.7 ± 1.2	4.3 ± 2.0	5.0 ± 2.3	3.8 ± 1.3
	NaCl—1.1 g L^{-1}	3.9 ± 1.2	5.1 ± 3.3	2.6 ± 1.4	2.8 ± 1.5
	Na_2SO_4—0.40 g L^{-1}	4.0 ± 1.6	4.8 ± 2.5	2.5 ± 1.1	1.7 ± 0.7
Orenburgskaya 10	control	4.8 ± 1.8	3.5 ± 1.7	5.4 ± 2.0	1.3 ± 0.5
	NaCl—1.1 g L^{-1}	4.7 ± 0.9	4.1 ± 2.0	2.8 ± 1.3	0.9 ± 0.3
	Na_2SO_4—0.40 g L^{-1}	5.9 ± 1.6	4.9 ± 2.2	4.7 ± 2.5	2.5 ± 1.2
Zolotaya	control	5.1 ± 1.3	5.5 ± 2.3	4.2 ± 2.0	1.7 ± 0.7
	NaCl—1.1 g L^{-1}	4.2 ± 0.8	4.4 ± 2.6	2.3 ± 0.5	0.8 ± 0.3
	Na_2SO_4—0.40 g L^{-1}	3.9 ± 1.2	5.2 ± 2.9	4.8 ± 0.6	1.6 ± 0.6
LSD_{05}		0.62	0.08	0.67	0.72
LSD_{05}A (Cultivar)		0.51	0.74	0.55	0.53
LSD_{05}B (Experiment variant)		0.44	0.62	0.48	0.42
LSD_{05}AB		0.44	0.62	0.48	0.42

LSD—least significant difference test ($p = 0.05$). Arithmetic mean ± standard deviation, $p = 0.05$.

Under the influence of various salinity options, the number of adventitious roots decreased in cultivars Ulyanovskaya 105, Orenburgskaya 10, and Zolotaya. Salt exposure caused a decrease in the number of nodal roots in the cv Ulyanovskaya 105.

Differences in the suppressive effect of chloride and sulfate salinization are insignificant and are not clearly marked, taking into account the least significant difference (LSD) for the 5% significance level.

A decrease in plant height as a factor in the inhibitory effect of salt stress was manifested in three cultivars (Table 2). In cultivars Orenburgskaya 23 and Ulyanovskaya 105, it was noted under the influence of Na_2SO_4 treatment. In the cv Orenburgskaya 10, both types of salinity contributed to the decrease in plant height. The decrease in the length of the spike in cultivars Orenburgskaya 23, Ulyanovskaya 105, and Orenburgskaya 10 was largely due to the negative effect of chloride salinity. For the cv Zolotaya, no negative effects of additional salinization on the length of the spike were noted, and even a positive effect of chloride salinization on this indicator was noted.

Table 2. Influence of different salinity variants on the morphological parameters of spring wheat cultivars in the earing phase.

Cultivar	Experiment Variant	Plant Height, cm	Spike Length, cm
Orenburgskaya 23	control	46.7 ± 6.8	8.4 ± 0.6
	NaCl—1.1 g L^{-1}	49.1 ± 6.2	6.5 ± 0.7
	Na_2SO_4—0.40 g L^{-1}	43.1 ± 6.5	8.0 ± 1.1
Ulyanovskaya 105	control	55.6 ± 10.4	6.8 ± 0.9
	NaCl—1.1 g L^{-1}	57.6 ± 8.3	6.2 ± 1.4
	Na_2SO_4 − 0.40 g L^{-1}	51.0 ± 6.9	7.6 ± 0.9
Orenburgskaya 10	control	60.0 ± 9.9	6.6 ± 0.7
	NaCl—1.1 g L^{-1}	52.0 ± 7.9	5.3 ± 0.9
	Na_2SO_4—0.40 g L^{-1}	47.0 ± 6.6	5.8 ± 0.8
Zolotaya	control	61.6 ± 6.0	6.8 ± 0.6
	NaCl—1.1 g L^{-1}	63.4 ± 10.5	7.8 ± 0.8
	Na_2SO_4—0.40 g L^{-1}	62.0 ± 9.8	6.6 ± 1.0
LSD $_{05}$		3.82	0.41
LSD $_{05}$A (Cultivar)		3.12	0.33
LSD $_{05}$B (Experiment variant)		2.70	0.29
LSD $_{05}$AB		2.70	0.29

LSD—least significant difference test (p = 0.05). Arithmetic mean ± standard deviation, p = 0.05.

2.2. *Effect of Salts on Fulfilling (Plumpness) of Kernels*

The study revealed that out of 251 kernels examined in cv Zolotaya plants, only 50 kernels were incompletely matured or had damage. In the control variant, the percentage of unmatured kernels was 33% from the sampling (38 out of 114 kernels), while in the variants with the treatment by salts, the result differed in a positive direction—10% with Na_2SO_4 and 7% with NaCl. When comparing the percentage of matured, well-filled kernels among the left and right kernels in spikelet, it turned out that the left kernels have a greater percentage in all salinity options: 73% in the control (61% of the maturing in the right kernels), 94% when exposed to NaCl (89% of the right kernels), and 92% under exposure to Na_2SO_4 (88% of the right kernels).

A similar result in maturation was noted in the cv Ulyanovskaya 105. Thus, in the variant with application of NaCl, 98% of the kernels (130 out of 133 kernels) were matured, under Na_2SO_4 exposure 91% of the kernels (150 out of 164 kernels) were matured, and in the control variant 72%. In total, 466 kernels were examined in the study, and 401 of them (86%) were maturing (well-filled). Upon a detailed examination of the left and right grains, it should be noted that in the experiment with NaCl exposure, 100% of the left grains and 95% of the right kernels were maturing, while under exposure to Na_2SO_4—93% of the right kernels and 88% of the left kernels, and in the control 73% of the left kernels and only 65% of the right kernels were maturing.

Cultivar Orenburgskaya 23 matured better in the variant with exposure to NaCl (92% of matured or 96 of 104 kernels), in the experiment with exposure to Na_2SO_4—88% (78 of 89 kernels), and in the control variant 83% (95 of 115 kernels). In total, in the sample of 308 kernels examined, 269 matured, well-filled kernels (87%) were noted. The proportion of the distribution of matured grains relative to their location in the spikelet after exposure to salts on wheat plants was as follows: in the variant with NaCl, 93% of the right and 93% of the left kernels were maturing; in the variant with Na_2SO_4,—90% of the right and 83% of the left kernels; and in the control variant without salt impact—77% right and 84% left kernels were maturing.

Approximately the same result in maturation of kernels is observed in the cv Orenburgskaya 10 in the control variant and the variant with the use of NaCl—69% and 68%, respectively. However, the best result was obtained in the experiment with Na_2SO_4, where fully matured kernels accounted for 78%. In general, the sampling consisted of 147 kernels, of which 42 were damaged or unmatured. Salt exposure caused greater maturation of the right kernels in spikelets (71% of the right and 57% of the left ones in the variant with NaCl, and 86% of the right and 69% of the left ones in the variant with Na_2SO_4), while in the control variant more left kernels in spikelets were matured (72% of the left and only 58 % right ones).

2.3. Effect of Salts on the Main Parameters of Kernels

The Kruskal–Wallis test confirmed the hypothesis of the effect of salts on the weight, area, perimeter, and width of the kernel, except for the length of the kernels in the cv Zolotaya (Table 3). The weight of kernels in the cv Zolotaya plants exposed to NaCl (average kernel weight 0.0305 ± 0.0123 g) and Na_2SO_4 (0.0281 ± 0.0143 g) does not have significant differences; however, it is significantly higher than in the control variant (0. 0199 ± 0.0141 g). The differences with the control in terms of kernel area are similar: the average area in the control variant was 15.22 ± 0.54 mm^2, while when exposed to NaCl it was 18.73 ± 0.76 mm^2, and to Na_2SO_4 was 17.41 ± 0.59 mm^2 (Figure 1). According to the "perimeter" parameter, there are significant differences in kernels only between the control (23.16 ± 0.53 mm) and plants exposed to Na_2SO_4 (25.20 ± 0.55 mm), in contrast to NaCl (24.84 ± 0.57 mm), and in the kernel width—only between the control (2.92 ± 0.07 mm) and NaCl (3.41 ± 0.09 mm)-exposed plants.

Table 3. Confirmation of the hypothesis about the significant effect of salts on the parameters of kernels in different wheat cultivars.

	Kernel Weight	Area	Perimeter	Length	Width
Orenburgskaya 10	–	+	+	+	+
Zolotaya	+	+	+	–	+
Ulyanovskaya 105	+	+	+	+	+
Orenburgskaya 23	–	–	+	–	–

Confirmation was made by the Kruskal–Wallis analysis of variance (p = 0.05), where "+" there is influence, "–" no influence.

Exposure to salts did not have a significant effect on the weight of kernels of the cv Orenburgskaya 10 (control—0.0262 ± 0.0024 g, NaCl—0.0254 ± 0.0017 g, Na_2SO_4—0.0311 ± 0.0024 g); however, an effect on other parameters was noted. Significant differences were observed both in the area of kernels due to the effect of salts (NaCl—17.94 ± 0.67 mm^2, Na_2SO_4—20.97 ± 0.89 mm^2), and in width (NaCl—3.07 ± 0.89, Na_2SO_4—3.47 ± 0.12). The values of the kernel perimeter in the control variant significantly differ from the results with salt exposure (control—23.97 ± 0.69 mm, NaCl—27.60 ± 0.65 mm, Na_2SO_4—26.12 ± 0.55 mm). A shorter kernel length was noted in the control variant—7.07 ± 0.16 mm, and a longer length in the experiment with Na_2SO_4—7.61 ± 0.13 mm; there were significant differences between these variants.

The differences between the control and NaCl in kernel weight (control 0.0232 ± 0.0008 g, NaCl 0.0266 ± 0.0008 g) were noted in the cv Ulyanovskaya 105. According to the parameter "kernel area", the values of the control variant and exposure to Na_2SO_4 were similar (control—16.21 ± 0.35 mm^2, Na_2SO_4—16.15 ± 0.28 mm^2), but both values were less than in the variant with NaCl (17.60 ± 0.32 mm^2). There were also no differences between the control and NaCl in terms of such an indicator as the perimeter of the kernels (control—20.40 ± 0.29 mm and variant with NaCl—20.32 ± 0.31 mm), while a greater result was achieved in the experiment with Na_2SO_4 (21.77 ± 0.32 mm). The kernel length values differed in the experiment with Na_2SO_4 (5.94 ± 0.05 mm) and NaCl (6.22 ± 0.06 mm), and salinity factors did not affect the kernel width (control—3.38 ± 0.05 mm, NaCl—3.57 ± 0.04 mm, and Na_2SO_4—3.42 ± 0.04 mm).

The least influence of salts on the parameters of kernels was noted in the cv Orenburgskaya 23. Thus, there was no significant effect on the weight of the kernel (control—0.0231 ± 0.0010 g, NaCl—0.0251 ± 0.0009 g, Na_2SO_4—0.0248 ± 0.0011 g), transverse section area, length, and width of kernels. However, there was an effect on the perimeter of the kernel transverse section, where the values differed significantly between the control (21.26 ± 0.29 mm) and the variant with NaCl (22.91 ± 0.38 mm).

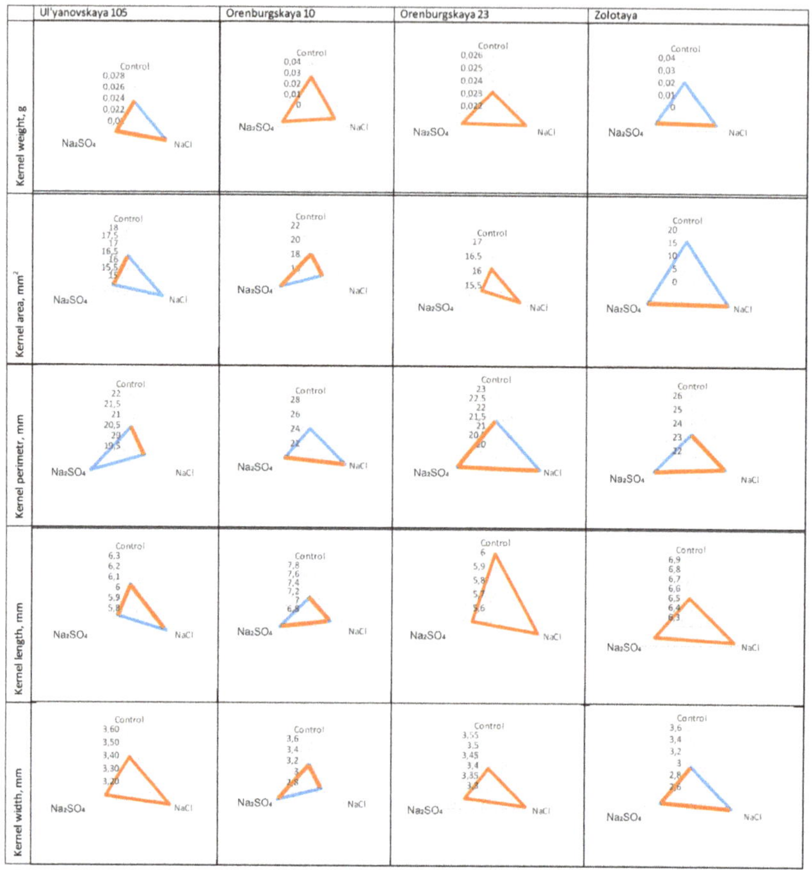

Figure 1. The examined kernel parameters in various wheat cultivars exposed to salt treatment in field conditions. The results are represented as radar charts. Significant differences are indicated in blue, non-significant differences in orange.

2.4. Effect of Salts on the General Asymmetry of Kernels

It has been established that the average value of the general (fluctuating) asymmetry in the cv Zolotaya was higher in the control variant; moreover, more even kernels were observed in the variants with Na_2SO_4 and NaCl exposure (Table 4). The Kruskal–Wallis method revealed significant differences between the control variant and Na_2SO_4 exposure on this trait.

Table 4. Average values of the general (fluctuating) asymmetry of kernels in wheat cultivars with different salt exposure variants.

	Control	NaCl	Na_2SO_4
Zolotaya	0.140 ± 0.055	0.111 ± 0.047	0.095 ± 0.05
Orenburgskaya 10	0.115 ± 0.009	0.109 ± 0.009	0.134 ± 0.011
Ulyanovskaya 105	0.109 ± 0.008	0.106 ± 0.007	0.107 ± 0.007
Orenburgskaya 23	0.111 ± 0.009	0.099 ± 0.008	0.097 ± 0.008

Arithmetic mean ± standard deviation, p = 0.05.

2.5. Effect of Salts on the Asymmetry of Left, Right, and Middle Kernels

In the control variant, cv Zolotaya had no significant differences (Kruskal–Wallis method) in the overall asymmetry between the left, right, and middle kernels (Figure 2). In the variant with NaCl exposure, the left kernels (0.131 ± 0.042) did not differ from the right ones; however, the left kernels significantly differed from the middle ones (0.068 ± 0.038). The result of the experiment with Na_2SO_4 exposure did not reveal any differences between the left kernels (0.109 ± 0.060) and the right ones (0.103 ± 0.044); however, both left and right kernels differed from the middle ones (0.060 ± 0.021). In general, kernels looked more symmetrical in the experiment with Na_2SO_4.

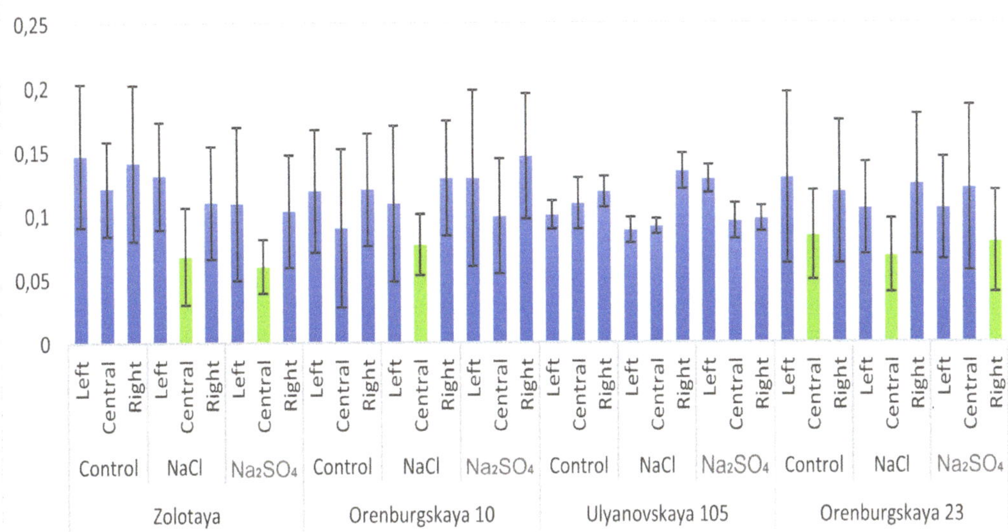

Figure 2. Average values of the total (fluctuating) asymmetry of kernels for wheat cultivars with different salinity variants, depending on the location in the spikelet.

In the cv Orenburgskaya 10, there were no differences between the left, right, and middle kernels in terms of overall asymmetry in the control variant, as well as in the experiment with Na_2SO_4, where the asymmetry level was 0.129 ± 0.069 for the left kernel, 0.146 ± 0.049 for the right kernel, and 0.099 ± 0.045 for the middle one. In the experiment with NaCl, the middle (0.077 ± 0.024) and right (0.129 ± 0.045) kernels differed, but did not differ from the left ones.

The right, left, or middle kernels in the spikelet were evaluated (arithmetic mean ± standard deviation, p = 0.05). Light green color indicates variants that have significant differences with other experimental variants that were established after applying the Kruskal–Wallis method.

The Kruskal–Wallis analysis of variance did not reveal the effect of the kernel arrangement in the cv Ulyanovskaya 105 and Orenburgskaya 23 on the asymmetry level in the control variant and in the experiment with Na_2SO_4 exposure. Therefore, there were no differences between kernels. However, in the cv Ulyanovskaya 105, the asymmetry of the left (0.088 ± 0.010) and right (0.134 ± 0.014) kernels in the experiment with NaCl has significant differences, and in the cv Orenburgskaya 23 the asymmetry of the middle kernels (0.068 ± 0.029) significantly differs from the extreme ones. The most symmetrical kernels of the cv Ulyanovskaya 105 were noted in the right row of the spikelet in experiment with Na_2SO_4.

2.6. Scale of Fluctuating Asymmetry

Fluctuating asymmetry coefficients range from 0.016 to 0.315 in this study (Table 5). In a recent study of the fluctuating asymmetry in wheat cultivars Zlata, Agata, and Rubezhnaya, a scale for measuring the strength (degree) of asymmetry was proposed [7]. However, the upper interval ended with asymmetry values from 0.150 to 0.184, so it could not include newly obtained data. Using the initial data from the previous work (278 asymmetry coefficients for three experiment variants) and the current 461 coefficients for all experiment variants, the following measurements were proposed, which are shown in Tables 5 and 6.

Table 5. Scale of indicators of overall asymmetry.

Asymmetry Range	Degree of Asymmetry
0.010–0.070	Very low
0.071–0.131	Low
0.132–0.192	Middle
0.193–0.253	High
0.254–0.314	Very high

The arithmetic means were calculated for five asymmetry indices in 739 kernels of cv Zlata, Agata, Rubezhnaya, Ulyanovskaya 105, Orenburgskaya 10, Orenburgskaya 23, Zolotaya, and average asymmetry values obtained were divided into 5 classes manifesting the degree of asymmetry.

Table 6. Distribution of the number of kernels in the sampling according to the degree of asymmetry.

Cultivar	Salinity Variant	Degree of Fluctuating Asymmetry					Total Kernels, Pieces
		Very Low	Low	Middle	High	Very High	
Zolotaya	Control	3	15	10	6	2	36
	NaCl	9	16	11	1	0	37
	Na_2SO_4	15	18	5	2	0	40
Total kernels, pieces		**27**	**49**	**26**	**9**	**2**	**113**
Orenburgskaya 10	Control	6	13	9	3	0	31
	NaCl	11	12	9	1	1	34
	Na_2SO_4	3	14	4	5	2	28
Total kernels, pieces		**20**	**39**	**22**	**9**	**3**	**93**
Ulyanovskaya 105	Control	10	18	9	3	1	41
	NaCl	7	24	7	4	0	42
	Na_2SO_4	7	25	9	3	0	44
Total kernels, pieces		**24**	**67**	**25**	**10**	**1**	**127**
Orenburgskaya 23	Control	10	20	10	3	1	44
	NaCl	13	21	6	1	3	44
	Na_2SO_4	16	14	3	4	0	37
Total kernels, pieces		**39**	**55**	**19**	**8**	**4**	**125**

Bold type indicates the numbers corresponding to the total number of kernels in the experiment variant.

2.7. Determination of the Main Factors of Sampling Variability

Principal component analysis identified four main factors (components) that explain the cumulative variance of sampling. The sampling variability of 51.98 % is explained by the influence of the first component, which consists of kernel size parameters: transverse section area, transverse section perimeter, length, width, length of the symmetry axis (kernel thickness), and index 4. The second component (14.77 %) is responsible for the left-sided shape displacement (index 2 and index 5), and the third one (10.06 %) is responsible for the right-sided displacement (index 1 and index 6). The fourth component consists of index 3, which is responsible for the displacement of the grain width under the hollow triangle of the groove, and explains 9.34 % of variance of the sampling. Altogether, four components explain the variability of 86.15 % from the sampling size.

3. Discussion

The study of the effect of salt stress on the parameters of roots and aboveground biomass is consistent with data from other studies. As in our case, salt treatment led to a decrease in aboveground biomass and root length of wheat [25,26]. However, it was unexpected that with the suppression of the vegetative organs in comparison with the control variant, we will achieve an improvement in the studied parameters of the grain. When simulating salt stress in the studied varieties, a greater yield of the proportion of mature kernels was observed, and the values for weight, area, perimeter, width, and symmetry of wheat caryopses were also significantly higher.

There are basically two types of changes in the degree of asymmetry in response to stress—either an increase in asymmetry, or no reaction at all. In our case, low salt concentrations in 5 out of 36 variants (Figure 3) "worked uniquely"—significantly reduced the asymmetry. Previously, such an effect was observed only when studying the leaves of different cohorts of *Quercus ilex*, it turned out that plants living in more stressful places (excessive moisture conditions) are more symmetrical [27]. In rodents *Peromyscus maniculatus*, it was found that after a natural disaster (destruction of a forest reserve by a tornado), fluctuating asymmetry in the length of the femur became lower [28].

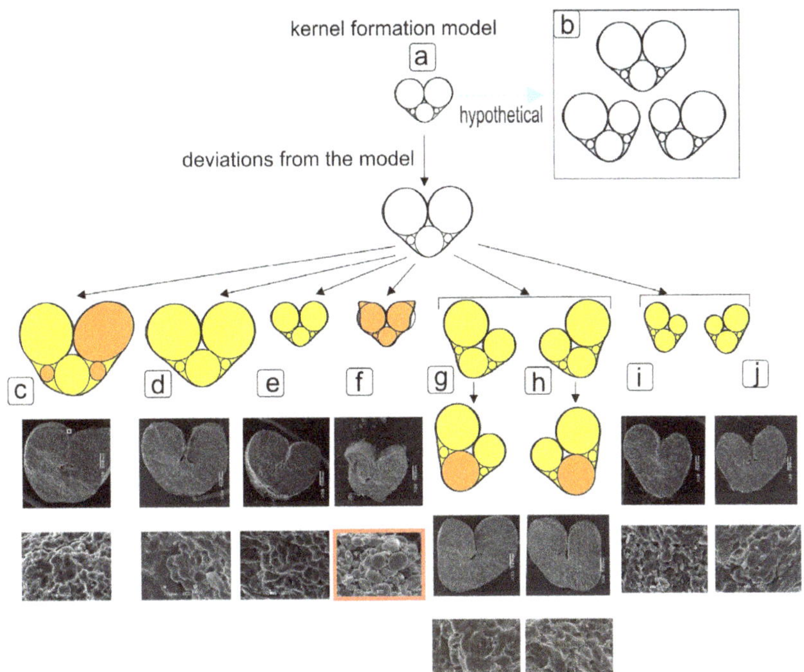

Figure 3. Scheme for testing the hypothesis about the stability of the dimension of conditional elements of the kernel during the formation of a kernel with manifested asymmetry, as well as hypothesis about the relationship of low-quality kernels (insufficient plumpness (fulfillment)) with the asymmetry of kernels. Designations: a—initial stage of kernel (caryopsis) formation, after the formation of the aleurone layer and endosperm; b—idealized model for the formation of symmetrical and asymmetric kernels in spikelets; c—large kernel from the lower tier with a manifested asymmetry; d—medium-sized kernel from the upper layer with a relatively symmetrical shape; e—small kernel of the upper tier with a relatively symmetrical shape; f—a small kernel from the upper tier with a manifested asymmetry and a lesion of the smoothed shape of the surface; g, h—asymmetric kernels from the lower tier of the spikelet with a manifested increase in one half and a decrease in the second one; i, j—small asymmetric kernels from the upper tier of the spikelet with a manifested increase in one half and a decrease in the second one.

Often, external stresses led to an increase in the asymmetry of individual plant organs. Thus, the fluctuating asymmetry of birch leaves is affected by environmental pollution due to emissions from copper smelters [29], height of growth [30], excess nitrogen on *Betula pubescens* [31], interspecific hybridization on the leaves of *Betula nana*, *B. pubescens*, and *B. pendula* [32]. Leaf asymmetry is enhanced by exposure to electromagnetic fields in soybean when placed under high-voltage power lines generating pulsed magnetic fields from 3 to 50 mG [33], in *Salix borealis* due to exposure to leaf beetles *Melasoma lapponica* around the copper smelter [34], in a population of *Clarkia tembloriensis* due to inbreeding [35], in *Phaseolus vulgaris* due to water deficiency [36], in *Dimorphotheca sinuata* due to UV exposure [37], in *Lythrum salicaria* due to excess nutrients [38].

There are also works that have shown that the above dependence is not always the case. The impact of stresses (water deficit, pathogen attack, and competition) on *Salix sericea* and *Salix eriocephala* slightly changed the fluctuating leaf asymmetry. However, this study found a strong negative correlation between plant biomass and fluctuating asymmetry, i.e., plants with symmetrical leaves are better able to protect themselves from stress [39]. Mountain birch does not have asymmetry associated with insect damage [28] and pollution concentration, but is sensitive to cold conditions of the year [40]. No relationship has been found between a wide NaCl concentration gradient on *Glycine max* and fluctuating asymmetry [41], nor between asymmetry and fitness components in *Brassica cretica* [42].

Heterospermia of cereal grains is an important quality indicator that can significantly affect the properties and composition of the resulting product and products of its processing [43]. The reasons that cause the appearance of different-quality seeds (grains) can be genetic, that is, characteristic of the variety, but can be induced by cultivation features: humidity, temperature, nutrient composition, acidity, predominant air movement, and the spectral composition of light and its intensity. Additionally, a similar effect should be expected from features associated with the location: gravitropism or exposure to a magnetic field or atmospheric pressure. A special factor may be the impact of pests and diseases, in which the parameters of symmetry and asymmetry of grains (kernels) in the spike will be significantly distorted.

Meanwhile, the predictability of qualitative indicators is extremely important for creating modern foundations for predictive phenotyping and analysis of factors causing changes in the qualitative and quantitative characteristics of grain, which is a prospect for the near future [44,45]. If the heterogeneity of the upper, smaller kernels in each spikelet, and the lower, respectively, larger kernels, can be used, for example, by separating small and large kernels using sieves or using differences in their mass, then taking into account and eliminating changes in characteristics associated with symmetry and asymmetry of kernels conditionally upper and conditionally lower in each spikelet mechanically is currently not possible. Nevertheless, qualitative differences between both small and large kernels and between relatively symmetrical and asymmetric kernels are obvious. They are caused by differences in the ratio of the volume of the germ, endosperm, and aleurone layer, which allows us to predict the quality of flour for specific purposes [46]. Obviously, in small grains, the ratio of protein to starch will be higher than in large ones, since the larger the volume of the endosperm, the more starch in the grain and the smaller the number of cells of the aleurone layer, which is one layer of cells around the endosperm [47]. It is likely that the noted effects may be accompanied by a change in the biosynthesis or distribution of some plant hormones, for example, auxins [5]. With pronounced genetic or adverse effect-induced asymmetry, unevenness already occurs at the level of each kernel (grain), which can create problems not only during processing, but also during storage. Thus, a grain with cavities can have altered respiration [48], and consequently lipid and protein oxidation [49], which can affect the duration of viability during storage and subsequent germination [50]. The development of kernels (grains), which are characterized by a significant degree of asymmetry, causes disturbances in the development of the root system [51] and subsequently can affect the plant as a whole.

The contribution of climate change and increased effects of environmental stressors have had and will continue to affect the development and productivity of plants. The need for a transition to the digitalization of agriculture, especially for such a fundamental crop as wheat, is very urgent [52,53]. Therefore, identifying the effects of each individual factor is critical to the development of projected grain production. By analyzing the relationships between traits and environment conditions, 3D X-ray micro computed tomography (CT) images find that the grain-to-grain distance, aspect ratio, and porosity are more likely affected by the genome than environment [44].

Earlier, we made sure that the degree of manifestation of asymmetry indices is expressed differently in different varieties of wheat [7]. The phenomenon of peculiarities in the formation of right and left seeds is widespread. In the present work, we studied which parts of the kernel can be used to understand the structural basis for changes of ideal morphology. For this, an idealized model based on conditional cylinders, reflected in the diagram in the form of circles, was considered (Supplementary Material Figures S1 and S2). In cases where no differences in the shape and diameter of the circles from the idealized model were noted in the studied specimen, yellow coloring was used. Where significant deviations from the idealized model were noted, this is indicated in orange in those circles that had differences. Although some kernels could be considered symmetrical and corresponding to the idealized model, slight deformation in the region of the groove (crease) was observed. Thus, it was found that larger kernels were characterized by an increase in area due to an enlargement in the circle area in the upper part of the corresponding part of the kernels (Supplementary Material Figure S1). In asymmetric kernels of medium size, there was an increase in the area of the circle in the lower part of the kernel (Figure 3 and Supplementary Material Figure S2). Asymmetric kernels from the upper parts of the spikelets were smaller, but generally consistent with the idealized model.

Interestingly, in unfulfilled kernels, the asymmetry of the parts is preserved. We also raised the issue of comparing the asymmetry factor and kernel fulfillment. Asymmetry did not have a significant effect on the preservation of all components of the kernel fine structure in comparison with symmetrical kernels. While a looser distribution of starch grains was observed in unfulfilled kernels, non-binding components of the cytoplasm and the formation of cavities were observed. Such a structure improves the access of oxygen to internal structures and, accordingly, makes it problematic and very limited to use accelerated selection using immature grains, since their storage even for a short time is unlikely to be possible due to the rapid loss of germination.

In this paper, we expected to see the negative effect of salinity, a priori assuming that salinity is undesirable, as widely reported in long-term studies [11,54,55]. However, it was found that the characteristics of the kernels studied were rather good than bad. Meanwhile, it is worth considering the probable causes of the noted effect, which should be taken into account when looking for ways to reduce the negative effects of salinity, usually mitigated by the use of salt-tolerant varieties and the application of specific fertilizers and agrichemicals [56]. A possible reason could be that the salt concentration was not high enough to cause a noticeable drop in yield. In addition, the influence of salt ions and their interaction with plant roots and the soil-absorbing complex could cause the priming effect, i.e., the triggering of a cascade of protective reactions in plants, leading to their better adaptation to the high temperatures and drought characteristic of this region. It is also necessary to take into account the soil and root-associated microflora, which, with the addition of NaCl, could be modified in a way that provides benefits for the growth and development of the root system in this particular case.

This study of the differences in wheat kernel will be greatly helpful for accelerated breeding if we can automatically discover wheat phenotype in a nondestructive and fast manner in perspective [44]. Salinity, as a damaging factor that causes oxidative, osmotic, and toxic effects, is a rather complex stress. Further experiments will likely require simplifying the models inducing osmotic and oxidative stresses, for example using paraquat or hydrogen peroxide to trigger oxidative effects [57] and PEG to simulate drought [58].

4. Materials and Methods

4.1. Plant Material

Materials for the study were obtained from the Federal State Budgetary Scientific Institution "Federal Scientific Center for Biological Systems and Agrotechnologies of the Russian Academy of Sciences" (Orenburg).

The following varieties of wheat were taken as objects of study:

- Zolotaya—spring hard wheat, a variety of hordeiforme;
- Orenburgskaya 10—spring hard wheat, a variety of hordeiforme;
- Orenburgskaya 23—soft spring wheat, Lutescens variety;
- Ulyanovskaya 105—soft spring wheat, Lutescens variety.

The experiment was carried out in the central zone of the Orenburg region (fields coordinates—51°46′41.928000″ N, 55°19′14.292000″ E, 51°46′43.074000″ N, 55°19′14.598000″ E, 51°46′43.338000″ N, 55°19′22.620000″ E 51°46′42.120000″ N, 55°19′22.404000″ E).

The soil is southern carbonate low-humus heavy loamy chernozem, pH of the soil solution is 7.0–8.0 (pH meter brand—Ecotest, Moscow, Russia).

Sowing varieties were carried out on a black fallow. Site preparation consisted of non-moldboard loosening after harvesting the previous crop, early spring harrowing, four-fold cultivation during the summer, and autumn deep loosening of the fallow. In the year of sowing the experiment, spring harrowing of the site and pre-sowing cultivation to the depth of seeding were carried out.

Sowing was carried out using a seeder SN-16 in an ordinary way to a depth of 6–8 cm with a seeding rate of 4.5 million seeds per ha, followed by post-sowing rolling with ring-spur rollers. The area of plots under each variety was 50 m^2, each variety was sown in three repetitions (Supplementary Materials Figure S3).

To start the experiment on salt stress, in two non-adjacent repetitions for each variety, microplots of 0.50 m^2 were selected, which included three rows of sowing with a row spacing of 15 cm with a row length of 111 cm. The total number of microplots in the experiment for each variety was 30 (Supplementary Materials Figure S3).

For watering, the solutions of two salts, NaCI and Na_2SO_4, were used. Salt solutions were prepared by dissolving NaCl in water at a concentration of 1.10 g L^{-1}, and Na_2SO_4 at a concentration of 0.40 g L^{-1}. The plots were irrigated once, in the phase of full shoots. The control was irrigated with water, simultaneously with the watering of the experimental variants. The water consumption rate was 30 L per m^2, which corresponds to the monthly rainfall rate in the region. After watering, the plots were mulched with dry soil.

In the tillering phase, 10 plants were selected for each wheat cultivar to analyze the root system of wheat plants. In the earing phase, 10 plants of each wheat cultivar were selected for accounting the height of the plants and the length of the spike.

In the phase of full maturing of the wheat grain in the plots, 5 spikes were randomly selected, the kernels of which were used for further analysis (Supplementary Materials Figure S4).

4.2. Preparation of Material for Primary Analysis

All available spikes were disassembled into simple ones, keeping their placement order on the main spike axis. Next, the kernels were extracted and glued due to double-sided tape onto a white sheet of A4 paper while maintaining the order of the kernels in the spike. Each sheet of paper contained kernels of only one cultivar with a specific salt exposure variant. Images were scanned by a professional Epson Perfection V550 Photo scanner with a resolution of 600 dpi. The resulting images were processed in the ImageJ program [59]. The transverse section area (mm^2), transverse section perimeter (mm), length (mm), and width (mm) of kernels were measured by the program. Then, the mass of each kernel was measured on electronic scales (p = 0.0001 g). In total, 1173 kernels were examined in the study (Supplementary Material Table S1).

A sufficient sample size was collected to compare fluctuating asymmetry indices according to the algorithm proposed earlier for horticultural crops [60].

It was assumed that the volume of the general population of kernels tends to infinity $\hat{N} \rightarrow \infty$, therefore, Student's criterion was taken as $t_{st}05 = 1.96$. It was assumed that the minimum value of the asymmetry index in general population (x_{min}) can be equal to zero, which corresponds to a completely symmetrical kernel, and the maximum value (x_{max}) is approx. 0.2 (based on the information on the fluctuating asymmetry scale proposed earlier [7]. Using the 6-sigma rule, the value of the standard deviation $\hat{\sigma}$ was determined:

$$\hat{\sigma} = (x_{max} - x_{min})/6 = (0.2 - 0)/6 = 0.333$$

Allowable error $\Delta = 0.035$ (corresponds to the class interval on the fluctuating asymmetry scale) or $\Delta = 0.017$ (corresponds to half of the class interval on the fluctuating asymmetry scale) proposed earlier [7].

Allowable accuracy (k) was calculated by the following formula:

$$k = \frac{\Delta}{\hat{\sigma}}$$

$$k = 0.035/0.333 = 1.05 \text{ or } k = 0.017/0.333 = 0.51$$

The sample size (N) was calculated using the following formula:

$$N = {t_{st}}^2/k^2 \quad (1)$$

$N = (1.96 \times 1.96)/(1.05 \times 1.05) = 3.69$, the minimum sample size for a sample to obtain a reliable value is 4 kernels.

Or

$N = (1.96 \times 1.96)/(0.51 \times 0.51) = 14.77$, i.e., this ideal sample size, minimizing the experimental error, is 15 kernels.

After data on kernel parameters were entered into the SPSS Statistics 25 program, the "random number generator" function was turned on and the SPSS program randomly selected 15 matured left, right, and middle kernels for each of the four varieties in three salinity options (the algorithm is given in the Supplementary Material Table S1 about the number of kernels) where the number of mature kernels exceeded 15 pieces. If the number of mature kernels was less than 15, for example, 6 pieces (Zolotaya, control, middle from the Supplementary Material Table S1), then all 6 kernels were taken.

458 transverse slices of kernels were prepared, of which 113 sections were from the cv Zolotaya, 93 sections were from the cv Orenburgskaya 10, 125 sections were from the cv Orenburgskaya 23, and 127 sections were from the cv Ulyanovskaya 105. Sections were made from the middle part of the kernels. The sections were glued with double-sided adhesive tape on a white A4 sheet; then, the images were scanned on an Epson Perfection V550 Photo with a resolution of 600 dpi. In the ImageJ program by means of the "segmented line" tool the length of the central axis of the kernel cut, the distance from the control points of the left and right sides of the kernel to the axis to establish asymmetry indices were measured, in millimeters. Index 1 is the ratio of the distance from the top of the kernel to the central axis to the total distance between the tops. Index 2 is the ratio of the distance along the widest part to the central axis to the total distance of width. Index 3 is the ratio of the distance from the edge of the kernel under the triangle to the central axis to the total distance. Index 4 is the ratio of the distance from the edge of the triangle to the axis of symmetry to the total length of the triangle. Index 5 is the ratio of each segment extending 45° from the axis of symmetry to the length segment from the triangle to the bottom (Figure 4).

Fluctuating asymmetry (FA) indices were calculated by the formula of Wilsey et al. (1998) [30].

$$FA = \frac{|L_1 - L_1'|}{L_1 + L_1'} + \frac{|L_2 - L_2'|}{L_2 + L_2'}$$

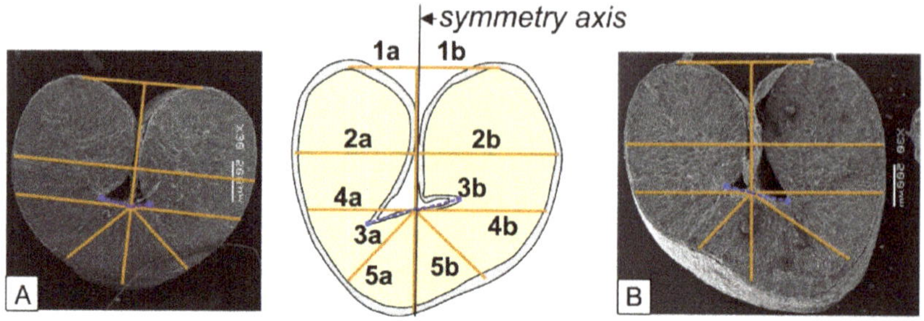

Figure 4. Distances between the control points of the left and right sides of the kernel to the axis of symmetry. Fluctuating asymmetry indices: 1a,b—arm lengths from the intersection of the perpendicular to the tangent (axis of symmetry) to the shares of halves of the wheat kernel on the median section; 2a,b—arm lengths of the perpendicular to the widest sections of the shares of halves of the wheat kernel; 3a,b—segment lengths of maximum distance of the inner corners of the crease (ventral furrow) from the intersection point of the symmetry axis with the flat part of the ventral furrow between the halves of the shares of wheat kernel; from 4a,b—the arm lengths of the perpendicular at the point of intersection of the symmetry axis with the flat part of the ventral furrow between the halves of the wheat kernel shares; 5a,b—the lengths of the bisectors extending from the perpendicular at the point of intersection of the symmetry axis with the flat part of the ventral furrow between the halves of the wheat kernel shares.

L_1 corresponds to the length from the left top of the kernel to the axis of symmetry, and L_1' corresponds to the length from the right top of the kernel to the axis of symmetry. The measurements L_1 and L_1' are obtained for calculating index 1 (described above), L_2 and L_2' for index 2, and so on.

All statistic calculations were obtained by IBM SPSS Statistics 25 program. Two-way ANOVA was used to determine the effect of salts on embryonic, adventitious, and nodal roots, as well as on plant height and spike length. The difference between the variants was determined by LSD (least significant difference test). The compliance of the sample with the normal distribution law was verified using Kolmogorov–Smirnov method. Since the data in most cases did not conform the normal distribution law, despite the significant sampling size, Kruskal–Wallis analysis of variance for non-parametric criteria ($p = 0.05$) was used to establish differences between the experimental variants. This method allows us to check the statement that salt exposure affects the parameters of wheat kernels. The principal component method was used to decrease the dimension.

5. Conclusions

A slight salt effect of NaCl and Na_2SO_4 on wheat plants of four cultivars caused an increase in the weight of kernels, as well as a number of other parameters of kernel (perimeter of transverse section, area of transverse section, length, and width). Methods for assessing fluctuating asymmetry revealed a trend towards an increase in the number of symmetrical kernels in a spike. However, this trend was reliably confirmed only in the case of exposure to NaCl for the central kernels in the spikelet in the cultivars Zolotaya, Orenburgskaya 10, and Orenburgskaya 23. Additionally, similar results were obtained for the central kernels in spikelets in the cv Zolotaya as a result of exposure to Na_2SO_4, and for the left kernels in spikelets in the cv Ulyanovskaya 105 exposed to NaCl. Thus, the methods proposed in this research for assessing fluctuating asymmetry in wheat kernels under field conditions were able to reliably confirm the observed deviations as a result of exposure to low salt concentrations not for all the studied specimens. It is possible that when using the proposed indices for assessing fluctuating asymmetry at high salt concentrations corresponding to natural salt stress, both an increase and a decrease in asymmetry can be detected, which requires further study.

Supplementary Materials: The following supporting information can be downloaded at: https://www.mdpi.com/article/10.3390/plants12050980/s1, Figure S1: Scanning electron microscopy of transverse sections of the central part of wheat kernels with manifested asymmetry; Figure S2: Scanning electron microscopy of transverse sections of the central part of wheat kernels with different degrees of fulfillment; Figure S3: Wheat plants with root system, experimental plots, scheme of experiment; Figure S4: Appearance of mature spikes of wheat; Table S1: The number of kernels involved in the experiment.

Author Contributions: Conceptualization, T.S.A. and E.N.B.; validation, T.S.A. and V.A.K.; formal analysis, T.S.A. and E.N.B.; investigation, resources, I.N.B., T.S.A. and E.N.B.; data curation, S.V.L., N.S.R., I.N.B., A.A.P., A.A.G., T.S.A. and E.N.B.; writing—original draft preparation, T.S.A. and E.N.B.; writing—review and editing, T.S.A., A.A.G. and E.N.B.; visualization, T.S.A. and E.N.B.; supervision, E.N.B. All authors have read and agreed to the published version of the manuscript.

Funding: The reported study was supported by assignments 0431-2022-0003 (to E.N.B., A.A.G.—ARRIAB RAS); 122011400178-7 (to E.N.B., T.S.A., V.A.K.—MBG RAS) and 0526-2022-0014 (to S.V.L., N.S.R., I.N.B., A.A.P.—FSC BSA) of the Ministry of Science and Higher Education of the Russian Federation.

Data Availability Statement: Not applicable.

Acknowledgments: The authors thank Anatoly G. Bogdanov and the Center for Electron Microscopy in the Life Sciences, a Unique Scientific Installation for Three-Dimensional Electron Microscopy and Spectroscopy (Biological Faculty of M.V. Lomonosov Moscow State University).

Conflicts of Interest: The authors declare no conflict of interest.

References

1. Afzal, I.; Basra, S.M.A.; Hameed, A.; Farooq, M. Physiological enhancements for alleviation of salt stress in wheat. *Pak. J. Bot.* **2006**, *38*, 1649–1659.
2. Koppolu, R.; Chen, S.; Schnurbusch, T. Evolution of inflorescence branch modifications in cereal crops. *Curr. Opin. Plant Biol.* **2022**, *65*, 102168. [CrossRef]
3. Frølich, W.; Åman, P. Whole grain for whom and why? *Food Nutr. Res.* **2010**, *54*, 5056. [CrossRef]
4. Chaban, I.A.; Gulevich, A.A.; Smirnova, E.A.; Baranova, E.N. Morphological and ultrastructural features of formation of the skin of wheat (*Triticum aestivum* L.) kernel. *Plants* **2021**, *10*, 2538. [CrossRef]
5. Fischer, C.; Neuhaus, G. Influence of auxin on the establishment of bilateral symmetry in monocots. *Plant J.* **1996**, *9*, 659–669. [CrossRef]
6. Zhou, H.; Riche, A.B.; Hawkesford, M.J.; Whalley, W.R.; Atkinson, B.S.; Sturrock, C.J.; Mooney, S.J. Determination of wheat spike and spikelet architecture and grain traits using X-ray Computed Tomography imaging. *Plant Methods* **2021**, *17*, 1–9. [CrossRef]
7. Baranova, E.N.; Aniskina, T.S.; Kryuchkova, V.A.; Shchuklina, O.A.; Khaliluev, M.R.; Gulevich, A.A. Evaluation of the heterogeneity of wheat kernels as a traditional model object in connection with the asymmetry of development. *Symmetry* **2022**, *14*, 1124. [CrossRef]
8. Boshnakian, S. The mechanical factors determining the shape of the wheat kernel. *J. Amer. Soc. Agron.* **1918**, *10*, 503–512. [CrossRef]
9. Shewry, P.R.; Evers, A.D.; Bechtel, D.B.; Abecassis, J. Development, structure, and mechanical properties of the wheat grain. In *Wheat Chemistry and Technology*, 4th ed.; Khan, K., Shewry, P., Eds.; AACC International: St. Paul, MN, USA, 2009; pp. 51–96.
10. Cossani, C.M.; Sadras, V.O. Symmetric response to competition in binary mixtures of cultivars associates with genetic gain in wheat yield. *Evol. Appl.* **2021**, *14*, 2064–2078. [CrossRef]
11. El Sabagh, A.; Islam, M.S.; Skalicky, M.; Ali Raza, M.; Singh, K.; Anwar Hossain, M.; Hossain, A.; Mahboob, W.; Iqbal, M.A.; Ratnasekara, D.; et al. Salinity stress in wheat (*Triticum aestivum* L.) in the changing climate: Adaptation and management strategies. *Front. Agron.* **2021**, *3*, 661932. [CrossRef]
12. Rengasamy, P. World salinization with emphasis on Australia. *J. Exp. Bot.* **2006**, *57*, 1017–1023. [CrossRef]
13. Kaushal, S.S.; Groffman, P.M.; Likens, G.E.; Belt, K.T.; Stack, W.P.; Kelly, V.R.; Band, L.E.; Fisher, G.T. Increased salinization of fresh water in the northeastern United States. *Proc. Natl. Acad. Sci. USA* **2005**, *102*, 13517–13520. [CrossRef] [PubMed]
14. Löfgren, S. The chemical effects of deicing salt on soil and stream water of five catchments in southeast Sweden. *Water Air Soil Pollut.* **2001**, *130*, 863–868. [CrossRef]
15. Manca, F.; Capelli, G.; Tuccimei, P. Sea salt aerosol groundwater salinization in the Litorale Romano natural reserve (Rome, Central Italy). *Environ. Earth Sci.* **2015**, *73*, 4179–4190. [CrossRef]
16. Chamberlain, S.D.; Hemes, K.S.; Eichelmann, E.; Szutu, D.J.; Verfaillie, J.G.; Baldocchi, D.D. Effect of drought-induced salinization on wetland methane emissions, gross ecosystem productivity, and their interactions. *Ecosystems* **2020**, *23*, 675–688. [CrossRef]
17. Abuduwaili, J.; Liu, D.W.; Wu, G.Y. Saline dust storms and their ecological impacts in arid regions. *J. Arid Land* **2010**, *2*, 144–150. [CrossRef]
18. Herbert, E.R.; Boon, P.; Burgin, A.J.; Neubauer, S.C.; Franklin, R.B.; Ardón, M.; Hopfensperger, K.N.; Lamers, L.P.M.; Gell, P. A global perspective on wetland salinization: Ecological consequences of a growing threat to freshwater wetlands. *Ecosphere* **2015**, *6*, 1–43. [CrossRef]

19. Zeeshan, M.; Lu, M.; Sehar, S.; Holford, P.; Wu, F. Comparison of biochemical, anatomical, morphological, and physiological responses to salinity stress in wheat and barley genotypes deferring in salinity tolerance. *Agronomy* **2020**, *10*, 127. [CrossRef]
20. Kononenko, N.; Baranova, E.; Dilovarova, T.; Akanov, E.; Fedoreyeva, L. Oxidative damage to various root and shoot tissues of durum and soft wheat seedlings during salinity. *Agriculture* **2020**, *10*, 55. [CrossRef]
21. Al-Ashkar, I.; Alderfasi, A.; Ben Romdhane, W.; Seleiman, M.F.; El-Said, R.A.; Al-Doss, A. Morphological and genetic diversity within salt tolerance detection in eighteen wheat genotypes. *Plants* **2020**, *9*, 287. [CrossRef]
22. Semenova, G.; Fomina, I.; Ivanov, A. Combined effect of water deficit and salt stress on the structure of mesophyll cells in wheat seedlings. *CellBio* **2014**, *3*, 14–24. [CrossRef]
23. Aldesuquy, H.S. Impact of seawater salinity on ultrastructure of chloroplasts and oleosomes in relation to fat metabolism in flag leaf of two wheat cultivars during grain-filling. *Adv. Crop Sci. Technol.* **2015**, *4*, 1–7. [CrossRef]
24. Radi, A.A.; Farghaly, F.A.; Hamada, A.M. Physiological and biochemical responses of salt-tolerant and salt-sensitive wheat and bean cultivars to salinity. *J. Biol. Earth Sci.* **2013**, *3*, 72–88.
25. Zhao, D.; Gao, S.; Zhang, X.; Zhang, Z.; Zheng, H.; Rong, K.; Zhao, W.; Khan, S.A. Impact of saline stress on the uptake of various macro and micronutrients and their associations with plant biomass and root traits in wheat. *Plant Soil Environ.* **2021**, *67*, 61–70. [CrossRef]
26. Quan, X.; Liang, X.; Li, H.; Xie, C.; He, W.; Qin, Y. Identification and characterization of wheat germplasm for salt tolerance. *Plants* **2021**, *10*, 268. [CrossRef]
27. Hódar, J.A. Leaf fluctuating asymmetry of Holm oak in response to drought under contrasting climatic conditions. *J. Arid Environ.* **2002**, *52*, 233–243. [CrossRef]
28. Hopton, M.E.; Cameron, G.N.; Cramer, M.J.; Polak, M.; Uetz, G.W. Live animal radiography to measure developmental instability in populations of small mammals after a natural disaster. *Ecol. Indic.* **2009**, *9*, 883–891. [CrossRef]
29. Kozlov, M.V.; Wilsey, B.; Koricheva, J.; Haukioja, E. Fluctuating asymmetry of birch leaves increase under pollution impact. *J. Appl. Ecol.* **1996**, *33*, 1489–1495. [CrossRef]
30. Hagen, S.B.; Ims, R.A.; Yoccoz, N.G.; Sørlibråten, O. Fluctuating asymmetry as an indicator of elevation stress and distribution limits in mountain birch (*Betula pubescens*). *Plant Ecol.* **2008**, *195*, 157–163. [CrossRef]
31. Lappalainen, J.H.; Martel, J.; Lempa, K.; Wilsey, B.; Ossipov, V. Effects of resource availability on carbon allocation and developmental instability in cloned birch seedings. *Int. J. Plant Sci.* **2000**, *161*, 119–125. [CrossRef]
32. Wilsey, B.; Haukioja, E.; Koricheva, J.; Sulkinoja, M. Leaf fluctuating asymmetry increases with hybridization and elevation in three-line birches. *Ecology* **1998**, *79*, 2092–2099. [CrossRef]
33. Freeman, D.C.; Graham, J.H.; Tracy, M.; Emlen, J.M.; Alados, C.L. Developmental instability as a means of assessing stress in plants: A case study using electromagnetic fields and soybeans. *Int. J. Plant Sci.* **1999**, *160* (Suppl. 6), S157–S166. [CrossRef]
34. Zvereva, E.L.; Kozlov, M.V.; Niemela, P.; Haukioja, E. Delayed induced resistance and increase in leaf fluctuating asymmetry as responses of *Salix borealis* to insect herbivory. *Oecologia* **1997**, *109*, 368–373. [CrossRef]
35. Sherry, R.A.; Lord, E.M. Developmental stability in leaves of *Clarkia tembloriensis* (*Onagraceae*) as related to population outcrossing rates and heterozygosity. *Evolution* **1996**, *50*, 80–91. [CrossRef]
36. Souza, G.M.; Viana, J.O.F.; Oliveira, R.F. Asymmetrical leaves induced by water deficit show asymmetric photosynthesis in common bean. *Braz. J. Plant Physiol.* **2005**, *17*, 223–227. [CrossRef]
37. Midgley, G.F.; Wand, S.J.E.; Musil, C.F. Repeated exposure to enhanced UV-B radiation in successive generations increases developmental instability (leaf fluctuating asymmetry) in a desert annual. *Plant Cell Environ.* **2002**, *21*, 437–442. [CrossRef]
38. Milligan, J.R.; Krebs, R.A.; Mal, T.K. Separating developmental and environmental effects on fluctuating asymmetry in *Lythrum salicaria* and *Penthorum sedoides*. *Int. J. Plant Sci.* **2008**, *169*, 625–630. [CrossRef]
39. Hochwender, C.G.; Fritz, R.S. Fluctuating asymmetry in a *Salix hybrid* system: The importance of genetic versus environmental causes. *Evolution* **1999**, *53*, 408–416. [CrossRef]
40. Valkama, J.; Kozlov, M.V. Impact of climatic factors on the developmental stability of mountain birch growing in a contaminated area. *J. Appl. Ecol.* **2001**, *38*, 665–673.
41. Anne, P.; Mawri, F.; Gladstone, S.; Freeman, D.C. Is fluctuating asymmetry a reliable biomonitor of stress? A test using life history parameters in soybean. *Int. J. Plant Sci.* **1998**, *159*, 559–565. [CrossRef]
42. Rao, G.-Y.; Andersson, S.; Widen, B. Flower and cotyledon asymmetry in *Brassica cretica*: Genetic variation and relationships with fitness. *Evolution* **2002**, *56*, 690–698. [PubMed]
43. Matilla, A.; Gallardo, M.; Puga-Hermida, M.I. Structural, physiological and molecular aspects of heterogeneity in seeds: A review. *Seed Sci. Res.* **2005**, *15*, 63–76. [CrossRef]
44. Xiong, B.; Wang, B.; Xiong, S.; Lin, C.; Yuan, X. 3D Morphological Processing for Wheat Spike Phenotypes Using Computed Tomography Images. *Remote Sens.* **2019**, *11*, 1110. [CrossRef]
45. Faralli, M.; Williams, K.S.; Han, J.; Corke, F.M.; Doonan, J.H.; Kettlewell, P.S. Water-saving traits can protect wheat grain number under progressive soil drying at the meiotic stage: A phenotyping approach. *J. Plant Growth Regul.* **2019**, *38*, 1562–1573. [CrossRef]
46. Shewry, P.R.; Wan, Y.; Hawkesford, M.J.; Tosi, P. Spatial distribution of functional components in the starchy endosperm of wheat grains. *J. Cereal Sci.* **2020**, *91*, 102869. [CrossRef]
47. Nuttall, J.G.; O'leary, G.J.; Panozzo, J.F.; Walker, C.K.; Barlow, K.M.; Fitzgerald, G.J. Models of grain quality in wheat—A review. *Field Crops Res.* **2017**, *202*, 136–145. [CrossRef]

48. Mukhtarovna, N.R.; Kizi, R.D.T.; Kholdorovich, A.K.; Botiraliyevich, U.N. Breathing of grain during storage and factors affecting the intensity of respiration. *Int. Multidisc. Res. J.* **2021**, *11*, 290–296. [CrossRef]
49. Bouchard, J.; Malalgoda, M.; Storsley, J.; Malunga, L.; Netticadan, T.; Thandapilly, S.J. Health benefits of cereal grain-and pulse-derived proteins. *Molecules* **2022**, *27*, 3746. [CrossRef]
50. Chen, X.; Yin, G.; Börner, A.; Xin, X.; He, J.; Nagel, M.; Liu, X.; Lu, X. Comparative physiology and proteomics of two wheat genotypes differing in seed storage tolerance. *Plant Physiol. Biochem.* **2018**, *130*, 455–463. [CrossRef]
51. Baranova, E.N.; Gulevich, A.A. Asymmetry of plant cell divisions under salt stress. *Symmetry* **2021**, *13*, 1811. [CrossRef]
52. Sarkar, A.; Fu, B.X. Impact of quality improvement and milling innovations on durum wheat and end products. *Foods* **2022**, *11*, 1796. [CrossRef] [PubMed]
53. Mondejar, M.E.; Avtar, R.; Diaz, H.L.B.; Dubey, R.K.; Esteban, J.; Gómez-Morales, A.; Hallam, B.; Mbungu, N.T.; Okolo, C.C.; Prasad, K.A.; et al. Digitalization to achieve sustainable development goals: Steps towards a smart green planet. *Sci. Total Environ.* **2021**, *794*, 148539. [CrossRef] [PubMed]
54. Abbas, G.; Saqib, M.; Rafique, Q.; Rahman, A.U.; Akhtar, J.; Haq, M.A.U.; Nasim, M. Effect of salinity on grain yield and grain quality of wheat (*Triticum aestivum* L.). *Pak. J. Bot.* **2013**, *50*, 185–189.
55. Price, P.B.; Parsons, J. Distribution of lipids in embryonic axis, bran-endosperm, and hull fractions of hulless barley and hulless oat grain. *J. Agric. Food Chem.* **1979**, *27*, 813–815. [CrossRef]
56. Zheng, Y.; Xu, X.; Simmons, M.; Zhang, C.; Gao, F.; Li, Z. Responses of physiological parameters, grain yield, and grain quality to foliar application of potassium nitrate in two contrasting winter wheat cultivars under salinity stress. *J. Plant Nutr. Soil Sci.* **2010**, *173*, 444–452. [CrossRef]
57. Angelova, M.B.; Pashova, S.B.; Spasova, B.K.; Vassilev, S.V.; Slokoska, L.S. Oxidative stress response of filamentous fungi induced by hydrogen peroxide and paraquat. *Mycol. Res.* **2005**, *109*, 150–158. [CrossRef]
58. Dong, B.; Zheng, X.; Liu, H.; Able, J.A.; Yang, H.; Zhao, H.; Zhang, M.; Qiao, Y.; Wang, Y.; Liu, M. Effects of drought stress on pollen sterility, grain yield, abscisic acid and protective enzymes in two winter wheat cultivars. *Front. Plant Sci.* **2017**, *8*, 1008. [CrossRef]
59. Broeke, J.; Pérez, J.M.M.; Pascau, J. *Image Processing with ImageJ: Extract and Analyze Data from Complex Images with ImageJ, the World's Leading Image Processing Tool*, 2nd ed.; Community Experience Distilled; Packt Publishing Open Source: Birmingham Mumbai, UK, 2015. Available online: http://dspace.library.uu.nl/handle/1874/204900 (accessed on 8 March 2021).
60. Isachkin, A.B.; Kryuchkova, B.A. Algoritmi opredelenia dostatochnykh ob'emov vyborok (na primere sadovykh rastenii). *Bulleten Glavnogo botanicheskogo sada* **2020**, *4*, 68–78. (In Russian)

plants

MDPI

Article

Genome-Wide Identification and Expression Profiling of Glutathione *S*-Transferase Gene Family in Foxtail Millet (*Setaria italica* L.)

Linlin Wang [1], Hongbo Fu [2], Juan Zhao [1], Jiagang Wang [3], Shuqi Dong [1], Xiangyang Yuan [1], Xiaorui Li [1,*] and Mingxun Chen [4,*]

1 State Key Laboratory of Sustainable Dryland Agriculture (in preparation), College of Agronomy, Shanxi Agricultural University, Taiyuan 030031, China
2 Key Laboratory for Research and Utilization of Characteristic Biological Resources in Southern Yunnan, College of Biological and Agricultural Sciences, Honghe University, Mengzi 661100, China
3 National Laboratory of Minor Crops Germplasm Innovation and Molecular Breeding (in preparation), Shanxi Agricultural University, Taiyuan 030031, China
4 College of Agronomy, Northwest A&F University, Yangling 712100, China
* Correspondence: lixiaorui@sxau.edu.cn (X.L.); cmx786@nwafu.edu.cn (M.C.)

Abstract: Glutathione *S*-transferases (GSTs) are a critical superfamily of multifunctional enzymes in plants. As a ligand or binding protein, GSTs regulate plant growth and development and detoxification. Foxtail millet (*Setaria italica* (L.) P. *Beauv*) could respond to abiotic stresses through a highly complex multi-gene regulatory network in which the GST family is also involved. However, *GST* genes have been scarcely studied in foxtail millet. Genome-wide identification and expression characteristics analysis of the foxtail millet GST gene family were conducted by biological information technology. The results showed that 73 *GST* genes (*SiGSTs*) were identified in the foxtail millet genome and were divided into seven classes. The chromosome localization results showed uneven distribution of *GSTs* on the seven chromosomes. There were 30 tandem duplication gene pairs belonging to 11 clusters. Only one pair of *SiGSTU1* and *SiGSTU23* were identified as fragment duplication genes. A total of ten conserved motifs were identified in the GST family of foxtail millet. The gene structure of *SiGSTs* is relatively conservative, but the number and length of exons of each gene are still different. The cis-acting elements in the promoter region of 73 *SiGST* genes showed that 94.5% of *SiGST* genes possessed defense and stress-responsive elements. The expression profiles of 37 *SiGST* genes covering 21 tissues suggested that most *SiGST* genes were expressed in multiple organs and were highly expressed in roots and leaves. By qPCR analysis, we found that 21 *SiGST* genes were responsive to abiotic stresses and abscisic acid (ABA). Taken together, this study provides a theoretical basis for identifying foxtail millet GST family information and improving their responses to different stresses.

Keywords: foxtail millet; glutathione *S*-transferase (GSTs); expression analysis; stress response

Citation: Wang, L.; Fu, H.; Zhao, J.; Wang, J.; Dong, S.; Yuan, X.; Li, X.; Chen, M. Genome-Wide Identification and Expression Profiling of Glutathione *S*-Transferase Gene Family in Foxtail Millet (*Setaria italica* L.). *Plants* **2023**, *12*, 1138. https://doi.org/10.3390/plants12051138

Academic Editor: Cristina Crosatti

Received: 8 February 2023
Revised: 27 February 2023
Accepted: 28 February 2023
Published: 2 March 2023

1. Introduction

Glutathione *S*-transferases (GSTs), a superfamily of enzymes encoded by multiple genes and having multiple functions, are ubiquitous in animals, plants, and microorganisms. Glutathione binds to harmful heterologous substances or oxidation products through GSTs, thereby promoting the metabolism, compartmentalization, or elimination of such substances [1]. The classical reaction mode is that GSTs catalyze the binding of glutathione to various hydrophobic and electrophilic electronic groups to form soluble S-glutathionylated products [2]. Fourteen categories have been identified according to the amino acid sequence similarity, among which eight categories are extensive, including eight subclasses: tau (U type), phi (F type), lambda (L type), theta (T type), zeta (Z type), γ-subunit of translation elongation factor (EF1G), dehydroascorbate reductase (DHAR), and tetrachloro hydroquinone dehalogenase (TCHQD) [3–5]. Among these classes of GSTs, tau, phi, lambda, and TCHQD are endemic to plants, and the tau and phi classes are the most abundant GST types in plants [6,7]. Although the sequence homology of the GST gene is low (about 25%), it has been found, by studying the structure of a large number of GSTs proteins [8,9], that these proteins have highly conserved structural characteristics.

Since the first discovery of glutathione *S*-transferases in maize in the 1970s, GSTs have been identified as a multigene family [10], and GSTs have been found in many plants. Genome-wide analyses revealed that there are 55 *GST* genes in *Arabidopsis thaliana* [11,12], 79 in *Oryza sativa* [13,14], 330 in *Triticum aestivum* [15,16], 84 in *Hordeum vulgare* [17], 59 in *Gossypium Raimondi* [18], 141 in *Brassica napus* [19], and 52 in *Malus domestica* [20]. In addition, the *GST* genes were identified in these plants, laying the foundation for isolating new *GST* genes from other plants.

Many studies have shown that the plant *GST* gene family can regulate the adaptability of plants to various kinds of stresses through electrophilic substitution, detoxification, and peroxide scavenging [21–24]. The expression of this gene family is not just affected by drought [25], saline-alkali [26], low temperature [27], pathogen infection [28], herbicides [29], heavy metals [30], and other stresses. It is also subject to ABA [31], auxins (IAA), ethylene [32], salicylic acid (SA) [32], jasmonic acid (JA) [33] and other plant hormones.

The expression of *TaGSTU39* was significantly up-regulated throughout the treatment period under drought and salt stress treatments. *TaGSTU62* of wheat could be down-regulated by gibberellin (GA) and up-regulated by ABA [15]. In *A. thaliana*, overexpression of grape *GSTF13* could increase resistance to drought, salt, and methyl viologen stresses [34]. Multiple photoreceptors regulated the expression of *AtGSTU17* and various development of *A. thaliana* seeds, including hypocotyl elongation and anthocyanin accumulation [35]. *MdGSTF12* [20] and *MdGSTU12* [36] were strongly induced by aminolevulinic acid (ALA), and they play an essential role in ALA-induced anthocyanin accumulation in apples. *GmGSTU10* was induced explicitly by soybean mosaic virus and might have a highly efficient catalytic role in soybean [37].

A growing world population challenges global food and nutrition security [38]. In order to find suitable staple foods to overcome these difficult situations, millets are one of the potential candidates [39]. Foxtail millet is the oldest cultivated crop in the world, including China, and it is also the characteristic food crop in arid and semi-arid areas in northern China. Foxtail millet, as the main cultivated crop in dry green farming, has the characteristics of small genome size, short life cycle, self-pollination, drought resistance, and is able to grow in low fertility conditions, making it a model plant for studying C_4 cereal crops [10,40–42]. However, there is no relevant report on the type, quantity, structure, and function of GST family in foxtail millet. Therefore, this study used genome and bioinformatics methods to analyze the *GST* gene family of foxtail millet transcription factors. A total of 73 *GST* genes were identified and divided into seven classes. The protein physicochemical properties, chromosomal distribution, gene structure, conserved motifs, cis-acting element, and gene expression of members of the GST family in foxtail millet were analyzed. We also exhibited the syntenic correlation between foxtail millet and *A. thaliana* genes. In addition, we investigated expression profiling of *GST* genes in different tissues and detected the expression of *GST* genes after various biotic and abiotic stresses in foxtail millet. Our study suggested that *GST* genes may play a role in regulating development and responding to various biotic and abiotic stresses and to ABA. Overall, this study provided a comprehensive identification of foxtail millet GST family members and provided a theoretical basis for further research on the functional analysis, gene editing, and genetic engineering of *GST* genes in foxtail millet.

2. Results

2.1. Identification of the Foxtail Millet GST Proteins and Analysis of Phylogenetic Relationship

The GST protein sequences of *A. thaliana* were searched against the protein sequences of foxtail millet to identify the GST proteins in foxtail millet. A total of 73 SiGST proteins were identified from the foxtail millet genome, based on the conserved GST-N and GST-C domains from these proteins (Figure S1). These GSTs were divided into seven distinct classes according to their conserved domains: tau, lambda, zeta, phi, DHAR, TCHQD, and MGST, and theta is absent in foxtail millet (Figure 1; Table S1). Among SiGST proteins, there are 44 tau proteins and 18 phi proteins, accounting for the vast majority, just as there are more tau and phi proteins in most plant GST families [43]. Next is lambda, which contains six members. The DHAR class had two members, and the zeta, TCHQD, and MGST classes had the least number, all of which had only one member.

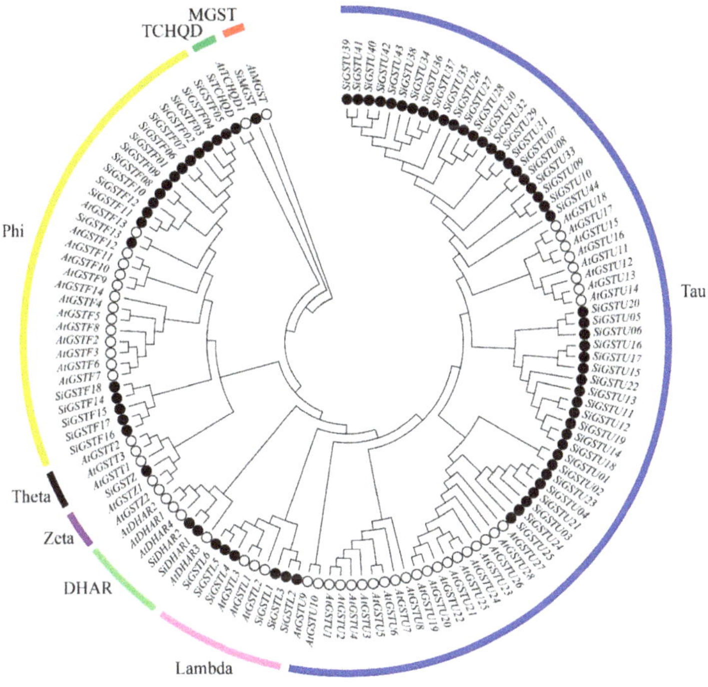

Figure 1. Phylogenetic tree of GST proteins among foxtail millet and *A. thaliana*. The GST protein sequences of 73 foxtail millet and 55 *A. thaliana* are divided into eight classes. Different subfamilies are marked as different colors.

The analysis of physicochemical property showed that the sequence length of SiGST proteins varied from 182 (*SiGSTF05*) to 320 (*SiGSTF18*) amino acid residues, and the molecular weight (MW) was 20,177.45 (*SiGSTF05*)—36,490.24 (*SiGSTF18*) Da. The isoelectric point (pI) values were changed from 4.74 (*SiGSTL1*) to 9.05 (*SiTCHQD*). The instability index of proteins ranged from 23.76 (*SiDHAR2*) to 60.81 (*SiGSTL5*), of which 31 were instability proteins with an instability index greater than 40. According to the correlation principle of the grand average of hydropathicity (GRAVY), the amphiphilic protein is between -0.5 and 0.5, the positive is hydrophobic protein, and the negative is hydrophilic protein. *SiGSTF14*, *SiGSTF18*, and *SiGSTU02* are hydrophilic proteins, and other proteins are amphotropic proteins, among which *SiGSTU08* has a maximum value of 0.185 and *SiGSTF18* has a minimum value of -1.085. The detailed data information for 73 SiGST protein sequences was tabulated (Table S2).

2.2. *Chromosome Location and Gene Replication of SiGST Genes in Foxtail Millet*

The chromosomal localization of 73 *SiGST* genes in foxtail millet revealed that SiGSTs were unevenly distributed on seven chromosomes. A high-density region containing GSTs was found on chromosomes III, V, and IX (Figure 2; Table S3). Among them, chromosome IX with 30 *SiGST* genes included the most members, followed by 22 on chromosome V. Eight *SiGST* genes were distributed on chromosome II, and nine *SiGST* genes were distributed on chromosome III. Only two *SiGST* genes were distributed on chromosome IV, and both chromosomes VII and VIII with one *SiGST* gene contained the least members. There is no *SiGST* gene on chromosomes I and VI.

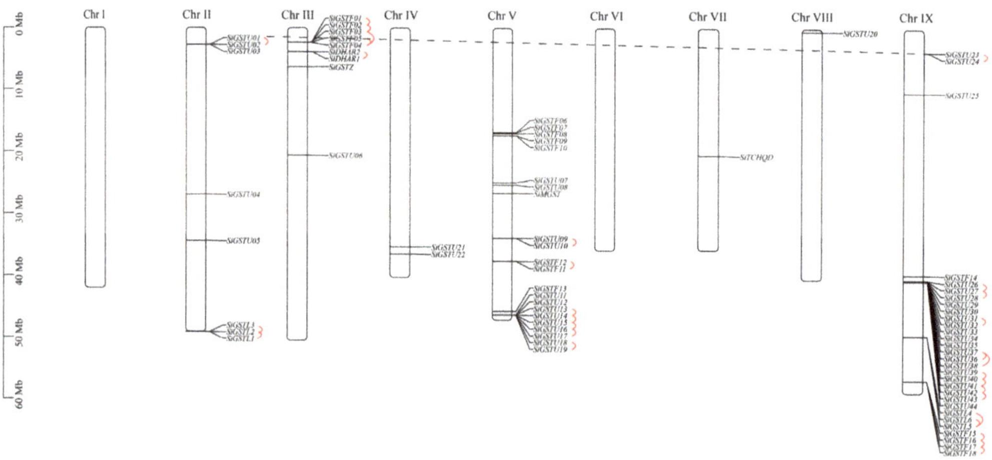

Figure 2. Chromosome mapping of *GST* genes in foxtail millet. A total of 30 tandem duplication gene pairs belonging to 11 clusters were present (highlighted by full red lines). Only one pair of gene segmental duplication event was found (highlighted by dashed black line).

Segmental duplication and tandem duplication are considered to be two important factors for gene family expansion [44]. Among the 73 *SiGST* genes, a total of 30 gene pairs of 11 clusters were identified as the tandem duplication type (Figure 2). Among them, one pair of tandem duplication in DHAR class, four pairs in lambda class, eight pairs in phi class, and 17 pairs in tau class, indicating that a tandem duplication event contributed more to the expansion of the phi and tau classes. Only a pair of gene segmental duplication events (*SiGSTU1* and *SiGSTU23*) occurred in all classes on seven chromosomes. Further analysis of the evolution of *SiGST* genes revealed that there was no syntenic relationships between *SiGST* gene and *AtGST* gene (Figure S2).

2.3. *Conserved Motif and Gene Structure Analysis of SiGSTs*

In order to better demonstrate the diversity and similarity of the SiGST motifs, the conserved structure of amino acids in the foxtail millet GST family was analyzed using the MEME database (Figure 3a,b). The results showed that ten conserved motifs were identified in the foxtail millet GST family, and each conserved motif length ranged from 11 to 50 amino acids (Figure S3). Tau and phi have more members, with motifs 1, 2, 3, 4, 5, and 6 found in 44 tau protein sequences, and motifs 4, 5, 7, and 8 found in 18 phi protein sequences. There were 72 SiGST proteins that contained motif 5, while 70 SiGST proteins contained motifs 2, 4, and 6. In lambda classes, there were coexistent motifs. In addition, some class members had completely identical motifs, such as DHAR, which contains motifs 1, 2, 4, 5, and 6.

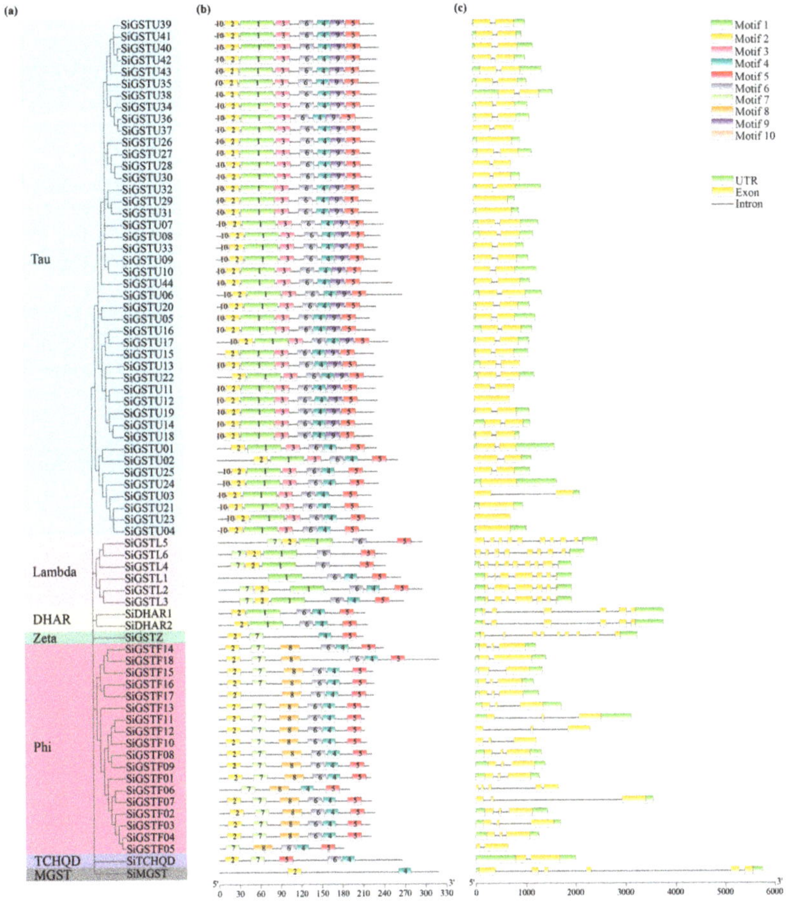

Figure 3. Gene structure and conserved motif of foxtail millet *GSTs*. (**a**) Neighbor-joining (NJ) phylogenetic tree based on SiGST protein sequences. (**b**) Conserved motif of *SiGSTs*. (**c**) Gene structure of *SiGSTs*.

Gain or loss of introns can alter gene structure and play a crucial role in the evolution of gene families [45]. Gene structure analysis showed that *SiGSTU11*, *SiGSTU12*, *SiGSTU23*, *SiGSTU28*, *SiGSTU37*, *SiGSTF05*, *SiGSTF06*, *SiGSTF10*, and *SiGSTF12* did not contain upstream and downstream regulatory regions (UTR). *SiGSTU02*, *SiGSTU03*, *SiGSTU10*, *SiGSTU25*, *SiGSTU29*, *SiGSTU30*, *SiGSTL5*, and *SiGSTF07* did not contain the upstream regulatory region outside. *SiGSTU13* did not contain the downstream regulatory region outside. The other 55 members all have an upstream and downstream regulatory region (Figure 3c). At the same time, the gene sequence structure of the members of the GST family was analyzed, and the "exon-intron" structure diagram was obtained. Of these, most tau, phi, and TCHQD classes had one to three exons, while a phi member consisted of five exons. The DHAR and MGST classes exhibited six exons, while the zeta and lambda classes contained more exons than other classes (6–10 exons). The number of introns of the 73 *GST* genes in foxtail millet was less than ten, the most had nine, and the least had none. In addition, each member of the same subfamily has the same or similar gene structure. For example, most tau classes contained one intron; phi classes contained one to two introns, but *SiGSTF06* contained four introns; TCHQD classes contained one intron; DHAR and MGST classes contained five introns; and zeta classes contained eight introns.

2.4. Prediction of Cis-Acting Elements in Promoter Region of GSTs Gene Family Members in Foxtail Millet

Using PlantCARE to analyze the 5′-upstream promoter (2.0 kb) region of 73 *SiGSTs*, we found that the cis-acting elements in the promoter region of 73 *SiGST* genes mainly included 20 kinds, including defense and stress responsive elements involved in salt, drought, low-temperature and anaerobic, light responsive element, hormone-responsive element associated with IAA, ABA, Methyl Jasmonate, GA, SA, and other elements related to growth regulation and circadian control, including meristem expression element, cell cycle regulation element, endosperm expression element, seed-specific regulation element, root specific element and MYB (MER), MYBHv1 binding site (Figures 4 and S4; Table S4). The promoter regions of 67 *SiGST* genes presented in defense and stress response elements, and the cis-acting elements of 67 *SiGST* gene promoters contained hormone responsive elements. In addition, more than 40 cis-acting elements were identified in the promoters of the *SiGSTF07*, *SiGSTF13*, *SiGSTU21*, and *SiTCHQD* transcripts. However, only two cis-elements were identified in the promoters of *SiGSTU39*.

Figure 4. Predicted cis-acting element in *SiGST* promoter of foxtail millet.

2.5. Relative Expression Patterns (FPKM Value) of GST Genes in Different Tissues

To explore the expression specificity of *GST* genes in foxtail millet at different tissues and developmental stages, we analyzed the expression profiles of 37 *SiGST* genes covering 21 tissues at different growth stages (Figure 5; Table S5). The results demonstrated partial differences in the expression of *SiGST* genes in different tissues of the same class. Additionally, most of the *SiGST* genes were expressed at higher levels in roots, which indicates that they may first play a role when the roots sense adversity. The transcription abundance of individual *SiGST* genes was low in all tissues and organs, such as *SiGSTF07*, which was weakly expressed in all tissues. However, some members had high transcript richness in all tissues and organs. *SiGSTF11*, *SiGSTF12*, and *SiGSTF13* were strongly expressed in all tissues. There were tissue-specific expression characteristics in different *SiGSTs* members, such as *SiGSTF01*, *SiGSTF03*, *SiGSTU05*, and *SiGSTU19*, which are highly expressed in the leaves and roots of foxtail millet during grain fillinfigure s1 stage. The expression of *SiGSTU06* in developing seeds and spikelets is relatively high, especially in panicle, which have the highest expression level. *SiGSTL4* was slightly or even not expressed in different tissues. In addition, the genes were differentially expressed in different tissues and organs at the same developmental stage. *SiGSTF11* and *SiGSTF12* were highly expressed in roots during the filling stage and *SiGSTF13* was highly expressed in neck-panicle-internodes during the filling stage. These results indicated that the relative expression pattern of *SiGST* genes in different tissues predicted its complex roles in foxtail millet growth and development.

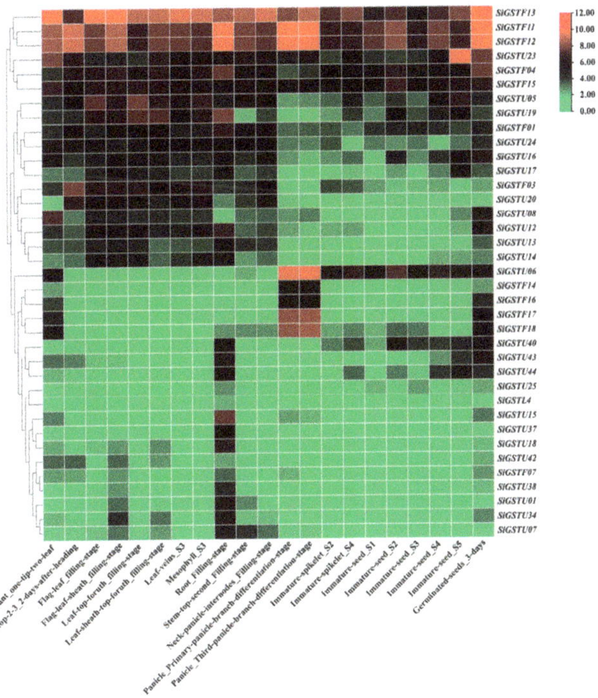

Figure 5. Relative expression patterns (FPKM value) of *GST* genes involved in 21 tissues of foxtail millet. The FPKM values of *SiGST* genes in different tissues were obtained from the *Setrari italica* Multiomics database (http://foxtail-millet.biocloud.net/home, accessed on 8 June 2022). The color bar represents $\log_2$ expression levels (FPKM), with red indicating high gene expression levels and green indicating low gene expression levels.

2.6. Relative Expression Patterns of 21 SiGSTs under Abiotic Stresses and ABA Treatments in Foxtail Millet

To understand the responses of *SiGST* genes to ABA and other abiotic stresses, we selected 21 genes to analyze their expression in foxtail millet treated with osmotic (20% PEG 6000), salt (200 mM NaCl), cold stress (4 °C temperature), and 100 μM ABA, respectively (Figure 6). In general, *SiGST* gene expression did not show consistent characteristics under stress and hormone treatments. Under osmotic stress treatment, the expression of *SiGSTF11*, *SiGSTF14*, *SiGSTU01*, *SiGSTU05*, *SiGSTU13*, *SiGSTU14*, *SiGSTU20*, *SiGSTU24*, and *SiGSTU42* were significantly up-regulated, and the expression level of *SiGSTU1* was remarkably up-regulated. Its expression peaked at 12 h, which was 56.5 times that of 0 h. The expression of *SiGSTF07*, *SiGSTU16*, *SiGSTU20*, and *SiGSTU23* peaked at 6 h; *SiGSTU13* at 12 h; *SiGSTF12*, *SiGSTF16*, *SiGSTU14*, and *SiGSTU17* at 24 h; and *SiGSTF03* and *SiGSTU44* at 48 h (Figure 6a).

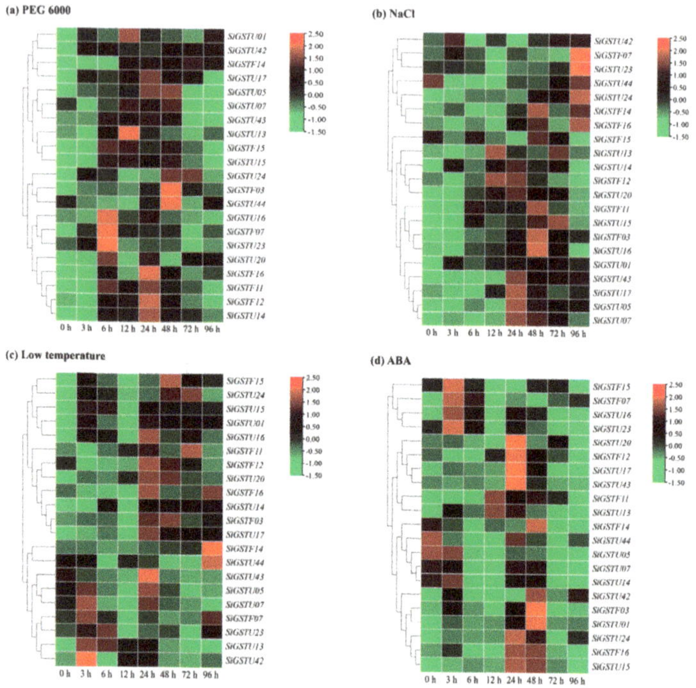

Figure 6. Relative expression patterns of 21 *SiGSTs* in leaves of foxtail millet under abiotic stresses and ABA treatments. Relative expression patterns of 21 *SiGST* genes were analyzed with qPCR under osmotic (20% PEG 6000) (**a**), salt (200 mM NaCl) (**b**), cold stress (4 °C temperature) (**c**), and 100 μM ABA (**d**). The relative expression levels of each gene were calculated by the $2^{-\Delta\Delta Ct}$ method. The heat map of SiGSTs expression condition was constructed by TBtools, expression value was standardized by Log2. The expression level of target gene at 0 h was used as control.

Under salt stress treatment, the *SiGSTF03*, *SiGSTF11*, *SiGSTU01*, *SiGSTU15*, *SiGSTU24*, and *SiGSTU43* genes were induced more significantly, and the expression was higher than 0 h. The expression peak for *SiGSTF12* and *SiGSTU13* appeared at 12 h and up-regulated 7.2-fold and 6.7-fold compared to the control group. The expression levels of *SiGSTU05*, *SiGSTU07*, *SiGSTU14*, *SiGSTU17*, *SiGSTU20*, and *SiGSTU43* genes were the highest at 24 h, and the *SiGSTF03*, *SiGSTF11*, *SiGSTF14*, *SiGSTF15*, and *SiGSTU16* genes showed an expression peak at 48 h. The expression peak of *SiGSTF07*, *SiGSTF16*, *SiGSTU23*, and *SiGSTU24* appeared at 96 h, and *SiGSTU24* expression at 96 h, which is about 81 times that of 0 h (Figure 6b).

The expressions of *SiGSTF11*, *SiGSTF12*, *SiGSTF16*, and *SiGSTU43* were inhibited and lower than those of the control group in the early stage of 4 °C temperature treatment. However, *SiGSTU15*, *SiGSTU17*, *SiGSTU20*, and *SiGSTU24* expression levels were significantly higher than the control group during the whole 4 °C temperature treatment, and the *SiGSTU20* induced more remarkably. The expression levels of *SiGSTF12*, *SiGSTF16*, *SiGSTU05*, *SiGSTU16*, *SiGSTU17*, *SiGSTU20*, and *SiGSTU43* were significantly higher than the control group at 24 h. At 48 h, *SiGSTF12* and *SiGSTU20* were still significantly higher than in control group (Figure 6c).

Additionally, they could respond to ABA. For instance, the expression of *SiGSTU5*, *SiGSTU07*, and *SiGSTU44* were down-regulated by ABA; and the expression of *SiGSTF07*, *SiGSTU01*, *SiGSTU17*, *SiGSTU24*, and *SiGSTU42* could be up-regulated by ABA. The expression of *SiGSTF12*, *SiGSTU15*, *SiGSTU17*, *SiGSTU20*, *SiGSTU24*, and *SiGSTU43* peaked at 24 h. The expression peak for *SiGSTF03*, *SiGSTU01*, and *SiGSTU42* appeared at 48 h, up-regulated 16.1-fold, 19.2-fold, and 74.6-fold compared to the control group, respectively (Figure 6d).

3. Discussion

3.1. Identification and Analysis of GST Genes in Foxtail Millet

GSTs comprise a large and diverse gene family that is ubiquitous in a wide variety of organisms. Currently, the *GST* gene family has been confirmed to be involved in regulating the growth and development, stress resistance, and other processes of plants, such as *A. thaliana* [11,12] and rice [13], which has important biological significance. However, the identification and analysis of GST gene family in foxtail millet are still lacking. A total of 73 *SiGST* genes were identified from the foxtail millet genome, which were divided into seven categories. Among them, tau and phi contain the most SiGST family members, with 44 and 18 *SiGST* genes, respectively, which are the same as the findings of other plants such as soybean [46], rice [13], and pepper [47]. Tau and phi, known as plant-specific GSTs, had the most members and abundant content compared with other subfamilies [47]. A possible reason is that, compared with animals, in addition to detoxification, GST in higher plants also has unique functions such as transporting flavonoids, and regulating plant growth and development [30,48].

Gene structure is of great significance to the study of gene evolution [49]. The rearrangements and fusions of different chromosomal fragments may result in the gain or loss of an exon or intron, which play an important role in the diversification of multi-gene families [45]. We found that the number of *GST* exons is generally conserved within the same class in foxtail millet. GSTUs had one or two exons, GSTFs had two or three exons (except for *SiGSTF06*), DHAR and MGST had six exons, zeta had nine exons, and TCHQDs had two exons (Figure 3c). Similarly, the structural characteristics of *GST* genes were conserved in wheat, apple, and melon [15,20,50]. In addition, previous studies have shown that the earliest gene had the fewest number of introns, and as the replication progressed, the number of gene members continued to increase, and the number of introns gradually increased. It is pointed out that some of the origins of introns in the gene structure exist within themselves, and some are generated or increased with the insertion of transposons during gene replication [51,52]. In this study, through the analysis of the structure and conserved motifs of GST, it was found that most of the tau and phi classes contained one intron, the DHAR and MGST classes contained five introns, and the zeta class contained eight introns (Figure 3a,b). Therefore, according to the number of members of each subfamily and the number of introns in the gene, it is speculated that tau and phi may first appear in the GST family of foxtail millet. This was similar to the structure of GST family genes reported in wheat, apple, and radish [15,20,53].

The expansion of plant gene families is mainly achieved through different gene duplication methods (including fragment duplication, cascade duplication, gene transfer, and genome duplication, etc.) [18,54]. In addition, the expansion of GST family genes in plants was mainly completed by the tandem duplication of genes of tau and phi classes, making plant-specific tau and phi the two classes with the most members [6,47,55]. GST protein sequences of foxtail millet were compared, and the phylogenetic tree was constructed. It was divided into eight large groups (tau, phi, lambda, zeta, theta, DHAR, TCHQD, and MGST) (Figure 1). There were 28 *A. thaliana* genes in tau, accounting for 50.9% of the *A. thaliana* GST family. However, there were 44 members of the foxtail millet GST family in tau, accounting for 60.2%. This amplification in tau may be related to gene duplication during evolution. Indeed, we found that only one pair of segment duplication genes were in foxtail millet, and 30 pairs of genes were tandem duplications, and tandem duplications were more frequent than segment duplications (Figure 2). Similar phenomena have been observed in rice [13], apple [20], and *Capsella rubella* [56]. Gene duplication not only expands the content of the genome, but is also important for generating new gene functions, thereby enabling organisms to further adapt to complex environments [44]. Collinearity analysis indicated that strong selection pressure was subjected to during the evolution of *GST* gene family in foxtail millet.

3.2. The Expressions of SiGST Genes in Foxtail Millet

Tissue expression profiling analysis of *SiGST* genes indicated that most *SiGST* genes were expressed in multiple organs and highly expressed in roots and leaves, indicating that most *SiGST* genes may play a role in roots and leaves (Figure 5). We also found functional differentiation of *GSTs* in foxtail millet, and *SiGSTs* were expressed differently in different tissue parts of foxtail millet. For example, *SiGSTF03* was highly expressed in parietal leaf at heading stage. *SiGSTU06*, *SiGSTF17*, and *SiGSTF18* were highly expressed in panicle_primary-panicle-branch-differentiation-stage and panicle_third-panicle-branch-differentiation-stage. *SiGSTL4*, *SiGSTU15*, and *SiGSTU18* were specifically expressed in the underground part of foxtail millet. In some tandem repeat pairs, the expression levels of the two genes differed, suggesting that the retention of gene duplication may be related to the process of tissue expression differences [57,58]. For example, *SiGSTF15* was highly expressed in 21 different tissues, while *SiGSTF07* was less expressed.

Plant growth and development are affected by various abiotic stresses, which activate the molecular mechanisms of plants to adapt to adverse conditions [59]. Drought, high salinity, and extreme temperatures limit the geographical distribution of plants because they cause dehydration and, eventually, cell death [53]. The *GSTs* promoter region of foxtail millet contains a large number of light response elements and stress response elements (low temperature, drought, stress defense, and stress), and the cis-acting elements in promoter regions of *SiGST* genes were found to be involved in the response to diverse biotic and abiotic stresses, as well as hormones (Figure 4 and Figure S4; Table S4). The 94.5% of *SiGST* genes possessed defense and stress responsive elements, and only six of them had no hormones responsive elements.

Previous studies have shown that the GST gene family is induced by abiotic stresses and hormones. For example, *MdGSTF12* [20] and *MdGSTU12* [36] were strongly induced by ALA. *TaGSTU62* could be induced by osmotic stress, salt stress, and ABA [15]. However, *OsGSTU4* could only be induced by salt stress [60]. The expression profiles of 21 *SiGST* genes under three abiotic stresses and one hormone (osmotic, salt, cold stress, and ABA) were analyzed by qPCR, showing that *SiGST* genes could be induced by abiotic stresses and ABA, and they might play a key role in abiotic stress response through corresponding hormone-dependent pathways (Figure 6). This study found that the *SiGSTU17*, *SiGSTU24*, and *SiGSTF03* could be induced by abiotic stresses and ABA. Although these results indirectly proved that *SiGST* genes are involved in the response of Jingu 21 to stress, further analysis is still needed to confirm its ability to resist stresses.

4. Materials and Methods

4.1. GST Gene Identification, Phylogenetic Analysis and Physicochemical Properties of Foxtail Millet

To identify the SiGST proteins, the *Setaria italica* genome data were downloaded from the NCBI database (https://www.ncbi.nlm.nih.gov/genome/?term=Setaria+italica+, accessed on 14 August 2021). HMMER 3.0 software was used to identify *SiGST* genes, and the *GST* gene domain sequence (PF02798 and PF00043) was downloaded from the Pfam database (https://pfam.xfam.org/, accessed on 7 June 2022). The National Center for Biotechnology Information conserved domain database (www.ncbi.nlm.nih.gov/Structure/cdd/wrpsb.cgi, accessed on 7 June 2022) was used to detect GST domains which were then mapped to the conserved domain. Neighbor-joining (NJ) phylogenetic tree was constructed using MEGA 7 software by the neighbor-joining method (bootstraps = 1000). The GST protein sequences of *A. thaliana* (55 numbers) were downloaded from The *Arabidopsis* Information Resource (TAIR, http://www.arabidopsis.org, accessed on 27 October 2020). Length of the protein sequence, protein molecular weight (MW), genomic position, isoelectric point (pI), instability index, and aliphatic index were predicted using ExPASy-ProParam (https://web.expasy.org/protparam/, accessed on 9 June 2022).

4.2. Distribution and Duplication Analysis of SiGSTs

TBtools was used to display the distribution of *GST* genes on *Setaria italica* chromosomes. Two or more *GSTs* separated by no more than three genes on the chromosome are called *GST* gene clusters. The gene sequences were aligned using BLASTp to determine the form of gene replication with an e-value of 1×10^{-1}.

4.3. Gene Structure, Motif Compositions, and Gene Synteny of SiGST Genes

The gene structure map was produced, and an intron-exon map was compiled based on the *S. italica* genome annotation information (v 2.0). The MEME database (http://meme-suite.org/tools/meme, accessed on 9 June 2022) was used to conduct protein motif analysis. Chromosomal positions of *SiGST* genes were analyzed, and MCScanX was used to detect collinear regions between *SiGST* genes, as well as collinear blocks of *SiGST* genes with *A. thaliana* genes. The *A. thaliana* genome data were downloaded from the NCBI database. All above were visualized using TBtools [61].

4.4. Prediction of Cis-Acting Elements in the Promoter of SiGST Genes

Using the genome sequence of *SiGSTs* obtained in the phytozome database, the 2.0 kb DNA sequence upstream of *SiGST* genes was submitted to the PlantCARE database (http://bioinformatics.psb.ugent.be/webtools/plantcare/html/, accessed on 17 June 2022) to predict cis-acting element. Promoter cis-acting regulatory element predictions were performed and visualized in TBtools.

4.5. Analysis of the Expression Pattern of SiGST Genes in Different Tissues

The FPKM values of *SiGST* genes, in the Multi-omics Database for *Setrari italica* (http://foxtail-millet.biocloud.net/home, accessed on 8 June 2022) in different tissues such as roots, stems, leaves, and flowers were extracted, and the TBtools software was used to draw gene expression heat maps for visualization.

4.6. Plant Materials and Treatments

The foxtail millet seeds of Jingu 21 were in a germination box with a temperature of 26 °C, a relative humidity of 65%, and a light cycle of 16 h/8 h. Two days later, the seedlings were transplanted into a plastic container with a modified half-strength Hoagland nutrient medium for further cultivation. For osmotic stress, salt stress, and ABA treatments, the 12-day-old seedlings were allowed to grow in 1/2 Hoagland solution containing 20% PEG 6000, 200 mM NaCl, and 100 μM ABA, respectively. For cold stress treatment, the 12-day-old seedlings were placed in a low temperature (4 °C) incubator to grow. The seedlings in 1/2 Hoagland solution without treatment at 26 ° C were regarded as controls. The leaves of the seedlings were sampled at 0 h, 3 h, 6 h, 12 h, 24 h, 48 h, 72 h, and 96 h after 20% PEG 6000, 200 mM NaCl, and 100 μM ABA treatment. All samples were immediately frozen in liquid nitrogen and stored in a −80 °C refrigerator.

4.7. RNA Isolation and qPCR

The acquisition of RNA from leaves was accomplished with TRIzol kit (Accurate Biology, Changsha, China), cDNA was synthesized using reverse transcription kit (Accurate Biology, Changsha, China), and real-time PCR was performed with SYBR Green dye method (Accurate Biology, Changsha, China). PCR primers were designed using Primer Premier 5 software (Table S6). The qRT-PCR reaction was performed in a Bio-Rad CFX system, a 20 μL reaction system containing 10 μL 2× SYBR Green *Pro Taq HS* premix, 2 μL cDNA, 0.4 μL each of forward primer and reverse primer, 7.2 μL RNase free water. The following cycling conditions were used: 95 °C for 30 s, followed by 40 cycles of 95 °C for 5 s, and 60 °C for 30 s. The *SiActin* (*SETIT_026509mg*) was used as internal standard, and the expression levels of each gene were calculated by the $2^{-\Delta\Delta Ct}$ method. The heat map of SiGSTs expression condition was constructed by TBtools, expression value was standardized by Log2. Each independent experiment was repeated at least three times.

5. Conclusions

Through the appraisal of foxtail millet *SiGST* gene families, system evolution, and expression analysis, identified 73 millet *SiGST* genes, and the chromosomal location, the physicochemical properties, gene structure, and conservative structure domain analysis, this study forecast the *SiGSTs* promoter segment in response to ABA treatment, several abiotic stresses, and developmental stages. The expression profile spectrum analyses of *SiGST* genes showed that most of the *SiGST* genes were highly expressed in the roots and leaves. The qPCR analyses of 21 *SiGST* genes confirmed that *SiGST* genes were widely involved in stress and hormone responses such as drought, salt, low temperature, and ABA. These results provide a reference for further research on the gene function of foxtail millet molecular breeding and mining potential genetic resources.

Supplementary Materials: The following supporting information can be downloaded at: https://www.mdpi.com/article/10.3390/plants12051138/s1, Figure S1: Conserved domain of foxtail millet *GSTs*. Figure S2: Collinearity analysis of *GSTs* in *Setaria italica* and *Arabidopsis thaliana*. Figure S3: The putative motifs of SiGST proteins in foxtail millet. Figure S4: Predicted cis-acting element in *SiGST* promoter of foxtail millet. Table S1: GST protein sequences of foxtail millet and *A. thaliana* used to construct the phylogenetic tree. Table S2: The detailed information of 73 *SiGST* genes, including gene name, accession number, number of amino acids, MW, chromosomal location, theoretical PI, instability index, aliphatic index, and grand average of hydropathicity. Table S3: The distributions of *SiGST* class members on foxtail millet chromosomes. Table S4: Putative cis-acting elements identified in the promoter regions of *SiGSTs*. Table S5: The expression levels (FPKM value) of *SiGST* genes involved in 21 tissues. Table S6: Primers used for qPCR.

Author Contributions: Conceptualization, X.L. and M.C.; Methodology, L.W., H.F. and X.L.; Software, L.W., H.F. and X.L.; Formal analysis, L.W., X.L. and M.C.; Investigation, L.W., H.F., J.Z., S.D. and H.F.; Resources, S.D., X.L. and H.F.; Data curation, L.W., J.W., J.Z., S.D, X.Y. and X.L.; Writing—original draft preparation, L.W.; Writing—Review & Editing, L.W., H.F., X.L. and M.C.; Visualization, X.L. and M.C.; Supervision, S.D., X.Y. and X.L.; Project Administration, J.W., S.D., X.Y., X.L. and M.C. All authors have read and agreed to the published version of the manuscript.

Funding: This work was supported by State Key Laboratory of Sustainable Dryland Agriculture (in preparation), Shanxi Agricultural University (YJHZKF2104 and 202003-5), the Shanxi Province Basic Research Program Project (20210302124699), the Doctoral Research Start Project of Shanxi Agricultural University (2021BQ20), the Shanxi Province Doctoral Graduates and Postdoctoral Researchers to Work Award Fund Research Project (SXBYKY2021044), the China Agriculture Research System of MOF and MARA (CARS-06-14.5-A28), the National Natural Science Foundation of China (32272229 and 32200222), and the National Laboratory of Minor Crops Germplasm Innovation and Molecular Breeding (in preparation) (K462202040-32).

Institutional Review Board Statement: Not applicable.

Informed Consent Statement: Not applicable.

Data Availability Statement: All GST protein sequences are provided in Table S1.

Acknowledgments: We thank Hongbo Fu (College of Biological and Agricultural Sciences, Honghe University) for his great contribution to the data analysis.

Conflicts of Interest: The authors declare no conflict of interest.

References

1. Nutricati, E.; Miceli, A.; Blando, F.; De Bellis, L. Characterization of two *Arabidopsis thaliana* glutathione S-transferases. *Plant Cell Rep.* **2006**, *25*, 997–1005. [CrossRef] [PubMed]
2. Abdul Kayum, M.; Nath, U.K.; Park, J.I.; Biswas, M.K.; Choi, E.K.; Song, J.Y.; Kim, H.T.; Nou, I.S. Genome-Wide Identification, Characterization, and Expression Profiling of Glutathione S-Transferase (GST) Family in Pumpkin Reveals Likely Role in Cold-Stress Tolerance. *Genes* **2018**, *9*, 84. [CrossRef] [PubMed]
3. Lallement, P.A.; Brouwer, B.; Keech, O.; Hecker, A.; Rouhier, N. The still mysterious roles of cysteine-containing glutathione transferases in plants. *Front. Pharmacol.* **2014**, *5*, 192. [CrossRef] [PubMed]
4. Dixon, D.P.; Edwards, R. Glutathione Transferases. In *The Arabidopsis Book, Biologists;* American Society of Plant Biologists: Rockville, MD, USA, 2010; Volume 8, pp. 1–15.
5. Sasan, M.; Maryam, E.; Fateme, M.; Maryam, S.; Babak, S.; Hassan, M. Plant glutathione S-transferase classification, structure and evolution. *Afr. J. Biochem. Res.* **2011**, *10*, 8160–8165.
6. Islam, S.; Rahman, I.A.; Islam, T.; Ghosh, A. Genome-wide identification and expression analysis of glutathione S-transferase gene family in tomato: Gaining an insight to their physiological and stress-specific roles. *PLoS ONE* **2017**, *12*, e0187504. [CrossRef]
7. Dixon, D.P.; Lapthorn, A.; Edwards, R. Plant glutathione transferases. *Genome Biol.* **2002**, *3*, 1–10. [CrossRef]
8. Brian McGonigle, S.J.K.; Sze-Mei, C.L.; Koeppe, M.K.; O'Keefe, D.P. A Genomics Approach to the Comprehensive Analysis of the Glutathione S-Transferase Gene Family in Soybean and Maize. *Plant Physiol.* **2000**, *124*, 1105–1120. [CrossRef]
9. Thom, R.; Dixon, D.P.; Edwards, R.; Cole, D.J.; Lapthorn, A.J. The Structure of a Zeta Class Glutathione S-Transferase from *Arabidopsis thaliana*: Characterisation of a GST with Novel Active-site Architecture and a Putative Role in Tyrosine Catabolism. *J. Mol. Biol.* **2001**, *308*, 949–962. [CrossRef]
10. Shimabukuro, R.H.; Swanson, H.R.; Walsh, W.C. Glutathione Conjugation. *Plant Physiol.* **1970**, *46*, 103–107. [CrossRef]
11. Sappl, P.G.; Carroll, A.J.; Clifton, R.; Lister, R.; Whelan, J.; Harvey Millar, A.; Singh, K.B. The *Arabidopsis glutathione* transferase gene family displays complex stress regulation and co-silencing multiple genes results in altered metabolic sensitivity to oxidative stress. *Plant J.* **2009**, *58*, 53–68. [CrossRef]
12. Wagner, U.; Edwards, R.; Dixon, D.P.; Mauch, F. Probing the Diversity of the Arabidopsis glutathione S-Transferase Gene Family. *Plant Mol. Biol.* **2002**, *49*, 515–532. [CrossRef] [PubMed]
13. Jain, M.; Ghanashyam, C.; Bhattacharjee, A. Comprehensive expression analysis suggests overlapping and specific roles of rice glutathione S-transferase genes during development and stress responses. *BMC Genom.* **2010**, *11*, 73. [CrossRef] [PubMed]
14. Soranzo, N.; Sari Gorla, M.; Mizzi, L.; De Toma, G.; Frova, C. Organisation and structural evolution of the rice glutathione S-transferase gene family. *Mol. Genet. Genom.* **2004**, *271*, 511–521. [CrossRef] [PubMed]
15. Wang, R.; Ma, J.; Zhang, Q.; Wu, C.; Zhao, H.; Wu, Y.; Yang, G.; He, G. Genome-wide identification and expression profiling of glutathione transferase gene family under multiple stresses and hormone treatments in wheat (*Triticum aestivum* L.). *BMC Genom.* **2019**, *20*, 986. [CrossRef] [PubMed]
16. Hao, Y.; Xu, S.; Lyu, Z.; Wang, H.; Kong, L.; Sun, S. Comparative Analysis of the Glutathione S-Transferase Gene Family of Four Triticeae Species and Transcriptome Analysis of GST Genes in Common Wheat Responding to Salt Stress. *Int. J. Genom.* **2021**, *2021*, 6289174. [CrossRef]
17. Rezaei, M.K.; Shobbar, Z.S.; Shahbazi, M.; Abedini, R.; Zare, S. Glutathione S-transferase (GST) family in barley: Identification of members, enzyme activity, and gene expression pattern. *J. Plant Physiol.* **2013**, *170*, 1277–1284. [CrossRef]
18. Dong, Y.; Li, C.; Zhang, Y.; He, Q.; Daud, M.K.; Chen, J.; Zhu, S. Glutathione S-Transferase Gene Family in Gossypium raimondii and G. arboreum: Comparative Genomic Study and their Expression under Salt Stress. *Front. Plant Sci.* **2016**, *7*, 139. [CrossRef]
19. Wei, L.; Zhu, Y.; Liu, R.; Zhang, A.; Zhu, M.; Xu, W.; Lin, A.; Lu, K.; Li, J. Genome wide identification and comparative analysis of glutathione transferases (GST) family genes in Brassica napus. *Sci. Rep.* **2019**, *9*, 9196. [CrossRef]
20. Fang, X.; An, Y.; Zheng, J.; Shangguan, L.; Wang, L. Genome-wide identification and comparative analysis of GST gene family in apple (*Malus domestica*) and their expressions under ALA treatment. *3 Biotech* **2020**, *10*, 307. [CrossRef]
21. Enayati, A.A.; Ranson, H.; Hemingway, J. Insect glutathione transferases and insecticide resistance. *Insect Mol. Biol.* **2005**, *14*, 3–8. [CrossRef]
22. Hayes, J.D.; Flanagan, J.U.; Jowsey, I.R. Glutathione transferases. *Annu. Rev. Pharmacol. Toxicol.* **2005**, *45*, 51–88. [CrossRef] [PubMed]
23. Townsend, D.M.; Tew, K.D. The role of glutathione-S-transferase in anti-cancer drug resistance. *Oncogene* **2003**, *22*, 7369–7375. [CrossRef] [PubMed]
24. Sheehan, G.M.D.; Foley, V.M.; Dowd, C.A. Structure, function and evolution of glutathione transferases: Implications for classification of non-mammalian members of an ancient enzyme superfamily. *Biochem. J.* **2001**, *360*, 1–16. [CrossRef] [PubMed]
25. George, S.; Venkataraman, G.; Parida, A. A chloroplast-localized and auxin-induced glutathione S-transferase from phreatophyte Prosopis juliflora confer drought tolerance on tobacco. *J. Plant Physiol.* **2010**, *167*, 311–318. [CrossRef]
26. Ji, W.; Zhu, Y.; Li, Y.; Yang, L.; Zhao, X.; Cai, H.; Bai, X. Over-expression of a glutathione S-transferase gene, GsGST, from wild soybean (*Glycine soja*) enhances drought and salt tolerance in transgenic tobacco. *Biotechnol. Lett.* **2010**, *32*, 1173–1179. [CrossRef]
27. Huang, C.; Guo, T.; Zheng, S.C.; Feng, Q.L.; Liang, J.H.; Li, L. Increased cold tolerance in *Arabidopsis thaliana* transformed with Choristoneura fumiferana glutathione S-transferase gene. *Biol. Plant.* **2009**, *53*, 183–187. [CrossRef]

28. Yan, Y.; Jia, H.; Wang, F.; Wang, C.; Liu, S.; Guo, X. Overexpression of GhWRKY27a reduces tolerance to drought stress and resistance to Rhizoctonia solani infection in transgenic Nicotiana benthamiana. *Front. Physiol.* **2015**, *6*, 00265. [CrossRef]
29. Lo Cicero, L.; Madesis, P.; Tsaftaris, A.; Lo Piero, A.R. Tobacco plants over-expressing the sweet orange tau glutathione transferases (CsGSTUs) acquire tolerance to the diphenyl ether herbicide fluorodifen and to salt and drought stresses. *Phytochemistry* **2015**, *116*, 69–77. [CrossRef]
30. Moons, A. Regulatory and Functional Interactions of Plant Growth Regulators and Plant Glutathione S-Transferases (GSTs). *Vitam. Horm.* **2005**, *72*, 155–202.
31. Dixon, D.P.; Cole, D.J.; Edwards, R. Purification, regulation and cloning of a glutathione transferase (GST) from maize resembling the auxin-inducible type-III GSTs. *Plant Mol. Biol.* **1998**, *36*, 75–87. [CrossRef]
32. Han, Q.; Chen, R.; Yang, Y.; Cui, X.; Ge, F.; Chen, C.; Liu, D. A glutathione S—transferase gene from Lilium regale Wilson confers transgenic tobacco resistance to Fusarium oxysporum. *Sci. Hortic.* **2016**, *198*, 370–378. [CrossRef]
33. Kumar, S.; Asif, M.H.; Chakrabarty, D.; Tripathi, R.D.; Dubey, R.S.; Trivedi, P.K. Expression of a rice Lambda class of glutathione S-transferase, OsGSTL2, in Arabidopsis provides tolerance to heavy metal and other abiotic stresses. *J. Hazard. Mater.* **2013**, *248–249*, 228–237. [CrossRef] [PubMed]
34. Xu, J.; Zheng, A.Q.; Xing, X.J.; Chen, L.; Fu, X.Y.; Peng, R.H.; Tian, Y.S.; Yao, Q.H. Transgenic Arabidopsis Plants Expressing Grape Glutathione S-Transferase Gene (VvGSTF13) Show Enhanced Tolerance to Abiotic Stress. *Biochemistry* **2018**, *83*, 755–765. [CrossRef] [PubMed]
35. Jiang, H.W.; Liu, M.J.; Chen, I.C.; Huang, C.H.; Chao, L.Y.; Hsieh, H.L. A glutathione S-transferase regulated by light and hormones participates in the modulation of Arabidopsis seedling development. *Plant Physiol.* **2010**, *154*, 1646–1658. [CrossRef] [PubMed]
36. Zhao, Y.W.; Wang, C.K.; Huang, X.Y.; Hu, D.G. Genome-Wide Analysis of the Glutathione S-Transferase (GST) Genes and Functional Identification of MdGSTU12 Reveals the Involvement in the Regulation of Anthocyanin Accumulation in Apple. *Genes* **2021**, *12*, 1733. [CrossRef] [PubMed]
37. Skopelitou, K.; Muleta, A.W.; Papageorgiou, A.C.; Chronopoulou, E.; Labrou, N.E. Catalytic features and crystal structure of a tau class glutathione transferase from Glycine max specifically upregulated in response to soybean mosaic virus infections. *Biochim. Biophys. Acta* **2015**, *1854*, 166–177. [CrossRef]
38. Saxena, R.; Vanga, S.; Wang, J.; Orsat, V.; Raghavan, V. Millets for Food Security in the Context of Climate Change: A Review. *Sustainability* **2018**, *10*, 2228. [CrossRef]
39. Chanwala, J.; Khadanga, B.; Jha, D.K.; Sandeep, I.S.; Dey, N. MYB Transcription Factor Family in Pearl Millet: Genome-Wide Identification, Evolutionary Progression and Expression Analysis under Abiotic Stress and Phytohormone Treatments. *Plants* **2023**, *12*, 355. [CrossRef]
40. Muthamilarasan, M.; Prasad, M. Advances in *Setaria* genomics for genetic improvement of cereals and bioenergy grasses. *Theor. Appl. Genet.* **2015**, *128*, 1–14. [CrossRef]
41. Yang, Z.; Zhang, H.; Li, X.; Shen, H.; Gao, J.; Hou, S.; Zhang, B.; Mayes, S.; Bennett, M.; Ma, J.; et al. A mini foxtail millet with an Arabidopsis-like life cycle as a C4 model system. *Nat. Plants* **2020**, *6*, 1167–1178. [CrossRef]
42. Bennetzen, J.L.; Schmutz, J.; Wang, H.; Percifield, R.; Hawkins, J.; Pontaroli, A.C.; Estep, M.; Feng, L.; Vaughn, J.N.; Grimwood, J.; et al. Reference genome sequence of the model plant *Setaria*. *Nat. Biotechnol.* **2012**, *30*, 555–561. [CrossRef] [PubMed]
43. Edwards, R.; Dixon, D.P. Plant Glutathione Transferases. *Meth. Enzymol.* **2005**, *401*, 169–186.
44. Freeling, M. Bias in plant gene content following different sorts of duplication: Tandem, whole-genome, segmental, or by transposition. *Annu. Rev. Plant Biol.* **2009**, *60*, 433–453. [CrossRef]
45. Xu, G.; Guo, C.; Shan, H.; Kong, H. Divergence of duplicate genes in exon-intron structure. *Proc. Natl. Acad. Sci. USA* **2012**, *109*, 1187–1192. [CrossRef]
46. Liu, H.J.; Tang, Z.X.; Han, X.M.; Yang, Z.L.; Zhang, F.M.; Yang, H.L.; Liu, Y.J.; Zeng, Q.Y. Divergence in Enzymatic Activities in the Soybean GST Supergene Family Provides New Insight into the Evolutionary Dynamics of Whole-Genome Duplicates. *Mol. Biol. Evol.* **2015**, *32*, 2844–2859. [CrossRef] [PubMed]
47. Islam, S.; Sajib, S.D.; Jui, Z.S.; Arabia, S.; Islam, T.; Ghosh, A. Genome-wide identification of glutathione S-transferase gene family in pepper, its classification, and expression profiling under different anatomical and environmental conditions. *Sci. Rep.* **2019**, *9*, 9101. [CrossRef]
48. Frova, C. Glutathione transferases in the genomics era: New insights and perspectives. *Biochem. Eng.* **2006**, *23*, 149–169. [CrossRef]
49. Kang, L.; Teng, Y.; Cen, Q.; Fang, Y.; Tian, Q.; Zhang, X.; Wang, H.; Zhang, X.; Xue, D. Genome-Wide Identification of R2R3-MYB Transcription Factor and Expression Analysis under Abiotic Stress in Rice. *Plants* **2022**, *11*, 1928. [CrossRef]
50. Wang, J.; Zhang, Z.; Wu, J.; Han, X.; Wang-Pruski, G.; Zhang, Z. Genome-wide identification, characterization, and expression analysis related to autotoxicity of the GST gene family in *Cucumis melo* L. *Plant Physiol. Biochem.* **2020**, *155*, 59–69. [CrossRef]
51. Tijet, N.; Helvig, C.; Feyereisen, R. The cytochrome P450 gene superfamily in Drosophila melanogaster: Annotation, intron-exon organization and phylogeny. *Gene* **2001**, *262*, 189–198. [CrossRef]
52. Jo, B.S.; Choi, S.S. Introns: The Functional Benefits of Introns in Genomes. *Genom. Inform.* **2015**, *13*, 112–118. [CrossRef] [PubMed]
53. Gao, J.; Chen, B.; Lin, H.; Liu, Y.; Wei, Y.; Chen, F.; Li, W. Identification and characterization of the glutathione S-Transferase (GST) family in radish reveals a likely role in anthocyanin biosynthesis and heavy metal stress tolerance. *Gene* **2020**, *743*, 144484. [CrossRef] [PubMed]

54. Mu, D.; Chen, W.; Shao, Y.; Wilson, I.W.; Zhao, H.; Luo, Z.; Lin, X.; He, J.; Zhang, Y.; Mo, C.; et al. Genome-Wide Identification and Expression Analysis of WRKY Transcription Factors in Siraitia siamensis. *Plants* **2023**, *12*, 288. [CrossRef]
55. Ma, L.; Zhang, Y.; Meng, Q.; Shi, F.; Liu, J.; Li, Y. Molecular cloning, identification of GSTs family in sunflower and their regulatory roles in biotic and abiotic stress. *World J. Microbiol. Biotechnol.* **2018**, *34*, 109. [CrossRef]
56. He, G.; Guan, C.N.; Chen, Q.X.; Gou, X.J.; Liu, W.; Zeng, Q.Y.; Lan, T. Genome-Wide Analysis of the Glutathione S-Transferase Gene Family in Capsella rubella: Identification, Expression, and Biochemical Functions. *Front. Plant Sci.* **2016**, *7*, 1325. [CrossRef]
57. Huerta-Cepas, J.; Dopazo, J.; Huynen, M.A.; Gabaldon, T. Evidence for short-time divergence and long-time conservation of tissue-specific expression after gene duplication. *Brief. Bioinform.* **2011**, *12*, 442–448. [CrossRef] [PubMed]
58. Ganko, E.W.; Meyers, B.C.; Vision, T.J. Divergence in expression between duplicated genes in Arabidopsis. *Mol. Biol. Evol.* **2007**, *24*, 2298–3309. [CrossRef] [PubMed]
59. Filyushin, M.A.; Kochieva, E.Z.; Shchennikova, A.V. ZmDREB2.9 Gene in Maize (*Zea mays* L.): Genome-Wide Identification, Characterization, Expression, and Stress Response. *Plants* **2022**, *11*, 3060. [CrossRef] [PubMed]
60. Moons, A. Osgstu3andosgtu4, encoding tau class glutathioneS-transferases, are heavy metal- and hypoxic stress-induced and differentially salt stress-responsive in rice roots1. *FEBS Lett.* **2003**, *553*, 427–432. [CrossRef]
61. Chen, C.; Chen, H.; Zhang, Y.; Thomas, H.R.; Frank, M.H.; He, Y.; Xia, R. TBtools: An Integrative Toolkit Developed for Interactive Analyses of Big Biological Data. *Mol. Plant* **2020**, *13*, 1194–1202. [CrossRef]

plants

MDPI

Article

Does Potassium Modify the Response of Zinnia (*Zinnia elegans* Jacq.) to Long-Term Salinity?

Hanna Bandurska *, Włodzimierz Breś, Małgorzata Zielezińska and Elżbieta Mieloszyk

Department of Plant Physiology, Poznań University of Life Sciences, Wołyńska 35, 60-637 Poznań, Poland
* Correspondence: hanna.bandurska@up.poznan.pl

Abstract: Salinity is one of the major abiotic stress factors hindering crop production, including ornamental flowering plants. The present study examined the response to salt stress of *Zinnia elegans* 'Lilliput' supplemented with basic (150 mg·dm^{-3}) and enhanced (300 mg·dm^{-3}) potassium doses. Stress was imposed by adding 0.96 and 1.98 g of NaCl per dm^{-3} of the substrate. The substrate's electrical conductivity was 1.1 and 2.3 dS·m^{-1} for lower potassium levels and 1.2 and 2.4 dS·m^{-1} for higher potassium levels. Salt stress caused a significant and dose-dependent reduction in leaf RWC, increased foliar Na and Cl concentrations, and reduced K. About 15% and 25% of cell membrane injury at lower and higher NaCl doses, respectively, were accompanied by only slight chlorophyll reduction. Salt stress-induced proline increase was accompanied by increased P5CS activity and decreased PDH activity. More than a 25% reduction in most growth parameters at EC 1.1–1.2 dS·m^{-1} but only a slight decrease in chlorophyll and a 25% reduction in the decorative value (number of flowers produced, flower diameter) only at EC 2.3–2.4 dS·m^{-1} were found. Salt stress-induced leaf area reduction was accompanied by increased cell wall lignification. An enhanced potassium dose caused a reduction in leaf Na and Cl concentrations and a slight increase in K. It was also effective in membrane injury reduction and proline accumulation. Increasing the dose of potassium did not improve growth and flowering parameters but affected the lignification of the leaf cell walls, which may have resulted in growth retardation. *Zinnia elegans* 'Lilliput' may be considered sensitive to long-term salt stress.

Keywords: salinity; decorative value; membrane injury; proline; lignins

Citation: Bandurska, H.; Breś, W.; Zielezińska, M.; Mieloszyk, E. Does Potassium Modify the Response of Zinnia (*Zinnia elegans* Jacq.) to Long-Term Salinity? *Plants* **2023**, *12*, 1439. https://doi.org/10.3390/plants12071439

Academic Editor: Mingxun Chen

Received: 27 January 2023
Revised: 17 March 2023
Accepted: 21 March 2023
Published: 24 March 2023

1. Introduction

Soil salinity is one of the major stress factors affecting plants growing in natural environments and those cultivated by humans [1]. It is widespread in arid, semi-arid, and coastal regions with poor soil water resources caused by low rainfall and high evapotranspiration [2]. However, it also arises from the natural weathering of saline rocks, anthropogenic activities such as inappropriate irrigation practices with salt water, and persistently changing climatic conditions [3–5]. Nowadays, almost 10% of total land and 50% of irrigated land are affected by salinity [6]. Soils are considered saline when the electrical conductivity (EC) of the saturated soil extract is 4 dS m^{-1} or higher. It is an equivalent of 40 mM NaCl and gives an osmotic potential of approximately −0.2 MPa [4,7]. Soil salinity is also an increasing threat to the ornamental plants used in gardening and landscaping and as cut flowers [8]. This is a significant handicap for urban green belts because soil salinity in these areas is common due to the high salt content caused by the use of NaCl in the deicing of roads and sidewalks [9–11]. Salt-affected soil is becoming a serious problem in landscaping and gardening because of the diminishing sources of high-quality water, which requires the re-use of wastewater for the irrigation of green areas in the urban environment as well as in gardens and fields [8,12]. Salt stress is also an issue in the production and cultivation of potted ornamental plants. Large amounts of salt may often be present in potting substrates made from a mixture of different materials [8].

Moreover, bedding plants produced in containers with regular watering usually have large leaf areas and are root-bound. When they grow in saline soil, they quickly use the water from the root ball but are frequently unable to obtain water from the surrounding soil and suffer from dehydration [13]. Ornamental plants used in landscaping and gardening significantly differ in their susceptibility to salt stress [12,14].

Zinnia (*Zinnia elegans* Jacq.) is one of the most popular annual ornamental plants. It has relatively low requirements, can adjust to most soil conditions (clay, chalk, sand, loam), and tolerates alkaline, neutral, and even acid soils [15,16]. This species is cultivated worldwide as bedding plants in gardening and landscaping and as cut flowers. Moreover, the low requirement and decorative value of zinnia (lively and uniform flower colors) make it applicable in vertical gardens coverage and garden roofs. The wide use of zinnia and more common salt stress in the cultivation of ornamental plants makes salt resistance an important feature. The assessment of salt resistance in ornamental plants should include both quantitative data (biomass production, growth parameters) and qualitative analyses, including flower production and overall plant appearance [17]. The existing literature data show a great cultivar variability in resistance to salinity between zinnia species and cultivars. Villarino and Mattos [14] revealed that *Zinnia angustifolia* 'Star Gold' exhibited significant growth inhibition at an EC of 6 dS m^{-1} and 100% mortality at an EC of 14.2 dS m^{-1}. Similarly, the cultivars of *Zinnia marylandica* and *Zinnia maritima* were shown to be sensitive to salt stress based on growth reduction and mortality. The shoot dry weight in all examined cultivars was reduced at an EC of 4.2 dS m^{-1} by 50% to 56%, and plants died at an EC of 6.0 to 8.2 dS m^{-1} [18]. Some *Zinnia elegans* cultivars were slightly more resistant to salinity. Zinnia 'Dreamland' irrigated with varying concentrations of NaCl for 5 weeks showed reduced growth without affecting flowering. Plants can survive NaCl concentrations that produce EC of the potting medium at 12 dS m^{-1} but show severe leaf injury and water stress symptoms [19]. However, *Zinnia elegans* 'Magellan' was resistant to moderate salinity (EC below 10 dS m^{-1}), had flowers reduced by 75% at an EC of 10 dS m^{-1}, and died at an EC of 15 dS m^{-1} [20]. Escalona et al. [21] showed that salt stress at an EC level of 6 dS m^{-1} caused a decrease in biomass in *Zinnia elegans* but did not affect flower production and toxicity symptoms in leaves. The decline in growth parameters was shown in *Zinnia elegans* 'Benary's Giant Salmon Rose' and 'Benary's Giant Golden Yellow' irrigated with saline water at an EC of 10 dS m^{-1}. However, marketable cut flowers based on stem length were produced [22].

The detrimental effect of soil salinity is water deficit in plants caused by the reduction in water availability (physiological drought) as well as the toxic effect of ions, especially sodium (Na^+) and chloride (Cl^-), caused by their excessive uptake [23]. Thus, plant responses to salt stress were divided into two phases. The first phase is caused by the restriction of water availability, takes place within minutes to days, and includes stomatal closure and the inhibition of cell expansion, mainly in the shoot, leading to growth reduction. The second phase takes place over days or even weeks and is caused by excessive toxic ions in plant cells, which harm metabolic processes [24]. The response to salinity is complex and involves numerous adjustments at morphological (early flowering, growth inhibition, prevention of lateral shoot development, root adaptations), physiological (stomatal closure, osmotic adjustment, Na/K discrimination), and biochemical levels (antioxidant activity, change in hormone level, increased proline level), which help plants cope with stress [23–25].

Growth inhibition and a reduction in the total leaf area under salinity conditions are the effects of lower cell turgor and cell wall extensibility caused by increased lignin deposition [26]. The inhibition of leaf growth in drought- and salt-stressed plants can be considered a component of a stress coping strategy. It is an avoidance mechanism that reduces water loss via transpiration under saline conditions and may prolong plant survival [27]. In the production of ornamental plants, it could be beneficial for nursery growers, especially when they want to produce compact plants and avoid using plant growth regulators [28].

Proline accumulation is another example of the salt stress coping strategy [25,29]. This amino acid may be involved in stress avoidance and stress tolerance strategies. Proline accumulation under salinity stress conditions decreases cells' osmotic potential and allows for water absorption. Besides acting as an osmolyte, proline is a membrane and protein stabilizer, as well as a free radical scavenger. It protects plant cells against the detrimental effects of dehydration and the accumulation of toxic ions [25,30]. The proline level in plant cells is controlled by several cellular mechanisms responsible for its synthesis and degradation [25,29]. This amino acid can be biosynthesized through glutamate and ornithine pathways in the cytosol. In salt-stressed plants, proline is primarily synthesized from glutamate that is converted to glutamic-γ-semialdehyde (GSA) by Δ^1-pyrroline-5-carboxylate synthetase (P5CS). GSA is spontaneously cyclized to Δ^1-pyrroline-5-carboxylate (P5C), which is reduced to proline by pyrroline-5-carboxylate reductase (P5CR). The proline degradation pathway occurs in mitochondria and is catalyzed by two enzymes. Proline dehydrogenase (ProDH) catalyzes proline oxidation to P5C, which is converted into glutamic acid by P5C dehydrogenase (P5CDH).

Salinity affected by a high concentration of NaCl in the soil solution resulted in an excessive accumulation of Na and Cl and reduced the uptake of other elements, including potassium, which plays an essential role in many physiological and biochemical processes [31]. One of the methods of alleviating the negative effect of salt stress caused by excess sodium chloride is fertilization with an increased dose of potassium [32]. A hypothesis that potassium application can reduce the harmful effects of salinity on plant growth and development was proposed by Ben-Hayyim et al. [33], and it was confirmed by, among others, Tzortzakis [34], Umar et al. [35], Amjad et al. [36], and Hashi et al. [37]. However, no effect of K fertilization under saline conditions or even a negative effect of K nutrition on salt resistance were reported [38,39].

The present study investigates whether the application of an increased potassium dose could ameliorate the harmful effect of salinity on *Zinnia elegans* 'Lilliput.' The assessment was made based on measurements of growth parameters, morphological features, and physiological and biochemical parameters. The choice of this variety was not accidental. Plants of this cultivar grow to a height of 20–25 cm, so they are very suitable for growing on flower beds in urban areas, especially along streets and sidewalks. The application rates of sodium were determined based on earlier studies and our preliminary experiment. For example, in Poland's urban areas, soil salinity rarely exceeds an EC of 3 dS m^{-1} [40–43], while in Canada, 50% of the electrical conductivity of soil along roads exceeded 2 dS m^{-1} [44]. According to Kotuby-Amacher [45], the growth of sensitive ornamental plants and grasses may be restricted when the soil EC exceeds 2 dS m^{-1} (e.g., China aster, geranium, annual bluegrass, Kentucky bluegrass).

2. Results

2.1. Results of the Preliminary Experiment

The results of the preliminary experiment showed that zinnia 'Lilliput' is sensitive to salinity. One-way ANOVA results showed a significant effect of salt stress on the examined growth parameters (Table 1). A gradual and substantial reduction in all examined growth parameters was observed as the NaCl dose increased (Figure 1). A negative effect of sodium chloride on the plant height, leaf area, fresh weight of the aboveground part, and diameter of the inflorescence on the main shoot (i.e., the features determining the ornamental value) was found just at the substrate EC of 1.67 dS m^{-1} (1.47 g NaCl).

Table 1. One-way ANOVA results for the effect of salt doses on growth parameters.

Treatments	*Df*	Main Shoot Height		Leaf Area		Aboveground Part Fresh Weight		Inflorescence Diameter	
		F	*p*	*F*	*p*	*F*	*p*	*F*	*p*
NaCl	6	24.369	**0.0000**	95.89	**0.0000**	130.435	**0.0000**	41.89	**0.0000**

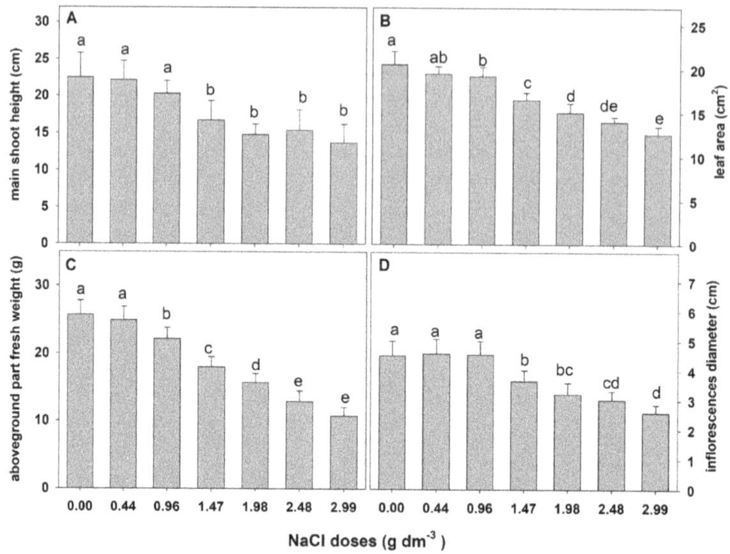

Figure 1. The effect of salinity doses on growth parameters in zinnia. Different letters show statistically significant differences between means. Values are expressed as the means (n = 15) ± standard deviation.

Moreover, a significant percentage of plants grown in the substrate with higher doses of sodium chloride (1.98, 2.48, and 2.99 g dm^{-3}) showed typical visual symptoms of saline injury such as scorching and necrosis around the leaf margins, extending to the whole blade over time and ultimately leading to leaf drop. The first symptoms of injury were observed after 22, 19, and 18 days of cultivation in substrates supplemented with 1.98, 2.48, and 2.99 g NaCl, respectively. The defoliation and final death of some plants were noted after 26, 25, and 24 days. Finally, 30 to 40% of the plants died. In the control treatment and in combinations with the three lowest doses of NaCl, no damage symptoms on the plant were observed. The presented results were the basis for selecting the range of salt concentrations used in the next experiments.

2.2. *Results of Main Experiments*

2.2.1. Water Management Parameters

A statistically significant effect of substrate salinity on water loss from containers was shown, as well as the effect of the date (days of the experiment) and potassium dose and their interaction (Table 2).

Table 2. ANOVA results for water loss from containers and relative water content (RWC) in zinnia leaves exposed to salinity with two potassium levels in the peat substrate.

Treatments	*Df*	Water Loss		*Df*	RWC	
		F	*p*		*F*	*p*
Potassium	1	13.21	**0.0003**	1	5.468	**0.0221**
NaCl	2	1381.32	**0.0000**	2	133.083	**0.0000**
Date	2	399.73	**0.0000**	2	69.872	**0.0000**
Date × potassium	25	5.86	**0.0000**	2	1.889	0.1587
NaCl × potassium	25	8.11	**0.0003**	2	0.925	0.4012
NaCl × date	50	12.97	**0.0000**	4	2.236	0.0735
NaCl × date × potassium	50	2.90	**0.0000**	4	0.977	0.4256

The water loss from containers was significantly higher in the control combination than in both stress combinations (Figure 2). Significantly less water evapotranspired from the containers with plants exposed to a higher NaCl dose. This shows that plants grown in the saline substrate can uptake less water than those produced without salt.

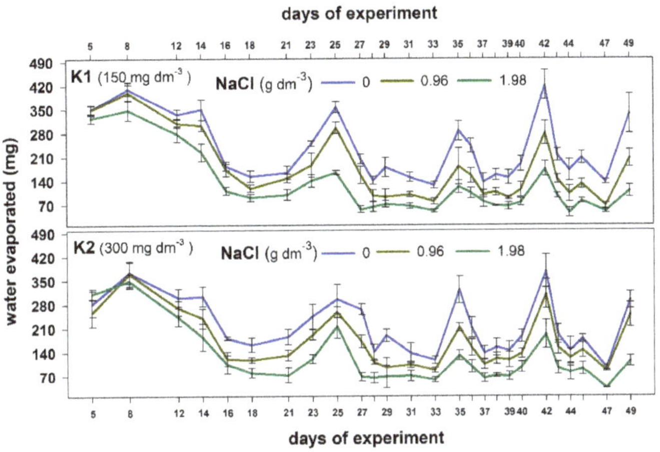

Figure 2. The effect of NaCl doses on water evaporated during the experiment from the containers with plants supplemented with two potassium doses (K1 and K2). Values are expressed as the means (n = 5) ± standard deviation.

Increasing the dose of potassium resulted in a slight reduction in water evaporation, especially at the beginning of the experiment in plants grown without salt and with a lower NaCl dose (Figure 3).

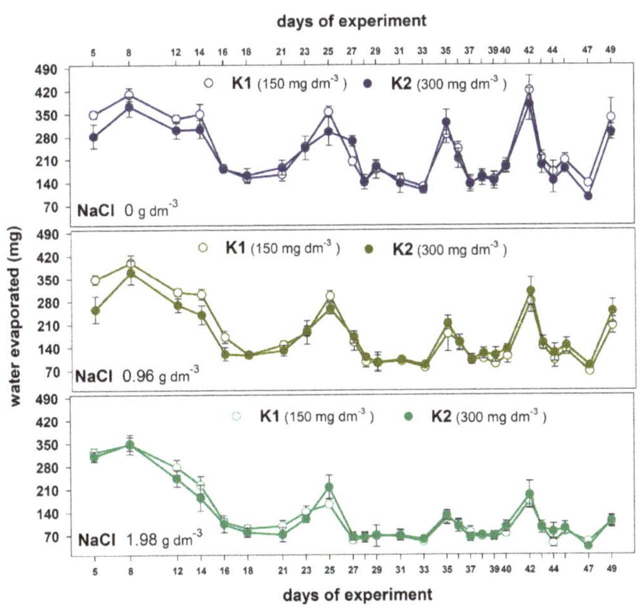

Figure 3. The effect of potassium dose (K1 and K2) on water evaporated during the experiment from the containers with plants grown in the substrate without salt and substrate supplemented with two NaCl doses. Values are expressed as the means (n = 5) ± standard deviation.

The restriction of water availability in salt-stressed plants affected the water balance, as evidenced by the statistically significant NaCl dose- and time-dependent changes in RWC. An interaction between treatments in terms of RWC was not found (Table 1). The greater the salt dose, the greater the leaf RWC reduction (Figure 4). Statistically significant differences in RWC were also noted between the dates of measurements. In both the control and salt-treated plants, the RWC was higher on the second and third sampling dates compared to the first one. Plants supplemented with higher potassium doses showed a statistically significant higher decrease in RWC, except those grown with the highest NaCl dose on the third sampling date. On this date, the leaf RWC in plants supplemented with a higher potassium level was higher than that in those supplemented with a lower dose.

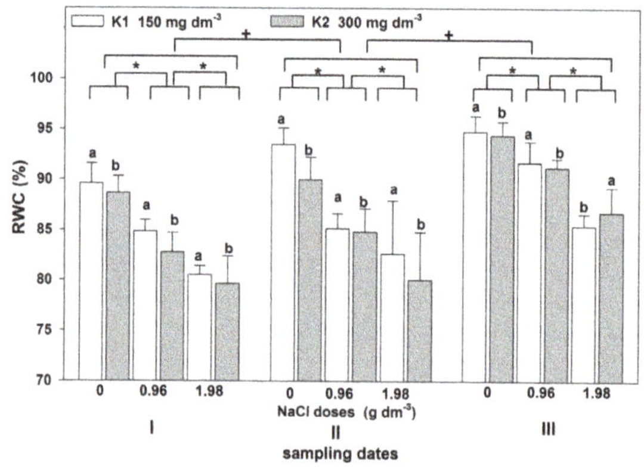

Figure 4. The effect of NaCl doses on RWC in leaves of zinnia grown in the peat substrate supplemented with two potassium doses. Letters a and b show a significant effect of potassium, an asterisk (*) shows a significant effect of salt doses, and + shows a significant effect of the sampling date (I, II, III). Values are expressed as the means (n = 5) ± standard deviation.

2.2.2. Na, Cl, and K Concentrations in Leaves

The two-way ANOVA showed that adding NaCl to the substrate caused statistically significant changes in leaf Na, Cl, and K concentrations (Table 3). In the leaves of control plants, the Na was at about 0.13% and was much lower than the Cl concentration (about 1.5%). The applied salt stress caused a significant and dose-dependent increase in Na and Cl concentrations (Figure 5A,B). In leaves of plants growing with the higher NaCl dose, the Na concentration increased threefold, and Cl increased about twofold. The concentration of K in control plants was about 5.75%. Salt stress caused a statistically significant and dose-dependent decrease in K in leaves (Figure 5C). About a 5% and 11% decrease was shown with the lower and higher doses of NaCl, respectively. As a result of these changes, salt stress caused a reduction in the K/Na ratio in leaves from 48.16 to 17.38.

Table 3. ANOVA results for sodium, chlorine, and potassium concentrations in zinnia leaves exposed to salinity stress with two potassium levels in the peat substrate.

Treatments	*Df*	K		Na		Cl		Na/K	
		F	*p*	*F*	*p*	*F*	*p*	*F*	*p*
NaCl	2	25.911	**0.0000**	102.867	**0.0000**	360.356	**0.0000**	**200.123**	**0.0000**
Potassium	1	11.509	**0.0020**	9.628	**0.0002**	8.597	**0.0064**	**16.122**	**0.0004**
NaCl × potassium	2	0.243	0.7856	7.276	**0.0007**	1.477	0.2444	0.288	0.7521

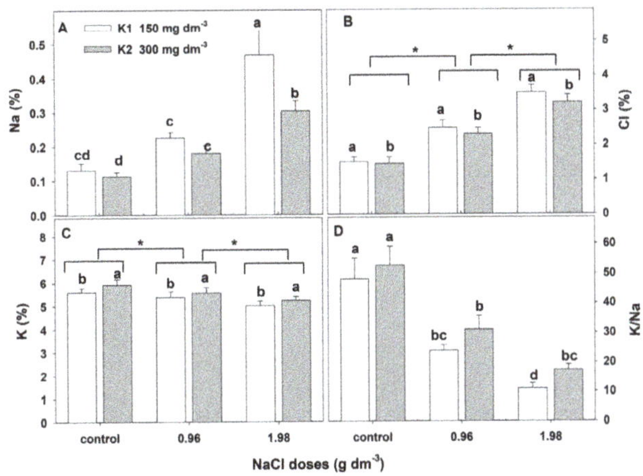

Figure 5. The effect of salinity doses on the concentration (%DM) of Na, Cl, K, and K/Na in the leaves of zinnia grown with two potassium doses. In (**A**,**D**), different letters show statistically significant differences between combinations; in (**B**,**C**), letters a and b show a significant effect of K doses, and an asterisk (*) shows a significant effect of salt doses. Values are expressed as the means (n = 6) ± standard deviation.

The addition of a higher potassium dose to the substrate caused a significant but slight increase in this element in the plant leaves from all combinations (Figure 5C) and an increase in the K/Na ratio. The concentration of Cl in the leaves of plants grown with the higher potassium dose was significantly but slightly lower than that in those grown with a lower dose (Figure 5B). However, the higher potassium dose was effective in the reduction in Na accumulation in leaves. Na was reduced by about 22% and 37% in plants exposed to lower and higher NaCl doses, respectively (Figure 5A). The interactive effect of NaCl and potassium on the concentration of Na in leaves was also observed (Table 3).

2.2.3. Growth Parameters

Analyses of variance indicated a significant effect of salinity on the estimated growth parameters (Table 4). A significant effect of the potassium dose on the changes in the main shoot height, lateral shoot number, leaf area, and inflorescence diameter was found. However, the interaction of NaCl and potassium treatment was found only in the main shoot height and leaf area changes.

Table 4. ANOVA results for growth parameters in zinnia exposed to salinity with two potassium doses in the peat substrate.

Treatments	*Df*	Main Shoot Height		Lateral Shoot Number		Leaf Area	
		F	*p*	*F*	*p*	*F*	*p*
NaCl	2	145.342	**0.0000**	7.9817	**0.0022**	385.693	**0.0000**
Potassium	1	29.244	**0.0000**	13.1460	**0.0013**	55.3951	**0.0000**
NaCl × potassium	2	5.9135	**0.0082**	0.3426	0.7133	11.8915	**0.0003**
Treatments	*Df*	**Aboveground Part Fresh Weight**		**Inflorescence Diameter**		**Number of Inflorescences**	
		F	*p*	*F*	*p*	*F*	*p*
NaCl		180.543	**0.0000**	13.779	**0.0001**	17.996	**0.0000**
Potassium		3.121	0.0900	13.390	**0.0012**	3.378	0.0785
NaCl × potassium		0.4259	0.7224	0.848	0.4408	1.4145	0.2626

Salt stress affected the reduction in all growth parameters (Figure 6). The reduction was greater with the increasing salt dose applied (Figure 6A–F). Higher potassium doses significantly reduced the main shoot height, the number of lateral shoots, the leaf blade area, and the inflorescence diameter (Figure 6A–E). In addition, the potassium-induced reduction in the mean shoot height and leaf area was higher in plants supplemented with a higher NaCl dose (Figure 6A,C).

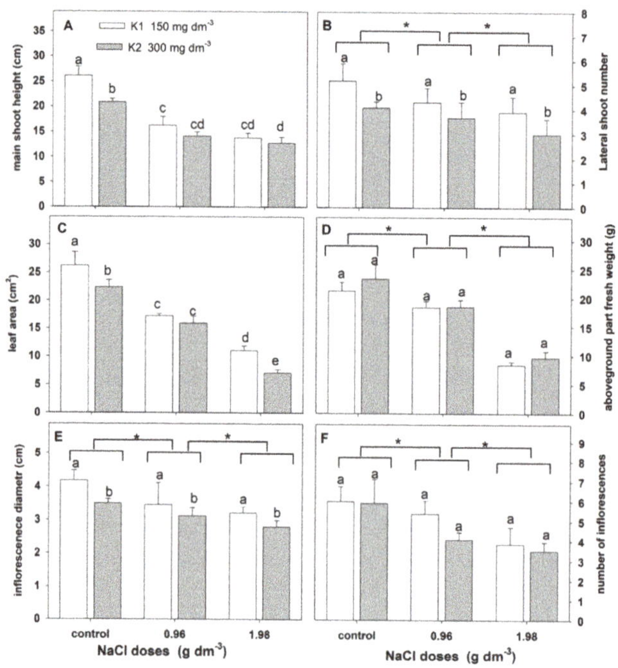

Figure 6. The effect of salinity doses on growth parameters in zinnia grown in the peat substrate supplemented with two potassium doses. In (**A**,**C**), different letters show statistically significant differences between combinations. In (**B**,**D**–**F**), letters a and b show a significant effect of potassium doses, and an asterisk (*) shows a significant effect of salt doses. Values are expressed as the means (n = 15) ± standard deviation.

2.2.4. Chlorophyll, Lignin, and Membrane Injury Index

The two-way ANOVA indicated a significant effect of salt stress on the content of leaf chlorophyll and lignin as well as the membrane injury index (Table 5). Potassium significantly affected the lignin content and membrane injury index but not the chlorophyll content. The interaction of these factors on the content of chlorophyll and lignin and membrane injury was not revealed.

Table 5. ANOVA results for leaf chlorophyll and lignin content and membrane injury index in zinnia exposed to salinity with two potassium levels in the peat substrate.

Treatments	*Df*	Chlorophyll		Lignin		Membrane Injury	
		F	*p*	*F*	*p*	*F*	*p*
NaCl	2	4.859	**0.0169**	19.683	**0.0000**	65.818	**0.0000**
Potassium	1	1.323	0.2615	40.929	**0.0000**	9.187	**0.0080**
NaCl × potassium	1	0.871	0.4313	1.877	0.1749	0.766	0.3943

An increase in salinity led to a decrease in the leaf chlorophyll content (Figure 7A). The lignin content in the cell wall increased significantly with the increase in NaCl. Higher lignin levels were found in the leaves of plants supplemented with higher potassium doses (Figure 7B). In addition, membrane injury increased with increasing salinity. However, it was lower in plants supplemented with higher potassium doses (Figure 7C).

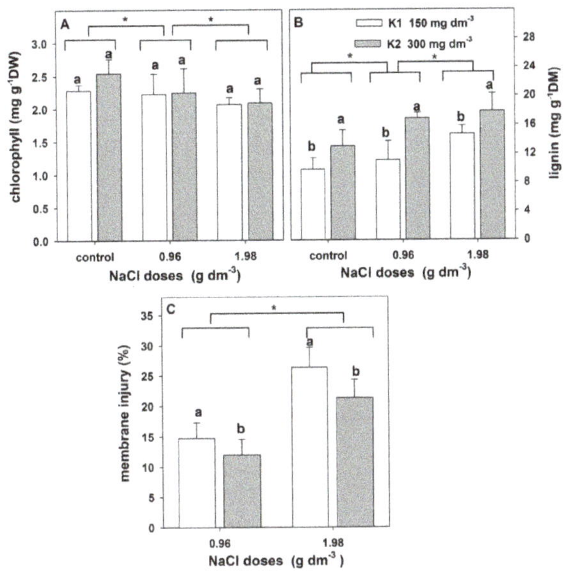

Figure 7. The effect of salinity doses on chlorophyll levels, lignin levels, and membrane injury index in the leaves of zinnia grown in the peat substrate supplemented with two doses of potassium. Letters a and b show a significant effect of K doses, and an asterisk (*) shows a significant effect of salt doses. Values are expressed as the means (n = 5) ± standard deviation.

2.2.5. Proline Content and Enzyme Activity

The three-way ANOVA showed a significant effect of the salinity, the sampling date, and their interaction on the proline content and the activity of both enzymes (Table 6). Moreover, a significant effect of potassium and the interaction between NaCl × potassium and NaCl × date × potassium on proline content and PDH activity was revealed.

Table 6. ANOVA results for proline content and P5CS and PDH activity in the leaves of zinnia exposed to salinity with two potassium doses in the peat substrate.

Treatments	*Df*	Proline		P5CS		PDH	
		F	*p*	*F*	*p*	*F*	*p*
NaCl	2	129.330	**0.0000**	40.582	**0.0000**	191.364	**0.0000**
Date	2	69.634	**0.0000**	15.330	**0.0000**	17.461	**0.0000**
Potassium	1	32.680	**0.0000**	1.270	0.2635	6.218	**0.0149**
NaCl × date	4	28.069	**0.0000**	4.043	**0.0052**	3.866	**0.0067**
NaCl × potassium	2	41.251	**0.0000**	0.805	0.4512	10.916	**0.0000**
Date × potassium	2	1.689	0.1920	0.233	0.7930	0.974	0.3826
NaCl × date × potassium	4	9.329	**0.0000**	0.441	0.7787	11.756	**0.0000**

Salt stress affected the time, dose, and potassium-dependent increase in leaf proline content (Figure 8A). On the first date, the proline content increased with the salinity increase and was the highest in plants grown with higher NaCl (1.98 g dm^{-3}) and potassium doses

(300 g dm^{-3}). A much smaller proline increase was shown on the second and third dates. However, on the second date, the elevated proline level was maintained in plants grown with higher NaCl and potassium doses. An increase in salinity resulted in a rise in P5CS activity (Figure 8B). The largest increases dependent on the salt dose were revealed on the first date. The potassium dose did not affect the activity of this enzyme. Salinity affected the decrease in PDH activity (Figure 8C). This decrease was dependent on the date and salt. A greater decrease was found with higher salinity on the third date. This decrease was slightly alleviated by the increased potassium dose in plants grown with lower salinity on the first and second dates.

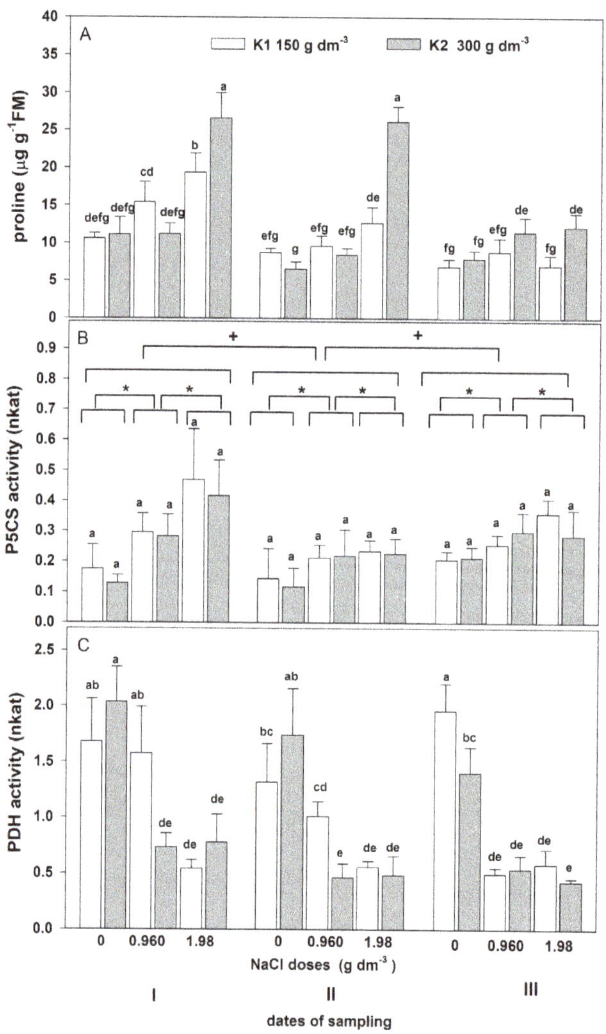

Figure 8. The effect of salinity doses on (**A**) proline content, (**B**) P5CS activity, and (**C**) PDH in the leaves of zinnia grown in the peat substrate with two doses of potassium. In (**A**,**C**), different letters show significant effects of NaCl and K doses and sampling dates. In (**B**), letter "a" shows the lack of a significant effect of the K^+ level, an asterisk (*) shows a significant effect of salt doses, and + shows a significant effect of the sampling dates (I, II, III). Values are expressed as the means (n = 5) ± standard deviation.

3. Discussion

Crops' resistance to salinity is an important feature affecting the productivity and final yield. It is not only the ability to survive stress but also the ability to reduce the adverse effects of stress on growth and yielding. The traditional methods of assessing plants' resistance to salinity are based on estimated growth parameters (fresh or dry weight of aboveground parts) and traits of agronomic interest, such as the yield of harvest organs [46]. However, the aesthetic value is also important for horticultural crops, especially ornamental ones [17,47]. Miyamoto et al. [47] proposed a multipurpose method of evaluating the salt stress resistance of horticultural crops. This method is based on establishing the electrical conductivity (EC) of the soil solution, which results in a 25 to 50% reduction in growth or causes at least 25% damage to leaves. Plants are classified into five categories: sensitive (0–3 dS m^{-1}), moderately sensitive (3–6 dS m^{-1}), moderately tolerant (6–8 dS m^{-1}), tolerant (8–10 dS m^{-1}), and highly tolerant (10 dS m^{-1}). According to this criterion, *Zinnia elegans* 'Lilliput' belongs to the category of sensitive plants, as it has over 25% reduction in most growth parameters at EC 1.1–1.2 dS·m^{-1}. To test salt resistance in ornamental plants, de Oliviera [17] proposed another method that considers both growth data and visual analyses. The cumulative reduction in the shoot biomass production and the overall appearance of plants (chlorophyll index, flower production) were used to estimate the ornamental index (Orn index) of four ornamental species. The results confirmed that growth inhibition is inadequate for measuring salinity resistance in all ornamental plants. In some plants, slight or moderate reductions in growth can be compensated by morphological traits that favor the aesthetical quality [18]. For example, *Catharanthus roseus* showed a rise in flower production at the salinity of EC 3.0 dS m^{-1}, even with a 10% reduction in shoot dry matter and a 15% reduction in the chlorophyll index [17]. The increase in flower production and about a 20% decrease in the chlorophyll index and shoot dry matter at a salinity higher than 8.0 dS m^{-1} were reported in *Ixora coccinea*. This species was considered moderately tolerant (resistant) to salinity [17]. So, the green color of leaves and flower production (plant appearance) and the assessment of the degree of growth reduction should be recommended to evaluate salinity resistance in ornamental species, especially in garden flowers of commercial interest [17]. According to the criteria proposed by Oliviera et al. [17], our results show that *Zinnia elegans* 'Lilliput' appears to be sensitive to salinity. It demonstrated a more than 25% reduction in most growth parameters at EC 1.1–1.2 dS·m^{-1} and a slight decrease in chlorophyll (about 10%) at EC 2.3–2.4 dS·m^{-1}. A reduction in the decorative value (flower production, flower diameter) of at least 25% was found at the EC 2.3–2.4 dS·m^{-1}. We did not find any adverse effects of salinity on the appearance of leaves, such as the tip and marginal leaf burn in 2016. However, in the preliminary experiment (2015), some initial damage symptoms were observed in plants at EC 2.3 dS·m^{-1}, along with typical saline injury such as scorching and necrosis around the leaf margins, leading to the defoliation and death of some plants at EC 3.1 and 3.3 dS·m^{-1}.

One of the reasons for the inhibition of leaf growth could be a decrease in their hydration level (RWC), which was noticeable on the first sampling date. Leaf growth results from the irreversible expansion of cells due to water absorption and cell wall extension capacity [48]. The presented studies found an increase in lignins in the cell walls of plants grown using both doses of salt. This increase was the highest on the last date. The reduction in the leaf area increased with the duration of stress. On the other hand, the reduction in RWC on the last date was lower than that on the first date, indicating the involvement of cell wall lignification in limiting leaf growth. Plant cells adjust to salt stress by accumulating lignin and thickening the cell wall [49]. Lignification results in a loss of wall extensibility. As a result, the cell absorbs less water, and growth is inhibited [50]. Cell wall thickening, especially in leaf tissues, has desirable effects on the capacity for turgor maintenance [51]. Moreover, the accumulation of lignin is beneficial for salt-stressed plants because it provides the mechanical strengthening of the cell walls and the protection of membrane integrity [52]. However, salt stress significantly reduced growth parameters in maize without lignin level alteration [53].

The harmful effect of salinity on plants is also caused by the accumulation of toxic sodium ions in cells and disturbances in the uptake of potassium ions [32,54]. This makes it difficult to maintain an optimal potassium (K^+)–sodium (Na^+) ratio, which is essential for enzymatic reactions in the cytoplasm, which are necessary for the maintenance of plant growth and yield development [55]. Moreover, an appropriate potassium level is important for maintaining membrane integrity [56]. Our results indicate that such disturbances in the nutritional status of plants (significant increase in leaf sodium and decrease in potassium concentration) caused the reduction in all yield parameters, affected membrane injury, and caused some decline in chlorophyll.

As mentioned in the introduction chapter, the negative effect of salt stress caused by excess sodium chloride concentration may be mitigated by increasing potassium fertilization. In our experiment, the increase in the potassium dose significantly reduced the accumulation of sodium ions in the leaves. There was a slightly smaller but statistically significant increase in potassium levels. A positive effect of the increased potassium dose on membrane integrity was also found. This alleviating effect of the increased potassium dose on cell membrane integrity could be related to the lignification of the walls, leading to their strengthening [52]. A higher dose of potassium had no effect on the chlorophyll content in the leaves of plants grown in the salt-free substrate and at both salinity levels. Similarly, the potassium dose had no impact on the aboveground parts and the number of flowers produced. On the other hand, an increase in potassium decreased the diameter of flowers, the number of lateral shoots, and the leaf area. One reason for leaf growth inhibition by the increased potassium dose could be the increased cell wall lignification. Additionally, in plants growing in the salt-free substrate with an increased dose of potassium, a slight but statistically significant inhibition of leaf growth and an increase in lignin levels were found.

Proline accumulation is a well-known plant strategy for coping with salt stress [57,58]. This amino acid plays an important role in the avoidance of osmotic stress (water deficit) as well as in stress tolerance (water deficit, accumulation of toxic ions) strategies [59,60]. The literature indicates that not only the accumulation of this amino acid but also its metabolism (synthesis, oxidation) may play an important role in stress-coping mechanisms [61]. However, there are limited data on the effect of salt stress on the proline concentration and the mechanism responsible for its accumulation in ornamental plants. Proline accumulation in Jerusalem artichoke plantlets under short-term salinity stress is caused by activating its synthesis and inhibiting its oxidation [62]. We found a significant effect of salinity on the accumulation of free proline in zinnia leaves on the first sampling date at a higher salt dose. Changes in the activity of proline metabolism enzymes indicate that the accumulation of this amino acid resulted from an increase in the activity of the enzyme catabolizing its synthesis (P5CS) and a decrease in the activity of the enzyme responsible for its oxidation (PDH). The proline level did not change at the lower salt dose on the first date. This may have been due to a slight increase in P5CS activity and no change in the activity of PDH involved in proline oxidation. On another date, the proline level did not change significantly under the influence of NaCl doses despite a slight increase in P5CS activity and a decrease in PDH activity. Similarly, in chrysanthemum leaves, proline accumulation decreased with the duration of salinity stress, but it was consistent with the changes in P5CS activity [63]. In cadmium-treated pea plants, proline accumulation corresponded closely with the expression of genes encoding proline synthesis (P5CS) and proline degradation (PDH) enzymes, but the level of accumulated proline was lower in older leaves than in younger ones [64]. In our experiment, leaves were taken from the fourth node (from the top of the plant) on each date. So, on subsequent dates, they were getting older.

Increasing the potassium dose resulted in a significant proline increase in combination with a higher NaCl concentration on the first and second dates and a slightly lower increase on the third date. This greater proline accumulation may also be involved in mitigating damage to cell membranes observed in plants supplemented with higher potassium doses. The involvement of potassium in proline accumulation has already been shown in salt-stressed *Avena sativa*, *Pereskia bleo*, and Pearl Millet [65–67]. Potassium-induced proline accumulation in maize was caused by the enhanced conversion of arginine into proline [68]. Given that potassium-induced proline accumulation in salt-stressed zinnia was not triggered by the changes in the activity of the glutamate-proline metabolic pathway (P5CS, PDH), it can be assumed that the synthesis from arginine may have been activated.

In conclusion, our findings indicate that, based on the growth and flowering parameters, *Zinnia elegans* 'Lilliput' may be considered sensitive to salinity. Long-term salt stress led to about a 25% reduction in most growth parameters at EC 1.1–1.2 dS·m^{-1} and a similar reduction in the decorative value (amount of flower production, flower diameter) only at EC 2.3–2.4 dS·m^{-1}. No signs of damage, such as leaf yellowing and browning, or necrotic spots were observed for any of the used salt concentrations. Salt-induced proline accumulation was the highest at the beginning of stress and was consistent with the change in the activity of P5CS and PDH. Higher potassium doses affected the increase in the K/Na ratio and aided the effective plant adjustment to salinity. Potassium improves *Zinnia elegans'* functions in saline conditions by lessening the accumulation of sodium and chlorine ions, diminishing membrane injury, and triggering proline accumulation. However, potassium did not improve growth and flowering parameters; on the contrary, it affected cell wall lignification in leaves, leading to growth restriction.

4. Materials and Methods

4.1. Plant Cultivation and Treatment

Three greenhouse independent experiments were conducted during three vegetation seasons (2015, 2016, 2017) from early March to early May. The greenhouse was equipped with a climate computer Hortimax GPK2000; therefore, growth conditions in both experiments were similar. Klasmann high moor peat (pH 3.91) was used to prepare the substrate. Zinnia ('Lilliput') seeds were sown to the limed peat (pH 6) enriched with multicomponent fertilizer PGMix (0.5g·dm^{-3}). Four-week-old seedlings were re-planted into a container filled with 7 dm^3 of limed peat substrate supplemented with the following final doses of macro- and micronutrients (mg dm^{-3} of the substrate): N-100, P-75, K-150, Ca-1245, Mg-160, Fe-75, Mn-35, Zn-30, Cu-10, B-2, and Mo-1.

In the first preliminary experiment (2015), plants were cultivated with the following doses of sodium chloride: 0 (control), 0.44, 0.96, 1.47, 1.98, 2.48, and 2.99 g·dm^{-3} of the substrate. The salts were used separately: once at the beginning of the experiment in one complete dose per container. The initial electrical conductivity (EC) of the substrates was 0.3 (control) 0.75, 1.1, 1.67, 2.3, 3.1, and 3.9 dS m^{-1}, respectively. The effect of salinity on the growth parameters (height of the main shoot, leaf blade area, fresh weight of the aboveground part of plants, and the diameter of the inflorescence of the main shoot) was evaluated.

The investigation conducted in 2016 and 2017 aimed to examine the effect of increased potassium doses on the response of zinnia to long-term salinity. Based on the results of a preliminary experiment, we used two doses of salt in these experiments, which had different effects on growth parameters. The experimental scheme for this study was based on a completely randomized design with two levels of sodium chloride (0.96 and 1.98 g·dm^{-3} of the substrate) and two levels of potassium (150 and 300 mg·dm^{-3} of the substrate in form of K_2SO_4). The level of other elements was the same as that in the preliminary experiment. Each combination contained five containers filled with the same volume of the substrate and three plants (Figure 9). After adding all the components, the electrical conductivity (EC) of the substrates with lower potassium levels reached 1.1 and 2.3 dS·m^{-1} and 1.2 and 2.4 for lower and higher NaCl doses, respectively. The control combination consisted of plants growing in the substrate with no added salt. The EC of the peat substrate without NaCl addition amounted to 0.3 and 0.4 mS m^{-1} for lower and higher potassium levels, respectively.

The EC of the growing medium in each experiment was measured using an Orion Benchtop Conductivity Meter (Thermo Electron Corporation). The water content in the substrate was kept at about 60% of the capillary water capacity. The plants were regularly irrigated using water purified by reverse osmosis. There were no holes in the bottom of the containers to prevent water from flowing out.

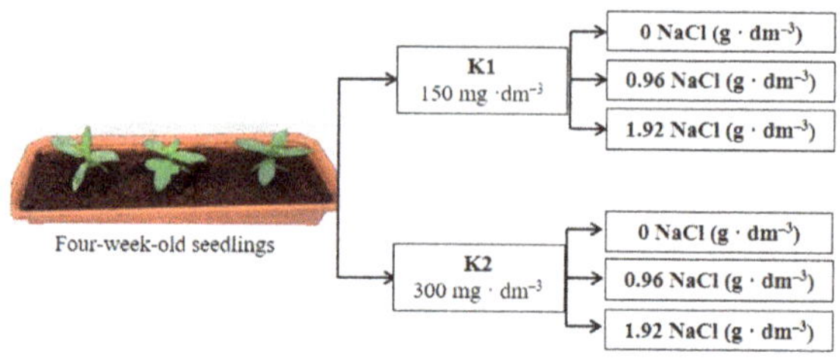

Figure 9. The schematic diagram of experiments; K1 and K2—potassium levels.

In the experiment conducted in 2016, the effect of salinity and an increased potassium dose on the content of Na, Cl, and K in leaves and growth parameters was examined. The height of the main shoot, the number of lateral shoots, the fresh weight of the aboveground part of plants, the leaf blade area, the number of flowering inflorescences, and the diameter of the inflorescence of the main shoot as well as the Na, Cl, and K concentrations were measured at the end of the experiment (after 49 days of growth in the saline substrate). In the next experiment (2017), the effect of salinity and an increased potassium dose on biochemical and physiological parameters were evaluated. Plant material was harvested three times during growth in the saline substrate (after 21, 35, and 49 days) in order to estimate the leaf water content, proline content, activity of Δ^1-pyrroline-5-carboxylate synthetase (P5CS), and activity of proline dehydrogenase (PDH). At the end of this experiment (last harvest), the membrane injury index, chlorophyll content, and lignin level in the cell wall were also estimated.

4.2. Water Loss Measurements

The total amount of water lost in the containers via transpiration and evaporation from the substrate surface was estimated. The containers were weighed every two or three days, and the water losses were supplemented to maintain soil moisture at about 60% of the capillary water capacity. On other days, water losses were supplemented by adding the same water volume to all containers. We measured the differences between the weight of containers after watering to 60% of capillary water capacity and their weight before watering. The volume of water used for plant watering in the meantime was also included. The increase in plant biomass during the experiments was taken into account, and the weight of the containers was considered. The results presented are the means from five replications (containers).

4.3. Growth Measurements

At the stage of full blooming (51 day of growth in the saline substrate), the height of the main shoot, the number of lateral shoots, the fresh matter of the aboveground parts, the average area of the leaf blades, the number of flowering inflorescences, and the diameter of the inflorescence of the main shoot were determined for 15 plants in each combination. The leaf area was determined with a Leaf Area Meter CI-202 (CID BioSciences Inc., Camas, WA, USA). For this purpose, leaves located at the fourth node (from the top of the plant) were collected.

4.4. Na, Cl, and K Measurements

Fully mature leaves were pre-dried at a temperature of 105 °C for 48 h and ground in a mixer. The plant material was then mineralized with a mixture of H_2SO_4 and H_2O_2 (2:1). The Na and K concentrations were determined by flame emission spectrophotometry. The plant material was mineralized at a temperature of 500 °C to establish chlorine levels. Next, the residue was dissolved in hot deionized water, and after sedimentation, the content of

Cl was determined in a clear solution by the nephelometric method [69]. All analyses were carried out in six biological replicates. Each replicate was derived from a separate sample of randomly chosen plants. The measurement results are expressed as a percentage of dry matter (% DM).

4.5. Physiological and Biochemical Parameter Measurements

Two or three fully mature leaves taken from the third node (counted from the top of the plant) of five randomly chosen plants were used for the estimation of relative water content (RWC), chlorophyll, lignin, and proline content, as well as the activity of Δ^1-pyrroline-5-carboxylate synthetase (P5CS) and proline dehydrogenase (PDH) and the membrane injury index. The tissue water content and the membrane injury index were estimated immediately after harvest. Plant material for estimating other parameters was frozen in liquid nitrogen and stored at 20 °C until analysis. The analyses were carried out using five independent biological replicates. Each replicate was derived from different samples of plant material.

4.6. Water Content in Leaves

The water content in leaves was estimated by measuring the relative water content (RWC) according to the standard method developed by Weatherly [70], with some modifications [71]. It was calculated using the following formula:

$$RWC = \frac{\text{fresh matter} - \text{dry matter}}{\text{fresh matter at full turgor} - \text{dry mater}} \cdot 100$$

4.7. Proline

The proline content was measured according to Bates et al. [72]. The plant material (200 mg FW) was homogenized with 4 cm^3 of 5% TCA. The homogenate was centrifuged at 5000× *g* for 15 min. The supernatant was used for proline determination by measuring the quantity of the colored reaction product of proline with ninhydrin acid. Absorbance was read at 520 nm. The amount of proline in the sample was calculated from a standard curve and expressed in milligrams per gram of dry matter (mg g^{-1} DM).

4.8. Enzyme Extraction and Assays

Frozen samples (0.5 g) were ground in a chilled mortar and pestle at 4 °C with 2.5 cm^3 of extraction buffer (50 mM phosphate buffer, pH 7.2, containing 1 mM phenylmethanesulfonyl fluoride (PMSF), 1mM EDTA-K_2, and 1% PVP). The homogenate was centrifuged at 14,000× *g* for 15 min at 4 °C. The supernatant was used to determine Δ^1-pyrroline-5-carboxylate synthetase (P5CS) and proline dehydrogenase (PDH).

The activity of P5CS was determined according to Zhang et al. [73], with some modifications. The reaction mixture in a final volume of 1 cm^3 contained 50 mM glutamic acid, 10 mM ATP, 20 mM $MgCl_2$, 100 mM hydroxamate-HCl, 50 mM Tris-HCL (pH 7.0), and 0.1 cm^3 enzyme extract. This mixture was incubated at 37 °C for 15 min. Next, the reaction was stopped by adding 2 cm^3 of the stop buffer (2.5% $FeCl_3$ (*w*/*v*) and 6% TCA (*w*/*v*) dissolved in 2.5 M HCl). The precipitated proteins were removed by centrifugation at 15,000× *g* for 30 min, and the supernatant's absorbance was measured at 535 nm. The amount of γ-glutamyl hydroxamate produced was calculated from the molar absorption coefficient (ε_{534} = 0.25 mM^{-1} cm^{-1}) of the Fe^{3+}-hydroxamate complex. The enzyme activity was expressed in nkat · mg^{-1} protein.

PDH activity was determined using a modified Rahnama and Ebrahimzadeh method [74]. The reaction mixture enclosed 100 mM Na CO_3-HCl buffer (pH 10.3), 20 mM proline, 10 mM NAD^{+}, and 0.100 cm^3 of enzyme extract in the final volume of 1 cm^3. The reduction in NAD^+ was measured by the absorbance increase at 340 nm through/over 3 min. The amount of reduced NAD $^+$ was calculated from the molar absorption coefficient (ε_{340} = 6.2 mM^{-1} cm^{-1}). Enzyme activity was expressed in nkat·mg^{-1} protein.

4.9. Protein Concentration

The total protein concentration was determined according to Bradford [75], using bovine serum albumin as a standard.

4.10. Chlorophyll Content

The total chlorophyll content was estimated according to the Hiscox and Israelstam method [76]. Leaf samples (100 mg) were cut into pieces, and pigments were extracted at 65 °C using 5 cm^3 of dimethyl sulfoxide (DMSO). The optical density of the extract was measured at 649 and 665 nm. The total chlorophyll content was calculated following the modified Arnon equations [77] and expressed in milligrams per gram of dry matter ($mg \cdot g^{-1}$ DM).

4.11. Lignin Determination

Lignin content was measured by a slightly modified Syros et al. method [78]. Leaf samples were air-dried at 70 °C, and 0.1 g of dry matter was extracted three times with 3 cm^3 80% (v/v) ethanol at 80 °C for 1.5 h. Ethanol was decanted, and the residue was extracted with 3 cm^3 chloroform at 62 °C. Chloroform was removed, and samples were dried at 50 °C. The dried material was digested in 2.6 cm^3 of 25% (v/v) acetyl bromide solution in acetic acid containing 2.7% (v/v) perchloric acid. After 1h, 100 μl of each sample was added to 580 μl of a solution containing 17.24% (v/v) 2N sodium hydroxide and 82.76% (v/v) acetic acid, and 20 μL of 7.5 M hydroxylamine hydrochloride was added to ensure termination of the reaction. Finally, the volume was corrected to 2 cm^3 with acetic acid. The absorbance of samples was recorded at 280 nm. The lignin content was calculated using a linear calibration curve with commercial lignin alkali (Sigma, St. Louis, MO, USA) and expressed in mg·g dry matter (DM).

4.12. Membrane Injury Index

The effect of salinity on cell membrane injury was determined according to Premachendra et al. [79], with modifications [71]. Leaf pieces (five pieces with 1.5 cm diameters for one replication) from control plants and plants treated with different NaCl concentrations were washed quickly three times in 10 cm^3 of deionized water to remove surface electrolytes. Then, leaf pieces were put into a 50 cm^3 flask, submerged in 10 cm^3 of deionized water, and kept at 10 °C for 24 h. After warming to 25 °C and shaking, the electrical conductivity of the effusate was measured. Next, tissues were killed by autoclaving for 15 min and cooled down to 25 °C, and the electrical conductivity of the effusate was measured again. Membrane injury (%) was evaluated according to the formula:

$$MI = 1 - \frac{1 - (T1/T2)}{1 - (C1/C2)} \times 100\%$$

where $C1$ and $C2$ represent the conductivity of the control samples before and after autoclaving, respectively; $T1$ and $T2$ represent the conductivity of the samples treated with NaCl before and after autoclaving, respectively.

4.13. Statistical Analysis

All data were analyzed statistically using STATISTICA 13.3 (StatSoft, Inc., Tulsa, OK, USA). The effect of two factors (salinity and potassium) on the growth parameters (15 replicates), the concentration of Na Cl and K (3 replicates) in leaves, the content of chlorophyll and lignin, as well as the membrane injury index (5 replicates) was determined using a two-way analysis of variance (ANOVA). A three-way ANOVA was used to determine whether the salinity, potassium, and estimation date significantly affected the water loss from containers, the RWC, the proline content, and the activity of P5CS and PDH. Because the results concerning some traits (leaf area, diameter of inflorescence, sodium content in leaves, RWC, P5CS, PDH) did not meet the ANOVA assumptions, they were transformed before performing the analysis using the Box–Cox method [80]. A post hoc Tukey's simultaneous comparison test was performed if significant differences were found between individual means of the treatment groups in each experiment.

Author Contributions: H.B.—data analysis and visualization, original draft preparation, manuscript revision; W.B.—experimental design, planning and conducting experiments, statistical analysis of data, original draft preparation, manuscript revision; M.Z.—plant care and watering, biochemical measurements, development, description of methods used, evapotranspiration measurement; E.M.—measurement of Na, Cl, and K ions and growth parameters. All authors have read and agreed to the published version of the manuscript.

Funding: This publication was co-financed within the framework of the Polish Ministry of Science and Higher Education's program: "Regional Initiative Excellence" in the years 2019–2022 (No. 005/RID/2018/19).

Institutional Review Board Statement: Not applicable.

Informed Consent Statement: Not applicable.

Data Availability Statement: Not applicable.

Conflicts of Interest: The authors declare no conflict of interest.

References

1. Mishra, A.; Tanna, B. Halophytes: Potential resources for salt stress tolerance genes and promoters. *Front. Plant Sci.* **2017**, *8*, 829. [CrossRef] [PubMed]
2. Ibrahim, E.A. Seed priming to alleviate salinity stress in germinating seeds. *J Plant Physiol.* **2016**, *192*, 38–46. [CrossRef] [PubMed]
3. Rengasamy, P. World salinization with emphasis on Australia. *J. Exp. Bot.* **2006**, *57*, 1017–1023. [CrossRef]
4. Yadav, S.; Irfan, M.; Ahmad, A.; Hayat, A. Causes of salinity and plant manifestations to salt stress: A review. *J. Environ. Biol.* **2011**, *32*, 667–685. [PubMed]
5. Singh, M.; Singh, A.; Prasad, S.M.; Singh, R.K. Regulation of plants metabolisms in response to salt stress: An omics approach. *Acta Physiol. Plant.* **2017**, *39*, 48. [CrossRef]
6. Guo, J.; Shan, C.; Zhang, Y.; Wang, X.; Tian, H.; Han, G.; Zhang, Y.; Wang, B. Mechanisms of Salt Tolerance and Molecular Breeding of Salt-Tolerant Ornamental Plants. *Front. Plant Sci.* **2022**, *13*, 854116. [CrossRef] [PubMed]
7. Singh, M.; Nara, U.; Kumar, A.; Choudhary, A.; Singh, H.; Thapa, S. Salinity tolerance mechanisms and their breeding implications. *J. Genet. Eng. Biotechnol.* **2021**, *19*, 173. [CrossRef]
8. García-Caparrós, P.; Lao, M.T. The effects of salt stress on ornamental plants and integrative cultivation practices. *Sci. Hort.* **2018**, *240*, 430–439. [CrossRef]
9. Devecchi, M.; Remotti, D. Effect of Salts on Ornamental Ground Covers for Green Urban Areas. *Acta Hortic.* **2004**, *643*, 153–156. [CrossRef]
10. Cunningham, M.A.; Snyder, E.; Yonkin, D.; Ross, M.; Elsen, T. Accumulation of deicing salts in soils in urban environment. *Urban Ecosyst.* **2008**, *11*, 17–31. [CrossRef]
11. Ordóñez-Barona, C.; Sabetski, V.; Millward, A.A.; Steenberg, J. De-icing salt contamination reduces urban tree performance in structural soil cells. *Environ. Pollut.* **2018**, *253*, 562–571. [CrossRef] [PubMed]
12. Cassaniti, C.; Romano, D.; Flowers, T.J. The response of ornamental plants to saline irrigation water. In *Irrigation–Water Management, Pollution and Alternative Strategies*; Garcia-Garizabal, I., Ed.; InTechOpen: London, UK, 2012; pp. 131–139. [CrossRef]
13. Reid, M.S.; Jiang, C.Z. Postharvest biology and technology of cut flowers and potted plants. *Hortic. Rev.* **2012**, *40*, 1–54.
14. Villarino, G.H.; Mattson, N.S. Assessing tolerance to sodium chloride salinity in fourteen floriculture species. *HortTechnology* **2011**, *21*, 539–545. [CrossRef]
15. Esringü, A.; Ekinci, M.; Turan, M. Effects of Different Growing Media on Growth Parameters of Zinnia (*Zinnia elegans*). Yuzuncu Yil University. *J. Agric. Sci.* **2022**, *32*, 175–185. [CrossRef]
16. Iannotti, M. How to plant and care for *Zinnia* flowers. The Spruce 2022. Available online: https://www.thespruce.com/zinnias-a-burst-of-hot-flower-colors-from-zinnias-1402919 (accessed on 15 December 2022).
17. Oliveira, E.V.; de Lacerda, C.F.; de Neves, A.L.R.; Gheyi, H.R.; Oliveira, D.R.; Oliveira, F.Í.F.; de Viana, T.V.A. A new method to evaluate salt tolerance of ornamental plants. *Theor. Exp. Plant Physiol.* **2018**, *30*, 173–180. [CrossRef]
18. Niu, G.; Wang, M.; Rodriges, D. Response of *Zinna* plants to saline water irrigation. *HortScience* **2012**, *478*, 793–797. [CrossRef]
19. Macherla, K.; McAvoy, R.J. The effect of salinity on the growth and nutrient status of zinnia grown under short- and long-cycle subirrigation management. *HortScience* **2017**, *52*, 770–773. [CrossRef]
20. Bizhaini, S.; Jowkar, A.; Abdolmaleki, M. Growth and antioxidant response of *Zinnia elegans* under salt stress conditions. *Techn. J. Eng. Appl. Sci.* **2013**, *13*, 1285–1292.

21. Escalona, A.; Salas-Sanjuán, M.C.; Dos Santos, C.; Guzmán, M. The effect of water salinity on growth and ionic concentration and relation in plant tissues in *Zinnia elegans* and *Tagetes erecta* for use in urban landscaping. *ITEA* **2014**, *110*, 325–334. [CrossRef]
22. Carter, C.T.; Grieve, C.M. Growth and nutrition of two cultivars of *Zinnia elegans* under saline conditions. *HortScience* **2010**, *45*, 1058–1063. [CrossRef]
23. Van Zelm, E.; Zhang, Y.; Testerink, C. Salt tolerance mechanisms of plants. *Annu. Rev. Plant Biol.* **2020**, *71*, 403–433. [CrossRef] [PubMed]
24. Isayenkov, S.V.; Maathuis, F.J. Plant salinity stress: Many unanswered questions remain. *Front. Plant Sci.* **2019**, *10*, 80. [CrossRef] [PubMed]
25. Singh, P.; Choudhary, K.K.; Chaudhary, N.; Gupta, S.; Sahu, M.; Tejaswini, B.; Sarkar, S. Salt stress resilience in plants mediated through osmolyte accumulation and its crosstalk mechanism with phytohormones. *Front. Plant Sci.* **2022**, *13*, 1006617. [CrossRef] [PubMed]
26. Munns, R.; Tester, M. Mechanisms of salinity tolerance. *Ann. Rev. Plant Biol.* **2008**, *59*, 651–681. [CrossRef] [PubMed]
27. Acosta-Motos, J.R.; Ortuño, M.F.; Bernal-Vicente, A.; Diaz-Vivancos, P.; Sanchez-Blanco, M.J.; Hernandez, J.A. Plant Responses to Salt Stress: Adaptive Mechanisms. *Agronomy* **2017**, *7*, 18. [CrossRef]
28. Rademacher, W. Plant growth regulators: Backgrounds and uses in plant production. *J. Plant Growth Regul.* **2015**, *34*, 845–872. [CrossRef]
29. Meena, M.; Divyanshu, K.; Kumar, S.; Swapnil, P.; Zehra, A.; Shukla, V.; Yadav, M.; Upadhyay, R.S. Regulation of l-proline biosynthesis, signal transduction, transport, accumulation and its vital role in plants during variable environmental conditions. *Heliyon* **2019**, *5*, e02952. [CrossRef]
30. Szabados, L.; Savourè, A. Proline: A multifunctional amino acid. *Trends Plant Sci.* **2009**, *15*, 89–97. [CrossRef]
31. Abbasi, H.; Jamil, M.; Haq, A.; Ali, S.; Ahmad, R.; Malik, Z.; Parveen. Salt Stress Manifestation on Plants, Mechanism of Salt Tolerance and Potassium Role in Alleviating It: A Review. *Zemdirbyste-Agriculture* **2016**, *103*, 229–238. [CrossRef]
32. Sayed, S.A.A.E.; Hellal, F.A.; El–Rab, N.G.; Zewainy, R.M. Ameliorative Effects of Potassium on the Salinity Stress in Plants: A Review. *Asian J. Soil Sci. Plant Nutr.* **2019**, *4*, 1–15. [CrossRef]
33. Ben-Hayyim, G.; Spiegel-Roy, P.; Neumann, H. Relation between ion accumulation of salt-sensitive and isolated stable salttolerant cell lines of *Citrus aurantium*. *Plant Physiol.* **1985**, *178*, 144–148. [CrossRef] [PubMed]
34. Tzortzakis, N.G. Potassium and calcium enrichment alleviate salinity-induced stress in hydroponically grown endives. *Hort. Sci.* **2010**, *37*, 155–162. [CrossRef]
35. Umar, S.; Diva, I.; Anjum, N.; Iqbal, M.; Ahmad, I.; Pereira, E. Potassium-induced alleviation of salinity stress in *Brassica campestris* L. *Open Life Sci.* **2011**, *6*, 1054–1063. [CrossRef]
36. Amjad, M.; Akhtar, J.; Haq, M.; Aanwar-ul-Haq, M.; Imran, S.; Jacobsen, S.E. Soil and foliar application of potassium enhances fruit yield and quality of tomato under salinity. *Turk. J. Biol.* **2014**, *38*, 7. [CrossRef]
37. Hashi, U.S.; Karim, A.; Saikat, H.M.; Islam, R.; Islam, M.A. Effect of Salinity and Potassium Levels on Different Morpho-Physiological Characters of two Soybean (*Glycine max* L.) Genotypes. *J. Rice Res.* **2015**, *3*, 143. [CrossRef]
38. Henry, E.E.Y.; Kinsou, E.; Mensah, A.C.G.; Komlan, F.A.; Gandonou, C.B. Response of tomato (*Lycopersicon esculentum* mill.) plants cultivated under salt stress to exogenous application of calcium and potassium. *J. Appl. Biosci.* **2021**, *159*, 16363–16370. [CrossRef]
39. Yurtseven, E.; Kesmez, G.D.; Ünlükara, A. The effects of water salinity and potassium levels on yield, fruit quality and water consumption of a native central anatolian tomato species (*Lycopersicon esculantum*). *Agric. Water Manage.* **2005**, *78*, 128–135. [CrossRef]
40. Kleiber, T. Nutritional resources of soil in the localities of monumental large-leaved linden (*Tilia platyphyllos f. aurea*) alleys. *Ecol. Chem. Eng.* **2009**, *16*, 277–286.
41. Gałuszka, A.; Migaszewski, Z.M.; Podlaski, R.; Dołęgowska, S.; Michalik, A. The influence of chloride deicers on mineral nutrition and the health status of roadside trees in the city of Kielce, Poland. *Environ. Monit. Assess.* **2011**, *176*, 451–464. [CrossRef]
42. Wilkaniec, B.; Breś, W.; Fružyńska-Jóźwiak, D.; Borowiak-Sobkowiak, B.; Wilkaniec, A. The assessment of chemical properties of soil, the chemical composition of leaves and the occurrence of diseases on *Acer platanoides* and *Tilia cordata* in selected sites of urban greenery in Poznań. *Phytopathology* **2012**, *65*, 19–28.
43. Breś, W.; Kozłowska, M.; Kupska, A. The salinity of soils located along the selected streets of Poznan. Current trends in the horticultural plants cultivation. In *National Scientific Conference*; Department of Soil Cultivation and Horticultural Plant Fertilization: Lublin-Susiec, Poland, 2014; p. 40. (In Polish)
44. Equiza, M.A.; Calvo-Polanco, M.; Cirelli, D.; Señorans, J.; Wartenbe, M.; Saunders, C.; Zwiazek, J.J. Long-term impact of road salt (NaCl) on soil and urban trees in Edmonton, Canada. *Urban For. Urban Green.* **2017**, *21*, 16–28. [CrossRef]
45. Kotuby-Amacher, J.; Koenig, R.; Kitchen, B. Salinity and Plant Tolerance 2000, All Archived Publications. Paper 43. Available online: https://digitalcommons.usu.edu/extension_histall/43 (accessed on 15 December 2022).
46. Maas, E.V.; Hoffman, G.J. Crop salt tolerance-current assessment. *ASCE J. Irrig. Drain. Divsion* **1977**, *10*, 115–134. [CrossRef]
47. Miyamoto, S.; Martinez, I.; Portillo, M.; Ornelas, A. Landscape Plant Lists for Salt Tolerance Assessment. USDI Bureau of Reclamation. Texas Agricultural Extension Station, El Paso. 2004. Available online: http://agrilife.org/elpaso/files/2011/10/Landscape-Plant-Lists-for-Salt-Tolerance-Assessment.pdf (accessed on 15 December 2022).
48. Cosgrove, D.J. Expansive growth of plant cell walls. *Plant Physiol. Biochem.* **2000**, *38*, 109–124. [CrossRef] [PubMed]

49. Zhao, S.; Zhang, Q.; Liu, M.; Zhou, H.; Ma, C.; Wang, P. Regulation of Plant Responses to Salt Stress. *Int. J. Mol. Sci.* **2021**, *22*, 4609. [CrossRef]
50. Xie, M.; Zhang, J.; Tschaplinski, T.J.; Tuskan, G.A.; Chen, J.-G.; Muchero, W. Regulation of Lignin Biosynthesis and Its Role in Growth-Defense Tradeoffs. *Front. Plant Sci.* **2018**, *9*, 1427. [CrossRef]
51. Neuman, P.M. The role of cel wall adjustment in plant resistance to watre deficyt. *Crop. Sci.* **1995**, *35*, 1258–1266. [CrossRef]
52. Liu, J.; Zhang, W.; Long, S.; Zhao, C. Maintenance of CellWall Integrity under High Salinity. *Int. J. Mol. Sci.* **2021**, *22*, 3260. [CrossRef]
53. Oliveira, D.M.; Mota, T.R.; Salatta, F.V.; Sinzker, R.C.; Končitíková, R.; Kopečný, D.; Simister, R.; Silva, M.; Goeminne, G.; Morreel, K.; et al. Cell wall remodeling under salt stress: Insights into changes in polysaccharides, feruloylation, lignification, and phenolic metabolism in maize. *Plant Cell Environ.* **2020**, *43*, 2172–2191. [CrossRef]
54. Tavakkoli, E.; Fatehi, F.; Coventry, S.; Rengasamy, P.; McDonald, G.K. Additive effects of Na^+ and Cl^- ions on barley growth under salinity stress. *J. Exp. Bot.* **2011**, *62*, 2189–2203. [CrossRef]
55. Wakeel, A. Potassium-sodium interactions in soil and plant under saline-sodic conditions. *J. Plant Nutr. Soil Sci.* **2013**, *176*, 344–354. [CrossRef]
56. Wang, M.; Zheng, Q.; Shen, Q.; Guo, S. The Critical Role of Potassium in Plant Stress Response. *Int. J. Mol. Sci.* **2013**, *14*, 7370–7390. [CrossRef]
57. Hayat, S.; Hayat, Q.; Alyemeni, M.N.; Wani, A.S.; Pichtel, J.; Ahmad, A. Role of proline under changing environments: A review. *Plant Signal Behav.* **2012**, *7*, 1456–1466. [CrossRef] [PubMed]
58. Per, T.S.; Khan, N.A.; Reddy, P.S.; Masood, A.; Hasanuzzamn, M.; Khan, I.R.; Ajum, N.A. Approaches in modeling proline metabolism in plants for salt and drought tolerance: Phytohormones, mineral nutrients and transgenics. *Plant Physiol. Biochem.* **2017**, *115*, 126–140. [CrossRef] [PubMed]
59. Trovato, M.; Forlani, G.; Signorelli, S.; Funck, D. Proline Metabolism and Its Functions in Development and Stress Tolerance. In *Osmoprotectant-Mediated Abiotic Stress Tolerance in Plants*; Springer Nature: Cham, Switzerland, 2019. [CrossRef]
60. Moukhtari, A.E.; Cabass-Hourton, C.C.; Farissi, M.; Savouré, A. How does proline treatment promote salt stress tolerance during crop plant development? *Front. Plant Sci.* **2020**, *11*, 1127. [CrossRef] [PubMed]
61. Bhaskara, G.B.; Yang, T.H.; Verslues, P.E. Dynamic proline metabolism: Importance and regulation in water limited environments. *Front. Plant Sci.* **2015**, *6*, 484. [CrossRef] [PubMed]
62. Huang, Z.; Zhao, L.; Chen, D.; Liang, M.; Liu, Z.; Shao, H.; Long, X. Salt Stress Encourages Proline Accumulation by Regulating Proline Biosynthesis and Degradation in Jerusalem Artichoke Plantlets. *PLoS ONE* **2013**, *8*, e62085. [CrossRef]
63. Li, W.; Meng, R.; Liu, Y.; Chen, S.; Jiang, J.; Wang, L.; Zhao, S.; Wang, Z.; Fang, W.; Chen, F.; et al. Heterografted chrysanthemums enhance salt stress tolerance by integrating reactive oxygen species, soluble sugar, and proline. *Hort. Res.* **2022**, *9*, uhac073. [CrossRef]
64. Zdunek-Zastocka, E.; Grabowska, A.; Michniewska, B.; Orzechowski, S. Proline Concentration and Its Metabolism Are Regulated in a Leaf Age Dependent Manner But Not by Abscisic Acid in Pea Plants Exposed to Cadmium Stress. *Cells* **2021**, *10*, 946. [CrossRef]
65. Ahanger, M.A.; Agarwal, R.; Tomar, N.S.; Shrivastava, M. Potassium induces positive changes in nitrogen metabolism and antioxidant system of oat (*Avena sativa* L. cultivar Kent). *J. Plant Inter.* **2015**, *10*, 211–223. [CrossRef]
66. Sulandjari, A.; Sakya, T.; Wijayanti, R.N. Salinity and potassium fertilizer on growth and proline of the medicinal plant *Pereskia bleo*. In *IOP Conference Series: Earth and Environmental Science, Proceedings of the 6th International Conference on Climate Change, 25 May 2021, Surakarta, Indonesia*; IOP Publishing Ltd.: Bristol, UK, 2021; Volume 824. [CrossRef]
67. Heidari, M.; Jamshidi. Effects of Salinity and Potassium Application on Antioxidant Enzyme Activities and Physiological Parameters in Pearl Millet. *Agric. Sci. China* **2011**, *10*, 228–237. [CrossRef]
68. Rao, R.C.N.; Krishnasastry, K.S.; Udayakumar, M. Role of potassium in proline metabolism. I. Conversion of precursors into proline under stress conditions in K-sufficient and K-deficient plants. *Plant Sci. Lett.* **1981**, *23*, 327–334. [CrossRef]
69. Kalra, Y.P. *Handbook of Reference Methods for Plant Analysis*; CRC Press, Taylor & Francis Group: Boca Raton, FL, USA, 1998; p. 287.
70. Weatherly, P.E. Studies in water relation of cotton plants. The measurement of water deficits in leaves. *New Phytol.* **1950**, *49*, 81–97. [CrossRef]
71. Bandurska, H. Does proline accumulated in leaves of water deficit stressed barley plants confine cell membrane injury? I. Free proline accumulation and membrane injury index in drought and osmotic stressed plants. *Acta Physiol. Plantarum* **2000**, *22*, 409–415. [CrossRef]
72. Bates, L.S.; Waldren, R.P.; Teare, J.D. Rapid determination of proline for water stress studies. *Plant Soil* **1973**, *39*, 205–207. [CrossRef]
73. Zhang, C.-S.; Lu, Q.; Verma, D.P.S. Removal of feedback inhibition of Δ^1-pyrroline-carboxylate synthetase, a bifunctional enzyme catalyzing the first two steps of proline biosynthesis in plants. *J. Biol. Chem.* **1995**, *270*, 20491–20496. [CrossRef]
74. Rahnama, H.; Ebrahimzadeh, H. The effect of NaCl on proline accumulation in potato seedlings and calli. *Acta Physiol. Plant.* **2004**, *26*, 263–270. [CrossRef]
75. Bradford, M.M.A. A rapid and sensitive method for the quantitation of microgram quantities of protein utilizing the principle of protein-dye binding. *Anal. Biochem.* **1976**, *72*, 248–254. [CrossRef] [PubMed]

76. Hiscox, J.C.; Israelstam, G.F. A method for the extraction of chlorophyll from tissue without maceration. *Can. J. Bot.* **1979**, *57*, 1332–1334. [CrossRef]
77. Wellburn, A.R. The spectral determination of chlorophylls a and b, as well as total carotenoids, using various solvents with spectrophotometers of different resolution. *J. Plant Physiol.* **1994**, *144*, 307–313. [CrossRef]
78. Syros, T.; Yupsanis, T.; Zafiriadis, H.; Economou, A. Activity and isoforms of peroxidases, lignin and anatomy, during adventitious rooting in cuttings of *Ebenus cretica* L. *J. Plant Physiol.* **2004**, *161*, 69–77. [CrossRef]
79. Premachandra, G.S.; Saneoka, G.S.; Ogata, H. Cell membrane stability, as indicator of drought tolerance, as affected by applied nitrogen and soybean. *J. Agric. Sci.* **1990**, *115*, 63–66. [CrossRef]
80. Sakia, R.M. The Box-Cox Transformation Technique: A Review. *J. R. Statist. Soc. Ser. D* **1992**, *41*, 169–178. [CrossRef]

plants

MDPI

Article

Linum usitatissimum AccD Enhances Seed Fatty Acid Accumulation and Tolerance to Environmental Stresses during Seed Germination in *Arabidopsis thaliana*

Rui Du, Xinye Li, Huan Hu, Yu Zhao, Mingxun Chen and Zijin Liu *

National Yangling Agricultural Biotechnology & Breeding Center, Shaanxi Key Laboratory of Crop Heterosis and College of Agronomy, Northwest A&F University, Yangling 712100, China; durui000@nwafu.edu.cn (R.D.); lixinye0324@163.com (X.L.); huhuan@nwafu.edu.cn (H.H.); nwafu_zy@163.com (Y.Z.); cmx786@nwafu.edu.cn (M.C.)

* Correspondence: liuzijin@nwafu.edu.cn

Abstract: Flax (*Linum usitatissimum* L.), as an important oil-producing crop, is widely distributed throughout the world, and its seeds are rich in polyunsaturated fatty acids (FAs). Previous studies have revealed that *Arabidopsis thaliana* ACETYL-CoA CARBOXYLASE (AtACCase) is vital for FA biosynthesis. However, the functions of *L. usitatissimum AccD* (*LuAccD*) on FA accumulation and seed germination remain unclear. In the present study, we cloned the LuAccD coding sequence from the flax cultivar 'Longya 10', identified conserved protein domains, and performed a phylogenetic analysis to elucidate its relationship with homologs from a range of plant species. Ectopic expression of *LuAccD* in *A. thaliana* wild-type background enhanced seed FA accumulation without altering seed morphological characteristics, including seed size, 1000-seed weight, and seed coat color. Consistently, the expression of key genes involved in FA biosynthesis was greatly up-regulated in the developing seeds of *LuAccD* overexpression lines. Additionally, we demonstrated that LuAccD acts as a positive regulator of salt and mannitol tolerance during seed germination in *A. thaliana*. These results provide important insights into the functions of *LuAccD*, which facilitates the oil quantity and abiotic stress tolerance of oil-producing crops through genetic manipulation.

Keywords: *LuAccD*; fatty acids; salt stress; mannitol stress; seed germination

Citation: Du, R.; Li, X.; Hu, H.; Zhao, Y.; Chen, M.; Liu, Z. *Linum usitatissimum AccD* Enhances Seed Fatty Acid Accumulation and Tolerance to Environmental Stresses during Seed Germination in *Arabidopsis thaliana*. *Plants* **2023**, *12*, 3100. https://doi.org/10.3390/plants12173100

Academic Editor: Wei Ma

Received: 25 July 2023
Revised: 15 August 2023
Accepted: 27 August 2023
Published: 29 August 2023

1. Introduction

Flax (*Linum usitatissimum* L., 2n = 30) is a versatile annual plant with global production areas of approximately 12 million acres mainly in Kazakhstan, Russia, Canada, and China; primarily cultivated for its seed oil (oilseed flax) and stem fiber (fiber flax) [1]. Oilseed flax generally contains approximately 50% oil which is composed of five major fatty acids (FAs): palmitic acid (C16:0), stearic acid (C18:0), oleic acid (C18:1), linoleic acid (C18:2), and linolenic acid (C18:3) [2,3]. Among them, the percentage of C18:3 in the flaxseed oil ranges from 40% to 60%, which is significantly higher than that of *Zea mays* (~1%), *Glycine max* (~8%), and *Brassica napus* (~11%) [3]. As polyunsaturated FAs, C18:2 and C18:3 cannot be biosynthesized in the human body and are the precursors for long-chain polyunsaturated FAs, inclusive of arachidonic acid and eicosapentaenoic acid. These long-chain polyunsaturated FAs have a significant role in the prevention of a variety of diseases, including cancers, inflammatory, cardiovascular, and autoimmune diseases [4–6]. Therefore, together with a high amount of proteins (up to 18.29%), fiber (27.3%), vitamin B1, and lignans, particularly secoisolariciresinol diglucoside (294–700 mg/100 g) [7–9], flax serves as a predominant source which offers a wide range of nutritional and therapeutic applications. In the past decade, China has become the largest importer with the import of $31,108 million, which is equivalent to 26.8% of total global flax import in the year 2020 [10]. However, oilseed flax is mainly grown in the arid and semi-arid regions of the Northern and Northwestern China, which is one of the areas more vulnerable to global climate change [11]. Unpredictable environmental stresses, such as drought and salinity–alkalinity, pose a threat to biological diversity and the quality of oilseed flax. Therefore, identifying the key genes involved in seed FA accumulation and response to adversities in *L. usitatissimum* would provide potential targets for molecular breeding in oil-producing crops including *L. usitatissimum*.

In plants, FA biosynthesis starts with the provision of carbon from glycolysis. After glycolysis, pyruvate dehydrogenase catalyzes the conversion of pyruvate to acetyl-CoA, the initial substrate for de novo FA biosynthesis which occurs in the plastids [12]. Acetyl-CoA carboxylase (ACCase) converts acetyl-CoA and bicarbonate into malonyl-CoA, which is the first committed step in FA biosynthesis [13,14]. In the plastids of dicots and non-graminaceous monocots, ACCase mainly comprises four distinct subunits, namely biotin carboxylase, biotin carboxyl carrier protein, α-subunit of carboxyltransferase (CTα), and β-subunit of carboxyltransferase (CTβ) [15,16]. Studies have shown that increased activity of *A. thaliana* AtACCase in the tuber amyloplasts of *Solanum tuberosum* led to an increase of more than five times in the triacylglycerol content [17]. The mutation of *A. thaliana ACC1* (*AtACC1*), an essential gene encoding ACCase, significantly decreased the contents of long-chain FAs in leaves under cold treatment [18]. Overexpression of *AtACC1* in *B. napus* not only altered seed FA compositions, with the largest effect being an increase in C18:1, but also caused an increase of approximately 5% in seed oil content [19]. Meanwhile, overexpression of each subunit of *Gossypium hirsutum* ACCase effectively increased seed oil content in the transgenic plants of *G. hirsutum*. Among them, the oil content of *GhBCCP1* transgenic seeds was significantly increased by 21.92%, while that of *GhBC1* and *GhCTβ* transgenic seeds was elevated by ~17% [20]. The latest study showed that the interaction between α-CT and CARBOXYLTRANSFERASE INTERACTORs was enhanced by light, which in turn attenuates carbon flux into triacylglycerol accumulation in *A. thaliana* leaves [21]. Homologous expression of *NtAccD*, located in the plastid genome, raised the ACCase level and FA content in the resultant transgenic leaves in *Nicotiana tabacum* cv. *Xanthi* [22]. Semi-quantitative RT-PCR and quantitative real-time PCR (qRT-PCR) results showed that the expression level of *EgAccD* is positively correlated with the *Elaeis guineensis* productivity [23]. The functions of *AccD* genes from *A. thaliana* and other plants have been well characterized, but the roles of *AccD* from *L. usitatissimum* in the regulation of seed FA accumulation and in response to salt and osmotic stresses remain unclear.

For the sessile crops, environmental factors are crucial in determining crop growth and development. Of these, drought and salt are the most prevalent and detrimental constraints to agricultural production [24–27]. Previous studies have demonstrated that drought can negatively affect the yield potential, oil content and FA compositions, and fiber quality traits of flax [26,28,29]. Meanwhile, soil salinity–alkalinity can result in delayed germination, low seedling survival, irregular growth, and lower yield of flax [27]. In addition, drought can result in osmotic stress by altering water potential and cell turgor, and salt can induce osmotic stress and ion toxicity [30]. It is worth noting that the hyperosmotic signal caused by drought and salt stresses promotes the accumulation of phytohormone abscisic acid (ABA), which in turn triggers a series of adaptive responses in plants [31]. Therefore, ABA biosynthesis and signal transduction are of great importance for plants to resist abiotic stresses.

In this study, we cloned the *LuAccD* gene from the flax cultivar 'Longya 10′ and found that overexpression of *LuAccD* in *A. thaliana* wild-type plants significantly increased the accumulation of seed total FAs by boosting the transcription levels of several key genes involved in FA biosynthesis. We also demonstrated that LuAccD enhances tolerance to salt and mannitol stresses during seed germination via mediating the ABA biosynthesis and ABA-responsive pathway in *A. thaliana*.

2. Results

2.1. Sequence Analysis of LuAccD Protein

The protein sequence of AtAccD was applied to BLASTP in the Phytozome (https://phytozome-next.jgi.doe.gov/, accessed on 12 October 2020) database and one homologous polypeptide of Lus10002473 was identified from the *L. usitatissimum* genome, namely LuAccD. As shown in Figure 1A, LuAccD and AtAccD had 330 and 488 amino acids, respectively. A 58.4% identity in amino acid sequence was matched between LuAccD and AtAccD, and their carboxyltransferase domains shared 61.5% identity (Table S2). Phylogenetic analysis indicated that LuAccD presents a relatively distant relationship with AccD from other crops we selected (Figure 1B). These results suggested that LuAccD may have a similar function as AtAccD in some ways.

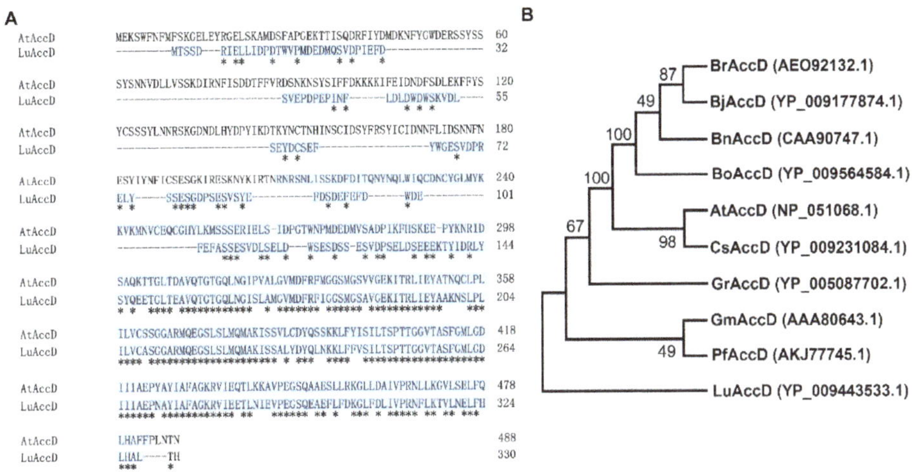

Figure 1. Protein sequence alignment and phylogenetic analysis of AccD. (**A**) Sequence alignment of amino acids from LuAccD and AtAccD. The asterisks represent strictly conserved amino acids. The crotonase-like superfamily domain, which was predicated by the NCBI Conserved Domain Database (https://www.ncbi.nlm.nih.gov/Structure/cdd/wrpsb.cgi, accessed on 20 October 2022), is highlighted in blue in the sequences. (**B**) Phylogenetic analysis of AccD proteins from *L. usitatissimum*, *A. thaliana*, and other crops. Numbers indicate the phylogenetic confidence of the tree topology and denote the bootstrap values on neighbor-joining analysis. Br: *Brassica rapa*, Bj: *Brassica juncea*, Bn: *Brassica napus*, Bo: *Brassica oleracea* var. *oleracea*, At: *Arabidopsis thaliana*, Cs: *Camelina sativa*, Gr: *Gossypium raimondii*, Gm: *Glycine max*, Pf: *Perilla frutescens*, Lu: *Linum usitatissimum*. The accession numbers of AccD are listed in parentheses.

2.2. LuAccD Increases the Seed FA Accumulation in A. thaliana

Studies have revealed that the loss of *AtAccD* function results in embryo lethality of *A. thaliana* [32]. To preliminarily investigate the functions of *LuAccD* on the accumulation of seed FAs, we introduced the overexpression construct of *35S: LuAccD–6HA* (Figure 2A) into the *A. thaliana* wild-type (Col-0) plants. We obtained six independent T_3 homozygous *Col-0 35S: LuAccD–6HA* transgenic lines (#1, #2, #4, #5, #6, and #11) and identified them by the analysis of PCR-based DNA genotyping (Figure 2B). Meanwhile, qRT-PCR results showed that the *LuAccD* expression is not detected in the Col-0, but highly present in the six transgenic lines (Figure 2C). Therefore, we selected *Col-0 35S: LuAccD–6HA#2* and *Col-0 35S: LuAccD–6HA#4* for follow-up experiments. The phenotype analysis showed that there are no significant differences in the seed coat color, seed size, and 1000-seed weight between Col-0 and *Col-0 35S: LuAccD–6HA* transgenic plants (#2 and #4) (Figure S1). However, the contents of seed total FAs and all major FA compositions were both significantly elevated in *Col-0 35S: LuAccD–6HA* plants compared to those in Col-0 (Figure 2D,E). These results suggested that ectopic expression of *LuAccD* promotes FA accumulation without affecting other measured agronomic traits in *A. thaliana* seeds.

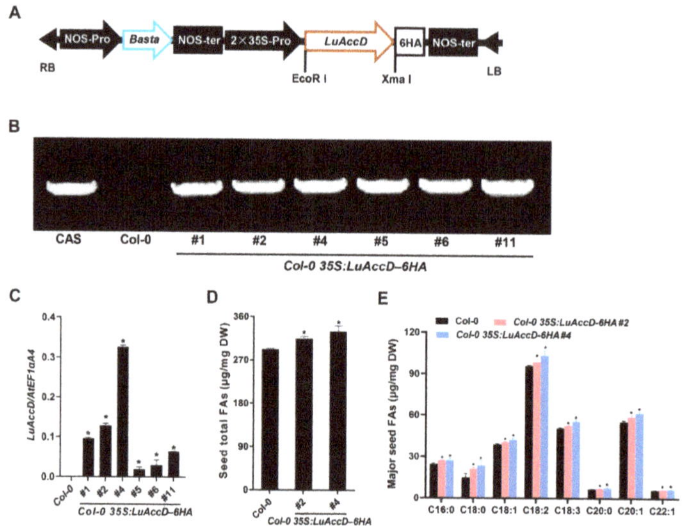

Figure 2. Overexpression of *LuAccD* increased the accumulation of seed FAs in *A. thaliana*. (**A**) Schematic illustration of the constitutive expression cassette of *LuAccD*. RB, right border; LB, left border; NOS-pro, nopaline synthase promoter; NOS-ter, nopaline synthase terminator; Basta, glyphosate; 35S-pro, CaMV 35S promoter. (**B**) PCR-based DNA genotyping of *Col-0 35S: LuAccD–6HA* transgenic plants. Cas, cassette. (**C**) Transcript levels of *LuAccD* in the wild-type (Col-0) and *Col-0 35S: LuAccD–6HA* developing seeds at 12 days after pollination measured by qRT-PCR. *AtEF1αA4* was used as an internal control. The values are presented as the mean ± SD (n = 3). (**D**) Comparisons of seed total FA content between Col-0 and *Col-0 35S: LuAccD–6HA* transgenic plants. (**E**) Comparison of the major seed FA compositions between the Col-0 and *Col-0 35S: LuAccD–6HA* transgenic plants. Values represent means ± SD and error bars denote SD. Three independent experiments were carried out and each biological replicate contains three technical replicates. Asterisks (*) indicate significant differences in the FA contents between *Col-0 35S: LuAccD–6HA* and Col-0 plants (two-tailed paired Student's *t*-test, $p \leq 0.05$).

2.3. *LuAccD* Increases the Expression Levels of Genes Contributing to Seed FA Accumulation

To further investigate how LuAccD controls seed FA accumulation at transcription level, several key genes inclusive of *AtBCCP1* (*BIOTIN CARBOXYL CARRIER PROTEIN ISOFORM1*), *AtBCCP2*, *AtMCAT* (*MALONYL COA-ACP MALONYLTRANSFERASE*), *AtKASI* (*3-KETOACYL-ACYL CARRIER PROTEIN SYNTHASE I*), *AtKASII*, *AtSSI2* (*SUPPRESSOR OF SA INSENSITIVE 2*), *AtFAD2* (*FATTY ACID DESATURASE2*), *AtFAD3*, and *AtPDAT2* (*PHOSPHOLIPID: DIACYLGLYCEROL ACYLTRANSFERASE2*), were selected for expression analysis. The expression levels of these genes were assessed by qRT-PCR using the developing seeds at 12 days after pollination (DAP) between Col-0 and *Col-0 35S: LuAccD–6HA#4* transgenic plants. The transcript levels of *AtBCCP1*, *AtBCCP2*, *AtMCAT*, *AtKASI*, *AtKASII*, *AtSSI2*, *AtFAD2*, *AtFAD3*, and *AtPDAT2* in the developing seeds of *Col-0 35S: LuAccD–6HA#4* transgenic plants were significantly higher than those of the Col-0 at 12 DAP (Figure 3). These results demonstrated that LuAccD contributes to seed FA accumulation by up-regulating the expression of *AtBCCP1*, *AtBCCP2*, *AtMCAT*, *AtKASI*, *AtKASII*, *AtSSI2*, *AtFAD2*, *AtFAD3*, and *AtPDAT2* during seed development in *A. thaliana*.

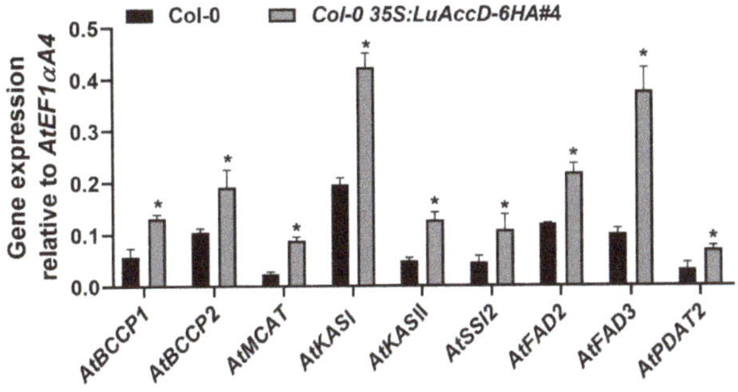

Figure 3. Expression analysis of genes contributing to FA accumulation in the wild-type (Col-0) and *Col-0 35S: LuAccD–6HA#4* developing seeds at 12 days after pollination. Results were normalized against the expression of *AtEF1αA4* as an internal control. Values are means ± SD (n = 3). Asterisks (*) represent significant differences between Col-0 and *Col-0 35S: LuAccD–6HA#4* transgenic plants determined by two-tailed paired Student's *t*-test ($p \leq 0.05$).

2.4. *LuAccD Promotes Seed Germination under Salt and Mannitol Stresses in A. thaliana*

To determine the effects of LuAccD in response to abiotic stresses, seed germination of Col-0 and *Col-0 35S: LuAccD–6HA* plants were observed on MS agar medium containing 150 mM NaCl or 300 mM mannitol. As shown in Figure 4, Col-0 and *Col-0 35S: LuAccD–6HA* lines displayed similar germination rates and seedling growth on the medium without stress treatment (Figure 4). However, the seed germination rate of *Col-0 35S: LuAccD–6HA* lines was higher than that of Col-0 under the stress of 150 mM NaCl or 300 mM mannitol (Figure 4). Therefore, we indicated that LuAccD positively regulates the resistance of salt and mannitol stresses during seed germination in *A. thaliana*.

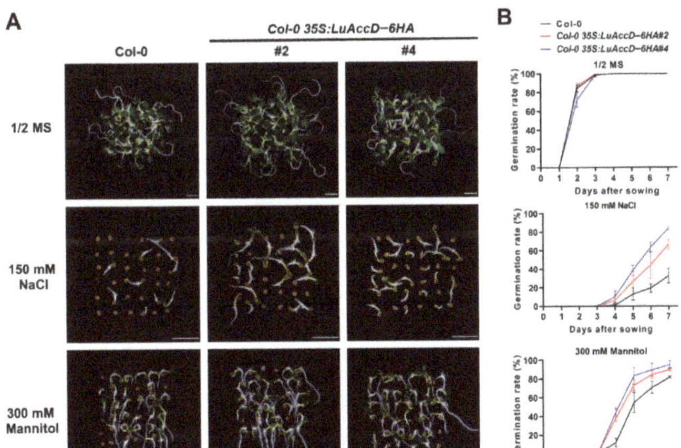

Figure 4. Response of wild-type (Col-0) and transgenic plants overexpressing *LuAccD* to NaCl and mannitol in seed germination. (**A**) Germination phenotype of seeds from the different lines grown on 1/2 MS plates or 1/2 MS plates with 150 mM NaCl or 300 mM mannitol for 7 days after sowing. Bar = 2 mm. (**B**) Germination rates of seeds from the different lines grown on 1/2 MS plates or 1/2 MS plates with 150 mM NaCl or 300 mM mannitol. Seed germination percentages were quantified every day from 1st day to the 7th day after sowing, and the embryonic axis protrusion was considered as seed germination. Date are the means ± SD (n = 3). Error bars denote SD. The value of each biological replicate was the average calculated over three technical replicates. For each technical replicate, we recorded the germination rates of 150 seeds from the same batch.

2.5. LuAccD Inhibits Expression Levels of Several Genes Contributing to ABA Biosynthesis and Signal Transduction

To better understand how LuAccD influences seed germination in response to salt and mannitol stresses, we assessed the expression of five ABA-related genes in Col-0 and *Col-0 35S: LuAccD–6HA#4* transgenic seeds at 12 h after sowing. As illustrated in Figure 5, there were no significant differences in the expression levels of *AtNCED3* (*NINE-CIS-EPOXYCAROTENOID DIOXYGENASE 3*), *AtAAO3* (*ABSCISIC ALDEHYDE OXIDASE 3*), *AtABI3* (*ABSCISIC ACID INSENSITIVE 3*), *AtEM1* (*EARLY METHIONINE-LABELED 1*), and *AtEM6* between Col-0 and *Col-0 35S: LuAccD-6HA#4* transgenic seeds under the normal condition. The treatment of 150 mM NaCl or 300 mM mannitol remarkably induced the expression of these genes in both Col-0 and *Col-0 35S: LuAccD–6HA#4* germinating seeds. But the expression levels of these genes were always lower in *Col-0 35S: LuAccD–6HA#4* transgenic lines than those in Col-0 (Figure 5). These results suggested that overexpression of *LuAccD* inhibits the expression of *AtNCED3*, *AtAAO3*, *AtABI3*, *AtEM1*, and *AtEM6*, which weakens the ABA biosynthesis and ABA signal transduction, thereby resulting in the low sensitivity of transgenic plants to salt and mannitol stresses during seed germination.

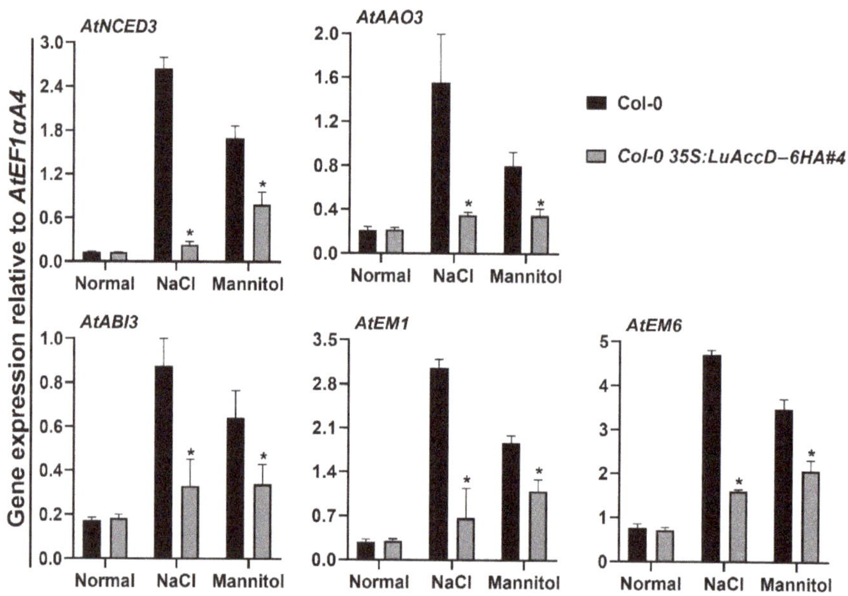

Figure 5. Comparison of relative transcript levels of ABA-related genes between wild-type (Col-0) and *Col-0 35S: LuAccD–6HA#4* transgenic plants. Total RNA was extracted from germinating seeds grown on the 1/2 MS or 1/2 MS containing 150 mM NaCl or 300 mM mannitol at 12 h after sowing. The expression levels of genes were calculated relative to that of the internal control *AtEF1αA4*. Values represent means ± SD (n = 3). Asterisks (*) represent significant differences between Col-0 and *Col-0 35S: LuAccD–6HA#4* lines (two-tailed paired Student's *t*-test, $p \leq 0.05$).

3. Discussion

Flax is an important oil-producing crop that has attracted interest due to its high content of C18:3. Flax with improved tolerance to stresses can also be used to expand cultivation into currently undeveloped and marginal lands [33]. Therefore, it is desirable to generate elite flax germplasm with a high oil content and resistance to environmental stresses, including salt and drought stresses. In this study, we found that *LuAccD* promotes the seed FA accumulation and facilitates seed germination under salt and mannitol stresses in *A. thaliana*.

The previous study showed that *AtAccD* is essential for FA biosynthesis [32]. Consistently, we found that ectopic expression of *LuAccD* significantly promotes the accumulation of seed total FAs and major FA compositions in *A. thaliana* (Figure 2D,E). The high percent identity of carboxyltransferase domains, which play an important role in FA biosynthesis [34], was observed between LuAccD and AtAccD (Figure 1A). Therefore, we inferred that LuAccD exhibits a conserved role with AtAccD in regulating the FA accumulation of *A. thaliana* seeds. Inconsistently, the overexpression of LuAccD in Col-0 did not alter the 1000-seed weight (Figure S1). This might be ascribed to the fact that other seed components affecting seed weight, such as storage proteins, offset the higher seed total FA content in *LuAccD* transgenic seeds. The exact explanation needs to be supported by further experimental results. Notably, seed coat color, seed length and width were also not altered (Figure S1). These results indicated that *LuAccD* can be regarded as a valuable potential for flax molecular breeding.

The collaborative expression of genes participating in FA biosynthesis is important for the oil accumulation in seeds [35–37]. Overexpression of *LuAccD* induced the transcript levels of several genes involved in oil biosynthetic processes, including FA biosynthesis and modification, and triacylglycerol deposition, which, in turn, contributes to oil accumulation in seeds (Figure 3). Of these enzymes, *BCCP1* and *BCCP2*, like *AccD*, also encode the subunit of ACCase, which functions as a sensor or gating system that controls the overall flux of FA biosynthesis [38,39]. MCAMT converts malonyl-CoA and ACYL CARRIER PROTEIN (ACP) into CoA and malonyl-ACP, which is a key building block for the FA biosynthesis [40]. Therefore, the up-regulated expression of *AtBCCP1*, *AtBCCP2*, and *AtMACT* by LuAccD should increase the overall flux of seed FAs at the early stage of the FA biosynthetic pathway in *A. thaliana*. Additionally, three separate condensing enzymes, or 3-ketoacyl-ACP synthases (KASI–KASIII), are essential for the production of C18 FAs. Among them, KASI participates in the conversion of acetyl-ACP to palmitoyl-ACP, whereas KASII mainly utilizes palmitoyl-ACP as the substrate to produce stearoyl-ACP [41]. Studies have shown that the deficiency of KASI leads to disrupted embryo development before the globular stage and noticeably decreases seed total FA content (~33.6% of the wild-type) in *A. thaliana* [42]. SSI2 (FAB2) encodes a stearoyl-acyl carrier protein desaturase that converts C18:0 into C18:1 [43]. FAD2 catalyzes the conversion of C18:1 to C18:2 which is further desaturated by FAD3 to form C18:3 [44–46]. Thus, the highly up-regulated expression of *AtKASI*, *AtKASII*, *AtSSI2*, *AtFAD2*, and *AtFAD3* in *Col-0 35S: LuAccD–6HA* would accelerate the accumulation of FAs in seeds at the middle stage of the biosynthetic pathway. *PDAT2* encoding a phospholipid: diacylglycerol acyl-transferase promotes triacylglycerol production [47]. Therefore, ectopic expression of *LuAccD* in *A. thaliana* could trigger multiple transcriptional regulatory events that affect FA accumulation in seeds.

Seed germination is a critical checkpoint for crop survival under adverse conditions, and ABA plays a critical role in affecting seed germination and seedling establishment, especially under abiotic stresses [48–51]. In our study, we found that overexpression of *LuAccD* in Col-0 weakens the sensitivity of the transgenic seeds to salt and mannitol during germination (Figure 4). At the cellular level, the transcript levels of five stress-response genes, *AtNCED3*, *AtAAO3*, *AtABI3*, *AtEM1*, and *AtEM6*, were higher in Col-0 germinating seeds than in the *Col-0 35S: LuAccD–6HA* under the NaCl or mannitol stress (Figure 5). *AtNCED3* encodes 9-cis-epoxy carotenoid dioxygenase which functions in osmotic stress-induced ABA biosynthesis in *A. thaliana* [52]. It is highly induced by salt and drought stresses, and its inactivation is responsible for enhanced germination upon salt stress [53–55]. *AtAAO3* encodes an enzyme that catalyzes the final step of ABA biosynthesis [56], and the *Oryza sativa OsAAO3* mutation exhibited earlier seed germination [57]. AtABI3 as a major downstream component of ABA signaling has been long recognized as a master regulator of seed dormancy and ABA inhibition of seed germination [50]. The higher percentage of seed germination was observed in Atabi3 mutant compared to wild-type when exposed to ABA, mannitol or NaCl treatments [58]. *AtEM1* and *AtEM6* encoding the late embryogenesis abundant proteins are ABA-responsive marker genes, which are induced by ABA, salt and osmotic stresses [59–61]. Owing to these results, we concluded that the lower expression of *AtNCED3*, *AtAAO3*, *AtABI3*, *AtEM1*, and *AtEM6* caused by the overexpression of *LuAccD* in *A. thaliana* attenuates the ABA biosynthesis and ABA signal transduction, thereby resulting in low sensitivity of *A. thaliana* to salt and mannitol stresses during seed germination.

4. Materials and Methods

4.1. Plant Materials and Growth Conditions

All *A. thaliana* materials used in this study were in the Columbia ecotype (Col-0) background, and were grown in a growth chamber at 22°C with a 16/8 h light/dark cycle, which has been reported in detail previously [62].

4.2. Gene Cloning and Plasmid Construction

The protein sequence of AtAccD (ATCG00500) was used for protein blast against the *L. usitatissimum* reference genome (https://phytozome-next.jgi.doe.gov/pz/portal.html, accessed on 12 October 2020). One identified highly conserved sequence Lus10002473 was named *LuAccD*. The template cDNA was synthesized from total RNA extracted from germinated seeds of oil flax cultivar 'Longya 10'. The full-length CDS of *LuAccD* without the stop codon was amplified using specific primers by PCR and was cloned into the pGreen-35S–6HA vector, forming the *35S: LuAccD–6HA* fusion vector. Primer information for the plasmid construction is given in Table S1.

4.3. Analysis of Protein Sequence and Phylogenetic Tree

The protein sequence of LuAccD was obtained from Phytozome (https://phytozome-next.jgi.doe.gov/pz/portal.html, accessed on 12 October 2020). Multiple sequence alignment of AtAccD and LuAccD proteins was carried out using MUSCLE website (https://www.ebi.ac.uk/Tools/msa/muscle/, accessed on 15 October 2022). The NCBI Conserved Domain Database (https://www.ncbi.nlm.nih.gov/Structure/cdd/wrpsb.cgi, accessed on 20 October 2022) was used to predicate the conserved domain of LuAccD. The phylogenetic tree was constructed by the Neighbor-Joining (NJ) method using MEGA 7.0 software with 1000 bootstrap replications and the *p*-distance model.

4.4. Generation of A. thaliana Transgenic Plants

The construct of *35S: LuAccD–6HA* was transformed into the *Agrobacterium tumefaciens* strain GV3101, which was then introduced into Col-0 via the floral dip method [63]. The T_1 transgenic plants were selected by Basta® (Bayer, Langenfeld, Germany) on soil and identified by using PCR in DNA level. T_2 and T_3 seeds were screened on 1/2 MS medium (pH 5.7, 1% sucrose, 1% agar) containing 10 μg/mL glufosinate-ammonium, and positive seedlings were transferred to soil. The T_3 generation homozygous plants were used for subsequent experiments after cultivation under similar conditions.

4.5. RNA Extraction and qRT-PCR Analysis

The total RNA samples were isolated using the MiniBEST Plant RNA extraction kit (Takara Bio, Dalian, China). RNA reverse reaction was carried out with the PrimeScript RT kit (Takara Bio, Dalian, China). qRT-PCR was performed using an SYBR Green Mix (Takara Bio, Dalian, China) on a Quant Studio 7 real-time system. The relative expression values were normalized to that of the internal control *AtEF1αA4*. Statistical data were obtained from three biological replicates. For each biological replicate, two technical repetitions were performed. Primer information for qRT-PCR is given in Table S1.

4.6. Microscopic Observation of A. thaliana Seed Traits

The *A. thaliana* seeds were harvested from the siliques at the basal part of the major inflorescences. The mature seeds were imaged under an SZ61 stereomicroscope (Olympus, Tokyo, Japan), and their length and width were determined with ImageJ 1.48v software. The 1000-seed weight was measured by using a 0.0001 precision test analytical balance

(BSA124S-CW, Sartorius, Beijing, China). Three independent biological replicates and three technical replicates were performed. For seed size measurement, each technical replicate contains 300 seeds.

4.7. Measurement of Seed FAs

Isolation and determination of FAs were performed according to previously described [64]. In brief, seed FAs were methylated in the 2.5% (v/v) H_2SO_4 solution diluted with methanol at 80 °C for 2 h. After cooling to room temperature, the solution was added with 2 mL of 0.9% (w/v) NaCl and 2 mL of hexane in due order, and the organic phase was analyzed by gas chromatography using GC-2010 plus instrument (Shimadzu, Kyoto, Japan) with a flame ionization detector and a 30 m (length) × 0.25 mm (internal diameter) × 0.5 μm (liquid membrane thickness) column (Supelco wax-10, Supelco, Shanghai, China). Methyl heptadecanoate was used as an internal standard. The initial column temperature was maintained at 160 °C for 1 min, increased by 4 °C min^{-1} to 240 °C, and held for 16 min at the final temperature. The peak for each FA composition was identified by their unique retention time, and their concentrations were calculated against the internal control.

4.8. Determination of Seed Germination

The *A. thaliana* seeds used for the germination analysis were harvested from plants grown under the same conditions at the same time and allowed to mature at room temperature for 3 months. The A. thaliana seeds were surface sterilized with 75% ethyl alcohol and were subsequently sown on 1/2 MS solid medium supplemented with or without 150 mM NaCl or 300 mM mannitol. The seeds were stratified at 4°C for 2 days in darkness and were then placed in the climate chamber. The germination (emergence of radicles) rate was scored daily. After 7 days, the seedlings were photographed. Seed germination percentages were quantified every day from 1st day to the 7th day after sowing, and the embryonic axis protrusion was considered as seed germination. Date are the means ± SD (n = 3). Error bars denote SD. The value of each biological replicate was the average calculated over three technical replicates. For each technical replicate, we recorded the germination rates of 150 seeds from the same batch.

5. Conclusions

In this study, our results demonstrated that LuAccD exhibits a conserved role with AtAccD in promoting the seed FA accumulation in *A. thaliana*. Meanwhile, LuAccD could enhance the tolerance to salt and mannitol stresses during seed germination in *A. thaliana*. In this regard, LuAccD can be utilized as a potential target for the breeding of flax varieties with high FA content and stress tolerance.

Supplementary Materials: The following supporting information can be downloaded at: https://www.mdpi.com/article/10.3390/plants12173100/s1, Figure S1. Morphological observation of mature *A. thaliana* seeds randomly selected from wild-type (Col-0) and overexpression transgenic plants carrying *LuAccD* (*Col-0 35S: LuAccD–6HA#2* and *#4*). Table S1. The primers used in this study. Table S2. Comparison of percent identity between the amino acid sequences of AtAccD and LuAccD.

Author Contributions: Z.L. conceived and designed the experiments. X.L. conducted the experiments and analyzed the data. R.D., H.H. and Y.Z. conducted parts of the experiments. R.D. wrote the draft of the manuscript, and M.C. and Z.L. revised the manuscript. All authors have read and agreed to the published version of the manuscript.

Funding: This work was supported by the Key Research and Development Program of Shaanxi Province (grant no. 2022NY-158 and 2021LLRH-07), the PhD Start-up Fund of Northwest A&F University (grant no. Z1090121052), and a grant from the Yang Ling Seed Industry Innovation Center (Grants no. K3031122024 and K3031123009).

Data Availability Statement: All data included in this study are available upon reasonable request by contact with the corresponding author.

Conflicts of Interest: The authors declare no conflict of interest.

References

1. Hall, L.M.; Booker, H.; Siloto, R.M.P.; Jhala, A.J.; Weselake, R.J. Flax (*Linum usitatissimum L.*). In *Industrial Oil Crops*; Elsevier Inc.: Amsterdam, The Netherlands, 2016; pp. 157–194. [CrossRef]
2. Hall, C., III; Tulbek, M.; Xu, Y. Flaxseed. *Adv. Food Nutr. Res.* **2006**, *51*, 1–97.
3. Radovanovic, N.; Thambugala, D.; Duguid, S.; Loewen, E.; Cloutier, S. Functional characterization of flax fatty acid desaturase FAD2 and FAD3 isoforms expressed in yeast reveals a broad diversity in activity. *Mol. Biotechnol.* **2014**, *56*, 609–620. [CrossRef]
4. Gogos, C.A.; Ginopoulos, P.; Salsa, B.; Apostolidou, E.; Zoumbos, N.C.; Kalfarentzos, F. Dietary omega-3 polyunsaturated fatty acids plus vitamin E restore immunodeficiency and prolong survival for severely ill patients with generalized malignancy. *Cancer* **1998**, *82*, 395–402. [CrossRef]
5. Shahidi, F.; Ambigaipalan, P. Omega-3 polyunsaturated fatty acids and their health benefits. *Annu. Rev. Food Sci. Technol.* **2018**, *9*, 345–381. [CrossRef]
6. Zhao, J.V.; Schooling, C.M. Role of linoleic acid in autoimmune disorders: A mendelian randomisation study. *Ann. Rheum. Dis.* **2019**, *78*, 711–713. [CrossRef]
7. Singh, K.K.; Mridula, D.; Rehal, J.; Barnwal, P. Flaxseed: A potential source of food, feed and fiber. *Crit. Rev. Food Sci. Nutr.* **2011**, *51*, 210–222. [CrossRef]
8. Goyal, A.; Sharma, V.; Upadhyay, N.; Gill, S.; Sihag, M. Flax and flaxseed oil: An ancient medicine & modern functional food. *J. Food Sci. Technol.* **2014**, *51*, 1633–1653.
9. Kajla, P.; Sharma, A.; Sood, D.R. Flaxseed-a potential functional food source. *J. Food Sci. Technol.* **2015**, *52*, 1857–1871. [CrossRef]
10. Yadav, B.; Kaur, V.; Narayan, O.P.; Yadav, S.K.; Kumar, A.; Wankhede, D.P. Integrated omics approaches for flax improvement under abiotic and biotic stress: Current status and future prospects. *Front Plant Sci.* **2022**, *13*, 931275. [CrossRef]
11. Chen, Y.N.; Yang, Q.; Luo, Y.; Shen, Y.J.; Pan, X.L.; Li, L.H.; Li, Z.Q. Ponder on the issue of water resources in the arid region of northwest China. *Arid Land Geogr.* **2012**, *35*, 1–9.
12. Yang, Y.Z.; Kong, Q.; Lim, A.R.Q.; Lu, S.P.; Zhao, H.; Guo, L.; Yuan, L.; Ma, W. Transcriptional regulation of oil biosynthesis in seed plants: Current understanding, applications, and perspectives. *Plant Commun.* **2022**, *3*, 100328. [CrossRef]
13. Turnham, E.; Northcote, D.H. Changes in the activity of acetyl-CoA carboxylase during rape-seed formation. *Biochem. J.* **1983**, *212*, 223–229. [CrossRef]
14. Simcox, P.D.; Garland, W.; DeLuca, V.; Canvin, D.T.; Dennis, D.T. Respiratory pathways and fat synthesis in the developing castor oil seed. *Can. J. Bot.* **1979**, *57*, 1008–1014. [CrossRef]
15. Kozaki, A.; Kamada, K.; Nagano, Y.; Iguchi, H.; Sasaki, Y. Recombinant carboxyltransferase responsive to redox of pea plastidic acetyl-CoA carboxylase. *J. Biol. Chem.* **2000**, *275*, 10702–10708. [CrossRef]
16. Sasaki, Y.; Konishi, T.; Nagano, Y. The compartmentation of acetyl-coenzyme a carboxylase in plants. *Plant Physiol.* **1995**, *108*, 445–449. [CrossRef]
17. Klaus, D.; Ohlrogge, J.B.; Neuhaus, H.E.; Dörmann, P. Increased fatty acid production in potato by engineering of acetyl-CoA carboxylase. *Planta* **2004**, *219*, 389–396. [CrossRef]
18. Amid, A.; Lytovchenko, A.; Fernie, A.; Warren, G.; Thorlby, G. The sensitive to freezing3 mutation of *Arabidopsis thaliana* is a cold-sensitive allele of homomeric acetyl-CoA carboxylase that results in cold-induced cuticle deficiencies. *J. Exp. Bot.* **2012**, *63*, 5289–5299. [CrossRef]
19. Roesler, K.; Shintani, D.; Savage, L.; Boddupalli, S.; Ohlrogge, J. Targeting of the *Arabidopsis* homomeric acetyl-coenzyme a carboxylase to plastids of rapeseeds. *Plant Physiol.* **1997**, *113*, 75–81. [CrossRef]
20. Cui, Y.; Liu, Z.J.; Zhao, Y.P.; Wang, Y.M.; Huang, Y.; Li, L.; Wu, H.; Xu, S.X.; Hua, J.P. Overexpression of heteromeric GhACCase subunits enhanced oil accumulation in upland cotton. *Plant Mol. Biol. Rep.* **2017**, *35*, 287–297. [CrossRef]
21. Ye, Y.; Nikovics, K.; To, A.; Lepiniec, L.; Fedosejevs, E.T.; Van Doren, S.R.; Baud, S.; Thelen, J.J. Docking of acetyl-CoA carboxylase to the plastid envelope membrane attenuates fatty acid production in plants. *Nat. Commun.* **2020**, *11*, 6191. [CrossRef]
22. Yuka, M.; Ken, T.; Junya, M.; Ikuo, N.; Yukio, N.; Yukiko, S. Chloroplast transformation with modified accD operon increases acetyl-CoA carboxylase and causes extension of leaf longevity and increase in seed yield in tobacco. *Plant Cell Physiol.* **2002**, *43*, 1518–1525.
23. Nakkaew, A.; Chotigeat, W.; Eksomtramage, T.; Phongdara, A. Cloning and expression of a plastid-encoded subunit, beta-carboxyltransferase gene (*accD*) and a nuclear-encoded subunit, biotin carboxylase of acetyl-CoA carboxylase from oil palm (*Elaeis guineensis* Jacq.). *Plant Sci.* **2008**, *175*, 497–504. [CrossRef]
24. Basu, S.; Ramegowda, V.; Kumar, A.; Pereira, A. Plant adaptation to drought stress. *F1000 Res.* **2016**, *5*, 1554. [CrossRef]
25. Li, Y.; Han, S.; Sun, X.; Khan, N.U.; Zhong, Q.; Zhang, Z.; Zhang, H.; Ming, F.; Li, Z.; Li, J. Variations in *OsSPL10* confer drought tolerance by directly regulating *OsNAC2* expression and ROS production in rice. *J. Integr. Plant Biol.* **2023**, *65*, 918–933. [CrossRef]
26. El-Fatah, A.A. Comparative study on some flax cultivars. *J. Plant Prod.* **2007**, *32*, 7111–7119.
27. Dubey, S.; Bhargava, A.; Fuentes, F.; Shukla, S.; Srivastava, S. Effect of salinity stress on yield and quality parameters in flax (*Linum usitatissimum* L.). *Not. Bot. Horti Agrobot.* **2020**, *48*, 954–966. [CrossRef]
28. Fofana, B.; Cloutier, S.; Duguid, S.; Ching, J.; Rampitsch, C. Gene expression of stearoyl-ACP desaturase and D12 fatty acid desaturase 2 is modulated during seed development of flax (*Linum usitatissimum*). *Lipids* **2006**, *41*, 705–720. [CrossRef]
29. Heller, K.; Byczyńska, M. The impact of environmental factors and applied agronomy on quantitative and qualitative traits of flax fiber. *J. Nat. Fibers.* **2015**, *12*, 26–38. [CrossRef]

30. Zhu, J.K. Abiotic Stress Signaling and Responses in Plants. *Cell* **2016**, *167*, 313–324. [CrossRef]
31. Zhu, J.K. Salt and drought stress signal transduction in plants. *Ann. Rev. Plant Biol.* **2002**, *53*, 247–273. [CrossRef]
32. Bryant, N.; Lloyd, J.; Sweeney, C.; Myouga, F.; Meinke, D. Identification of nuclear genes encoding chloroplast-localized proteins required for embryo development in *Arabidopsis*. *Plant Physiol.* **2010**, *155*, 1678–1689. [CrossRef]
33. Wang, N.; Lin, Y.; Qi, F.; Xiaoyang, C.; Peng, Z.; Yu, Y.; Liu, Y.; Zhang, J.; Qi, X.; Deyholos, M.; et al. Comprehensive analysis of differentially expressed genes and epigenetic modification-related expression variation induced by saline stress at seedling stage in fiber and oil flax, *Linum usitatissimum* L. *Plants* **2022**, *11*, 2053. [CrossRef]
34. Marchler-Bauer, A.; Bo, Y.; Han, L.; He, J.; Lanczycki, C.J.; Lu, S.; Chitsaz, F.; Derbyshire, M.K.; Geer, R.C.; Gonzales, N.R.; et al. CDD/SPARCLE: Functional classification of proteins via subfamily domain architectures. *Nucleic. Acids. Res.* **2017**, *45*, D200–D203. [CrossRef]
35. Ohlrogge, J.B.; Jaworski, J.G. Regulation of fatty acid synthesis. *Annu. Rev. Plant Physiol. Plant Mol. Biol.* **1997**, *48*, 109–136. [CrossRef]
36. Graham, I.A. Seed storage oil mobilization. *Annu. Rev. Plant Biol.* **2008**, *59*, 115–142. [CrossRef]
37. Li, D.; Jin, C.Y.; Duan, S.W.; Zhu, Y.N.; Qi, S.H.; Liu, K.G.; Gao, C.H.; Ma, H.L.; Zhang, M.; Liao, Y.C.; et al. MYB89 Transcription Factor Represses Seed Oil Accumulation. *Plant Physiol.* **2017**, *173*, 1211–1225. [CrossRef]
38. Thelen, J.J.; Ohlrogge, J.B. Both antisense and sense expression of biotin carboxyl carrier protein isoform 2 inactivates the plastid acetyl-coenzyme A carboxylase in *Arabidopsis thaliana*. *Plant J.* **2002**, *32*, 419–431. [CrossRef]
39. Mu, J.; Tan, H.; Zheng, Q.; Fu, F.; Liang, Y.; Zhang, J.; Yang, X.; Wang, T.; Chong, K.; Wang, X.J.; et al. LEAFY COTYLEDON1 is a key regulator of fatty acid biosynthesis in *Arabidopsis*. *Plant Physiol.* **2008**, *148*, 1042–1054. [CrossRef]
40. Jung, S.H.; Kim, R.J.; Kim, K.J.; Lee, D.H.; Suh, M.C. Plastidial and mitochondrial malonyl CoA-ACP malonyltransferase is essential for cell division and its overexpression increases storage oil content. *Plant Cell Physiol.* **2019**, *60*, 1239–1249. [CrossRef]
41. Shimakata, T.; Stumpf, P.K. Isolation and function of spinach leaf beta-ketoacyl-[acyl-carrier-protein] synthases. *Proc. Natl. Acad. Sci. USA* **1982**, *79*, 5808–5812. [CrossRef]
42. Wu, G.Z.; Xue, H.W. *Arabidopsis* b-ketoacyl-[acyl carrier protein] synthase I is crucial for fatty acid synthesis and plays a role in chloroplast division and embryo development. *Plant Cell* **2010**, *22*, 3726–3744. [CrossRef]
43. Kachroo, A.; Shanklin, J.; Whittle, E.; Lapchyk, L.; Hildebrand, D.; Kachroo, P. The *Arabidopsis* stearoyl-acyl carrier protein-desaturase family and the contribution of leaf isoforms to oleic acid synthesis. *Plant Mol. Biol.* **2007**, *63*, 257–271. [CrossRef]
44. Shah, S.; Xin, Z.; Browse, J. Overexpression of the FAD3 desaturase gene in a mutant of *Arabidopsis*. *Plant Physiol.* **1997**, *114*, 1533–1539. [CrossRef]
45. Choudhary, A.K.; Mishra, G. Functional characterization and expression profile of microsomal *FAD2* and *FAD3* genes involved in linoleic and alpha-linolenic acid production in *Leucas cephalotes*. *Physiol. Mol. Biol. Plants.* **2021**, *27*, 1233–1244. [CrossRef]
46. Wang, J.J.; Liu, Z.J.; Liu, H.; Peng, D.S.; Zhang, J.P.; Chen, M.X. *Linum usitatissimum FAD2A* and *FAD3A* enhance seed polyunsaturated fatty acid accumulation and seedling cold tolerance in *Arabidopsis thaliana*. *Plant Sci.* **2021**, *311*, 111014. [CrossRef]
47. Xue, P.; Fred, Y.P.; Randall, W. Genome-wide analysis of *PHOSPHOLIPID: DIACYLGLYCEROL ACYLTRANSFERASE* (*PDAT*) genes in plants reveals the eudicot-wide *PDAT* gene expansion and altered selective pressures acting on the core eudicot PDAT paralogs. *Plant Physiol.* **2015**, *167*, 887–904.
48. Pan, J.; Wang, H.; Hu, Y.; Yu, D. *Arabidopsis* VQ18 and VQ26 proteins interact with ABI5 transcription factor to negatively modulate ABA response during seed germination. *Plant J.* **2018**, *95*, 529–544. [CrossRef]
49. Luo, X.; Li, C.; He, X.; Zhang, X.; Zhu, L. ABA signaling is negatively regulated by *GbWRKY1* through JAZ1 and ABI1 to affect salt and drought tolerance. *Plant Cell Rep.* **2020**, *39*, 181–194. [CrossRef]
50. Ali, F.; Qanmber, G.; Li, F.; Wang, Z. Updated role of ABA in seed maturation, dormancy, and germination. *J. Adv. Res.* **2021**, *31*, 199–214. [CrossRef]
51. Li, Y.; Zhou, J.; Li, Z.; Qiao, J.; Quan, R.; Wang, J.; Huang, R.; Qin, H. SALT and ABA RESPONSE ERF1 improves seed germination and salt tolerance by repressing ABA signaling in rice. *Plant Physiol.* **2022**, *189*, 1110–1127. [CrossRef]
52. Iuchi, S.; Kobayashi, M.; Taji, T.; Naramoto, M.; Seki, M.; Kato, T.; Tabata, S.; Kakubari, Y.; Yamaguchi-Shinozaki, K.; Shinozaki, K. Regulation of drought tolerance by gene manipulation of 9-cis-epoxycarotenoid dioxygenase, a key enzyme in abscisic acid biosynthesis in *Arabidopsis*. *Plant J.* **2001**, *27*, 325–333. [CrossRef]
53. Ruggiero, B.; Koiwa, H.; Manabe, Y.; Quist, T.M.; Inan, G.; Saccardo, F.; Joly, R.J.; Hasegawa, P.M.; Bressan, R.A.; Maggio, A. Uncoupling the effects of abscisic acid on plant growth and water relations. Analysis of *sto1/nced3*, an abscisic acid-deficient but salt stress-tolerant mutant in *Arabidopsis*. *Plant Physiol.* **2004**, *136*, 3134–3147. [CrossRef]
54. Barrero, J.M.; Rodríguez, P.L.; Quesada, V.; Piqueras, P.; Ponce, M.R.; Micol, J.L. Both abscisic acid (ABA)-dependent and ABA-independent pathways govern the induction of *NCED3*, *AAO3* and *ABA1* in response to salt stress. *Plant Cell Environ.* **2010**, *29*, 2000–2008. [CrossRef] [PubMed]
55. Sato, H.; Takasaki, H.; Takahashi, F.; Suzuki, T.; Iuchi, S.; Mitsuda, N.; Ohme-Takagi, M.; Ikeda, M.; Seo, M.; Yamaguchi-Shinozaki, K.; et al. *Arabidopsis thaliana* NGATHA1 transcription factor induces ABA biosynthesis by activating *NCED3* gene during dehydration stress. *Proc. Natl. Acad. Sci. USA* **2018**, *115*, E11178. [CrossRef]
56. Seo, M.; Aoki, H.; Koiwai, H.; Kamiya, Y.; Nambara, E.; Koshiba, T. Comparative studies on the *Arabidopsis* aldehyde oxidase (AAO) gene family revealed a major role of *AAO3* in ABA biosynthesis in seeds. *Plant Cell Physiol.* **2004**, *45*, 1694–1703. [CrossRef]

57. Shi, X.; Tian, Q.; Deng, P.; Zhang, W.; Jing, W. The rice aldehyde oxidase *OsAO3* gene regulates plant growth, grain yield, and drought tolerance by participating in ABA biosynthesis. *Biochem. Biophys. Res. Commun.* **2021**, *548*, 189–195. [CrossRef]
58. Lin, J.H.; Yu, L.H.; Xiang, C.B. ARABIDOPSIS NITRATE REGULATED 1 acts as a negative modulator of seed germination by activating ABI3 expression. *New Phytol.* **2020**, *225*, 835–847. [CrossRef]
59. Morris, P.C.; Kumar, A.; Bowles, D.J.; Cuming, A.C. Osmotic stress and abscisic acid induce expression of the wheat Em genes. *FEBS J.* **2010**, *190*, 625–630. [CrossRef]
60. Li, Y.J.; Fang, Y.; Fu, Y.R.; Huang, J.G.; Wu, C.A.; Zheng, C.C. NFYA1 is involved in regulation of postgermination growth arrest under salt stress in *Arabidopsis*. *PLoS ONE* **2013**, *8*, e61289. [CrossRef]
61. Zhao, X.; Dou, L.R.; Gong, Z.Z.; Wang, X.F.; Mao, T.L. BES1 hinders ABSCISIC ACID INSENSITIVE5 and promotes seed germination in *Arabidopsis*. *New Phytol.* **2019**, *221*, 908–918. [CrossRef] [PubMed]
62. Chen, M.; Zhang, B.; Li, C.; Kulaveerasingam, H.; Chew, F.T.; Yu, H. TRANSPARENT TESTA GLABRA1 regulates the accumulation of seed storage reserves in *Arabidopsis*. *Plant Physiol.* **2015**, *169*, 391–402. [CrossRef] [PubMed]
63. Clough, S.J.; Bent, A.F. Floral dip: A simplified method for Agrobacterium-mediated transformation of *Arabidopsis thaliana*. *Plant J.* **1998**, *16*, 735–743. [CrossRef] [PubMed]
64. Chen, M.X.; Wang, Z.; Zhu, Y.; Li, Z.L.; Hussain, N.; Xuan, L.J.; Guo, W.L.; Zhang, G.P.; Jiang, L.X. The effect of TRANSPARENT TESTA2 on seed fatty acid biosynthesis and tolerance to environmental stresses during young seedling establishment in *Arabidopsis*. *Plant Physiol.* **2012**, *160*, 1023–1036. [CrossRef] [PubMed]

plants

MDPI

Article

Comparison of Salt Stress Tolerance among Two Leaf and Six Grain Cultivars of *Amaranthus cruentus* L.

Adrien Luyckx, Stanley Lutts and Muriel Quinet *

Groupe de Recherche en Physiologie Végétale, Earth and Life Institute-Agronomy, Université Catholique de Louvain, 1348 Louvain-la-Neuve, Belgium; adrien.luyckx@uclouvain.be (A.L.); stanley.lutts@uclouvain.be (S.L.)
* Correspondence: muriel.quinet@uclouvain.be

Abstract: Amaranths (*Amaranthus* L.) are multi-use crop species renowned for their nutritional quality and their tolerance to biotic and abiotic stresses. Since the soil salinity of croplands is a growing problem worldwide, we tested the salinity tolerance of six grain and two leaf cultivars of *Amaranthus cruentus* L. The plants were grown for 53 days under hydroponic conditions at 0, 50 and 100 mM NaCl. We investigated the growth rate, photosynthetic activity, mineral content, pigments and biochemical compounds involved in oxidative stress. Although 100 mM NaCl always decreased biomass production, we highlighted Don Leon and K91 as tolerant cultivars under moderate salt stress (50 mM NaCl). Under salinity, sodium accumulated more in the shoots than in the roots, particularly in the stems. Sodium accumulation in the plants decreased the net photosynthetic rate, transpiration rate and stomatal conductance but increased water use efficiency, and it decreased chlorophyll, betalain and polyphenol content in the leaves. It also decreased the foliar content of calcium, magnesium and potassium but not the iron and zinc content. The physiological parameters responded differently to sodium accumulation depending on the cultivar, suggesting a different relative importance of ionic and osmotic phases of salt stress among cultivars. Our results allowed us to identify the morpho-physiological traits of the cultivars with different salt tolerance levels.

Keywords: abiotic stress; amaranth; orphan crop; plant physiology; pseudocereal; salinity

Citation: Luyckx, A.; Lutts, S.; Quinet, M. Comparison of Salt Stress Tolerance among Two Leaf and Six Grain Cultivars of *Amaranthus cruentus* L. *Plants* **2023**, *12*, 3310. https://doi.org/10.3390/plants12183310

Academic Editors: Mingxun Chen, Lixi Jiang, Yuan Guo and Sylvia Lindberg

Received: 20 July 2023
Revised: 19 August 2023
Accepted: 11 September 2023
Published: 19 September 2023

1. Introduction

In a context of the increasing food demand and more adverse biotic and abiotic conditions for agricultural production, it is necessary to rely on robust crops that are able to thrive in stressful conditions, while limiting agriculture's contribution to global change [1]. Currently, soil salinity is one of the main abiotic stress threatening agricultural production worldwide [1].

Soil salinization, caused by an accumulation of soluble salts (mainly NaCl and Na_2SO_4) in the upper horizon of soil, is an expanding agronomic constraint for food production, especially in Asia, Africa, South America and Australasia in arid regions where precipitations are too low to leach excessive salts [2,3]. A soil is saline when its electrical conductivity is higher than 4 $dS \cdot m^{-1}$ and sodic when the exchangeable sodium percentage is higher than 6% [4]. Salinization can be primary when it is of natural origin (weathering of saline rocks, saline bedrock, atmospheric deposition) or secondary when it is human-induced. In the latter case, saline or sodic soils can be the result of bad irrigation practices (e.g., with brackish water), sea level rising or an excessive use or bad management of mineral fertilizers [5]. Saline and sodic soils are estimated to cover more than 800 Mha, but we are lacking recent data [6,7]. Most of the salt-contaminated topsoils (>400 Mha, 0–30 cm) are saline (85%), whereas a much lower proportion is sodic or saline–sodic (15%) [6]. Salt-affected subsoils are twice as common as salinized topsoils (>800 Mha) and a higher proportion of them are sodic or saline–sodic (38%) [6]. Salt-affected soils are in expansion in part because of climate change and this issue is particularly exacerbating under arid and semi-arid climates and on irrigated croplands [5].

Salt stress in plants occurs in two phases, namely the osmotic and the ionic phases [8]. First, the osmotic stress is driven by the low hydric potential of saline soils that decreases water uptake. Osmotic stress has a direct effect on plant growth. Then, the ionic stress is caused by the toxicity of salt ions entering the plant tissues and the competitive effect with important nutrients. Tolerance mechanisms consist of, depending on the species, salt exclusion or salt accumulation involving tissue partitioning and subcellular compartmentation, and the synthesis of organic osmolytes for osmotic adjustment [8–10]. These tolerance mechanisms consume energy, which impedes growth and development [11].

Most major crops, which provide most of the world's calories, are salt-sensitive even though extensive research is in progress to increase their salt tolerance [12–14]. Therefore, the use of so-called "orphan" or "indigenous" crops and wild crop relatives (WCR) has been flourishing in the last few decades [15–18]. It is a large research field with different approaches, including the introgression of genes from orphan crops or WCR to major crops, breeding of under-domesticated crops or de novo domestication of wild species [19–22]. This array of approaches is promising for breeding for salinity tolerance [22,23]. The amaranth genus (*Amaranthus* spp.) contains several orphan crops and WCR that has gained increasing attention in the past few decades because of its tolerance to abiotic and biotic stress, including salinity.

Amaranthus is a subcosmopolitan genus of 50–70 herbaceous plant species, most of which are annual plants, in the Amaranthaceae family [24]. Several species of amaranths are used as crops, either as leafy vegetables for their nutritious leaves or as pseudocereals for their quinoa-like seeds rich in high-quality proteins. Although some species were staple crops in several Mesoamerican civilizations, they fell into disuse for several centuries after european colonization of the Americas. However, a surge in interest for amaranths has arisen in the last few decades because of their nutritive qualities [25–27] and their tolerance to several biotic and abiotic stresses. Among grain species, studies have been conducted on salinity [28–34] and drought [35,36] in *A. cruentus*, on salinity in *A. caudatus* [37,38] and salt and drought in *A. hypochondriacus* [39,40]. The plant response to abiotic stress has also been investigated in some leaf species, such as *A. tricolor* [41–43] or *A. hybridus* and *A. albus* [44]. Most of these studies show that amaranth is highly tolerant to drought and salinity, particularly at the vegetative stage. These works report variability among species and among cultivars of each species. It has been shown that gas exchange regulation, antioxidant defense, ion transporter regulation and osmotic adjustment are involved in the stress response. Amaranths express the NAD malic-type C_4 photosynthetic pathway, which makes them more competitive in warm and/or dry environments by means of an higher water use efficiency compared to C_3 plants [45]. *Amaranthus cruentus* L. (red amaranth), domesticated by the Aztecs in Mesoamerica, is used either for leaves (in Africa and South-East Asia but also in America [46]) or for seed production (mainly in America and Asia, but also in Africa [47]), usually with distinct cultivars [48–50]. However, this species, along with the two other grain amaranths (*A. hypochondriacus* L. and *A. caudatus* L.), are promising as dual-use crops, with leaves and seeds harvested on the same plants [51–53]. In these three 'grain' species, the nutritional quality of both leaves and seeds is high. Seeds are rich in proteins, more than most true cereals, and minerals [54,55], while leaves are also rich in proteins, minerals and vitamins [56–58]. Grain amaranth can produce seed yields of 2000–3500 $kg \cdot ha^{-1}$ [59], whereas leaf amaranth can produce several dozens of tons per hectare of fresh leaves and young stems eaten as vegetables [60]. Despite their tolerance to salinity and their exceptional agronomic and nutritional value, amaranths have remained understudied crops. Indeed, the physiological mechanisms underlying abiotic stress tolerance are poorly known.

To deepen our understanding of the salt response in *Amaranthus cruentus* at the vegetative stage, a screening of the salt tolerance of six grain and two leaf cultivars was conducted under hydroponic conditions at moderate and strong levels of salt stress (50 mM and 100 mM NaCl, respectively [61]). Plant growth and photosynthetic activity were monitored. Na and K were quantified in leaves, stem and roots. In addition, other mineral contents (Ca, Fe, Mg and Zn) were determined in leaves since Na accumulation in plants is known to affect mineral nutrition [41]. The pigments (chlorophylls, betalains) and biochemical compounds involved in oxidative stress (malondialdehyde, polyphenols, flavonoids and ascorbate) were investigated in the leaves due to their importance in stress response [62]. The aims were to identify (1) contrasted cultivars regarding their tolerance to salt stress and (2) the main physiological mechanisms explaining variability in salt tolerance among leaf and grain cultivars of *A. cruentus*.

2. Results

Plants were subjected to 0, 50 and 100 mM NaCl for 53 days. The eight cultivars differed by their origin, morphology, color and food purpose (Table S1). The plants tolerated the salt stress well. Indeed, the mortality rate was low. Complete senescence was observed at 100 mM NaCl only in two plants of Montana 5 and two plants of Don Leon, for a total of 10 plants per cultivar and condition.

2.1. Biomass Production

Under the control conditions, the mean leaf dry weight of all cultivars was 1.79 ± 0.43 g. The most productive cultivars were Alegria Disciplinada and Don Armando, whereas the less productive ones were K91, Montana 5 and Locale (Figure 1). Total dry weight was strongly correlated with leaf dry weight (r = 0.98), with leaves accounting for 62.5 ± 3.8% of the total dry biomass, whereas stem and roots accounted for 28.6 ± 4% and 8.8 ± 3.5%, respectively.

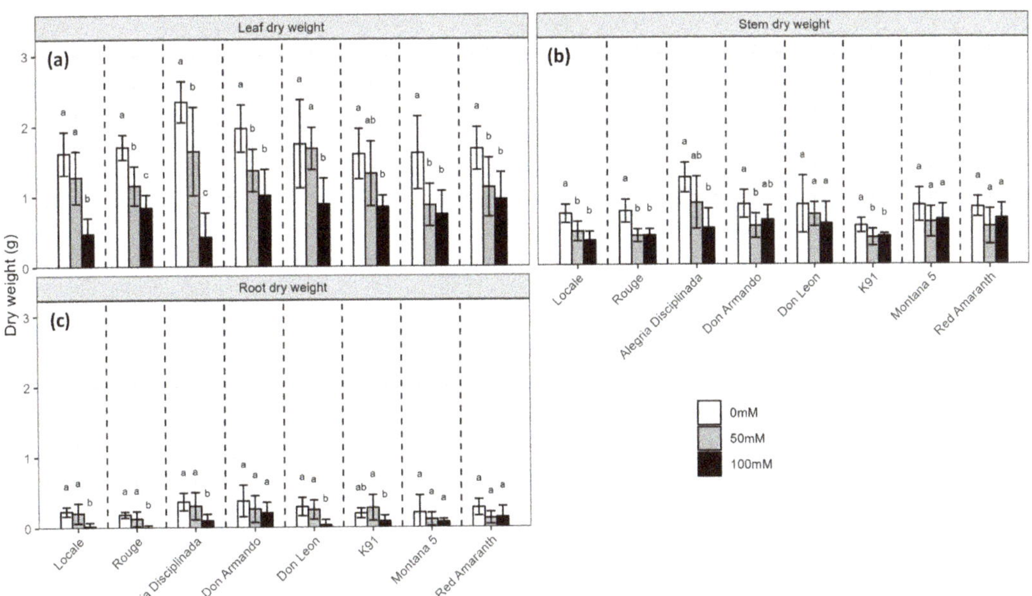

Figure 1. Effect of salinity (0, 50 and 100 mM NaCl) on the dry biomass production of eight *A. cruentus* cultivars after 53 days of growth. (**a**) Leaf dry weight; (**b**) stem dry weight; (**c**) root dry weight. Treatments followed by different letters for the same cultivar are significantly different ($p < 0.05$).

Salt decreased the dry weight of all organs ($p < 0.001$ for leaves, stem and roots) in all cultivars (Figure 1, Table S2). The leaf dry weight decreased by on average 57% at 100 mM NaCl (Figure 1a). The strongest effect was observed on Alegria Disciplinada (−82%), which was the most productive without salt. Red Amaranth and K91 were the least affected, with less than 50% of leaf weight loss. At 50 mM NaCl, salt caused an average decrease in leaf biomass of 27%, but with some variability among cultivars. Don Leon leaf dry weight was hardly affected by 50 mM NaCl, whereas Montana 5 leaf production dropped by 46%. Moreover, the leaf production of Don Armando, K91, Montana 5 and Red Amaranth was similar at 50 and 100 mM NaCl.

Stem dry weight decreased significantly at 50 mM NaCl only in Locale, Rouge, K91 and Don Armando, whereas it was not affected at 100 mM NaCl in Montana 5, Don Leon, Don Armando and Red Amaranth (Figure 1b). Root dry weight rarely decreased at 50 mM NaCl, but often strongly at 100 mM NaCl, by more than 70% in Alegria Disciplinada and even more than 80% in Don Leon (Figure 1c).

Salt stress slightly influenced the stem and root water content (p = 0.004 and p = 0.049, respectively), which were 95 ± 1% and 97 ± 3% in control conditions, respectively (Tables S2 and S3). The leaf water content was 88 ± 1% in control conditions. While no effect was observed at 50 mM NaCl, salt substantially decreased the leaf water content at 100 mM NaCl, which dropped to 81 ± 13%. This decrease was only significant in Montana 5 (−12.3%), Alegria Disciplinada (−16.2%) and Don Leon (−17.9%).

Based on the DW, the salt tolerance index (STI) was calculated (Table 1). At 50 mM NaCl, the most sensitive cultivar to salt was Montana 5, whereas Don Leon was the most tolerant. Cultivars behaved substantially differently at 100 mM NaCl, since Alegria Disciplinada was the most sensitive and Red Amaranth the most tolerant. Indeed, the tolerance index did not demonstrate any correlation at 50 mM and 100 mM NaCl (r = −0.11, p = 0.80).

Table 1. Salt tolerance index (STI) of the eight *A. cruentus* cultivars at 50 and 100 mM NaCl.

Cultivar	STI_{50mM}	STI_{100mM}
Locale	0.759	0.327
Rouge	0.639	0.476
Montana 5	0.595	0.546
Don Leon	0.913	0.521
K91	0.841	0.569
Alegria Disciplinada	0.717	0.264
Don Armando	0.681	0.580
Red Amaranth	0.631	0.617

2.2. Sodium Distribution in the Plant Organs

Salt stress significantly increased Na concentrations in the leaves (p < 0.001), stems (p < 0.001) and roots (p < 0.001) but no differences were observed among different cultivars (Figure 2, Table S4). The sodium content in stems and roots was similar between 50 mM and 100 mM NaCl, whereas the accumulation was proportional to the stress intensity in leaves, especially in Locale, Rouge and Alegria Disciplinada. Salt stress caused an accumulation of Na in leaves up to 5.45 ± 2.74 $mg{\cdot}g^{-1}$ DW in Montana 5 at 50 mM NaCl and up to 7.91 ± 1.61 $mg{\cdot}g^{-1}$ DW at 100 mM NaCl in the same cultivar (Figure 2a). Rouge had the lowest accumulation of Na in leaves at 50 mM NaCl with 1.99 ± 0.36 $mg{\cdot}g^{-1}$ DW; at 100 mM NaCl, the lowest accumulation was observed in K91 (3.95 ± 2.43 $mg{\cdot}g^{-1}$ DW). Sodium accumulated in stems at a higher magnitude than in leaves (up to 8.38 ± 1.07 $mg{\cdot}g^{-1}$ DW at 50 mM NaCl and up to 9.78 ± 1.17 $mg{\cdot}g^{-1}$ DW at 100 mM NaCl, in both cases in Don Leon, Figure 2b). In contrast to stems, Na accumulated in roots at lower concentrations than in leaves (up to 4.90 ± 1.23 $mg{\cdot}g^{-1}$ DW in K91 at 50 mM NaCl and up to 4.17 ± 0.60 $mg{\cdot}g^{-1}$ DW in Red Amaranth at 100 mM NaCl, Figure 2c).

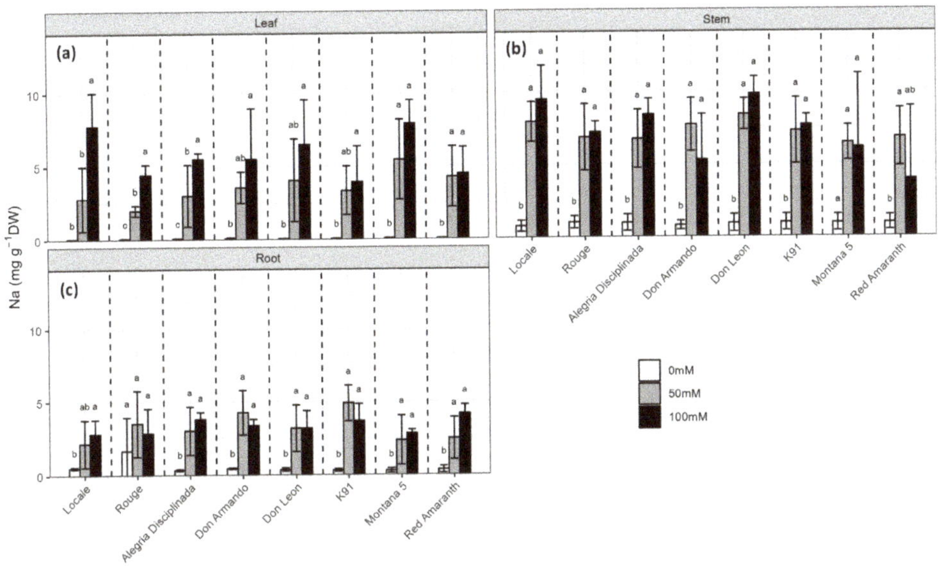

Figure 2. Effect of salinity (0, 50 and 100 mM NaCl) on the sodium content in (**a**) leaves; (**b**) stems and (**c**) roots of the eight *A. cruentus* cultivars. Treatments followed by different letters for the same cultivar are significantly different ($p < 0.05$).

2.3. *Overview of the Physiological Response of the Cultivars to Salt Stress*

In addition to plant growth, salinity affected the physiology of the eight cultivars. In order to identify the main physiological parameters involved in salt tolerance in *A. cruentus* and to differentiate the cultivars, principal component analysis (PCA) and correlation plots were used, as shown in Figures 3 and 4.

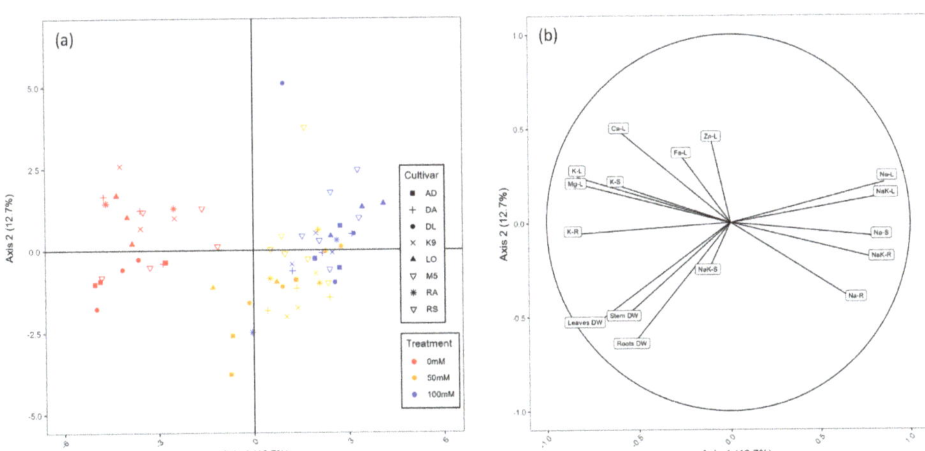

Figure 3. Principal component analysis (PCA) of the plant growth and mineral content of eight *A. cruentus* cultivars exposed to 0 mM, 50 mM and 100 mM NaCl. (**a**) Individual plot showing the eight cultivars position (LO, Locale; RO, Rouge; AD, Alegria Disciplinada; DA, Don Armando; DL, Don Leon; K9, K91; RA, Red Amaranth; M5, Montana 5) in the three salt treatments (red, 0 mM; yellow, 50 mM; blue, 100 mM). (**b**) Variable plot showing correlations between mineral content and biomass data (DW, dry weight; -L, leaf; -S, stem; -R, root).

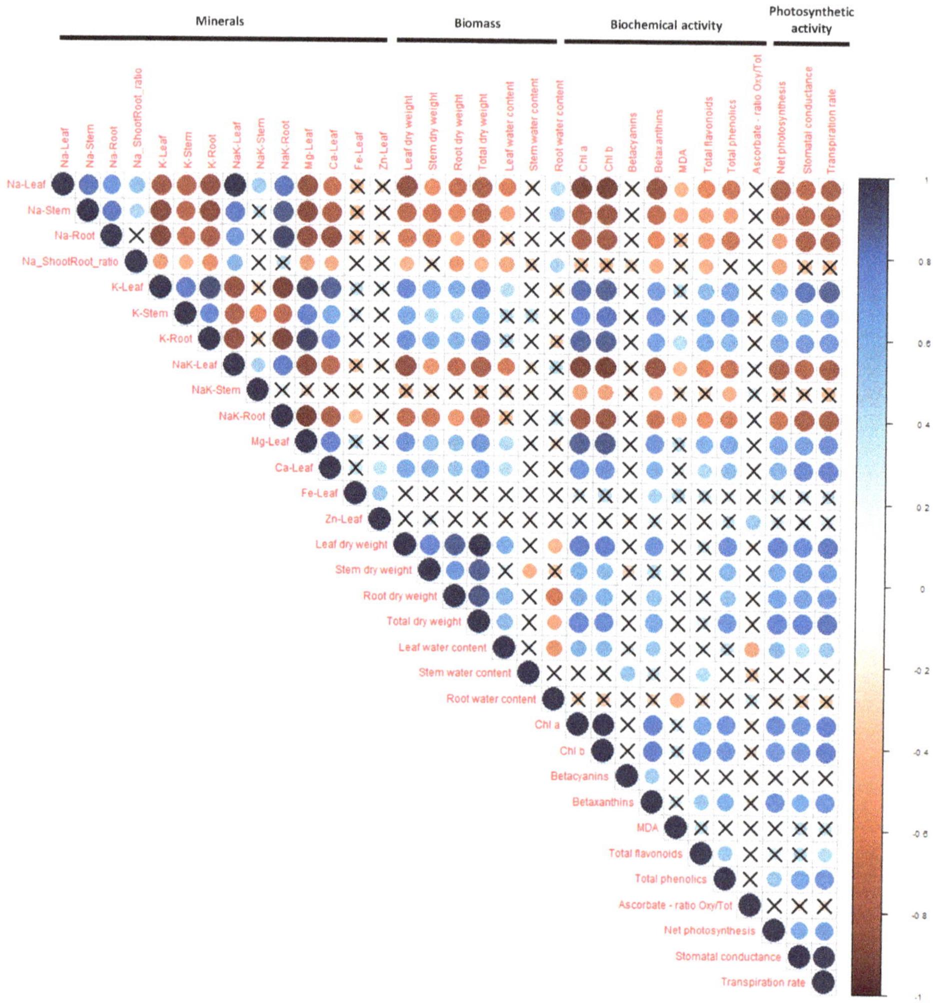

Figure 4. Correlation graph of all measured parameters, grouped in four categories: mineral content, biomass production, pigments, biochemical activity in leaves and photosynthetic activity. Non-significant ($p < 0.05$) correlations are crossed out. Negative correlations are colored in shades of red, whereas positive correlations are in blue (see the color legend on the right). Na_Shoot and Root_ratio, ratio between the quantity of sodium in the shoot to the quantity in the roots; NaK, Na/K ratio; MDA, malondialdehyde.

Axis 1 and axis 2 of the PCA explained 55.4% of the variance (Figure 3). Axis 1 of the PCA separated plants in control conditions from those exposed to salt (Figure 3a). The sodium contents in leaves, stem and roots and Na/K ratios in leaves and roots were good positive predictors of salt-treated plants, whereas K and Mg content were negatively correlated with Na content (Figure 3b). Axis 2 was mainly explained by growth parameters. In contrast to NaCl treatments, there was no clear discrimination between cultivars.

Sodium accumulation in the plants affected several parameters, as shown on the correlation matrix (Figure 4). The sodium content in all organs (leaves, stems and roots) was negatively correlated with K content in all organs, Mg and Ca content in leaves, biomass of all organs, pigments (chlorophylls and betaxanthins but not betacyanins), photosynthetic activity (net photosynthetic rate, transpiration rate and stomatal conductance) and slightly with polyphenols in leaves. The Na/K ratio in leaves and roots, but not in stems, was also negatively correlated with all these parameters.

The impacted physiological, biochemical and mineral parameters will be further analyzed below.

2.4. Mineral Content

Salt decreased the K content in all organs ($p < 0.001$), but similarly decreased the content at 50 mM and 100 mM NaCl (Figure 5, Table S4). However, this decrease was proportionally higher in stems and roots compared to leaves (Figure 5). In leaves, the highest K reduction was observed in Red Amaranth (−59%) while the lowest was observed in Don Leon (−47%) (Figure 5a). In stems, the reduction in K content ranged from −47% (in Alegria Disciplinada) to −93% (in Red Amaranth), while it ranged from −39% (in K91) to −74% (in Locale and Don Leon) in roots (Figure 5b,c). As a result, salt stress caused an increase in the Na/K ratio in all organs (Figure 5d–f, Table S4). This increase was similar among the cultivars for leaves and roots but depended on the cultivar in stems (Figure 5d–f, Table S4).

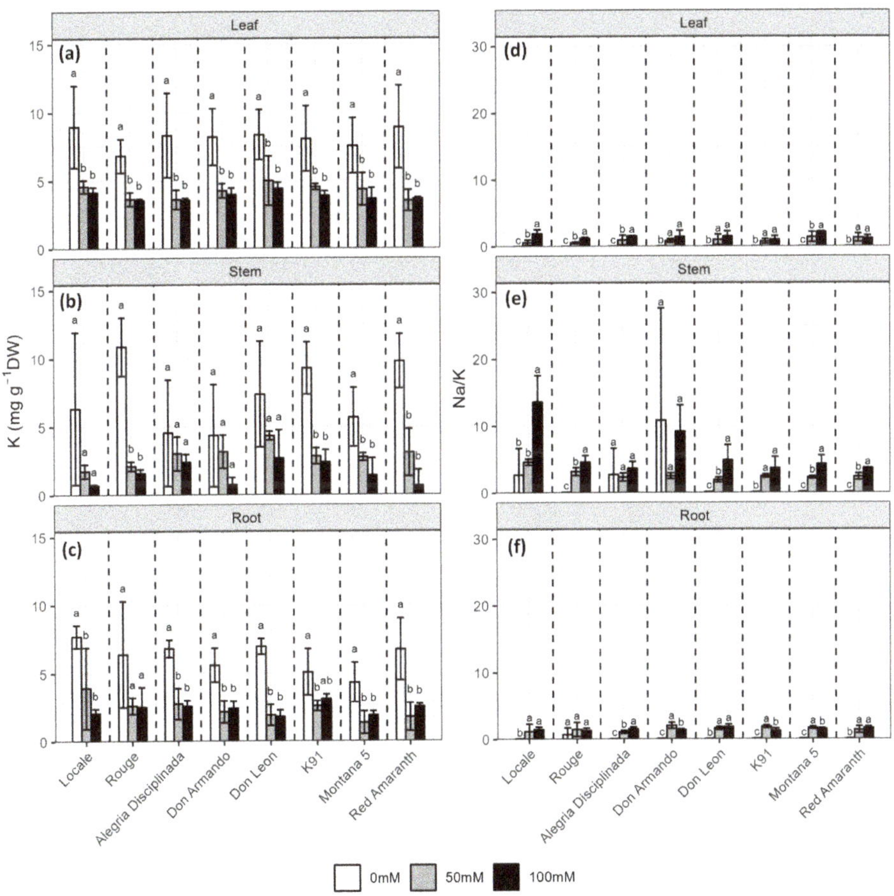

Figure 5. Effect of salinity (0, 50 and 100 mM NaCl) on the potassium content in (**a**) leaves; (**b**) shoot and (**c**) roots and the Na/K ratio in (**d**) leaves; (**e**) shoot and (**f**) roots of the eight *A. cruentus* cultivars. Treatments followed by different letters for the same cultivar are significantly different ($p < 0.05$).

Ca, Fe, Mg and Zn contents were measured in leaves only (Figure 6). A decrease in the Ca content was observed in response to salinity ($p < 0.001$, Figure 6a, Table S4), except in Montana 5. Magnesium content in the leaves of stressed plants decreased by more than 50% compared to the control plants ($p < 0.001$), but again there were no differences between 50 and 100 mM NaCl (Figure 6b, Table S4). Neither Fe ($p = 0.24$) nor Zn ($p = 0.87$) contents were affected by salinity (Figure 6c,d, Table S4).

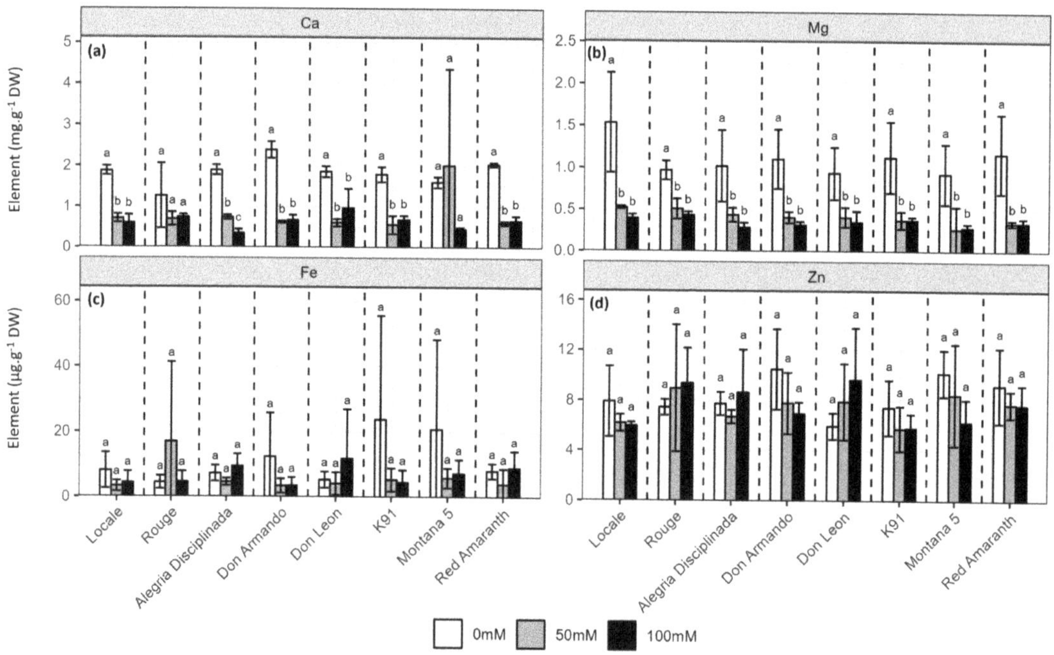

Figure 6. Effect of salinity (0, 50 and 100 mM NaCl) on the (**a**) calcium, (**b**) magnesium, (**c**) iron and (**d**) zinc contents in leaves of the eight *A. cruentus* cultivars. Treatments followed by different letters for the same cultivar are significantly different ($p < 0.05$).

2.5. Photosynthetic Activity in Relation to Sodium Accumulation in Leaves

Salt decreased net photosynthesis (A, $p = 0.0015$), stomatal conductance (g_s, $p < 0.001$) and net transpiration (E, $p < 0.001$), whereas it increased instantaneous water use efficiency (instWUE, $p < 0.001$) (Table S5). Figure 7 shows the photosynthetic parameters in relation to the sodium content in leaves.

The decrease in A was not always linked to Na content in the leaves, depending on the cultivar (Figure 7a). In Montana 5, the decrease in A was proportional to Na accumulation, whereas in most cultivars (Alegria Disciplinada, Don Armando, K91, Rouge), a sharp decrease occurred between 50 mM and 100 mM, despite the modest accumulation of Na in some cultivars (particularly in K91). In Red Amaranth, A was higher at 100 mM compared to 50 mM. In Locale and Don Leon, despite a significant accumulation of Na in the leaves at 100 mM compared to 50 mM NaCl, no decrease in A was observed.

The plant response was cultivar-dependent for E and g_s (Figure 7b,c, Table S5). Both parameters strongly correlated (r = 0.976). They decreased proportionally with salt accumulation in Don Armando, Don Leon and Montana 5, whereas they did not differ much between 50 and 100 mM NaCl in Alegria Disciplinada, Locale and Rouge. A sharp decrease in g_s and E, despite a modest accumulation of Na between 50 and 100 mM NaCl, occurred in Red Amaranth and K91.

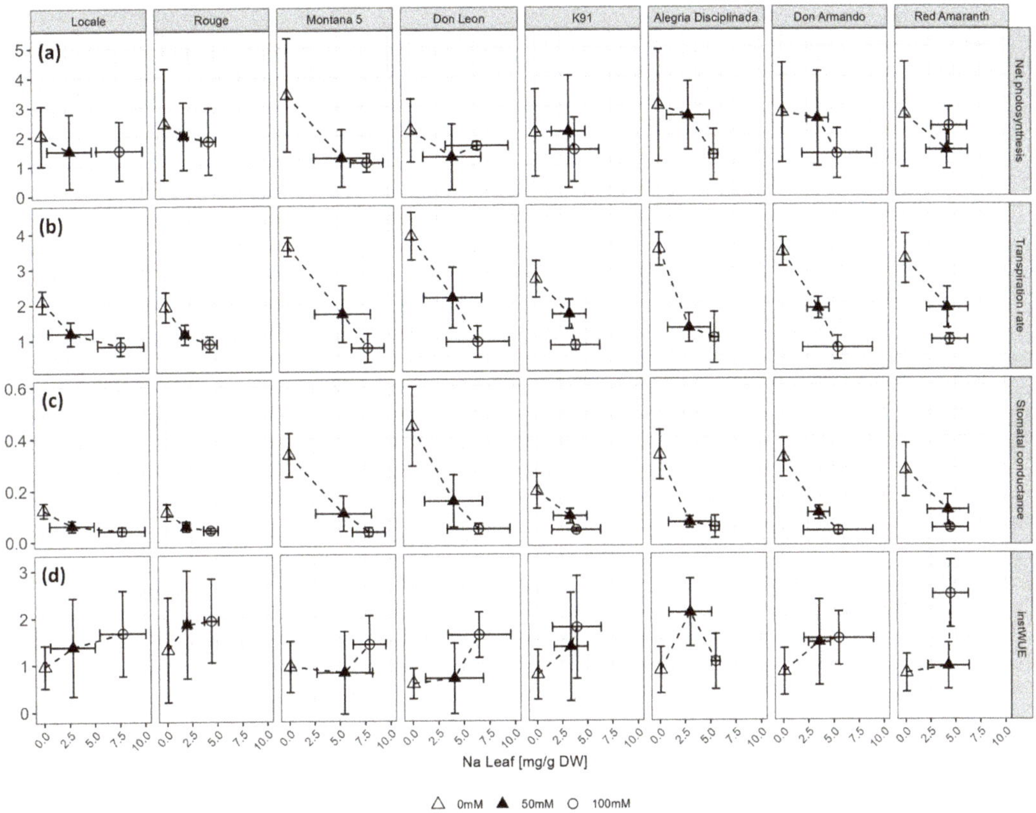

Figure 7. Response of photosynthetic activity to foliar Na accumulation. (**a**) Net photosynthesis (μmol CO_2 m^{-2} s^{-1}); (**b**) transpiration rate (mmol H_2O m^{-2} s^{-1}); (**c**) stomatal conductance (mmol H_2O m^{-2} s^{-1}); (**d**) intrinsic water use efficiency (μmol CO_2 mmol H_2O^{-1}).

The response of instWUE to salt was less conspicuous compared to A, E and g_s (Figure 7). Water use efficiency increased with salt accumulation in most of the cultivars, but only at 100 mM NaCl in Don Leon (Figure 7d). The increase was modest in Locale and Montana 5, whereas it was more marked in Rouge, K91 and Red Amaranth. In Alegria Disciplinada and Don Armando, instWUE increased at 50 mM NaCl but decreased at 100 mM and showed a high standard deviation value.

2.6. Biochemical Compound Contents in Relation to Sodium Accumulation in Leaves

Pigments (chlorophyll *a*, chlorophyll *b*, betaxanthins and betacyanins) and oxidative stress-related compounds (malondialdehyde (MDA), total flavonoids, total phenolics, ascorbate) were quantified (Figures 8 and 9, Table S6). Betacyanins, MDA and ascorbate were not affected by salt stress, while flavonoids ($p = 0.018$) were only slightly affected (Table S7). In contrast, the concentrations of chlorophyll *a* ($p < 0.001$), chlorophyll *b* ($p < 0.001$), betaxanthins ($p < 0.001$), and total phenolics ($p < 0.001$) strongly decreased with salt stress (Table S7). Since the content of chlorophyll *a* and *b* strongly correlated with one another ($r = 0.938$), only chlorophyll *a* is presented in Figure 8 and the data of chlorophyll *b* are shown in Table S6.

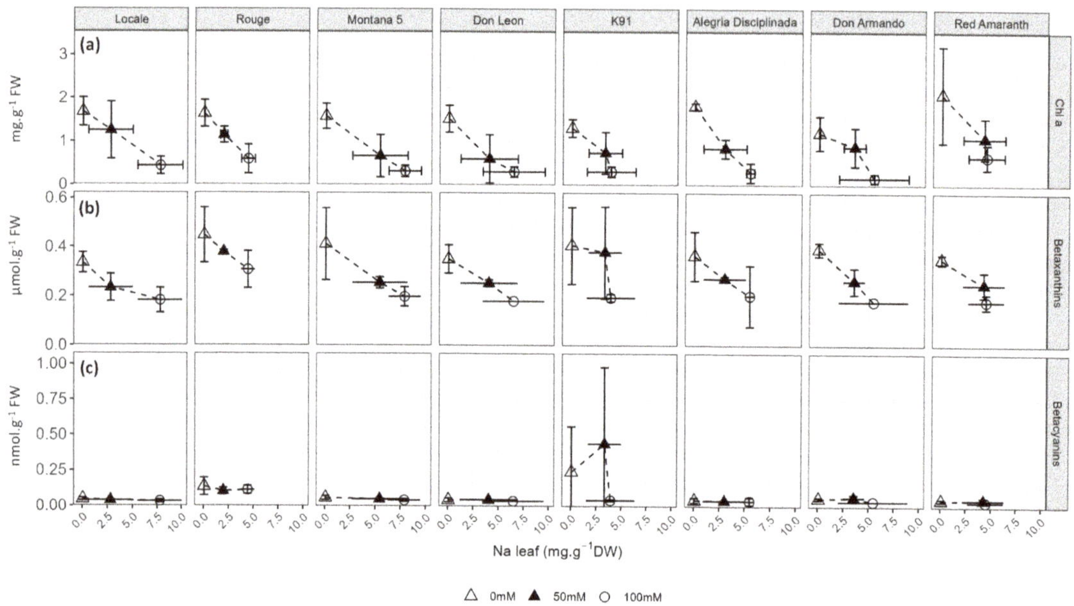

Figure 8. Response of foliar (**a**) chlorophyll *a* (mg·g^{-1} FW), (**b**) betaxanthins (µmol·g^{-1} FW); and (**c**) betacyanins (nmol·g^{-1} FW) to foliar sodium accumulation.

The content in chlorophylls was strongly reduced by salt in all cultivars at 100 mM NaCl, but only in Alegria Disciplinada, Don Leon and Montana 5 at 50 mM NaCl (Figure 8a, Table S6). Salinity decreased the betaxanthin content by more than 50% in Don Armando, K91 and Montana (Figure 8a). As a result, the chlorophyll and betaxanthin contents in leaves regularly decreased with Na content increase in most cultivars, with the exception of K91, Don Armando and Red Amaranth (Figure 8a,b). In contrast, the concentration of betacyanins was not affected by the Na concentration in the leaves, except in K91 (Figure 8c).

Although it was not affected by salt stress and did not significantly vary among cultivars (Table S7), MDA response to sodium accumulation in the leaves was cultivar-dependent (Figure 9a). The response of total polyphenol and flavonoid content to sodium concentrations in the leaves was similar in most cultivars, with the exception of Don Leon and Alegria Disciplinada (Figure 9b,c). For most cultivars, the concentration of polyphenols decreased proportionally to the Na content in leaves. Some exceptions were nevertheless observed. For example, in Red Amaranth, the foliar Na content was roughly the same at 50 and 100 mM NaCl but a decrease in phenolics was observed (Figure 9b). In Don Armando and Don Leon, the Na concentration in the leaves of plants treated at 50 mM NaCl had no effect or even a positive effect on the phenolic content, respectively (Figure 9b). The ratio between oxidized and total ascorbate was similar whatever the Na content in all cultivars (Figure 9d).

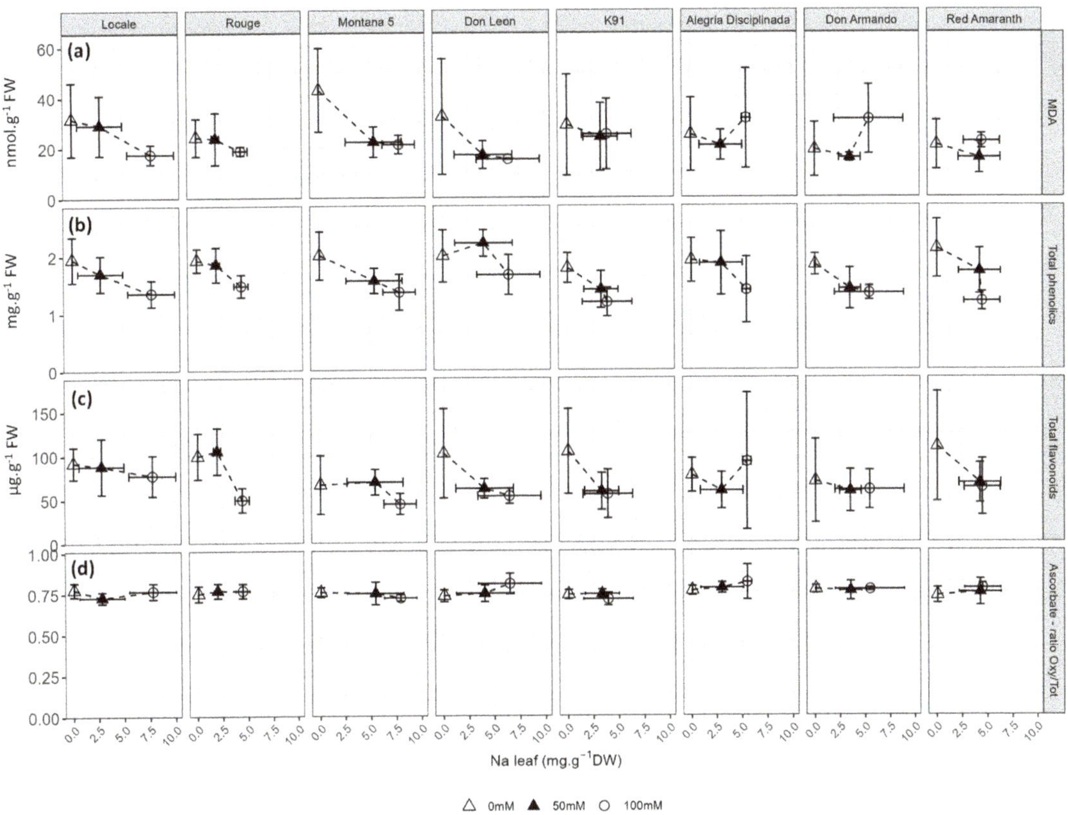

Figure 9. Response of the foliar (**a**) malondialdehyde (MDA, nmol·g^{-1} FW), (**b**) total phenolics (mg·g^{-1} FW), (**c**) total flavonoids (µg·g^{-1} FW) and (**d**) ratio between oxidized and total ascorbate to foliar sodium accumulation.

3. Discussion

In this study, we compared the tolerance of eight cultivars of *Amaranthus cruentus* to 50 and 100 mM NaCl in hydroponic conditions at the vegetative stage. Our results revealed different levels of tolerance and various physiological responses among the cultivars. Sodium accumulated in all plant organs regardless of the cultivar. Since the salt treatments were applied for four weeks, amaranth plants were subjected to both osmotic and ionic phases of salt stress. Indeed, it was previously observed that both phases of salt stress were detected a couple of days after stress imposition in amaranth [31].

3.1. Variability in Salt Tolerance among Leaf and Seed Cultivars of A. cruentus

Our results showed that *A. cruentus* plants were more affected by salt treatments than by the cultivars. This pattern had been observed in other studies screening genotypes for abiotic stresses. In a study of salinity resistance in 25 African rice cultivars (*Oryza glaberrima*), Prodjinoto et al. also found that plants were better discriminated by salt dose than by genotype [63]. Similarly, in a comparison of 12 Tartary buckwheat (*Fagopyrum tataricum*) cultivars, the response to water and heat stress was better explained by the differences between plants than the cultivar [64].

Despite this, differences between cultivars were highlighted in this study. Some cultivars such as Locale, Don Leon and K91 were tolerant to a moderate amount of salinity, given the similar biomass production at 0 and 50 mM NaCl. However, Montana 5, Red

Amaranth and Don Armando were nearly as tolerant at 50 mM than at 100 mM NaCl. We also observed that salt tolerance at 50 mM NaCl was not correlated with salt tolerance at 100 mM, meaning that a cultivar tolerant to a moderate salt dose is not always tolerant to a high salt dose in *A. cruentus*. In a previous study that investigated the salt tolerance of several leaf cultivars of *A. cruentus* after 2 weeks of NaCl treatment, Rouge was identified as salt-tolerant, while Locale was identified as more salt-sensitive compared to other *A. cruentus* cultivars [28]. Our results showed that Rouge accumulated less sodium in leaves and stems than Locale after 4 months of stress, which could explain its higher tolerance. However, both produced an equivalent shoot biomass in this experiment, despite being slightly higher in Rouge at 100 mM NaCl. Amaranth species are usually considered to be salt tolerant [28,29,39] and most cultivars tested in this study survived at 100 mM NaCl, demonstrating the value of growing *A. cruentus* in areas moderately affected by salt. The closely related species *Amaranthus hypochondriacus* was tested for drought and multi-salinity (NaCl, $CaCl_2$, KCl, $MgCl_2$, $MgSO_4$) tolerance in field conditions in South Italy [39,40]. It was shown that this grain amaranth can be grown under conditions of moderate combined drought and saline stress, at the cost of a decrease in seed nutritional quality.

Genetic diversity is an important prerequisite for breeding for salt tolerance [65]. Here, we report a noticeable variability in the physiological response (photosynthetic activity and biochemical activity) of *A. cruentus* cultivars to moderate salt stress, paving the way for developing salt-tolerant lines.

3.2. Putative Physiological Role of Sodium in Amaranth

Our results showed that the salt response of *A. cruentus* may differ according to the NaCl concentration. A previous study on *A. cruentus* leaf cultivars found that a low concentration of NaCl in hydroponic conditions could stimulate several parameters related to mineral content and oxidative status [34]. After two weeks of exposure, 30 mM NaCl had a positive effect on the plant growth and health compared to control conditions, whereas higher concentrations (60 and 90 mM) had detrimental effects [34]. After 4 weeks of exposure to 50 and 100 mM NaCl, the results of the present experiment differed substantially. Even though some parameters were not affected at 50 mM NaCl in some cultivars, or in some cases slightly up-regulated, generally, no positive effect of salt was observed on any parameter in our study. Previous works on *A. tricolor* demonstrated that similar NaCl concentrations (50–100 mM) had a positive effect on the nutritional quality (several minerals, macronutrients and phenolics) of leaves [41,42]. Often considered as a "functional nutrient" rather than as an essential nutrient in plants (e.g., possible substitution of K in some metabolic functions), Na is required in a small quantity in some NAD-ME-type C_4 plants for pyruvate transport and conversion of some metabolic intermediates [66–69]. It was also demonstrated that Na is important in amaranth (*A. tricolor*) besides its putative role in C_4 photosynthesis, for instance by stimulating N assimilation [70–73]. Further research using lower salt concentrations could determine the range of NaCl concentrations that stimulates *A. cruentus* growth and the nutritional quality of leaves.

3.3. Impact of Sodium Accumulation on Photosynthetic Activity

The eight cultivars investigated in this study could be distinguished based on the accumulation of Na in leaves, especially the accumulation difference between 50 mM and 100 mM NaCl, and its consequence on photosynthetic activity. The salt accumulation in leaves caused by salt stress was linked to an important decrease in stomatal conductance, transpiration rate and, to a lesser extent, carbon assimilation in some cultivars, whereas some others maintained their photosynthetic activity despite foliar sodium accumulation.

The significant water use efficiency increase in stressed plants was caused by a stronger decrease in the transpiration rate compared to the salt-induced decrease in net photosynthesis. Omamt et al. investigated the effect of saline stress on the WUE of various *A. cruentus*, *A. hypochondriacus* and *A. tricolor* genotypes [30]. They observed an increase of about 50% in instantaneous WUE in *A. cruentus* at 100 mM NaCl. However, the values recorded in their study were 2–3 times higher than what we observed in the current study, even in control conditions. The same authors demonstrated that the decrease in photosynthetic activity was, at least in part, due to salt-induced stomatal closure and decrease in stomatal density [30]. An increase in WUE in *A. cruentus* cv. Locale was also observed by Gandonou et al. [29]. An increase in WUE is considered as a salt-tolerance mechanism, improving the capacity of the plant to limit water loss despite the salinity toxic effects [74,75]. Liao et al. reported an increase in WUE in maize in combined water and salt stress conditions [76]. These authors identified salt-induced osmotic adjustment in stomata as the main mechanism of stomatal conductance regulation, resulting in an increased salt tolerance in this C_4 crop. In our study, the highest WUE under salinity was observed in K91 and Red Amaranth, suggesting that these cultivars were able to maintain photosynthesis and limit transpiration and thus showed a higher salt tolerance from the physiological point of view. In grasses, halophytism has been associated with C_4 photosynthesis, which could be explained by the high WUE provided by this type of carbon fixation [77]. This could be also true in amaranths, since all species use C_4 photosynthesis [45].

Another parameter often considered as a reliable physiological index for the tolerance to salt stress is the Na/K ratio, which significantly increased in all organs in our study [78]. In contrast to Na, K is an essential element used in different functions, such as in various steps of cell metabolism, photosynthesis and turgor pressure maintenance [79]. Maintaining a low Na/K ratio in plants and mainly in leaves is thus necessary. Red Amaranth and K91 showed the lowest Na/K ratios explained by their ability to restrict Na accumulation in leaves at 100 mM compared to 50 mM NaCl. Indeed, those cultivars were among the most tolerant at 100 mM NaCl.

3.4. Foliar Biochemical Activity Response to Salt Stress

Malondialdehyde is a by-product of lipid peroxidation, caused by reactive oxygen species, which are produced in response to stress, including salt stress [62,80,81]. In our experiment, foliar MDA content did not increase in response to salinity, suggesting low salt-induced oxidative stress on lipids in leaves of the selected *A. cruentus* cultivars. We observed that the total foliar polyphenol content (flavonoids also, to a lesser extent) decreased in plants exposed to NaCl. However, the phenylpropanoid pathway is often upregulated in response to many abiotic stresses because polyphenols have protective roles, for instance against oxidative stress [82,83]. Moreover, no difference in the ratio between oxidized and total ascorbate was observed between the control and stressed plants, although this metabolite is also involved in oxidative stress mitigation [84]. The low intensity of oxidative stress suggested by the low MDA content could explain why the plants exposed to salinity did not upregulate polyphenols and ascorbate production. Betalains, which include the two subfamilies of pigments betaxanthins and betacyanins, are also involved in plant tolerance to abiotic stress. Salinity-induced betalain accumulation has been described in two halophyte species of the Amaranthaceae family [85]. Sarker and Oba (2018) reported an increase in betacyanin and betaxanthin content in the leaves of *A. tricolor* exposed to 200 mM NaCl [86]. A photoprotective role of betacyanins was also described in *A. cruentus* [87]. Here, only the betaxanthin content significantly increased in response to stress.

3.5. Differences between Leaf and Seed Cultivars

There are two types of amaranth cultivars, depending on the harvested parts. African Rouge and Locale cultivars have been bred for leaf production, whereas the others are cultivated for their seeds. Even though the cultivar types are clearly distinguishable by several morphological traits (size and color of seeds, shape of leaves, branching pattern or inflorescence structure (see Table S1)), their physiological response to salinity stress did not differ at the vegetative stage. Only the transpiration rate was, on average, 1.5× higher in the grain cultivars than in leaf cultivars in the control conditions and at 50 mM NaCl, although it dropped at the same level at 100 mM NaCl. Comparing leaf and grain cultivars at the reproductive stage could reveal additional differences in response to salt stress. Although amaranths are usually used either for grain production or leaf production, the dual use of leaves and seeds on a single cultivar seems promising on the basis of several defoliation experiments [51–53,88–90]. To our knowledge, the tolerance of amaranth to defoliation under abiotic stress has not been investigated yet.

3.6. Effect of Salt of the Nutritional Quality of Amaranth

Amaranths are recognized for their exceptional nutritional value, both of their leaves and seeds [34,54,56,86,91]. Adverse environmental conditions could alter the nutritional quality of plants, although it was demonstrated that *A. tricolor* keeps its nutritional value under abiotic stress [86,92]. Although the effect of salinity on the nutritional quality was not a main objective of this work, we showed that salt stress caused a decrease in phenolic content in most cultivars. These metabolites, however, have nutritional and health benefits [93]. The decrease in the foliar content of several essential minerals (K, Ca, Mg) should also be highlighted. Given the outstanding nutritional value of this crop, understanding the effects of moderate salt stress on the nutritional quality of *A. cruentus* leaves is crucial and needs further research. A previous work on the two leaf cultivars demonstrated an increase in Mg, P, Fe, vitamin C, phenolic, α-tocopherol and carotenoid contents in leaves when exposed to 30 mM of NaCl, particularly in Rouge [34]. Salt stress and drought stress, the latter being in some respects similar to salt stress, can also improve the nutritional quality of amaranth leaves [42,94]. To our knowledge, the effect of abiotic stress, particularly salinity, on the nutritional quality of the seeds of grain amaranth has not been investigated extensively yet. More broadly, the effect of salt on the reproduction of amaranths requires further research.

4. Materials and Methods

4.1. Plant Material and Growth Conditions

Leafy cultivars Rouge and Locale were kindly provided by Dr. Christophe B. Gandonou (University of Abomey-Calavi, Cotonou, Benin) and selected based on previous works [28]. Since no information was available about the salt tolerance of grain cultivars, six grain cultivars differing by their origin were randomly selected and obtained from the Genebank of the Crop Research Institute (CRI, Prague, Cezch Republic) (see Table S1 for accession numbers).

Plants were cultivated in greenhouses (SeFy, UCLouvain) at 23–25 °C at day, 20–22 °C at night and 65% RH, under a 16 h photoperiod. When necessary, artificial light was provided by 650 W red-blue LumiGrow LED lights (minimum light intensity of 150 μmol m^{-2} s^{-1}). Seeds were sown in 2/3 peat compost (DCM, Amsterdam, The Netherlands) and 1/3 river sand (Mpro, Wavre, Belgium) (volume:volume). Two weeks later, seedlings were transplanted individually in 6 × 6 × 6 cm plastic pots in 2/3 peat compost +1/3 river sand. Two weeks later, they were transplanted in 15 L plastic tanks filled with Hoagland nutritive solution (5 mM KNO_3, 5.5 mM $Ca(NO_3)_2$, 1 mM $NH_4H_2PO_4$, 0.5 mM $MgSO_4$, 25 μM KCl, 10 μM H_3BO_4, 1 μM $MnSO_4$, 0.25 μM $CuSO_4$, 1 μM $ZnSO_4$, 10 μM $(NH_4)_6Mo_7O$ and 1.87 g L^{-1} Fe-EDTA, and pH 5.5–6), with 1 seedling of each cultivar per tank (9 plants/tank). Tanks were randomly assigned to 0, 50 or 100 mM NaCl, with 9 replicates (27 tanks). The NaCl was added in the Hoagland solution and salt stress started 9 days after the transfer to plastic tanks. The nutritive solution was renewed once a week. The experiment took place over 53 days in November and December 2020.

4.2. Biomass and Harvest

Plants were harvested 53 days after sowing for destructive measurements, mineral and biochemical analyses. For three plants per cultivar and salt treatment, three young but well-expanded leaves were harvested in liquid nitrogen and stored at −80 °C for further biochemical analyses (see below). For five other plants per cultivar and treatment, the stems, leaves and roots were separated, weighted (for fresh weight), dried at 60 °C for 72 h, then weighed again (for dry weight). Water content was calculated as (fresh weight − dry weight)/fresh weight. Dry material was used for mineral analyses (see below). The salt tolerance index was calculated as the ratio between the mean total biomass production in salt conditions relative to the total biomass production in control conditions [78].

4.3. Biochemical Analyses

4.3.1. Pigments

Chlorophyll *a* and *b* were quantified in the leaves of three plants per cultivar and treatment according to [95]. Briefly, 1.2 mL of 80% acetone (*v*/*v*) was added to 50 mg of finely ground (in liquid nitrogen) fresh leaves. After 60 min of incubation at 4 °C, tubes were centrifugated (10,000× *g*, 4 °C, 10 min). A second identical extraction was performed on the pellet; supernatants were combined. Absorbance of the supernatant was read at 663.2 and 646.8 nm (UV-1800 spectrophotomer, Shimadzu, Kyoto, Japan). Chlorophyll *a* and *b* contents were measured as follows: Chl *a* (mg/L) = 12.25 × $Abs_{663.2}$ − 2.79 × $Abs_{646.8}$ and Chl *b* (mg/L) = 21.50 × $Abs_{646.8}$ − 5.10 × $Abs_{663.2}$.

Betalains were extracted in deionized water overnight. Absorbance of the supernatant was read at 540 nm (betacyanins) and 475 nm (betaxanthins) (UV-1800 spectrophotomer, Shimadzu, Kyoto, Japan). Molar extinction coefficients of 62 × 10^6 cm^2 mol^{-1} and 48 × 10^6 cm^2 mol^{-1}, respectively, were used to quantify the pigment content [96,97].

4.3.2. Phenolics

Total phenolics and flavonoids were quantified in the leaves of three plants per cultivar and treatment according to [98,99]. After grounding in liquid nitrogen, 1.4 mL of 80% methanol was added to 100 mg of fresh material, before centrifugation (20,000× *g*, 20 min, 4 °C). The supernatant was stored at −20 °C until quantification.

A volume of supernatant was added to an equal volume of 2% $AlCl_3$ for total flavonoid quantification (adapted from [98]). After 10 min of incubation at room temperature in the dark, absorbance was read at 440 nm, with quercetin as the standard.

Total phenolics were quantified as follows: 200 µL of supernatant was added to 2.8 mL of deionized water and 200 µL of Folin–Ciocalteu reagent [99]. Three minutes later, 0.8 mL of 20% Na_2CO_3 was added before incubation in a water bath at 40 °C for 40 min. Absorbance was read at 760 nm with gallic acid as the standard.

4.3.3. Malondialdehyde

Malondialdehyde, a marker of lipid peroxidation [100], was quantified in the leaves of three plants per cultivar and treatment according to [101]. It was extracted in 250 mg of finely ground fresh leaves with 4 mL of 5% trichloroacetic acid with 1.25% glycerol. After 5 min of incubation at 4 °C, tubes were centrifuged for 10 min at 4 °C, 12,000× *g*. Then, 2 mL of supernatant was added to 2 mL of a 0.67% aqueous solution of thiobarbituric acid. Samples were incubated for 30 min in a water bath at 100 °C. Absorbance was read at 532 and 600 nm. Malondialdehyde concentration (mM) was calculated as (A_{532nm} − A_{600})/155 mM cm^{-1} [101].

4.3.4. Ascorbate

Ascorbate was quantified with some adaptations from [102], in the leaves of three plants per cultivar and treatment. Briefly, it was extracted in 250 mg of finely ground fresh leaves with 4 mL of 5% trichloroacetic acid (TCA). After 15 min of incubation on ice, samples were centrifugated (5 min, 4 °C, 10,000× *g*). Next, 200 µL of supernatant was added to 400 µL of phosphate buffer (0.2 M, pH 7.4). For oxidized ascorbate quantification, 0.4 mL of water was added. For total ascorbate quantification, 200 µL of 10 mM 2,2′-dithiothreitol was added, the samples were incubated 5 min at room temperature, then 200 µL of 0.5% N-ethylmaleimide was added. After 1 min of incubation at room temperature, 1 mL 10% TCA was added in all the tubes, then 0.8 mL 42.5% H_3PO_4 and 0.8 mL 4% dipyridyl were added. Finally, 400 µL of 3% $FeCl_3$ was added while agitating. Tubes were incubated

60 min in a water bath at 37 °C. Absorbance was read at 525 nm. Ascorbic acid was used as the standard.

4.3.5. Mineral Content

The concentrations of Na and K were quantified in the roots, stem and leaves of three plants per cultivar and treatment while the concentrations of Ca, Fe, Mg and Zn were quantified only in the leaves. For mineralization, 4 mL of 68% nitric acid was added to 50–100 mg of ground dry plant material. After one night of incubation, nitric acid was evaporated using a sand bath. Then, 1.5 mL of aqua regia (500 μL of 68% nitric acid and 1.5 mL of 37% hydrochloric acid) was added and incubated two minutes on the sand bath. The volume was adjusted to 10 mL with deionized water before filtration on Whatman Grade 1 paper. The concentration of Na, K, Mg, Ca, Fe and Zn was determined by atomic absorption spectroscopy (ICE 3300, Thermo Scientific, Waltham, MA, USA) after the required dilutions and addition of 1% $LaCl_3$ for Na and K quantification.

4.4. Photosynthetic Activity

The portable photosynthesis system LCpro-SD (ADC Bioscientific Ltd., Hoddesdon, United Kingdom) was used for photosynthetic activity analyses. Measurements were performed on three plants per cultivar and treatment, 51 days after sowing (27 days after stress). Net photosynthesis (A), stomatal conductance (g_s) and transpiration rate (E) were recorded in a young, well-expanded leaf after several minutes of stabilization, in conditions of ambient irradiance, carbon dioxide concentration, air humidity and temperature, similarly to [103]. Water use efficiency was calculated as A/E.

4.5. Statistical Analyses

All statistical analyses were performed in R, version 4.2.1 [104]. Normality of the data was verified based on histograms of residuals and homoscedasticity was verified with a Levene test. For each variable (plant growth, mineral content, physiological and biochemical parameters), analyses of variance (ANOVA 2) were performed with the salt treatment and cultivars as the fixed factors using the function "aov()" (base R). Detailed results of ANOVA 2 are presented in Tables S2, S4, S5 and S7. For each variable, Tukey's test from base R was used to perform multiple comparison tests to evaluate the differences between salt treatments within each cultivar. The package *ade4* was used to perform a principal component analysis to visualize the differences among the cultivars and salt treatments according to mineral and biomass data. Redundant variables with $r > 0.8$ were removed prior to analysis. Correlations between all the variables were quantified with the Pearson correlation coefficient (r) and significance tests were used to calculate the associated *p*-values with R package *corrplot*. Data are presented in figures and tables as the mean ± standard deviation, with adequate rounding.

5. Conclusions

Our results demonstrate a noticeable tolerance of *A. cruentus* cv. Don Leon and K91 at the vegetative stage under moderate salt stress (50 mM) in hydroponic conditions, since biomass production did not differ from the control conditions. Sodium accumulated in all organs regardless of the cultivar, mainly in stems. Different physiological responses to foliar sodium accumulation were observed among the cultivars, suggesting a predominance of the ionic phase when physiology was negatively affected in response to ion accumulation or of the osmotic phase when adverse effects on physiology were observed, despite no or a low sodium accumulation. Water use efficiency increased in response to salt because of an efficient decrease in stomatal conductance, despite a decrease in net photosynthesis. The lower transpiration rate of the leaf cultivars compared to the grain cultivars was the unique discriminating physiological trait between the two types of cultivars. Since salt stress did not increase neither MDA content nor the metabolites involved in protection against radical oxygen species such as polyphenols, betacyanins and ascorbate, we suggest that the oxidative stress was limited. This work provides a basis for further investigation of the physiological mechanisms underlying the variations in salt tolerance in red amaranth, a plant with a promising future in resilient agro-ecosystems under global change.

Supplementary Materials: The following supporting information can be downloaded at: https://www.mdpi.com/article/10.3390/plants12183310/s1, Table S1: Origin and main morphological characteristics of the eight *A. cruentus* cultivars used in this study; Table S2: Anova results of biomass production; Table S3: Water content of leaves, stem and roots of eight cultivars of *A. cruentus* exposed to 0 mM, 50 mM and 100 mM of NaCl; Table S4: Anova results of mineral content; Table S5: Anova results of photosynthetic activity; Table S6: Foliar content of chlorophyll *b* of eight cultivars of *A. cruentus* exposed to 0 mM, 50 mM and 100 mM of NaCl; Table S7: Anova results of biochemical data.

Author Contributions: Conceptualization, A.L., M.Q. and S.L.; methodology, A.L., M.Q. and S.L.; formal analysis, A.L.; investigation, A.L.; resources, M.Q. and S.L.; data curation, A.L.; writing—original draft preparation, A.L. and M.Q.; writing—review and editing, A.L., M.Q. and S.L.; visualization, A.L.; supervision, M.Q. and S.L.; project administration, M.Q. and S.L. All authors have read and agreed to the published version of the manuscript.

Funding: This research received no external funding.

Data Availability Statement: Data are contained within the article or Supplementary Material.

Acknowledgments: The authors are grateful to Baudouin Capelle and Brigitte Vanpée for their technical assistance, and to the Genebank of the Crop Research Institute (Prague, Cezch Republic) for obtaining the seeds.

Conflicts of Interest: The authors declare no conflict of interest. The funders had no role in the design of the study; in the collection, analyses, or interpretation of data; in the writing of the manuscript; or in the decision to publish the results.

References

1. Mukhopadhyay, R.; Sarkar, B.; Jat, H.S.; Sharma, P.C.; Bolan, N.S. Soil Salinity under Climate Change: Challenges for Sustainable Agriculture and Food Security. *J. Environ. Manag.* **2021**, *280*, 111736. [CrossRef]
2. Butcher, K.; Wick, A.F.; DeSutter, T.; Chatterjee, A.; Harmon, J. Soil Salinity: A Threat to Global Food Security. *Agron. J.* **2016**, *108*, 2189–2200. [CrossRef]
3. Montanarella, L.; Badraoui, M.; Chude, V.; Baptista Costa, I.D.S.; Mamo, T.; Yemefack, M.; Singh Aulakh, M.; Yagi, K.; Young Hong, S.; Vijarnsorn, P.; et al. *Status of the World's Soil Resources Main Report*; FAO: Rome, Italy, 2015; ISBN 978-92-5-109004-6.
4. Hassani, A.; Azapagic, A.; Shokri, N. Predicting Long-Term Dynamics of Soil Salinity and Sodicity on a Global Scale. *Proc. Natl. Acad. Sci. USA* **2020**, *117*, 33017–33027. [CrossRef] [PubMed]
5. Eswar, D.; Karuppusamy, R.; Chellamuthu, S. Drivers of Soil Salinity and Their Correlation with Climate Change. *Curr. Opin. Environ. Sustain.* **2021**, *50*, 310–318. [CrossRef]
6. FAO. *Global Map of Salt-Affected Soils: GSASmap v1.0*; FAO: Rome, Italy, 2021.
7. Hopmans, J.W.; Qureshi, A.S.; Kisekka, I.; Munns, R.; Grattan, S.R.; Rengasamy, P.; Ben-Gal, A.; Assouline, S.; Javaux, M.; Minhas, P.S.; et al. Critical Knowledge Gaps and Research Priorities in Global Soil Salinity. In *Advances in Agronomy*; Academic Press: Cambridge, MA, USA, 2021.
8. Munns, R.; Tester, M. Mechanisms of Salinity Tolerance. *Annu. Rev. Plant Biol.* **2008**, *59*, 651–681. [CrossRef] [PubMed]
9. Munns, R.; Passioura, J.B.; Colmer, T.D.; Byrt, C.S. Osmotic Adjustment and Energy Limitations to Plant Growth in Saline Soil. *New Phytol.* **2020**, *225*, 1091–1096. [CrossRef] [PubMed]
10. Maathuis, F.J.M. Sodium in Plants: Perception, Signalling, and Regulation of Sodium Fluxes. *J. Exp. Bot.* **2014**, *65*, 849–858. [CrossRef] [PubMed]
11. Munns, R.; Day, D.A.; Fricke, W.; Watt, M.; Arsova, B.; Barkla, B.J.; Bose, J.; Byrt, C.S.; Chen, Z.-H.; Foster, K.J.; et al. Energy Costs of Salt Tolerance in Crop Plants. *New Phytol.* **2019**, *225*, 1072–1090. [CrossRef]
12. Chourasia, K.N.; Lal, M.K.; Tiwari, R.K.; Dev, D.; Kardile, H.B.; Patil, V.U.; Kumar, A.; Vanishree, G.; Kumar, D.; Bhardwaj, V.; et al. Salinity Stress in Potato: Understanding Physiological, Biochemical and Molecular Responses. *Life* **2021**, *11*, 545. [CrossRef]
13. EL Sabagh, A.; Islam, M.S.; Skalicky, M.; Ali Raza, M.; Singh, K.; Anwar Hossain, M.; Hossain, A.; Mahboob, W.; Iqbal, M.A.; Ratnasekera, D.; et al. Salinity Stress in Wheat (*Triticum aestivum* L.) in the Changing Climate: Adaptation and Management Strategies. *Front. Agron.* **2021**, *3*, 661932. [CrossRef]
14. Cao, D.; Li, Y.; Liu, B.; Kong, F.; Tran, L.-S.P. Adaptive Mechanisms of Soybean Grown on Salt-Affected Soils. *Land Degrad. Dev.* **2018**, *29*, 1054–1064. [CrossRef]
15. Zaidi, S.S.-A.; Vanderschuren, H.; Qaim, M.; Mahfouz, M.M.; Kohli, A.; Mansoor, S.; Tester, M. New Plant Breeding Technologies for Food Security. *Science* **2019**, *363*, 1390–1391. [CrossRef] [PubMed]

16. Dawson, I.K.; Powell, W.; Hendre, P.; Bančič, J.; Hickey, J.M.; Kindt, R.; Hoad, S.; Hale, I.; Jamnadass, R. The Role of Genetics in Mainstreaming the Production of New and Orphan Crops to Diversify Food Systems and Support Human Nutrition. *New Phytol.* **2019**, *224*, 37–54. [CrossRef] [PubMed]
17. Curtin, S.; Qi, Y.; Peres, L.E.P.; Fernie, A.R.; Zsögön, A. Pathways to de Novo Domestication of Crop Wild Relatives. *Plant Physiol.* **2022**, *188*, 1746–1756. [CrossRef] [PubMed]
18. Dwyer, W.; Ibe, C.N.; Rhee, S.Y. Renaming Indigenous Crops and Addressing Colonial Bias in Scientific Language. *Trends Plant Sci.* **2022**, *27*, 1189–1192. [CrossRef] [PubMed]
19. Singh, A.; Dubey, P.K.; Chaurasia, R.; Dubey, R.K.; Pandey, K.K.; Singh, G.S.; Abhilash, P.C. Domesticating the Undomesticated for Global Food and Nutritional Security: Four Steps. *Agronomy* **2019**, *9*, 491. [CrossRef]
20. Ye, C.-Y.; Fan, L. Orphan Crops and Their Wild Relatives in the Genomic Era. *Mol. Plant* **2021**, *14*, 27–39. [CrossRef] [PubMed]
21. van Zonneveld, M.; Kindt, R.; McMullin, S.; Achigan-Dako, E.G.; N'Danikou, S.; Hsieh, W.; Lin, Y.; Dawson, I.K. Forgotten Food Crops in Sub-Saharan Africa for Healthy Diets in a Changing Climate. *Proc. Natl. Acad. Sci. USA* **2023**, *120*, e2205794120. [CrossRef]
22. Razzaq, A.; Saleem, F.; Wani, S.H.; Abdelmohsen, S.A.M.; Alyousef, H.A.; Abdelbacki, A.M.M.; Alkallas, F.H.; Tamam, N.; Elansary, H.O. De-Novo Domestication for Improving Salt Tolerance in Crops. *Front. Plant Sci.* **2021**, *12*, 1623. [CrossRef]
23. Massel, K.; Lam, Y.; Wong, A.C.S.; Hickey, L.T.; Borrell, A.K.; Godwin, I.D. Hotter, Drier, CRISPR: The Latest Edit on Climate Change. *Theor. Appl. Genet.* **2021**, *134*, 1691–1709. [CrossRef]
24. Stetter, M.G.; Schmid, K.J. Analysis of Phylogenetic Relationships and Genome Size Evolution of the *Amaranthus* Genus Using GBS Indicates the Ancestors of an Ancient Crop. *Mol. Phylogenet. Evol.* **2017**, *109*, 80–92. [CrossRef] [PubMed]
25. Marx, J.L. Speaking of Science: Amaranth: A Comeback for the Food of the Aztecs? *Sci. New Ser.* **1977**, *198*, 40.
26. Kauffman, C.S. Realizing the Potential of Grain Amaranth. *Food Rev. Int.* **1992**, *8*, 5–21. [CrossRef]
27. Stallknecht, G.F.; Schulz-Schaeffer, J.R. *Amaranth Rediscovered;* Wiley: New York, NY, USA, 1993; pp. 211–218.
28. Wouyou, A.; Gandonou, C.; Komlan, F.; Montcho, D.; Zanklan, A.; Lutts, S.; Gnancadja, S. Salinity Resistance of Five Amaranth (*Amaranthus cruentus*) Cultivars at Young Plants Stage. *Int. J. Plant Soil Sci.* **2017**, *14*, 1–11. [CrossRef]
29. Gandonou, C.B.; Prodjinoto, H.; Ahissou Zanklan, S.; Dossou Wouyou, A.; Lutts, S.; Hambada Montcho, D.; Assogba Komlan, F.; Clément Goudjo Mensah, A. Effects of Salinity Stress on Growth in Relation to Gas Exchanges Parameters and Water Status in Amaranth (*Amaranthus cruentus*). *Int. J. Plant Physiol. Biochem.* **2018**, *10*, 19–27. [CrossRef]
30. Omami, E.N.; Hammes, P.S.; Robbertse, P.J. Differences in Salinity Tolerance for Growth and Water-use Efficiency in Some Amaranth (*Amaranthus* spp.) Genotypes. *N. Z. J. Crop Hortic. Sci.* **2006**, *34*, 11–22. [CrossRef]
31. Huerta-Ocampo, J.A.; Barrera-Pacheco, A.; Mendoza-Hernández, C.S.; Espitia-Rangel, E.; Mock, H.-P.; Barba de la Rosa, A.P. Salt Stress-Induced Alterations in the Root Proteome of *Amaranthus cruentus* L. *J. Proteome Res.* **2014**, *13*, 3607–3627. [CrossRef]
32. Wouyou, A.; Prodjinoto, H.; Zanklan, A.S.; Vanpee, B.; Lutts, S.; Gandonou, C.B. Implication of Ions and Organic Solutes Accumulation in Amaranth (*Amaranthus cruentus* L.) Salinity Resistance. *Am. J. Plant Sci.* **2019**, *10*, 2335–2353. [CrossRef]
33. Barba de la Rosa, A.P.; Huerta-Ocampo, J.A.; González-Escobar, J.L.; Aguilar-Hernández, H.S.; Salcedo-Barrientos, G.; Espitia-Rangel, E. Differential Expression of Iron Transporters in *Amaranthus cruentus* Roots When Are Subjected to Salt Stress: The Influence of Root Endophytes. *Rhizosphere* **2022**, *24*, 100620. [CrossRef]
34. Luyckx, A.; Beghin, C.; Quinet, M.; Achadé, B.; Prodjinoto, H.; Gandonou, C.B.; Lutts, S. Salinity Differently Affects Antioxidant Content and Amino Acid Profile in Two Cultivars of *Amaranthus cruentus* Differing in Salinity Tolerance. *J. Sci. Food Agric.* **2021**, *101*, 6211–6219. [CrossRef]
35. Mlakar, S.G.; Bavec, M.; Jakop, M.; Bavec, F. The Effect of Drought Occurring at Different Growth Stages on Productivity of Grain Amaranth *Amaranthus cruentus* G6. *J. Life Sci.* **2012**, *6*, 283–286.
36. Babayev, H.; Mehvaliyeva, U.; Aliyeva, M.; Feyziyev, Y.; Guliyev, N. The Study of NAD-Malic Enzyme in *Amaranthus cruentus* L. under Drought. *Plant Physiol. Biochem.* **2014**, *81*, 84–89. [CrossRef] [PubMed]
37. Tebini, M.; Luu, D.T.; Mguis, K.; Ben Ahmed, H.; Meddich, A.; Zribi, F.; Chalh, A. Physiological Exploration of Intra-Specific Variability in Salinity Tolerance of Amaranth. *Russ. J. Plant Physiol.* **2022**, *69*, 59. [CrossRef]
38. Tebini, M.; Rabaoui, G.; M'Rah, S.; Luu, D.-T.; Ben Ahmed, H.; Chalh, A. Effects of Salinity on Germination Dynamics and Seedling Development in Two Amaranth Genotypes. *Physiol. Mol. Biol. Plants* **2022**, *28*, 1489–1500. [CrossRef] [PubMed]
39. Pulvento, C.; Sellami, M.H.; Lavini, A. Yield and Quality of *Amaranthus hypochondriacus* Grain Amaranth under Drought and Salinity at Various Phenological Stages in Southern Italy. *J. Sci. Food Agric.* **2022**, *102*, 5022–5033. [CrossRef] [PubMed]
40. Lavini, A.; Pulvento, C.; d'Andria, R.; Riccardi, M.; Jacobsen, S.E. Effects of Saline Irrigation on Yield and Qualitative Characterization of Seed of an Amaranth Accession Grown under Mediterranean Conditions. *J. Agric. Sci.* **2016**, *154*, 858–869. [CrossRef]
41. Sarker, U.; Islam, M.T.; Oba, S. Salinity Stress Accelerates Nutrients, Dietary Fiber, Minerals, Phytochemicals and Antioxidant Activity in *Amaranthus tricolor* Leaves. *PLoS ONE* **2018**, *13*, e0206388. [CrossRef]
42. Sarker, U.; Oba, S. Salinity Stress Enhances Color Parameters, Bioactive Leaf Pigments, Vitamins, Polyphenols, Flavonoids and Antioxidant Activity in Selected *Amaranthus* Leafy Vegetables. *J. Sci. Food Agric.* **2019**, *99*, 2275–2284. [CrossRef]
43. Guo, S.-H.; Hu, N.; Li, Q.-S.; Yang, P.; Wang, L.-L.; Xu, Z.-M.; Chen, H.-J.; He, B.-Y.; Zeng, E.Y. Response of Edible Amaranth Cultivar to Salt Stress Led to Cd Mobilization in Rhizosphere Soil: A Metabolomic Analysis. *Environ. Pollut.* **2018**, *241*, 422–431. [CrossRef]

44. Bellache, M.; Allal Benfekih, L.; Torres-Pagan, N.; Mir, R.; Verdeguer, M.; Vicente, O.; Boscaiu, M. Effects of Four-Week Exposure to Salt Treatments on Germination and Growth of Two *Amaranthus* Species. *Soil. Syst.* **2022**, *6*, 57. [CrossRef]
45. Sage, R.F.; Sage, T.L.; Pearcy, R.W.; Borsch, T. The Taxonomic Distribution of C_4 Photosynthesis in Amaranthaceae Sensu Stricto. *Am. J. Bot.* **2007**, *94*, 1992–2003. [CrossRef] [PubMed]
46. Mapes, C.; Basurto, F.; Bye, R. Ethnobotany of Quintonil: Knowledge, Use and Management of Edible Greens *Amaranthus* spp. (Amaranthaceae) in the Sierra Norte de Puebla, México. *Econ. Bot.* **1997**, *51*, 293–306. [CrossRef]
47. Ogwu, M.C. Value of *Amaranthus* [L.] Species in Nigeria. In *Nutritional Value of Amaranth*; IntechOpen: London, UK, 2020; ISBN 978-1-83880-084-0.
48. Kietlinski, K.D.; Jimenez, F.; Jellen, E.N.; Maughan, P.J.; Smith, S.M.; Pratt, D.B. Relationships between the Weedy *Amaranthus hybridus* (Amaranthaceae) and the Grain Amaranths. *Crop Sci.* **2014**, *54*, 220–228. [CrossRef]
49. Stetter, M.G.; Müller, T.; Schmid, K.J. Genomic and Phenotypic Evidence for an Incomplete Domestication of South American Grain Amaranth (*Amaranthus caudatus*). *Mol. Ecol.* **2017**, *26*, 871–886. [CrossRef] [PubMed]
50. Stetter, M.G.; Vidal-Villarejo, M.; Schmid, K.J. Parallel Seed Color Adaptation during Multiple Domestication Attempts of an Ancient New World Grain. *Mol. Biol. Evol.* **2019**, *37*, 1407–1419. [CrossRef]
51. Dinssa, F.F.; Yang, R.-Y.; Ledesma, D.R.; Mbwambo, O.; Hanson, P. Effect of Leaf Harvest on Grain Yield and Nutrient Content of Diverse Amaranth Entries. *Sci. Hortic.* **2018**, *236*, 146–157. [CrossRef]
52. Hoidal, N.; Díaz Gallardo, M.; Jacobsen, S.-E.; Alandia, G. Amaranth as a Dual-Use Crop for Leafy Greens and Seeds: Stable Responses to Leaf Harvest Across Genotypes and Environments. *Front. Plant Sci.* **2019**, *10*, 817. [CrossRef]
53. Hoidal, N.; Jacobsen, S.-E.; Odone, A.; Alandia, G. Defoliation Timing for Optimal Leaf Nutrition in Dual-Use Amaranth Production Systems. *J. Sci. Food Agric.* **2020**, *100*, 4745–4755. [CrossRef]
54. Martinez-Lopez, A.; Millan-Linares, M.C.; Rodriguez-Martin, N.M.; Millan, F.; Montserrat-de la Paz, S. Nutraceutical Value of Kiwicha (*Amaranthus caudatus* L.). *J. Funct. Foods* **2020**, *65*, 103735. [CrossRef]
55. José Rodríguez Gómez, M.; Maestro-Gaitán, I.; Calvo Magro, P.; Cruz Sobrado, V.; Reguera Blázquez, M.; Matías Prieto, J. Unique Nutritional Features That Distinguish *Amaranthus cruentus* L. and *Chenopodium Quinoa* Willd Seeds. *Food Res. Int.* **2023**, *164*, 112160. [CrossRef]
56. Villarreal, M.; Iturriaga, L.B. Amaranth: An Andean Crop with History, Its Feeding Reassessment in America. In *Traditional Foods*; Kristbergsson, K., Oliveira, J., Eds.; Springer: Boston, MA, USA, 2016; pp. 217–232. ISBN 978-1-4899-7646-8.
57. Ngugi, C.C.; Oyoo-Okoth, E.; Manyala, J.O.; Fitzsimmons, K.; Kimotho, A. Characterization of the Nutritional Quality of Amaranth Leaf Protein Concentrates and Suitability of Fish Meal Replacement in Nile Tilapia Feeds. *Aquac. Rep.* **2017**, *5*, 62–69. [CrossRef]
58. Tang, Y.; Li, X.; Chen, P.X.; Zhang, B.; Hernandez, M.; Zhang, H.; Marcone, M.F.; Liu, R.; Tsao, R. Lipids, Tocopherols, and Carotenoids in Leaves of Amaranth and Quinoa Cultivars and a New Approach to Overall Evaluation of Nutritional Quality Traits. *J. Agric. Food Chem.* **2014**, *62*, 12610–12619. [CrossRef] [PubMed]
59. Gimplinger, D.M.; Dobos, G.; Schönlechner, R.; Kaul, H.-P. Yield and Quality of Grain Amaranth (*Amaranthus* Sp.) in Eastern Austria. *Plant Soil Environ.* **2008**, *53*, 105–112. [CrossRef]
60. Dinssa, F.F.; Hanson, P.; Ledesma, D.R.; Minja, R.; Mbwambo, O.; Tilya, M.S.; Stoilova, T. Yield of Vegetable Amaranth in Diverse Tanzanian Production Environments. *HortTechnology* **2019**, *29*, 516–527. [CrossRef]
61. Bigot, S.; Fuksová, M.; Martínez, J.-P.; Lutts, S.; Quinet, M. Sodium and Chloride Accumulation and Repartition Differed between the Cultivated Tomato (*Solanum lycopersicum*) and Its Wild Halophyte Relative *Solanum Chilense* under Salt Stress. *Sci. Hortic.* **2023**, *321*, 112324. [CrossRef]
62. Hasanuzzaman, M.; Raihan, M.R.H.; Masud, A.A.C.; Rahman, K.; Nowroz, F.; Rahman, M.; Nahar, K.; Fujita, M. Regulation of Reactive Oxygen Species and Antioxidant Defense in Plants under Salinity. *Int. J. Mol. Sci.* **2021**, *22*, 9326. [CrossRef] [PubMed]
63. Prodjinoto, H.; Gandonou, C.; Lutts, S. Screening for Salinity Tolerance of *Oryza glaberrima* Steud. Seedlings. *Afr. J. Agric. Res.* **2018**, *13*, 561–573. [CrossRef]
64. Aubert, L.; Quinet, M. Comparison of Heat and Drought Stress Responses among Twelve Tartary Buckwheat (*Fagopyrum tataricum*) Varieties. *Plants* **2022**, *11*, 1517. [CrossRef]
65. Munns, R.; Hare, R.A.; James, R.A.; Rebetzke, G.J. Genetic Variation for Improving the Salt Tolerance of Durum Wheat. *Aust. J. Agric. Res.* **2000**, *51*, 69–74. [CrossRef]
66. Subbarao, G.V.; Ito, O.; Berry, W.L.; Wheeler, R.M. Sodium—A Functional Plant Nutrient. *Crit. Rev. Plant Sci.* **2003**, *22*, 391–416. [CrossRef]
67. Rao, X.; Dixon, R.A. The Differences between NAD-ME and NADP-ME Subtypes of C_4 Photosynthesis: More than Decarboxylating Enzymes. *Front. Plant Sci.* **2016**, *7*, 1525. [CrossRef] [PubMed]
68. Johnston, M.; Grof, C.P.L.; Brownell, P.F. The Effect of Sodium Nutrition on the Pool Sizes of Intermediates of the C_4 Photosynthetic Pathway. *Funct. Plant Biol.* **1988**, *15*, 749–760. [CrossRef]
69. Murata, S.; Kobayashi, M.; Matoh, T.; Sekiya, J. Sodium Stimulates Regeneration of Phosphoenolpyruvate in Mesophyll Chloroplasts of *Amaranthus tricolor*. *Plant Cell Physiol.* **1992**, *33*, 1247–1250. [CrossRef]
70. Ohta, D.; Yasuoka, S.; Matoh, T.; Takahashi, E. Sodium Stimulates Growth of *Amaranthus tricolor* L. Plants through Enhanced Nitrate Assimilation. *Plant Physiol.* **1989**, *89*, 1102–1105. [CrossRef] [PubMed]

71. Ohta, D.; Matoh, T.; Takahashi, E. Early Responses of Sodium-Deficient *Amaranthus tricolor* L. Plants to Sodium Application. *Plant Physiol.* **1987**, *84*, 112–117. [CrossRef] [PubMed]
72. Ohta, D.; Matoh, T.; Takahashi, E. Sodium-Stimulated NO_3^- Uptake in *Amaranthus tricolor* L. Plants. *Plant Physiol.* **1988**, *87*, 223–225. [CrossRef] [PubMed]
73. Matoh, T.; Ohta, D.; Takahashi, E. Effect of Sodium Application on Growth of *Amaranthus tricolor* L. *Plant Cell Physiol.* **1986**, *27*, 187–192. [CrossRef]
74. Yan, J.; Zhu, C.; Liu, W.; Luo, F.; Mi, J.; Ren, Y.; Li, J.; Sang, T. High Photosynthetic Rate and Water Use Efficiency of *Miscanthus lutarioriparius* Characterize an Energy Crop in the Semiarid Temperate Region. *GCB Bioenergy* **2015**, *7*, 207–218. [CrossRef]
75. Tarin, T.; Nolan, R.H.; Medlyn, B.E.; Cleverly, J.; Eamus, D. Water-Use Efficiency in a Semi-Arid Woodland with High Rainfall Variability. *Glob. Chang. Biol.* **2020**, *26*, 496–508. [CrossRef]
76. Liao, Q.; Gu, S.; Kang, S.; Du, T.; Tong, L.; Wood, J.D.; Ding, R. Mild Water and Salt Stress Improve Water Use Efficiency by Decreasing Stomatal Conductance via Osmotic Adjustment in Field Maize. *Sci. Total Environ.* **2022**, *805*, 150364. [CrossRef]
77. Bromham, L.; Bennett, T.H. Salt Tolerance Evolves More Frequently in C_4 Grass Lineages. *J. Evol. Biol.* **2014**, *27*, 653–659. [CrossRef] [PubMed]
78. Tao, R.; Ding, J.; Li, C.; Zhu, X.; Guo, W.; Zhu, M. Evaluating and Screening of Agro-Physiological Indices for Salinity Stress Tolerance in Wheat at the Seedling Stage. *Front. Plant Sci.* **2021**, *12*, 646175. [CrossRef] [PubMed]
79. Nieves-Cordones, M.; Al Shiblawi, F.R.; Sentenac, H. Roles and Transport of Sodium and Potassium in Plants. In *The Alkali Metal Ions: Their Role for Life*; Sigel, A., Sigel, H., Sigel, R.K.O., Eds.; Metal Ions in Life Sciences; Springer International Publishing: Cham, Switzerland, 2016; pp. 291–324. ISBN 978-3-319-21756-7.
80. Yalcinkaya, T.; Uzilday, B.; Ozgur, R.; Turkan, I.; Mano, J. Lipid Peroxidation-Derived Reactive Carbonyl Species (RCS): Their Interaction with ROS and Cellular Redox during Environmental Stresses. *Environ. Exp. Bot.* **2019**, *165*, 139–149. [CrossRef]
81. Alché, J.d.D. A Concise Appraisal of Lipid Oxidation and Lipoxidation in Higher Plants. *Redox Biol.* **2019**, *23*, 101136. [CrossRef]
82. Singh, A.; Roychoudhury, A. Role of Phenolic Acids and Flavonoids in the Mitigation of Environmental Stress in Plants. In *Biology and Biotechnology of Environmental Stress Tolerance in Plants: Volume 1: Secondary Metabolites in Environmental Stress Tolerance*; Apple Academic Press: Palm Bay, FL, USA, 2023; pp. 227–248. ISBN 978-1-00-334617-3.
83. Sharma, A.; Shahzad, B.; Rehman, A.; Bhardwaj, R.; Landi, M.; Zheng, B. Response of Phenylpropanoid Pathway and the Role of Polyphenols in Plants under Abiotic Stress. *Molecules* **2019**, *24*, 2452. [CrossRef]
84. Gill, S.S.; Tuteja, N. Reactive Oxygen Species and Antioxidant Machinery in Abiotic Stress Tolerance in Crop Plants. *Plant Physiol. Biochem.* **2010**, *48*, 909–930. [CrossRef] [PubMed]
85. Duarte, B.; Santos, D.; Marques, J.C.; Caçador, I. Ecophysiological Adaptations of Two Halophytes to Salt Stress: Photosynthesis, PS II Photochemistry and Anti-Oxidant Feedback—Implications for Resilience in Climate Change. *Plant Physiol. Biochem.* **2013**, *67*, 178–188. [CrossRef]
86. Sarker, U.; Oba, S. Augmentation of Leaf Color Parameters, Pigments, Vitamins, Phenolic Acids, Flavonoids and Antioxidant Activity in Selected *Amaranthus tricolor* under Salinity Stress. *Sci. Rep.* **2018**, *8*, 12349. [CrossRef]
87. Nakashima, T.; Araki, T.; Ueno, O. Photoprotective Function of Betacyanin in Leaves of *Amaranthus cruentus* L. under Water Stress. *Photosynthetica* **2011**, *49*, 497–506. [CrossRef]
88. Vargas-Ortiz, E.; Délano-Frier, J.P.; Tiessen, A. The Tolerance of Grain Amaranth (*Amaranthus cruentus* L.) to Defoliation during Vegetative Growth Is Compromised during Flowering. *Plant Physiol. Biochem.* **2015**, *91*, 36–40. [CrossRef]
89. Vargas-Ortiz, E.; Espitia-Rangel, E.; Tiessen, A.; Délano-Frier, J.P. Grain Amaranths Are Defoliation Tolerant Crop Species Capable of Utilizing Stem and Root Carbohydrate Reserves to Sustain Vegetative and Reproductive Growth after Leaf Loss. *PloS ONE* **2013**, *8*, e67879. [CrossRef] [PubMed]
90. Castrillón-Arbeláez, P.A.; Martínez-Gallardo, N.; Arnaut, H.A.; Tiessen, A.; Délano-Frier, J.P. Metabolic and Enzymatic Changes Associated with Carbon Mobilization, Utilization and Replenishment Triggered in Grain Amaranth (*Amaranthus cruentus*) in Response to Partial Defoliation by Mechanical Injury or Insect Herbivory. *BMC Plant Biol.* **2012**, *12*, 163. [CrossRef] [PubMed]
91. Achigan-Dako, E.G.; Sogbohossou, O.E.D.; Maundu, P. Current Knowledge on *Amaranthus* Spp.: Research Avenues for Improved Nutritional Value and Yield in Leafy Amaranths in Sub-Saharan Africa. *Euphytica* **2014**, *197*, 303–317. [CrossRef]
92. Sarker, U.; Oba, S. Response of Nutrients, Minerals, Antioxidant Leaf Pigments, Vitamins, Polyphenol, Flavonoid and Antioxidant Activity in Selected Vegetable Amaranth under Four Soil Water Content. *Food Chem.* **2018**, *252*, 72–83. [CrossRef] [PubMed]
93. Escudero, N.L.; Albarracín, G.J.; Lucero López, R.V.; Giménez, M.S. Antioxidant Activity and Phenolic Content of Flour and Protein Concentrate of *Amaranthus cruentus* Seeds: Antioxidant Activity of *Amaranthus cruentus*. *J. Food Biochem.* **2011**, *35*, 1327–1341. [CrossRef]
94. Sarker, U.; Oba, S. Drought Stress Enhances Nutritional and Bioactive Compounds, Phenolic Acids and Antioxidant Capacity of *Amaranthus* Leafy Vegetable. *BMC Plant Biol.* **2018**, *18*, 258. [CrossRef] [PubMed]
95. Lichtenthaler, H.K. Chlorophylls and Carotenoids: Pigments of Photosynthetic Biomembranes. In *Methods in Enzymology*; Plant Cell Membranes; Academic Press: Cambridge, MA, USA, 1987; Volume 148, pp. 350–382.
96. Howard, J.E.; Villamil, M.B.; Riggins, C.W. Amaranth as a Natural Food Colorant Source: Survey of Germplasm and Optimization of Extraction Methods for Betalain Pigments. *Front. Plant Sci.* **2022**, *13*, 932440. [CrossRef] [PubMed]
97. Sarker, U.; Hossain, M.M.; Oba, S. Nutritional and Antioxidant Components and Antioxidant Capacity in Green Morph *Amaranthus* Leafy Vegetable. *Sci. Rep.* **2020**, *10*, 1336. [CrossRef]

98. Woisky, R.G.; Salatino, A. Analysis of Propolis: Some Parameters and Procedures for Chemical Quality Control. *J. Apic. Res.* **1998**, *37*, 99–105. [CrossRef]
99. Ainsworth, E.A.; Gillespie, K.M. Estimation of Total Phenolic Content and Other Oxidation Substrates in Plant Tissues Using Folin–Ciocalteu Reagent. *Nat. Protoc.* **2007**, *2*, 875–877. [CrossRef]
100. Tsikas, D. Assessment of Lipid Peroxidation by Measuring Malondialdehyde (MDA) and Relatives in Biological Samples: Analytical and Biological Challenges. *Anal. Biochem.* **2017**, *524*, 13–30. [CrossRef]
101. Heath, R.L.; Packer, L. Photoperoxidation in Isolated Chloroplasts: I. Kinetics and Stoichiometry of Fatty Acid Peroxidation. *Arch. Biochem. Biophys.* **1968**, *125*, 189–198. [CrossRef]
102. Kampfenkel, K.; Vanmontagu, M.; Inze, D. Extraction and Determination of Ascorbate and Dehydroascorbate from Plant Tissue. *Anal. Biochem.* **1995**, *225*, 165–167. [CrossRef]
103. Bigot, S.; Leclef, C.; Rosales, C.; Martínez, J.-P.; Lutts, S.; Quinet, M. Comparison of the Salt Resistance of *Solanum lycopersicum* × *Solanum chilense* Hybrids and Their Parents. *Front. Hortic.* **2023**, *2*, 1130702. [CrossRef]
104. R Core Team. *R: A Language and Environment for Statistical Computing*; R Foundation for Statistical Computing: Vienna, Austria, 2022.

MDPI
St. Alban-Anlage 66
4052 Basel
Switzerland
www.mdpi.com

Plants Editorial Office
E-mail: plants@mdpi.com
www.mdpi.com/journal/plants

www.ingramcontent.com/pod-product-compliance
Lightning Source LLC
Chambersburg PA
CBHW061611220926
51850CB00005B/475
9783036590844